Scientific Computation

More information about this series at http://www.springer.com/series/718

Hiqmet Kamberaj

Molecular Dynamics Simulations in Statistical Physics: Theory and Applications

Springer

Hiqmet Kamberaj
Computer Engineering
International Balkan University
Skopje, Republic of North Macedonia

Advanced Computing Research Center
University of New York Tirana
Tirana, Albania

ISSN 1434-8322 ISSN 2198-2589 (electronic)
Scientific Computation
ISBN 978-3-030-35701-6 ISBN 978-3-030-35702-3 (eBook)
https://doi.org/10.1007/978-3-030-35702-3

This Springer imprint is published by the registered company Springer Nature Switzerland AG.
The registered company address is: Gewerbestrasse 11, 6330 Cham, Switzerland

To the memory of my mother and father

Preface

Computer simulations are used very often to understand and solve practical problems in the area of statistical physics and biophysics. With proper knowledge of classical mechanics, thermodynamics, and statistical physics, you will be able to understand and judge the content of this book.

This book aims to be a recipe for computer simulations with molecular dynamics techniques in statistical physics, where the main emphases are the macromolecular systems. Numerical methods introduced in the form of computer algorithms can be implemented in computers using any desired computer programing language, such as Fortran 90, C/C++, and others. This book applies some of the discussed numerical methods and their algorithms in the existing computer programing software of macromolecular systems, such as the CHARMM program.

In this book, you will find out some advanced concepts of computer simulation techniques used in statistical physics and a particular understanding of biological and physical systems. It discusses the molecular dynamics approach in details to help understand its use in statistical physics problems.

Chapters 1, 2, and 3 introduce the principles of classical mechanics, thermodynamics, and statistical physics, which are necessary concepts to know to better understand real problems in different fields, such as physics, chemistry, and biology, when we use the computer simulations for solving them, while Chap. 4 introduces the use of statistical thermodynamics in understanding biological phenomena.

In Chap. 5, the main theory used for understanding many useful techniques in a computer simulation, such as calculations by means of molecular dynamics simulations of the absolute free energy, solvation free energy, and binding free energy, which have a broad area of applications in physics, chemistry, and biology, is discussed.

Chapter 6 then introduces molecular dynamics techniques, such as describing the equations of motion in different statistical ensembles of interest to mimic real experimental conditions, while Chap. 7 presents molecular mechanics and represents the primary methods for parameterization of the force fields; in particular, it discusses traditional and automated force field parameterization methods and presents the perspectives on the force field developments.

Chapter 8 further discusses different methods used to determine the frequency spectrum of the motions in a macromolecular system, namely, the normal modes, principal component analysis, and the time-lagged auto-encoder machine learning approach, and an improved modified version of the artificial neural network, called Bootstrapping Swarm Artificial Neural Network. Besides, it introduces an approach of how to derive the equations of motion in the reduced essential subspace of the slow collective variables using the harmonic bath coupling of these variables with the environmental fast degrees of freedom of the system.

Chapter 9 introduces some elements of the information theory and discusses the connection between information theory measures and statistical thermodynamics.

Chapter 10 subsequently discusses some technical aspects of the use of molecular dynamics method in simulations of macromolecular systems. In particular, it describes the periodic boundary condition, treatment of the long-range interactions, spherical cutoffs, and equilibration of the molecular dynamics simulations.

Chapter 11 introduces numerical techniques used to solve molecular dynamics equations of motion using Liouville's formalism and Trotter factorization scheme. Besides, the stability of numerical schemes will be discussed by applying to real physical systems for which the analytical solutions are known.

Finally, in Chap. 12, generalized ensemble molecular dynamics simulation methods used to improve the sampling of lower-energy configurations are discussed. In particular, this chapter introduces the multicanonical sampling, Tsallis ensemble, Swarm particle intelligence molecular dynamics, and replica exchange.

This book is aimed to graduate students and research scientists working in the fields of theoretical and computational biophysics, physics, and chemistry. Also, the book can be used by graduate students of other branches, such as applied mathematics, computer sciences, and bioinformatics.

Skopje, North Macedonia Hiqmet Kamberaj
April 2019

Acknowledgments

I am grateful to the support of Eco/logical Learning and Simulation Environments in Higher Education (ELSE) under the grant IT02-KA203-048006 and of the International Balkan University. Finally, I thank my family for their continuous support: Nera (my wife), Jon (my son), and Lina (my daughter).

Contents

About the Author

Hiqmet Kamberaj is an Associate Professor at International Balkan University and the Dean of the Faculty of Engineering. He completed his Ph.D. in Computational Physics in 2005 from Manchester Metropolitan University and his post-doctoral studies from the University of Minnesota, Arizona State University, and the National Institute for Nanotechnology, University of Alberta, Edmonton, Alberta. He received his Bachelor of Science in Physics in 1996 from the University of Tirana and his Master of Science in Physics from the University of Siegen in 2000.

He has published around 20 articles and book chapters in reputed journals and has been serving as Editor in Chief, Editorial Board Member, and Ad Hoc Reviewer of Repute. His articles focus on understanding the structure, dynamics, and thermodynamics of macromolecular systems using the laws of physics and biochemistry, and applied mathematics.

He loves teaching and preparing lecture notes for his students and especially enjoys preparing resumes for his students who are taking the first steps of their careers.

Please email kamberaj@yahoo.co.uk to contact Hiqmet or by visiting his ResearchGate profile at https://www.researchgate.net/profile/Hiqmet_Kamberaj to learn more about how his research articles could contribute to your research and teaching activities.

Chapter 1
Principles of Classical Mechanics

In this chapter, we will present some important concepts of classical mechanics.

For further reading about the following discussion on classical mechanics, one should consider the book by Goldstein (2002).

Although Newton's equations of motion are often used as a start to understanding the basis of molecular dynamics method presented in the next chapters, many other advanced simulation techniques use the Lagrangian and Hamiltonian formalism of classical mechanics.

1.1 Mechanics of the System of Particles

The system consists of N particles under the external forces, $\mathbf{F}_i^{(e)}$, acting on the particle i, and internal forces, $\mathbf{F}_{ji}$, acting on every particle i due to any other particle j in the system. Newton's second law gives the equations of motion for this system as:

$$\sum_{j=1\neq i}^{N} \left(\mathbf{F}_{ji} + \mathbf{F}_i^{(e)}\right) = \dot{\mathbf{p}}_i, \quad i = 1, 2, \cdots, N \tag{1.1}$$

where $\mathbf{p}_i$ is the i-th particle momentum:

$$\dot{\mathbf{p}}_i = m_i \frac{d^2\mathbf{r}_i}{dt^2}$$

The sum on Eq. (1.1) runs over all particle of system, excluding the i-th particle. The forces $\mathbf{F}_{ji}$, based on the third law of Newton, are given as

$$\mathbf{F}_{ji} = -\mathbf{F}_{ij}$$

H. Kamberaj, *Molecular Dynamics Simulations in Statistical Physics: Theory and Applications*, Scientific Computation,
https://doi.org/10.1007/978-3-030-35702-3_1

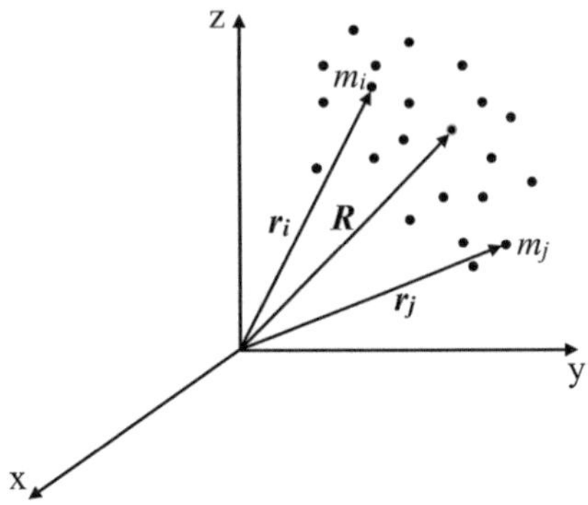

Fig. 1.1 The center of mass of a system of particles

By summing up over all particles, Eq. (1.1) takes the form:

$$\sum_{i=1}^{N} m_i \frac{d^2\mathbf{r}_i}{dt^2} = \sum_{i=1}^{N} \left(\sum_{j=1(j\neq i)}^{N} \mathbf{F}_{ji} + \mathbf{F}_i^{(e)} \right) \tag{1.2}$$

since $\mathbf{F}_{ji} + \mathbf{F}_{ij} = 0$. Determining the *center of mass* of system as (see also Fig. 1.1)

$$\mathbf{R} = \frac{\sum_{i=1}^{N} m_i \mathbf{r}_i}{\sum_{i=1}^{N} m_i} \tag{1.3}$$

then, the second derivative of **R** vector with respect to time is given as:

$$\frac{d^2\mathbf{R}}{dt^2} = \frac{\sum_{i=1}^{N} m_i \frac{d^2\mathbf{r}_i}{dt^2}}{\sum_{i=1}^{N} m_i} \tag{1.4}$$

Thus, Eq. (1.2) takes the form:

$$M\frac{d^2\mathbf{R}}{dt^2} = \mathbf{F}^{(e)} \tag{1.5}$$

Here, M gives the total mass of the system:

$$M = \sum_{i=1}^{N} m_i,$$

and $\mathbf{F}^{(e)}$ is the total external force acting on every particle of system:

$$\mathbf{F}^{(e)} = \sum_{i=1}^{N} \mathbf{F}_i^{(e)}.$$

For example, consider a system of N charged particles, each with charge q_i, interacting with each other via the Coulomb forces:

$$| \mathbf{F}_{ij} | = | \mathbf{F}_{ji} | = \frac{q_i q_j}{4\pi \varepsilon_0 r_{ij}^2}$$

where r_{ij} is the distance between the particles i and j and ε_0 is the permittivity of free space. Assume this system is placed in a uniform external electrical field $\mathbf{E}$, which is taken to be parallel with z-axis. The external force acting on each charge q_i is given as

$$\mathbf{F}_i^{(e)} = q_i \mathbf{E}$$

Then, the equation of motion of the center of the mass of system becomes

$$M \frac{d^2\mathbf{R}}{dt^2} = \mathbf{F}^{(e)} = \sum_{i=1}^{N} q_i \mathbf{E}$$

Projections along the x, y and z axis, respectively, give:

$$M \frac{d^2 X}{dt^2} = 0 \tag{1.6}$$

$$M \frac{d^2 Y}{dt^2} = 0 \tag{1.7}$$

$$M \frac{d^2 Z}{dt^2} = E \sum_{i=1}^{N} q_i \tag{1.8}$$

where $\mathbf{R} = (X, Y, Z)$. It can be seen that motions along x and y axes are without acceleration, on the other hand, the motion along z axis is with constant acceleration.

Moreover, if the system is neutral (i.e. $\sum_{i=1}^{N} q_i = 0$) or $\mathbf{E} = 0$, then

$$M \frac{d^2\mathbf{R}}{dt^2} = 0$$

hence, the center of mass is either at rest or moving with constant velocity.

Therefore, in general, we can say that if the total external force acting on the system is zero, then the center of mass is either at rest or moving with constant velocity. That is often the case of studying the dynamics of a system of N particles interacting via two bodies pair-wise potential energy function with no external forces. If we determine initially, $t = 0$, that the center of mass is at rest, then it remains so during the entire time of investigating the dynamics.

Denoting the total linear momentum of the system by

$$\mathbf{P} = \sum_{i=1}^{N} m_i \frac{d\mathbf{r}_i}{dt} = M\frac{d\mathbf{R}}{dt} \tag{1.9}$$

where $\mathbf{V} = d\mathbf{R}/dt$ is the center of mass velocity, then Eq. (1.5) can be written as

$$\frac{\mathbf{dP}}{dt} = \mathbf{F}^{(e)} \tag{1.10}$$

Eq. (1.10) can also be stated as the *Conservation Law for Linear Momentum* for a system of particles:

If the net external force acting on a system of particles is zero, the total linear momentum is conserved.

The *angular momentum* of the system is given by

$$\mathbf{L} = \sum_{i=1}^{N} \mathbf{r}_i \times \mathbf{p}_i \tag{1.11}$$

The *moment of force* or *torque* is defined as

$$\begin{aligned}\mathbf{T} = \frac{d\mathbf{L}}{dt} &= \sum_{i=1}^{N} (\dot{\mathbf{r}}_i \times \mathbf{p}_i + \mathbf{r}_i \times \dot{\mathbf{p}}_i) \\ &= \sum_{i=1}^{N} \mathbf{r}_i \times \dot{\mathbf{p}}_i \\ &= \sum_{i=1}^{N} \mathbf{r}_i \times \mathbf{F}_i^{(e)} + \sum_{i=1}^{N} \sum_{j=1 \neq i}^{N} \mathbf{r}_i \times \mathbf{F}_{ji} \\ &= \sum_{i=1}^{N} \mathbf{r}_i \times \mathbf{F}_i^{(e)}\end{aligned} \tag{1.12}$$

since the following two terms vanish:

$$\dot{\mathbf{r}}_i \times \mathbf{p}_i = 0; \quad \sum_{i=1}^{N} \sum_{j=1 \neq i}^{N} \mathbf{r}_i \times \mathbf{F}_{ji} = 0$$

Thus, the time derivative of the angular momentum, $\mathbf{L}$, equals the moment of the external force about a given point. That can also be formulated as the *Conservation Law for Total Angular Momentum*:

The angular momentum $\mathbf{L}$ is constant in time if the external torque is zero.

Consider a linear transformation on the particle coordinates as

$$\mathbf{r}'_i = \mathbf{r}_i + \boldsymbol{\ell}$$

where $\boldsymbol{\ell}$ is a constant vector. The following transformation of the angular momentum of the system of particles occurs:

$$\begin{aligned}\mathbf{L}' &= \sum_{i=1}^{N} \mathbf{r}'_i \times \mathbf{p}_i \\ &= \sum_{i=1}^{N} [\mathbf{r}_i \times \mathbf{p}_i + \boldsymbol{\ell} \times \mathbf{p}_i] \\ &= \mathbf{L} + \sum_{i=1}^{N} \boldsymbol{\ell} \times \mathbf{p}_i \\ &= \mathbf{L} + \boldsymbol{\ell} \times \mathbf{P}\end{aligned} \tag{1.13}$$

Thus, the total angular momentum on the new coordinates is the angular momentum on the old coordinates, plus the angular momentum of the vector $\boldsymbol{\ell}$. Only if total linear momentum of system is fixed to zero, then

$$\mathbf{L}' = \mathbf{L}$$

The momentum of the force on the new coordinates is

$$\begin{aligned}\mathbf{T}' &= \sum_{i=1}^{N} \mathbf{r}'_i \times \dot{\mathbf{p}}_i \\ &= \sum_{i=1}^{N} [\mathbf{r}_i \times \dot{\mathbf{p}}_i + \boldsymbol{\ell} \times \dot{\mathbf{p}}_i] \\ &= \mathbf{T} + \sum_{i=1}^{N} \boldsymbol{\ell} \times \dot{\mathbf{p}}_i \\ &= \mathbf{T} + \boldsymbol{\ell} \times \frac{d\mathbf{P}}{dt}\end{aligned} \tag{1.14}$$

It can be seen that the momentum of force is conserved only if the total linear momentum of system is a conserved quantity: $d\mathbf{P}/dt = 0$.

Such transformations could correspond to, for example, the position of particles with respect to the center of mass of the system. Or, transformations of the coordinates in a system with periodic boundary conditions.

Let calculate the work done by all forces in moving the system from initial configuration 1 to a final configuration 2:

$$
\begin{aligned}
W_{12} &= \sum_{i=1}^{N} \int_{1}^{2} \mathbf{F}_i \cdot d\mathbf{s}_i \\
&= \sum_{i=1}^{N} \int_{1}^{2} \mathbf{F}_i \cdot \frac{d\mathbf{s}_i}{dt} dt \\
&= \sum_{i=1}^{N} \int_{1}^{2} m_i \dot{\mathbf{v}}_i \cdot \mathbf{v}_i dt \\
&= \sum_{i=1}^{N} \int_{1}^{2} d(\frac{m_i v_i^2}{2}) \\
&= T_2 - T_1
\end{aligned} \tag{1.15}
$$

where

$$
T = \sum_{i=1}^{N} \frac{m_i v_i^2}{2}
$$

is the kinetic energy of the system. Equation (1.15) represents the work-kinetic energy law:

> The work done by all forces, W_{12}, in moving the system from the configuration 1 to a configuration 2 equals the increase in kinetic energy, $\Delta T = T_2 - T_1$.

On the other hand,

$$
\begin{aligned}
W_{12} &= \sum_{i=1}^{N} \int_{1}^{2} \mathbf{F}_i \cdot d\mathbf{s}_i \\
&= \sum_{i=1}^{N} \int_{1}^{2} \left[\mathbf{F}_i^{(e)} + \sum_{j=1 \neq i}^{N} \mathbf{F}_{ji} \right] \cdot d\mathbf{s}_i
\end{aligned} \tag{1.16}
$$

Assuming that the external and internal forces are conservative, hence they are derivative of some potential function V of coordinates. Thus,

$$
\mathbf{F}_i^{(e)} = -\nabla_i V
$$

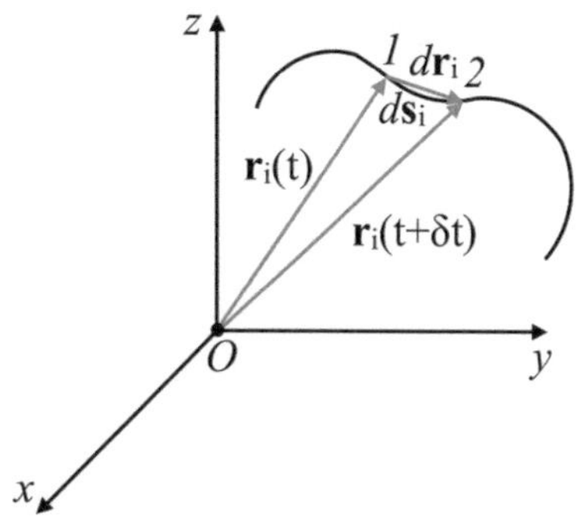

Fig. 1.2 The displacement of the particle i of the system due to the external force

and then the first term in Eq. (1.16) can be written as:

$$\sum_{i=1}^{N}\int_{1}^{2}\mathbf{F}_i^{(e)}\cdot d\mathbf{s}_i=\sum_{i=1}^{N}\int_{1}^{2}(-\nabla_i V)\cdot d\mathbf{s}_i=-\left[\sum_{i=1}^{N}V_i\right]_1^2 \tag{1.17}$$

where $\nabla_i=\partial/\partial\mathbf{r}_i$ and $d\mathbf{s}_i\approx d\mathbf{r}_i$ (see Fig. 1.2).

Furthermore, the mutual forces between two particles i and j can also be obtained from a potential function V_{ij} of the distance between the two particles $\mid\mathbf{r}_i-\mathbf{r}_j\mid$. For example, the force on the particle i due to particle j is:

$$\mathbf{F}_{ji}=-\nabla_i V_{ij}(\mid\mathbf{r}_i-\mathbf{r}_j\mid)$$

which satisfies the law of action and reaction (Newton's third law):

$$\mathbf{F}_{ji}=-\nabla_i V_{ij}(\mid\mathbf{r}_i-\mathbf{r}_j\mid)=+\nabla_j V_{ij}(\mid\mathbf{r}_i-\mathbf{r}_j\mid)=-\mathbf{F}_{ij}$$

where $\mathbf{F}_{ij}$ is the force acting on the particle j due to particle i. Furthermore, it can be written that

$$\nabla_i V_{ij}(\mid\mathbf{r}_i-\mathbf{r}_j\mid)=\frac{\partial\mid\mathbf{r}_i-\mathbf{r}_j\mid}{\partial\mathbf{r}_i}\frac{\partial V_{ij}(\mid\mathbf{r}_i-\mathbf{r}_j\mid)}{\partial\mid\mathbf{r}_i-\mathbf{r}_j\mid}=(\mathbf{r}_i-\mathbf{r}_j)f$$

where f is a scalar function of only the distance $\mid\mathbf{r}_i-\mathbf{r}_j\mid$, given as:

$$f\left(\mid\mathbf{r}_i-\mathbf{r}_j\mid\right)=\frac{1}{\mid\mathbf{r}_i-\mathbf{r}_j\mid}\frac{\partial V(\mid\mathbf{r}_i-\mathbf{r}_j\mid)}{\partial\mid\mathbf{r}_i-\mathbf{r}_j\mid}$$

This indicates that the forces $\mathbf{F}_{ij}$ (or $\mathbf{F}_{ji}$) lie along the direction connecting the two particles.

Note that, if V_{ij} will also depend on other difference vectors associated with the two particles, such as velocities, then the forces would still be equal and opposite, but not necessarily lie along the direction between them.

Equation (1.16) can be written as

$$
\begin{aligned}
W_{12} &= -\left[\sum_{i=1}^{N} V_i\right]_1^2 + \sum_{i=1}^{N}\sum_{j=1\neq i}^{N}\int_1^2 (-\nabla_i V_{ij})\cdot d\mathbf{s}_i \qquad (1.18)\\
&= -\left[\sum_{i=1}^{N} V_i\right]_1^2 + \sum_{i=1}^{N-1}\sum_{j=i+1}^{N}\int_1^2 -(\nabla_i V_{ij}\cdot d\mathbf{s}_i + \nabla_j V_{ij}\cdot d\mathbf{s}_j)\\
&= -\left[\sum_{i=1}^{N} V_i\right]_1^2 + \sum_{i=1}^{N-1}\sum_{j=i+1}^{N} -\int_1^2 (\nabla_{ij} V_{ij}\cdot d\mathbf{s}_i - \nabla_{ij} V_{ij}\cdot d\mathbf{s}_j)\\
&= -\left[\sum_{i=1}^{N} V_i\right]_1^2 + \sum_{i=1}^{N-1}\sum_{j=i+1}^{N} -\int_1^2 (d\mathbf{s}_i - d\mathbf{s}_j)\cdot\nabla_{ij} V_{ij}\\
&= -\left[\sum_{i=1}^{N} V_i\right]_1^2 + \sum_{i=1}^{N-1}\sum_{j=i+1}^{N} -\int_1^2 d\mathbf{r}_{ij}\cdot\nabla_{ij} V_{ij}\\
&= -\left[\sum_{i=1}^{N} V_i\right]_1^2 - \left[\sum_{i=1}^{N-1}\sum_{j=i+1}^{N} V_{ij}\right]_1^2\\
&= -\left[\sum_{i=1}^{N} V_i + \sum_{i=1}^{N-1}\sum_{j=i+1}^{N} V_{ij}\right]_1^2
\end{aligned}
$$

where ∇_{ij} represents

$$\nabla_{ij} \equiv \frac{\partial}{\partial \mathbf{r}_{ij}}$$

and $d\mathbf{r}_{ij} \approx d\mathbf{s}_i - d\mathbf{s}_j$. Denoting the total potential energy, V, of the system as

$$V = \sum_{i=1}^{N} V_i + \sum_{i=1}^{N}\sum_{j=i+1}^{N} V_{ij}$$

we obtain the work as

$$W_{12} = -(V_2 - V_1) \equiv -\Delta V \qquad (1.19)$$

Thus, both the external and internal forces are derivative of the total potential energy function. Equation (1.19) states that:

> The work done by the conservative forces equals the negative of the change in the potential energy associated with the forces, and the work done on a system of particles by the conservative forces does not depend on the path taken by the particles, but the work depends only on the particles initial and final configuration.

We can combine Eqs. (1.15) and (1.19) and obtain that

$$\Delta T = -\Delta V \tag{1.20}$$

which can be stated as:

> The potential energy implies the potential, or capability, of the system of particles of either gaining kinetic energy or doing work when it is at a certain configuration at some point in time under the influence of the conservative forces exerted on a member of a system by other members of the system.

Using Eq. (1.20), we obtain the *Conservation Law of Energy*:

$$T_1 + V_1 = T_2 + V_2$$

which states that:

> The total energy of the system of particles in the configuration 1 and configuration 2 are equal.

1.2 Generalized Coordinates for Unconstrained Systems

Here, we will discuss the formulation of a conservative unconstrained system in generalized coordinates. For that, consider n generalized coordinates q_j, which determine the system of the $3N$ Cartesian coordinates $\mathbf{r}_i$:

$$\mathbf{r}_i = \mathbf{r}_i\left(q_1, \cdots, q_n, t\right)$$

where it is assumed that $\mathbf{r}_i$ depends explicitly on time t. This expression represents a relationship between different descriptions of the same point in configuration space. Furthermore, the functions $\mathbf{r}_i\left(q_1, \cdots, q_n, t\right)$ are independent of the motion of any particle. We also assume that $\mathbf{r}_i$ and q_j are each a complete set of coordinates for the space, such that q_j are also functions of $\mathbf{r}_i$ and t:

$$q_j = q_j\left(\mathbf{r}_1, \cdots, \mathbf{r}_N, t\right) ,$$

where we have assumed an explicit dependence on time t.

Let us introduce $\{x_k\}$ the $3N$ Cartesian coordinates of the N 3-dimensional vectors $\mathbf{r}_i$:

$$\{x_1,\ x_2,\ \cdots,\ x_{3N}\} = \{x_1,\ y_1,\ z_1,\ \cdots,\ x_N,\ y_N,\ z_N\}$$

We also consider a small change in the coordinates of a particle in configuration space, which could be a change over a small time interval dt or a "virtual" change between the particles current position and the position it may be under slightly different circumstances, and describe it by the set of $\{\delta x_k\}_{k=1}^{3N}$ or $\{\delta q_j\}_{j=1}^{n}$. If we consider them to be the virtual changes at the same time, these are related by the chain rule as:

$$\delta x_k = \sum_{j=1}^{n} \frac{\partial x_k}{\partial q_j} \delta q_j\ , \tag{1.21}$$

$$\delta q_j = \sum_{k=1}^{3N} \frac{\partial q_j}{\partial x_k} \delta x_k\ , \quad (\text{for } \delta t = 0)\ .$$

In the case of the variations where $\delta t \neq 0$, a more general form can be obtained:

$$\delta x_k = \sum_{j=1}^{n} \frac{\partial x_k}{\partial q_j} \delta q_j + \frac{\partial x_k}{\partial t} \delta t\ , \tag{1.22}$$

$$\delta q_j = \sum_{k=1}^{3N} \frac{\partial q_j}{\partial x_k} \delta x_k + \frac{\partial q_j}{\partial t} \delta t\ , \quad (\text{for } \delta t \neq 0)\ .$$

A virtual displacement, with $\delta t = 0$, is the kind of variation needed to determine the forces described by a potential. Hence, the force is

$$F_k = -\frac{\partial U(x_1, \cdots, x_{3N})}{\partial x_k} = -\sum_{j=1}^{n} \frac{\partial U}{\partial q_j} \frac{\partial q_j}{\partial x_k} = \sum_{j=1}^{n} \frac{\partial q_j}{\partial x_k} Q_j \tag{1.23}$$

where

$$Q_j = -\frac{\partial U}{\partial q_j} \tag{1.24}$$

is the so-called *generalized force*, which can also be written as

$$Q_j = -\frac{\partial U}{\partial q_j} \tag{1.25}$$

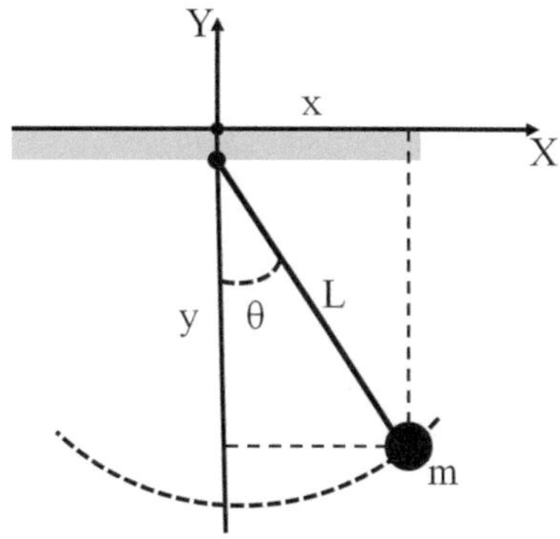

Fig. 1.3 The simple pendulum: A dimensionless mass m is connected to a non-extendable chord of length L. The position of mass m can be represented either by the Cartesian coordinates (x, y) or angle θ at any moment of the time t

$$= -\sum_{k=1}^{3N} \frac{\partial U}{\partial x_k} \frac{\partial x_k}{\partial q_j}$$

$$= \sum_{k=1}^{3N} F_k \frac{\partial x_k}{\partial q_j}$$

In general, the generalized forces depend explicitly on time t, in contrast to conservative forces, and hence they are not conservative. It is important to mention that the definition of the generalized forces in Eq. (1.25) holds even if the Cartesian forces F_k are not described by a potential.

The generalized coordinates q_k do not necessary have units of the distance. For example, q_k may be an angle, such as in the case of the simple pendulum shown in Fig. 1.3. The corresponding component of the generalized force will have the units of energy, and it might be considered as a torque rather than a force.

The potential can be considered as a function of the generalized coordinates using the following relation:

$$\tilde{U}(q_1, \cdots, q_n, t) = U(x_1(q_1, \cdots, q_n), \cdots, x_{3N}(q_1, \cdots, q_n), t) ,$$

and the generalized forces are given as derivative of the potential with respect to the corresponding generalized coordinate just as for ordinary Cartesian coordinates. For the entire system, the kinetic energy can be written as:

$$T = \frac{1}{2}\sum_{i=1}^{N} m_i \dot{\mathbf{r}}_i^2 = \frac{1}{2}\sum_{j=1}^{3N} m_j \dot{x}_j^2 ,$$

where m_j (for $j = 1, 2, \cdots, 3N$) are not exactly independent, considering that a particle has the same mass in all three directions. On the other hand, from the definition of velocity at time t, we obtain, for every $j = 1, 2, \cdots, 3N$, that

$$\dot{x}_j = \lim_{\Delta t \to 0} \frac{\Delta x_j}{\Delta t} = \frac{dx_j}{dt} \tag{1.26}$$
$$= \sum_{k=1}^{n} \left(\frac{\partial x_j}{\partial q_k} \right)_{q_{i \neq k}, t} \frac{dq_k}{dt} + \left(\frac{\partial x_j}{\partial t} \right)_{q_1, \cdots, q_n}$$
$$= \sum_{k=1}^{n} \left(\frac{\partial x_j}{\partial q_k} \right)_{q_{i \neq k}, t} \dot{q}_k + \left(\frac{\partial x_j}{\partial t} \right)_{q_1, \cdots, q_n}$$

where $(\cdots)_{q_{i \neq k}, t}$ means that t and q's other than q_k are held fixed, and the last term is due to the time dependence of the coordinates $x_i(q_1, \cdots, q_n, t)$ even for fixed values of q_i for $i = 1, 2, \cdots, n$.

Substituting this expression into the one for kinetic energy, we obtain that

$$T = \frac{1}{2} \sum_{j,k,l} m_j \left(\frac{\partial x_j}{\partial q_k} \frac{\partial x_j}{\partial q_l} \right)_{q_{i \neq k,l}} \dot{q}_k \dot{q}_l \tag{1.27}$$
$$+ \sum_{j,k} m_j \left(\frac{\partial x_j}{\partial q_k} \dot{q}_k \frac{\partial x_j}{\partial t} \right)_{q_{i \neq k}}$$
$$+ \frac{1}{2} \sum_{j} m_j \left[\left(\frac{\partial x_j}{\partial t} \right)_{q_1, \cdots, q_n} \right]^2 .$$

The first term arises if the relation between x and q is time independent. The second and the third terms are the sources of the $\dot{\mathbf{r}} \cdot (\boldsymbol{\omega} \times \mathbf{r})$ and $(\boldsymbol{\omega} \times \mathbf{r})^2$ terms in the kinetic energy if the rotation of coordinate systems is considered, with $\boldsymbol{\omega}$ being the angular velocity vector.

Equation (1.27) indicates that the kinetic energy as a function of the generalized coordinates and their velocities has a much more complicated expression compare to its form as a function of the Cartesian inertial coordinates, in which it is merely a diagonal quadratic functional form in the velocities, and it is independent on coordinates. On the other hand, in generalized coordinates, it is quadratic but not homogeneous in velocities (because it involves quadratic and lower order terms in the velocities), and with an arbitrary dependence on the coordinates.

Note that a time independent transformation of the coordinates will not in general avoid a dependence of the kinetic energy from the generalized coordinates. In addition, the quadratic form in the velocities may also include the off-diagonal terms. In the case of the time-independent situation, it can be written that

$$T = \frac{1}{2} \sum_{k,l} M_{kl} \left(q_1, \cdots, q_n \right) \dot{q}_k \dot{q}_l \ , \tag{1.28}$$

with

$$M_{kl}(q_1, \cdots, q_n) = \sum_j m_j \left(\frac{\partial x_j}{\partial q_k} \frac{\partial x_j}{\partial q_l} \right)_{q_{i \neq k,l}},$$

where M_{kl} is known as the *mass matrix*, and is always symmetric:

$$M_{kl}(q_1, \cdots, q_n) = M_{lk}(q_1, \cdots, q_n)$$

However, it is not necessary diagonal or coordinate independent.

1.3 Phase Space

By definition, the set of all possible configurations, $\mathbf{r}$, in space visited at any time t is called *trajectory*. If the trajectory of the system in configuration space, $\mathbf{r}(t)$, is known, the velocity as a function of time, $\mathbf{v}(t)$, is also determined as

$$\mathbf{v}(t) = \frac{d\mathbf{r}(t)}{dt}$$

Furthermore, since the mass of the particle is a constant quantity, the momentum

$$\mathbf{p}(t) = m\frac{d\mathbf{r}(t)}{dt}$$

contains the same information as the velocity. If we consider momentum $\mathbf{p}$ simply a function of time t, $\mathbf{p}(t)$, it does not give any information beyond the one contained in the trajectory. However, both $\mathbf{r}$ and $\mathbf{p}$, at any given time, can provide a complete set of initial conditions, which $\mathbf{r}$ alone does not provide.

The *phase space* is the set of possible positions and momenta for the system at some instant of time. Equivalently, it is the set of possible motions obeying the equations of motion, solved for a set of possible initial conditions. Note that as each initial conditions generate a unique trajectory of the system, there is an isomorphism between initial conditions and allowed trajectories. For a single particle in Cartesian coordinates, there are six coordinates in the phase space characterizing this particle: three components of $\mathbf{r}$ and three of $\mathbf{p}$. Thus, the system is represented by a point in this space at any moment, t, called the *phase point*, and that point moves with time according to the physical laws governing the dynamics of the system, for example, Newton's second law:

$$\begin{aligned} \frac{d\mathbf{r}}{dt} &= \frac{\mathbf{p}}{m}, \\ \frac{d\mathbf{p}}{dt} &= \mathbf{F}(\mathbf{r}, \mathbf{p}, t). \end{aligned} \tag{1.29}$$

It can be seen that these are first-order equations, which means that the motion of phase point representing the system in phase space is completely determined by its position, assuming that the force function is a well defined continuous function of its arguments. That is to be distinguished from the trajectory in configuration space, which is determined by both the initial position and its time derivative.

1.4 Dynamical Systems

The phase space of a single particle is represented by its coordinates **r** and momenta **p**, which from a mathematical point of view together give the coordinates of a phase point in phase space. This phase point is characterized by a six-dimensional vector for a single particle

$$\eta = (x, y, z, p_x, p_y, p_z)$$

A phase point moves through the phase space based on the physical laws, which determine at each point a velocity function of that phase point given as:

$$\frac{d\eta}{dt} = \mathbf{V}(\eta, t) \,, \tag{1.30}$$

This generalized form of the velocity contains two components. The first component is the usual velocity, while the other component represents the rate change of the momentum, i.e. the force. As the phase point moves in space, it traces a path in phase space, which is called the trajectory that forms a curve in phase space.

For a system of N particles in three dimensions, the complete set of initial conditions requires $3N$ spatial coordinates

$$(x_1, y_1, z_1, x_2, y_2, z_2, \cdots, x_N, y_N, z_N)$$

and $3N$ momenta

$$\left(p_{x1}, p_{y1}, p_{z1}, p_{x2}, p_{y2}, p_{z2}, \cdots, p_{xN}, p_{yN}, p_{zN}\right)$$

In this case, the phase space is represented by $6N$ degrees of freedom, $d = 6N$.

A system represented by a finite number of particles can use the first-order ordinary differential equation, Eq. (1.30), to describe its dynamics. In general, the complexity of this equation depends on the dimensions of the variable η and the form of the velocity function $\mathbf{V}(\eta, t)$. There are other systems, besides Newtonian dynamics, with dynamics governed by Eq. (1.30), with a suitable velocity function, such as Nosé-Hoover dynamics. All these systems are known as *dynamical systems*.

The order d of the differential equation for η in Eq. (1.30) is called the order of dynamical system. Note that a differential equation of order d in one independent variable may always equivalently be represented as d first-order differential equations. The space covered by these dependent variables is often called the phase space of the dynamical system. An even-order system can also characterize Newtonian systems because each spatial coordinate is paired with a momentum. For N particles unconstrained in three dimensions, the order of the dynamical system is $d = 6N$. That also is the case of the constrained Newtonian systems, because of the existence of pairs of coordinates and their conjugated momenta, which gives a restricting structure, called the symplectic structure, on phase space. If the force function does not depend explicitly on time, then the system is called *autonomous*. The velocity function has no explicit dependence on time, $\mathbf{V} = \mathbf{V}(\eta)$.

The paths taken by possible physical motions through the phase space of an autonomous system have an important property. Its initial velocity function completely determines the change on the rate and direction of a phase point from its initial point. Therefore, if the system returns to the same initial point, then it moves away from that point along the same path as it did previously. That is if the system at time T returns to a point in phase space that it was at time $t = 0$, then its subsequent motion must be just as it was, that is

$$\eta(T + t) = \eta(t)$$

In this case, the motion is called periodic with period T. That also implies that the trajectory of an object through phase space must be non-intersecting.

In the non-autonomous case, the force is explicitly depending on time t, $\mathbf{V} = \mathbf{V}(\eta, t)$. In this case, we can think of an extended phase space, a $6N + 1$ dimensional space with coordinates (η, t).

1.5 Lagrange Equations

In this section, we will consider a new formulation of Newtonian mechanics, the so-called *Lagrangian* mechanics, which naturally associates with configuration space, extended by time.

Lagrange method is, in general, an approach of treating dynamics in terms of generalized coordinates for configuration space.

The Lagrange's equation is derived as a simple change of coordinates in an unconstrained system, which is evolving according to Newton's laws with forces given by the gradient of some potential energy function. Note that the Lagrangian mechanics is also used to treat the systems under some constraints. Therefore, we are going to extend the formalism to this more general situation.

1.5.1 Lagrangian for Unconstrained Systems

For a system of N particles characterized by Cartesian coordinates $\mathbf{r}_1, \cdots, \mathbf{r}_N$ subject to conservative forces described by a potential $U(\mathbf{r}_1, \cdots, \mathbf{r}_N)$, the equations of motion in inertial Cartesian coordinates can be written

$$m_i \ddot{\mathbf{r}}_i = \mathbf{F}_i \ , \tag{1.31}$$

where

$$\mathbf{F}_i = -\frac{\partial U(\mathbf{r}_1, \cdots, \mathbf{r}_N)}{\partial \mathbf{r}_i} \equiv -\nabla_i U(\mathbf{r}_1, \cdots, \mathbf{r}_N)$$

and the left hand side is determined by the kinetic energy function

$$T(\mathbf{p}_1, \mathbf{p}_2, \cdots, \mathbf{p}_N) = \sum_{i=1}^{N} \frac{m_i \dot{\mathbf{r}}_i^2}{2} = \sum_{i=1}^{N} \frac{\mathbf{p}_i^2}{2m_i}$$

as the time derivative of the momentum:

$$m_i \ddot{\mathbf{r}}_i = \frac{d\mathbf{p}_i}{dt}$$

with

$$\mathbf{p}_i = \frac{\partial T}{\partial \dot{\mathbf{r}}_i} \ ,$$

Therefore, Eq. (1.31) can also be written as:

$$\frac{d}{dt}\left(\frac{\partial T}{\partial \dot{\mathbf{r}}_i}\right) = -\frac{\partial U}{\partial \mathbf{r}_i} \tag{1.32}$$

Since T is a function of only momentum $\mathbf{p}_1, \cdots, \mathbf{p}_N$ and it is independent of the coordinates $\mathbf{r}_i$ and U is independent of $\mathbf{p}_i$, the function $L = T - U$, which is the well known *Lagrangian* function, can be introduced, where L is a function of both the coordinates and their velocities. Then, from Eq. (1.32) we can obtain:

$$\frac{d}{dt}\left(\frac{\partial (T - U)}{\partial \dot{\mathbf{r}}_i}\right) = \frac{\partial (T - U)}{\partial \mathbf{r}_i} \tag{1.33}$$

or,

$$\frac{d}{dt}\left(\frac{\partial L}{\partial \dot{\mathbf{r}}_i}\right) - \frac{\partial L}{\partial \mathbf{r}_i} = 0 \ , \tag{1.34}$$

which is known as the *Lagrangian's equation*.

This equation can be projected along each degrees of freedom Cartesian coordinate, x_i, for $i = 1, 2, \cdots, f$ as:

$$\frac{d}{dt}\left(\frac{\partial L}{\partial \dot{x}_i}\right) - \frac{\partial L}{\partial x_i} = 0 \tag{1.35}$$

Here, f denotes the number of degrees of freedom.

The equation can often be generalized to any arbitrary coordinates q_i for $i = 1, \cdots, f$, as it will be shown in the following. Let $\{q_j\}$ be the set of generalized coordinates, which parameterize the coordinate space, such that each point may be described by $\{q_j\}_{j=1}^{f}$ or $\{x_i\}_{i=1}^{f}$, and therefore, each set may be considered as a function of the other set and time t as the following:

$$\begin{aligned} q_j &= q_j(x_1, \cdots, x_f, t) \\ x_i &= x_i(q_1, \cdots, q_f, t)\,. \end{aligned} \tag{1.36}$$

Using the chain rule

$$\frac{\partial L}{\partial \dot{x}_i} = \sum_j \frac{\partial L}{\partial q_j}\frac{\partial q_j}{\partial \dot{x}_i} + \sum_j \frac{\partial L}{\partial \dot{q}_j}\frac{\partial \dot{q}_j}{\partial \dot{x}_i}\,. \tag{1.37}$$

Since q_j does not depend on $\dot{x}_i$ (see Eq. (1.36)), the first term vanishes, and we get:

$$\frac{\partial L}{\partial \dot{x}_i} = \sum_j \frac{\partial L}{\partial \dot{q}_j}\frac{\partial \dot{q}_j}{\partial \dot{x}_i}\,. \tag{1.38}$$

Using Eq. (1.36) and the chain rule, we obtain:

$$\dot{q}_j = \sum_k \frac{\partial q_j}{\partial x_k}\dot{x}_k + \frac{\partial q_j}{\partial t} \tag{1.39}$$

By taking the partial derivative with respect to $\dot{x}_i$, we obtain

$$\begin{aligned} \frac{\partial \dot{q}_j}{\partial \dot{x}_i} &= \sum_k \frac{\partial q_j}{\partial x_k}\frac{\partial \dot{x}_k}{\partial \dot{x}_i} + \frac{\partial}{\partial \dot{x}_i}\frac{\partial q_j}{\partial t} \\ &= \sum_k \frac{\partial q_j}{\partial x_k}\delta_{ik} + \frac{\partial}{\partial t}\frac{\partial q_j}{\partial \dot{x}_i} \\ &= \frac{\partial q_j}{\partial x_i} \end{aligned} \tag{1.40}$$

where δ_{ik} is the Kronecker number and $\dfrac{\partial q_j}{\partial \dot{x}_i} = 0$.

Substituting Eq. (1.40) into Eq. (1.38), we obtain:

$$\frac{\partial L}{\partial \dot{x}_i} = \sum_j \frac{\partial L}{\partial \dot{q}_j} \frac{\partial q_j}{\partial x_i} \tag{1.41}$$

Lagrange's equation involves the time derivative of the above equation, which is the derivative along the path system takes as it moves through configuration space. It is also called the *stream derivative*, as in the fluid mechanics, giving the rate change of some property of a fixed element of the fluid defined throughout the fluid, $f(\mathbf{r}, t)$, as it flows as a whole. It represents the total derivative to indicate that the motion is followed rather than evaluating the rate of change at any specified point in space using the partial derivative.

The function $f(x, t)$ of extended configuration space has a total time derivative given by

$$\frac{df}{dt} = \sum_j \frac{\partial f}{\partial x_j} \dot{x}_j + \frac{\partial f}{\partial t} \tag{1.42}$$

Using the Leibnitz's rule on Eq. (1.41) and using the above relation in the second term, it can be found that

$$\frac{d}{dt}\frac{\partial L}{\partial \dot{x}_i} = \sum_j \left(\frac{d}{dt}\frac{\partial L}{\partial \dot{q}_j}\right) \frac{\partial q_j}{\partial x_i} + \sum_j \frac{\partial L}{\partial \dot{q}_j} \left(\sum_k \frac{\partial^2 q_j}{\partial x_i \partial x_k} \dot{x}_k + \frac{\partial^2 q_j}{\partial x_i \partial t}\right) \tag{1.43}$$

Using the chain rule again:

$$\frac{\partial L}{\partial x_i} = \sum_j \frac{\partial L}{\partial q_j} \frac{\partial q_j}{\partial x_i} + \sum_j \frac{\partial L}{\partial \dot{q}_j} \frac{\partial \dot{q}_j}{\partial x_i} \tag{1.44}$$

where the last term $\frac{\partial \dot{q}_j}{\partial x_i}$ is, in general, different from zero, since $\dot{q}_j$, in general, depends on both the coordinates and velocities.

Taking the derivative with respect to x_i of the expression in Eq. (1.39) gives

$$\frac{\partial \dot{q}_j}{\partial x_i} = \sum_k \frac{\partial^2 q_j}{\partial x_i \partial x_k} \dot{x}_k + \frac{\partial^2 q_j}{\partial x_i \partial t} \tag{1.45}$$

Hence,

$$\frac{\partial L}{\partial x_i} = \sum_j \frac{\partial L}{\partial q_j} \frac{\partial q_j}{\partial x_i} + \sum_j \frac{\partial L}{\partial \dot{q}_j} \left(\sum_k \frac{\partial^2 q_j}{\partial x_i \partial x_k} \dot{x}_k + \frac{\partial^2 q_j}{\partial x_i \partial t}\right). \tag{1.46}$$

Lagrange's equation in Cartesian coordinates indicates that the left hand sides of Eqs. (1.43) and (1.46) are equal, hence by subtracting them the second terms cancel, and hence

$$0 = \sum_j \left(\frac{d}{dt} \frac{\partial L}{\partial \dot{q}_j} - \frac{\partial L}{\partial q_j} \right) \frac{\partial q_j}{\partial x_i} . \tag{1.47}$$

The matrix $\partial q_j / \partial x_i$, is non-singular, because it has $\partial x_i / \partial q_j \neq 0$ as its inverse, therefore, the Lagrange's Equation in generalized coordinates can be derived:

$$\frac{d}{dt} \left(\frac{\partial L}{\partial \dot{q}_j} \right) - \frac{\partial L}{\partial q_j} = 0 \tag{1.48}$$

where $j = 1, 2, \cdots, f$. Thus, Lagrange's equations are invariant in form under the transformation from the Cartesian to generalized coordinates, used to represent the configuration of a system. It is primarily for this reason that the Lagrangian function in this particular and peculiar combination form of kinetic energy and potential energy is useful. Note that it is implicitly assumed the Lagrangian itself transformed like a scalar, in that its value at a given physical phase point in configuration space is independent of the choice of generalized coordinates that describe the point. The change of coordinates given by expression in Eq. (1.36) is called a *point transformation*.

1.5.2 Lagrangian for Constrained Systems

Now, the case of the constrained systems and non-conservative forces will be considered. The *holonomic* constraints are expressed as k real functions $\phi_\alpha(\mathbf{r}_1, \cdots, \mathbf{r}_N, t) = 0$, for $\alpha = 1, 2, \cdots, k$, which are "enforced" by constraint forces $\mathbf{F}_i^C$ on each particle i. In general, there may be other forces, which are called $\mathbf{F}_i^D$ that have a dynamical effect, which are possibly known functions of the configuration and time but not necessarily through the potential. In each of these cases, the full configuration space is $\Re^{3N}$, but the constraints restrict the motion to only an allowed subspace of the entire extended configuration space. It will also be assumed that the constraint forces, in general, satisfy the restriction no net *virtual* work is done by the forces of constraint for any possible virtual displacement.

From Newton's second law,

$$\dot{\mathbf{p}}_i = \mathbf{F}_i = \mathbf{F}_i^C + \mathbf{F}_i^D .$$

By multiplying by an arbitrary virtual displacement, $\delta \mathbf{r}_i$, summing up over all the particles, and rearranging the equation we obtain:

$$\sum_i \left(\mathbf{F}_i^D - \dot{\mathbf{p}}_i \right) \cdot \delta \mathbf{r}_i = - \sum_i \mathbf{F}_i^C \cdot \delta \mathbf{r}_i = 0$$

where the first equality would be true even if $\delta\mathbf{r}_i$ did not satisfy the constraints, but the second requires $\delta\mathbf{r}_i$ to be an allowed virtual displacement. Thus, we derived the so-called *D'Alembert's Principle*:

$$\sum_i \left(\mathbf{F}_i^D - \dot{\mathbf{p}}_i\right) \cdot \delta\mathbf{r}_i = 0$$

This equation determines the motion on the constrained subspace and does not involve the unspecified forces of constraint $\mathbf{F}^C$.

The constrained subspace is characterized by the vector

$$\mathbf{r}_i = \mathbf{r}_i(q_1, q_2, \cdots, q_f, t)$$

for $i = 1, 2, \cdots, N$, which are known functions of f independent $q_1, q_2, \cdots, q_f$ generalized coordinates. Since there are k holonomic constraints, and the number of degrees of freedom of unconstrained system is $3N$, then $f = 3N - k$. Furthermore,

$$\Delta\mathbf{r}_i = \sum_j \frac{\partial\mathbf{r}_i}{\partial q_j}\Delta q_j + \frac{\partial\mathbf{r}_i}{\partial t}\Delta t \;. \tag{1.49}$$

Dividing by Δt both sides of this equation and taking the limit when $\Delta t \to 0$, we get:

$$\mathbf{v}_i = \sum_j \frac{\partial\mathbf{r}_i}{\partial q_j}\dot{q}_j + \frac{\partial\mathbf{r}_i}{\partial t} \;, \tag{1.50}$$

where $\mathbf{v}_i = \lim_{\Delta t\to 0} \Delta\mathbf{r}_i/\Delta t$ is the particle velocity. On the other hand, for the *virtual* displacement, $\Delta t = 0$, it can be written that

$$\delta\mathbf{r}_i = \sum_j \frac{\partial\mathbf{r}_i}{\partial q_j}\delta q_j \;.$$

Differentiating Eq. (1.50) with respect to $\dot{q}_j$, it can be found:

$$\frac{\partial\mathbf{v}_i}{\partial\dot{q}_j} = \frac{\partial\mathbf{r}_i}{\partial q_j} \;,$$

and differentiating Eq. (1.50) with respect to q_j, yield

$$\frac{\partial\mathbf{v}_i}{\partial q_j} = \sum_k \frac{\partial^2\mathbf{r}_i}{\partial q_j \partial q_k}\dot{q}_k + \frac{\partial^2\mathbf{r}_i}{\partial q_j \partial t} = \frac{d}{dt}\frac{\partial\mathbf{r}_i}{\partial q_j} \;,$$

where Eq. (1.42) is used. The first term in Eq. (1.50) (D'Alembert's Principle) is

$$\sum_i \mathbf{F}_i \cdot \delta \mathbf{r}_i = \sum_j \sum_i \mathbf{F}_i \cdot \frac{\partial \mathbf{r}_i}{\partial q_j} \delta q_j = \sum_j Q_j \cdot \delta q_j \,. \tag{1.51}$$

The generalized force Q_j has the same form as in the unconstrained case (see Eq. (1.24)), but there are only as many of them as there are unconstrained degrees of freedom.

The second term in Eq. (1.50) can be expressed as the following:

$$\begin{aligned}
\sum_i \dot{\mathbf{p}}_i \cdot \delta \mathbf{r}_i &= \sum_i \frac{d\mathbf{p}_i}{dt} \sum_j \frac{\partial \mathbf{r}_i}{\partial q_j} \delta q_j \\
&= \sum_j \frac{d}{dt}\left(\sum_i \mathbf{p}_i \cdot \frac{\partial \mathbf{r}_i}{\partial q_j}\right) \delta q_j - \sum_{i,j} \mathbf{p}_i \cdot \left(\frac{d}{dt}\frac{\partial \mathbf{r}_i}{\partial q_j}\right) \delta q_j \\
&= \sum_j \frac{d}{dt}\left(\sum_i \mathbf{p}_i \cdot \frac{\partial \mathbf{v}_i}{\partial \dot{q}_j}\right) \delta q_j - \sum_{i,j} \mathbf{p}_i \cdot \frac{\partial \mathbf{v}_i}{\partial q_j} \delta q_j \\
&= \sum_j \left[\frac{d}{dt}\left(\sum_i m_i \mathbf{v}_i \cdot \frac{\partial \mathbf{v}_i}{\partial \dot{q}_j}\right) - \sum_i m_i \mathbf{v}_i \cdot \frac{\partial \mathbf{v}_i}{\partial q_j}\right] \delta q_j \\
&= \sum_j \left[\frac{d}{dt}\frac{\partial T}{\partial \dot{q}_j} - \frac{\partial T}{\partial q_j}\right] \delta q_j \,,
\end{aligned} \tag{1.52}$$

where T is the kinetic energy. Substituting this equation into the expression found for the D'Alembert's Principle, Eq. (1.51), we get

$$\sum_j \left[\frac{d}{dt}\frac{\partial T}{\partial \dot{q}_j} - \frac{\partial T}{\partial q_j} - Q_j\right] \delta q_j = 0 \,.$$

It was assumed a holonomic system and q_j, for $j = 1, 2, \cdots, f$, are independent, therefore, this equation holds for arbitrary virtual displacements δq_j, and hence

$$\frac{d}{dt}\frac{\partial T}{\partial \dot{q}_j} - \frac{\partial T}{\partial q_j} - Q_j = 0 \,. \tag{1.53}$$

In the cases when all the forces are conservative, we can write

$$\mathbf{F}_i = -\nabla_i U(\mathbf{r}_1, \mathbf{r}_2, \cdots, \mathbf{r}_N, t) \,,$$

or

$$Q_j = -\sum_i \frac{\partial \mathbf{r}_i}{\partial q_j} \cdot \nabla_i U = -\left(\frac{\partial \tilde{U}(q_1, \cdots, q_f, t)}{\partial q_j}\right)_t . \tag{1.54}$$

Notice that Q_j depends only on the value of U on the constrained surface. Since U is independent of $\dot{q}_i$, then

$$0 = \frac{d}{dt}\frac{\partial T}{\partial \dot{q}_j} - \frac{\partial T}{\partial q_j} + \frac{\partial U}{\partial q_j} = \frac{d}{dt}\frac{\partial (T-U)}{\partial \dot{q}_j} - \frac{\partial (T-U)}{\partial q_j}$$

or

$$\frac{d}{dt}\left(\frac{\partial L}{\partial \dot{q}_j}\right) - \frac{\partial L}{\partial q_j} = 0 \tag{1.55}$$

for $j = 1, 2, \cdots, f$. This is Lagrange's equation, which is derived in the more general case of constrained systems.

1.6 Hamilton's Principle

The configuration space of a system at any instant of time t can be represented by the generalized coordinates $q_i(t)$ for $i = 1, 2 \cdots, f$. The space characterized by the vector $(q_1, q_2, \cdots, q_f)$ is called *configuration space*. The trajectory or the motion of the point in configuration space gives the time evolution of the system as a function of time, specified by the time dependence of the vector $\big(q_1(t), q_2(t), \cdots, q_f(t)\big)$.

We can think that a system can take different paths for going from the state 1 at time t_1 to the state 2 at some other time t_2, which may or may not obey to the Newton's second law. However, only those paths for which $q_i(t)$ $(i = 1, 2, \cdots, f)$ are differentiable will be considered. Along such paths, an *action* can be defined as the following:

$$I = \int_{t_1}^{t_2} L\left(q(t), \dot{q}(t), t\right) dt \tag{1.56}$$

The action depends not only on the starting and ending points, respectively, $q(t_1)$ and $q(t_2)$, but its value also depends on the path, in contrast to what we know about the work of the conservative forces on a system moving in configuration space.

This action is also known as the *Hamilton's principle*, which states that the actual motion of the particle from $q(t_1) = q_i$ to $q(t_2) = q_f$ is along a path $q(t)$ for which the action is stationary. That means, for any small deviation of the path from the actual one, keeping the initial and final configurations fixed, the variation of the action vanishes to the first order in this deviation.

In general, to find out the stationary points of a differentiable function of one variable, we first differentiate it, and then solve the equation found by setting the derivative to zero. Here, let f be a differentiable function of several variables x_i, then the first-order variation of the function is given by

$$\Delta f = \sum_i (x_i - x_{0i}) \left(\frac{\partial f}{\partial x_i} \right)_{x_0}$$

Equalizing $\Delta f = 0$, then x_0 is a stationary point ($x_0 \neq x_i$), if

$$\left(\frac{\partial f}{\partial x_i} \right)_{x_0} = 0$$

for every i.

Therefore, let's consider a change $q(t) \to q(t) + \delta q(t)$, then the derivative will vary by

$$\delta \dot{q} = \delta \left[\frac{dq(t)}{dt} \right] = \frac{d}{dt} \left[\delta q(t) \right] ,$$

and the functional I will vary by

$$\delta I = \int_{t_1}^{t_2} \left(\frac{\partial L}{\partial q} \delta q + \frac{\partial L}{\partial \dot{q}} \delta \dot{q} \right) dt \tag{1.57}$$

Using the following relation:

$$\frac{d}{dt} \left(\frac{\partial L}{\partial \dot{q}} \delta q \right) = \frac{d}{dt} \left(\frac{\partial L}{\partial \dot{q}} \right) \delta q + \frac{\partial L}{\partial \dot{q}} \delta \dot{q}$$

we obtain

$$\frac{\partial L}{\partial \dot{q}} \delta \dot{q} = \frac{d}{dt} \left(\frac{\partial L}{\partial \dot{q}} \delta q \right) - \frac{d}{dt} \left(\frac{\partial L}{\partial \dot{q}} \right) \delta q \tag{1.58}$$

Replacing Eq. (1.58) into Eq. (1.57), we obtain:

$$\begin{aligned} \delta I &= \int_{t_1}^{t_2} \left(\frac{\partial L}{\partial q} \delta q + \frac{d}{dt} \left(\frac{\partial L}{\partial \dot{q}} \delta q \right) - \frac{d}{dt} \left(\frac{\partial L}{\partial \dot{q}} \right) \delta q \right) dt \\ &= \left(\frac{\partial L}{\partial \dot{q}} \delta q \right)_1^2 + \int_{t_1}^{t_2} \left[\frac{\partial L}{\partial q} - \frac{d}{dt} \frac{\partial L}{\partial \dot{q}} \right] \delta q dt , \end{aligned} \tag{1.59}$$

The first term in Eq. (1.59) is zero, because the boundary terms which have a factor of δq at the initial or final point from Hamilton' principle are hold at q_1 and q_2 fixed. Therefore, the functional is stationary, i.e. $\delta I = 0$, if and only if

$$\frac{d}{dt}\left(\frac{\partial L}{\partial \dot{q}}\right) - \frac{\partial L}{\partial q} = 0 \tag{1.60}$$

for $t \in (t_1, t_2)$. As it can be seen, Eq. (1.60) is the Lagrangian equation derived in the previous section.

1.7 Hamiltonian Method

Lagrange's equations do not form a dynamical system in the way we discussed in Sect. 1.4, because they implicitly contain second-order derivatives, $\ddot{\mathbf{q}}$. However, there exists the possibility to derive a system of the first-order equations from the second-order one.[1] This is done by doubling the dimensions of the phase space of time dependent variables, introducing the so-called generalized velocities $\mathbf{u}$ as independent of generalized coordinates. The dynamical system then becomes

$$\begin{aligned}\dot{\mathbf{q}} &= \mathbf{u} \\ \dot{\mathbf{u}} &= \ddot{\mathbf{q}}(\mathbf{q}, \mathbf{u}, t)\end{aligned} \tag{1.61}$$

with a phase space dimension of $2f$.

Assuming that the Lagrangian is a function of coordinates, velocities and time, as such $L = L(\mathbf{q}, \mathbf{u}, t)$, then

$$\frac{\partial L}{\partial u_i} \equiv \frac{\partial L}{\partial \dot{q}_i} = f(\mathbf{q}, \dot{\mathbf{q}}, t)$$

Then, we can evaluate the following derivative:

$$\begin{aligned}\frac{d}{dt}\left(\frac{\partial L}{\partial \dot{q}_i}\right) &= \frac{d}{dt} f(\mathbf{q}, \dot{\mathbf{q}}, t) \\ &= \sum_{j=1}^{f} \frac{\partial f}{\partial q_j}\frac{dq_j}{dt} + \sum_{j=1}^{f} \frac{\partial f}{\partial \dot{q}_j}\frac{d\dot{q}_j}{dt} + \frac{\partial f}{\partial t} \\ &= \sum_{j=1}^{f} \frac{\partial^2 L}{\partial \dot{q}_i \partial q_j}\dot{q}_j + \sum_{j=1}^{f} \frac{\partial^2 L}{\partial \dot{q}_i \partial \dot{q}_j}\ddot{q}_j + \frac{\partial^2 L}{\partial \dot{q}_i \partial t}\end{aligned} \tag{1.62}$$

[1]This method has been used frequently in numerical problems, because the standard numerical integrator methods require the problem to be represented in terms of systems of the first-order differential equations.

By replacing this expression into Eq. (1.55), we get

$$\sum_{j=1}^{f} \frac{\partial^2 L}{\partial \dot{q}_i \partial q_j} \dot{q}_j + \sum_{j=1}^{f} \frac{\partial^2 L}{\partial \dot{q}_i \partial \dot{q}_j} \ddot{q}_j + \frac{\partial^2 L}{\partial \dot{q}_i \partial t} - \frac{\partial L}{\partial q_i} = 0$$

Or, by determining the Hessian matrix, H_{ij}, which is a generalized mass tensor as

$$H_{ij} = \frac{\partial^2 L}{\partial \dot{q}_i \partial \dot{q}_j}$$

we get

$$\ddot{\mathbf{q}} = \mathbf{H}^{-1} \cdot \left[\frac{\partial L}{\partial \mathbf{q}} - \frac{\partial^2 L}{\partial \dot{\mathbf{q}} \partial \mathbf{q}} \dot{\mathbf{q}} - \frac{\partial^2 L}{\partial \dot{\mathbf{q}} \partial t} \right]$$

Thus, the dynamical system is

$$\begin{aligned} \dot{\mathbf{q}} &= \mathbf{u} \\ \ddot{\mathbf{q}} &= \mathbf{H}^{-1} \cdot \left[\frac{\partial L}{\partial \mathbf{q}} - \frac{\partial^2 L}{\partial \dot{\mathbf{q}} \partial \mathbf{q}} \dot{\mathbf{q}} - \frac{\partial^2 L}{\partial \dot{\mathbf{q}} \partial t} \right] \end{aligned} \tag{1.63}$$

which works for Hessian matrix $\mathbf{H}$ being non-singular, so that $\mathbf{H}^{-1}$ exists.

1.7.1 Hamilton Equations

In particular, we introduce a new variable

$$\mathbf{p} = \frac{\partial L(\mathbf{q}, \dot{\mathbf{q}}, t)}{\partial \dot{\mathbf{q}}}$$

which is called *generalized momentum* canonically conjugate to $\mathbf{q}$. From Eq. (1.55), we can easily find that

$$\dot{\mathbf{p}} = \frac{\partial L}{\partial \mathbf{q}}$$

Considering $L = L(\mathbf{q}, \dot{\mathbf{q}}, t)$, the differential of L is

$$\begin{aligned} dL &= \frac{\partial L}{\partial \mathbf{q}} \cdot d\mathbf{q} + \frac{\partial L}{\partial \dot{\mathbf{q}}} \cdot d\dot{\mathbf{q}} + \frac{\partial L}{\partial t} dt \\ &= \dot{\mathbf{p}} \cdot d\mathbf{q} + \mathbf{p} \cdot d\dot{\mathbf{q}} + \frac{\partial L}{\partial t} dt \end{aligned} \tag{1.64}$$

Defining the Hamiltonian $H(\mathbf{q}, \mathbf{p}, t)$ as Legendre transformation of L:

$$H(\mathbf{q}, \mathbf{p}, t) = \dot{\mathbf{q}} \cdot \mathbf{p} - L(\mathbf{q}, \dot{\mathbf{q}}, t) \tag{1.65}$$

We can calculate the differential of H as

$$\begin{aligned} dH &= d(\dot{\mathbf{q}} \cdot \mathbf{p}) - dL(\mathbf{q}, \dot{\mathbf{q}}, t) \\ &= d\dot{\mathbf{q}} \cdot \mathbf{p} + \dot{\mathbf{q}} \cdot d\mathbf{p} - \left(\dot{\mathbf{p}} \cdot d\mathbf{q} + \mathbf{p} \cdot d\dot{\mathbf{q}} + \frac{\partial L}{\partial t} dt \right) \\ &= \dot{\mathbf{q}} \cdot d\mathbf{p} - \dot{\mathbf{p}} \cdot d\mathbf{q} - \frac{\partial L}{\partial t} dt \end{aligned} \tag{1.66}$$

On the other hand, from Eq. (1.65), $H = H(\mathbf{q}, \mathbf{p}, t)$, therefore, dH is equal to

$$dH = \frac{\partial H}{\partial \mathbf{q}} \cdot d\mathbf{q} + \frac{\partial H}{\partial \mathbf{p}} \cdot d\mathbf{p} + \frac{\partial H}{\partial t} dt \tag{1.67}$$

By comparing Eq. (1.66) and Eq. (1.67), we get

$$\begin{aligned} \dot{\mathbf{p}} &= -\frac{\partial H}{\partial \mathbf{q}} \\ \dot{\mathbf{q}} &= \frac{\partial H}{\partial \mathbf{p}} \\ \frac{\partial H}{\partial t} &= -\frac{\partial L}{\partial t} \end{aligned} \tag{1.68}$$

For H which does not depend explicitly on time t, we get the so-called *Hamilton's equations of motion*

$$\begin{aligned} \dot{\mathbf{p}} &= -\frac{\partial H}{\partial \mathbf{q}} \\ \dot{\mathbf{q}} &= \frac{\partial H}{\partial \mathbf{p}} \end{aligned} \tag{1.69}$$

On the other hand, Hamilton's equations of motion, Eq. (1.69), form a dynamical system (as discussed in Sect. 1.4), where

$$\eta = (\mathbf{q}, \mathbf{p})$$

and

$$\mathbf{V} = \left(\frac{\partial H}{\partial \mathbf{p}}, -\frac{\partial H}{\partial \mathbf{q}} \right)$$

Then, Eq. (1.69) can equivalently be written as:

$$\dot{\eta} = \mathbf{V}(\eta)$$

1.8 Time Averages and Virial Theorem

The virial theorem is concerned with the time averages of some mechanical quantity of the system. For that, consider a system of N mass points particles with positions $\mathbf{r}_i$ and applied forces $\mathbf{F}_i$, which may also include external forces and constraints. The equations of motions governing the dynamics are given by Eq. (1.29). Consider the following quantity

$$G = \sum_{i=1}^{N} \mathbf{r}_i \cdot \mathbf{p}_i$$

Taking the total time derivative of the quantity G:

$$\frac{dG}{dt} = \sum_{i=1}^{N} \dot{\mathbf{r}}_i \cdot \mathbf{p}_i + \sum_{i=1}^{N} \mathbf{r}_i \cdot \dot{\mathbf{p}}_i \tag{1.70}$$

This expression can further be simplified as

$$\begin{aligned} \frac{dG}{dt} &= \sum_{i=1}^{N} \dot{\mathbf{r}}_i \cdot m_i \dot{\mathbf{r}}_i + \sum_{i=1}^{N} \mathbf{r}_i \cdot \mathbf{F}_i \\ &= 2T + \sum_{i=1}^{N} \mathbf{r}_i \cdot \mathbf{F}_i \end{aligned} \tag{1.71}$$

Here, T denotes the kinetic energy of the system.

The time average of the derivative $\frac{dG}{dt}$ in the interval from 0 to $\mathcal{T}$, $\langle \frac{dG}{dt} \rangle_t$, is obtained using Eq. (1.71) as the following:

$$\langle \frac{dG}{dt} \rangle_t = \frac{1}{\mathcal{T}} \int_0^{\mathcal{T}} \frac{dG}{dt} dt = \langle 2T \rangle_t + \langle \sum_{i=1}^{N} \mathbf{r}_i \cdot \mathbf{F}_i \rangle_t$$

where $\langle \cdots \rangle_t$ denotes the time average.

From here, it can be found that

$$\langle 2T\rangle_t + \langle\sum_{i=1}^{N} \mathbf{r}_i \cdot \mathbf{F}_i\rangle_t = \frac{1}{\mathcal{T}}\left[G(\mathcal{T}) - G(0)\right] \tag{1.72}$$

For the periodic motions, the coordinates will repeat after a certain time, and if $\mathcal{T}$ is the period, then the right-hand side of Eq. (1.72) vanishes. That also is true for non-periodic motions, given that all coordinates and velocities of particles remain finite so that for G to exist an upper limit. If $\mathcal{T}$ is sufficiently long, then the right-hand-side of Eq. (1.72) can be chosen to be very small since G is finite. Therefore, in both cases, it can be written that

$$\langle T\rangle_t = -\frac{1}{2}\langle\sum_{i=1}^{N} \mathbf{r}_i \cdot \mathbf{F}_i\rangle_t \tag{1.73}$$

which is known as *virial theorem*. The right-hand-side is known as the *virial of Clausius*.

Consider the general case when forces $\mathbf{F}_i$ are the sum of two terms, the non-frictional forces $\mathbf{F}'_i$ and frictional forces $\mathbf{f}_i$ (which are proportional to velocity). Hence,

$$\mathbf{F}_i = \mathbf{F}'_i + \mathbf{f}_i = \mathbf{F}'_i - \gamma\mathbf{v}_i$$

where γ is the proportionality constant. Substituting this expression into Eq. (1.73), we get

$$\langle T\rangle_t = -\frac{1}{2}\langle\sum_{i=1}^{N} \mathbf{r}_i \cdot \mathbf{F}'_i\rangle_t + \frac{\gamma}{2}\langle\sum_{i=1}^{N} \mathbf{r}_i \cdot \mathbf{v}_i\rangle_t \tag{1.74}$$

The second term vanishes, because:

$$\begin{aligned}
\langle\sum_{i=1}^{N} \mathbf{r}_i \cdot \mathbf{v}_i\rangle_t &= \langle\sum_{i=1}^{N} \mathbf{r}_i \cdot \frac{d\mathbf{r}_i}{dt}\rangle_t \\
&= \frac{1}{2}\langle\sum_{i=1}^{N} \frac{d(\mathbf{r}_i \cdot \mathbf{r}_i)}{dt}\rangle_t \\
&= \frac{1}{2}\sum_{i=1}^{N}\langle\frac{d(\mathbf{r}_i \cdot \mathbf{r}_i)}{dt}\rangle_t = 0
\end{aligned}$$

Therefore, there is no contribution from the frictional forces and the virial depends only on the non-frictional term $\mathbf{F}_i'$:

$$\langle T\rangle_t = -\frac{1}{2}\langle\sum_{i=1}^{N}\mathbf{r}_i\cdot\mathbf{F}_i'\rangle_t \tag{1.75}$$

Considering the forces are conservative, then they can be expressed as derivative of some potential V: $\mathbf{F}_i' = -\nabla_i V$. Thus, Eq. (1.75) can be written as

$$\langle T\rangle_t = \frac{1}{2}\langle\sum_{i=1}^{N}\mathbf{r}_i\cdot\nabla_i V\rangle_t \tag{1.76}$$

Assuming that the potential depends on the distance between two particles only, according to the power law:

$$V\left(\mid\mathbf{r}_i-\mathbf{r}_j\mid\right) = \alpha\mid\mathbf{r}_i-\mathbf{r}_j\mid^{-n}$$

where α is a real constant and n is an integer number.

Then, the virial theorem can be written as

$$\langle T\rangle_t = \frac{\alpha}{2}\langle\sum_{i=1}^{N}\sum_{j=1\neq i}^{N}\mathbf{r}_i\cdot\nabla_{ij}\mid\mathbf{r}_i-\mathbf{r}_j\mid^{-n}\rangle_t \tag{1.77}$$

or,

$$\langle T\rangle_t = -\frac{n}{2}\langle\sum_{i=1}^{N}\sum_{j=1\neq i}^{N}V\left(\mid\mathbf{r}_i-\mathbf{r}_j\mid\right)\rangle_t \tag{1.78}$$

Note that using the Euler's theorem for homogeneous functions, Eq. (1.78) holds for every homogeneous function in $\mid\mathbf{r}_i-\mathbf{r}_j\mid$ of V of degree n. For the special case when $n = 1$, then once obtains the virial theorem in well known form:

$$\langle T\rangle_t = -\frac{1}{2}\langle V\rangle_t$$

where

$$\langle V\rangle_t = \langle\sum_{i=1}^{N}\sum_{j=1\neq i}^{N}V\left(\mid\mathbf{r}_i-\mathbf{r}_j\mid\right)\rangle_t$$

1.9 Canonical Transformations

Consider a system where the Hamiltonian is a constant of motion. All the coordinates q_i are cyclic, and hence $H = H(\mathbf{p})$. Using these as conditions, all conjugate momenta p_i are constant, i.e.,

$$p_i = c_i$$

for $i = 1, 2, \cdots, n$.

Since the Hamiltonian is only a function of momenta and not of the time and cyclic coordinates (because it is conserved), it can be written that

$$H = H(c_1, c_2, \cdots, c_n)$$

The Hamilton's equation for the derivative of coordinates $\dot{q}_i$ can be written as

$$\dot{q}_i = \frac{\partial H}{\partial c_i} = \alpha_i \tag{1.79}$$

where α_i are functions of c_i only, and hence are constant in time. Solution of Eq. (1.79) gives

$$q_i = \alpha_i t + \beta_i \tag{1.80}$$

where β_i are constants of integration, and therefore are determined from the initial conditions:

$$\beta_i = q_i(0)$$

for $i = 1, 2, \cdots, n$.

That corresponds to motion with constant velocity for each particle of the system, where the relative distance between any two particles i and j of the system changes with time according to:

$$q_j - q_i = (\alpha_j - \alpha_i)t + (\beta_j - \beta_i)$$

If the velocities of the particles are the same (i.e., $\alpha_i = \alpha_j, \forall(i, j) : i \neq j$), then that would correspond to a motion where the entire system can move along a straight line as a rigid body (i.e., the relative distances between particles do not change with time, and they equal their initial separations, $\beta_j - \beta_i, \forall(i, j) : i \neq j$) with constant velocity of the center of mass, or the entire system can rotate about one of the axes with constant angular velocity. Another example of such motion could be the system of particles, such as photons, originating from different stars in space as seen from an observer on the Earth. The motion of these photons is with constant speed, and they follow separate streamlines which are curved because of the existence of gravitation waves in space. On the other hand, if the particles have different velocities (i.e., $\alpha_i \neq \alpha_j, \forall(i, j) : i \neq j$), then their separations vary linearly with time, but the relative velocity is constant. That corresponds to a flow motion of the particles of an ideal fluid along separate straight or curved streamlines. Here, the kinetic energy and the linear momentum of each particle are conserved, and hence the Hamiltonian function or total energy of the system is a constant of motion.

The number of cyclic coordinates depends on the choice of generalized coordinates. There may exist only one particular choice for each problem in which all the coordinates are cyclic. The only remaining problem is to find this set. The transformations considered above involve transformations from one set of coordinates, say q_i, to another set, say Q_i, by a transformation of the form:

$$Q_i = Q_i(q_1, \cdots, q_n, t), \quad i = 1, 2, \cdots, n \tag{1.81}$$

For example, these transformations can be an orthogonal transformation, or the change from Cartesian to planar polar coordinates. These transformations are known as *point transformations*. In the Hamiltonian formalism, the momenta are also independent variables, similarly to the generalized coordinates. Therefore, the equations of transformations from the independent coordinates and momenta q_i, p_i to the new set Q_i, P_i can be written, for $i = 1, 2, \cdots, n$, as

$$\begin{aligned} Q_i &= Q_i(q, p, t) \\ P_i &= P_i(q, p, t) \end{aligned} \tag{1.82}$$

which indicates that the new coordinates will be function of both, the old momenta and coordinates.

By definition, Eq. (1.81) determines a *point transformation* of configuration space, and Eq. (1.82) determines a *point transformation* of phase space. In Hamiltonian formulation, only the transformations in which the new set Q and P are canonical coordinates show an interest. This condition is satisfied if there exist a function $\mathcal{H}(Q, P, t)$ such that the equations of motion in the new set in the Hamiltonian formalism can be written as:

$$\begin{aligned} \dot{Q}_i &= \frac{\partial \mathcal{H}}{\partial P_i} \\ \dot{P}_i &= -\frac{\partial \mathcal{H}}{\partial Q_i} \end{aligned} \tag{1.83}$$

for $i = 1, 2, \cdots, n$. Here, the function $\mathcal{H}$ is considered the Hamiltonian in the new coordinate set. (Q, P) are canonical coordinates for all mechanical systems of the same number of degrees of freedom, and hence, these transformations are problem-independent. The new equations, Eq. (1.83), govern the dynamics of motion in the new momenta P and coordinates Q regardless of the form of old Hamiltonian H.

In order for Q_i and P_i to be canonical coordinates, they have to satisfy the modified Hamilton's principle of the following form:

$$\delta \int_{t_1}^{t_2} \left(\sum_{i=1}^{n} P_i \dot{Q}_i - \mathcal{H}(Q_1, \cdots, Q_n, P_1, \cdots, P_n, t) \right) dt = 0 \tag{1.84}$$

The same principle is also satisfied by the old coordinates and momenta:

$$\delta \int_{t_1}^{t_2} \left(\sum_{i=1}^{n} p_i \dot{q}_i - H(q_1, \cdots, q_n, p_1, \cdots, p_n, t) \right) dt = 0 \tag{1.85}$$

The general form of the Hamilton's principle provides zero variation at the endpoints, therefore both statements given by Eqs. (1.84) and (1.85) will be satisfied if the integrands are related to each other as:

$$\begin{aligned} &\lambda \left(\sum_{i=1}^{n} p_i \dot{q}_i - H(q_1, \cdots, q_n, p_1, \cdots, p_n, t) \right) \\ &= \sum_{i=1}^{n} P_i \dot{Q}_i - \mathcal{H}(Q_1, \cdots, Q_n, P_1, \cdots, P_n, t) + \frac{dF}{dt} \end{aligned} \tag{1.86}$$

where λ is a constant independent of the canonical coordinates and the time, and F is any function of the coordinates of phase space with continuous second derivative. This type of canonical transformation of coordinates is known as *scale transformation*.

For example, consider a change in units used to measure the coordinates and momenta, such that:

$$\begin{aligned} Q_i' &= \mu q_i \\ P_i' &= \nu p_i, \quad i = 1, 2, \cdots, n \end{aligned} \tag{1.87}$$

where μ and ν are two scaling factors of the size of the units of coordinates and momenta, respectively. Then, it can be seen that the Hamilton's equations given by Eq. (1.83) will be satisfied for a transformed Hamiltonian given as

$$H'(Q', P') = \mu\nu H(q, p)$$

On the other hand, the integrands of corresponding modified Hamilton's principles are related as:

$$\mu\nu\left(\sum_{i=1}^{n} p_i\dot{q}_i - H(q, p, t)\right) = \sum_{i=1}^{n} P_i'\dot{Q}_i' - H'(Q', P', t)$$

Comparing this expression with one given by Eq. (1.86), it can be seen that $\lambda = \mu\nu$. Here, the focus will be on these transformations of canonical coordinates for which $\lambda = 1$. For instance, if there is some transformation of canonical coordinates

$$(q, p) \rightarrow (Q', P')$$

for $\lambda \neq 1$, then there exists an intermediate set of canonical coordinates (Q, P) such that

$$\begin{aligned} Q_i &= \mu Q_i' \\ P_i &= \nu P_i', \quad i = 1, 2, \cdots, n \end{aligned} \tag{1.88}$$

where $\lambda = \mu\nu$.

Then, the transformations between two sets of canonical coordinates (q, p) and (Q, P) will satisfy Eq. (1.86) for $\lambda = 1$, that is

$$\sum_{i=1}^{n} p_i\dot{q}_i - H(q, p, t) = \sum_{i=1}^{n} P_i\dot{Q}_i - \mathcal{H}(Q, P, t) + \frac{dF}{dt} \tag{1.89}$$

The transformations of canonical coordinates with $\lambda \neq 1$ are called *extended canonical transformations* and those with $\lambda = 1$ are called *canonical transformations*.

If the transformations of canonical coordinates (see Eq. (1.82)) do not include time explicitly, i.e.

$$\begin{aligned} Q_i &= Q_i(q_1, \cdots, q_n, p_1, \cdots, p_n) \\ P_i &= P_i(q_1, \cdots, q_n, p_1, \cdots, p_n), \quad i = 1, 2, \cdots, n \end{aligned} \tag{1.90}$$

they are called *restricted canonical transformations*.

The term dF/dt in the expression given by Eq. (1.89) contributes to the variation of the action integral only at the endpoints. Therefore, the requirement that dF/dt vanishes at the endpoints is satisfied if F is a function of either (q, p, t), (Q, P, t) or any combination of the phase space coordinates, because they vanish at the endpoints.

Expressions in Eq. (1.82) indicate that F can be expressed partly in terms of old canonical coordinates (q, p) and partly of the new set (Q, P), as such, it provides a link between these two sets of canonical variables, and it is called the *generating function* of the transformation. To show that, consider F is given as

$$F \equiv F(q, Q, t)$$

Then, replacing this into expression given by Eq. (1.89), we get

$$\sum_{i=1}^{n} p_i\dot{q}_i - H(q,p,t) = \sum_{i=1}^{n} P_i\dot{Q}_i - \mathcal{H}(Q,P,t) + \frac{dF}{dt} \tag{1.91}$$
$$= \sum_{i=1}^{n} P_i\dot{Q}_i - \mathcal{H}(Q,P,t)$$
$$+ \frac{\partial F}{\partial t} + \sum_{i=1}^{n}\frac{\partial F}{\partial q_i}\dot{q}_i + \sum_{i=1}^{n}\frac{\partial F}{\partial Q_i}\dot{Q}_i$$

Because the new and old sets are separately independent, Eq. (1.91) can hold identically only if the coefficients before $\dot{q}_i$ and $\dot{Q}_i$ vanish, that is:

$$p_i = \frac{\partial F}{\partial q_i} \tag{1.92}$$
$$P_i = -\frac{\partial F}{\partial Q_i}$$

for $i = 1, 2, \cdots, n$.

Therefore,

$$\mathcal{H} = H + \frac{\partial F}{\partial t} \tag{1.93}$$

In Eq. (1.92) are n relations defining p_i as a function of q_i, Q_i, and t. Assume that the inverse of the function exists, then they can be solved to determine n other relations of Q_i in terms of q_i, p_i, and t, which give the first half of the transformations given by Eq. (1.82). After solving the relationship between Q_i and the old set (q, p), they can be substituted into second expression given by Eq. (1.92) to give in this way the n P_i expressions as functions of q_i, p_i and t, which form the second half of the transformation equations in Eq. (1.92). Finally, the Eq. (1.93) gives a relation between the new Hamiltonian and the old one, which is evaluated in this way: first the old canonical coordinates set q and p in expression of the old Hamiltonian H are expressed as functions of Q and P by calculating the inverses of Eq. (1.92), then, q_i and $\partial F/\partial t$ are expressed in terms of Q and P in a similar way and the two functions are added to give $\mathcal{H}(Q, P, t)$.

We showed here that if a generating function F is given, then the equations of canonical transformation can be obtained. The problem can also be inverted, that is, given a canonical transformation, how to obtain the generating function F. For that, first Eq. (1.82) are inverted to obtain expressions for p_i and P_i as functions of q, Q, and t. Then, the expressions in Eq. (1.92) are used by integrating them as a set of partial differential equations to derive the expression for F providing the transformation is canonical. This procedure will give a function F uncertain within an arbitrary additive function of t alone, which does not affect the equations of transformations.

It could also happen that it is not suitable to describe the canonical transformation by a generating function of the type $F(q, Q, t)$. For example, consider a transformation in which p_i cannot be written as functions of q, Q, and t, however they can be functions of q, P, and t. Then, we would try to derive a generating function F, which is a function of the old coordinates q and the new momenta P. It can be seen that Eq. (1.91) can be replaced by an equivalent relation involving $\dot{P}_i$ rather than $\dot{Q}_i$, for instance, by writing F in the following form:

$$F = F_2(q, P, t) - \sum_i Q_i P_i \tag{1.94}$$

Replacing this expression for F into Eq. (1.89) gives

$$\begin{aligned}
\sum_i \dot{q}_i p_i - H(q, p, t) &= \sum_i P_i \dot{Q}_i - \mathcal{H}(Q, P, t) \\
&\quad + \frac{dF_2(q, P, t)}{dt} - \sum_i \dot{Q}_i P_i - \sum_i Q_i \dot{P}_i \\
&= -\sum_i Q_i \dot{P}_i - \mathcal{H}(Q, P, t) + \frac{dF_2(q, P, t)}{dt} \\
&= -\sum_i Q_i \dot{P}_i - \mathcal{H}(Q, P, t) + \frac{\partial F_2(q, P, t)}{\partial t} \\
&\quad + \sum_i \frac{\partial F_2(q, P, t)}{\partial q_i} \dot{q}_i + \sum_i \frac{\partial F_2(q, P, t)}{\partial P_i} \dot{P}_i
\end{aligned} \tag{1.95}$$

In order that both sides to be equal, we should have the following equations:

$$\begin{aligned}
Q_i &= -\frac{\partial F_2(q, P, t)}{\partial P_i} \\
p_i &= \frac{\partial F_2(q, P, t)}{\partial q_i}
\end{aligned} \tag{1.96}$$

for $i = 1, 2, \cdots, n$, with

$$\mathcal{H}(Q, P, t) = H(q, p, t) + \frac{\partial F_2(q, P, t)}{\partial t} \tag{1.97}$$

First equation of Eq. (1.96) can be solved to obtain Q_i as a function of q_i, p_i, and t corresponding the first half of the transformations in Eq. (1.82). The second half can be obtained by solving P_i as functions of q_i, p_i, and t, using the second equation of Eq. (1.96).

There are four basic canonical transformations, which can be related to each other through Legendre transformations. For example, let's look at the transition from F_1 to F_2, which is equivalent to transformation from variables q, Q to q, P via the relation

$$-P_i = \frac{\partial F_1}{\partial Q_i}$$

The Legendre transformation for this change of variables is

$$F_2(q, P, t) = F_1(q, Q, t) + P_i Q_i$$

As an example, for a system with two degrees of freedom, a canonical transformation can be defined by a generating function of the form:

$$F'(q_1, p_2, P_1, Q_2, t)$$

which can be related to F according to

$$F = F'(q_1, p_2, P_1, Q_2, t) - Q_1 P_1 + q_2 p_2$$

Then, the equations representing the transformation of canonical coordinates are obtained as the following:

$$\begin{aligned} p_1 &= \frac{\partial F'}{\partial q_1}, \quad Q_1 = \frac{\partial F'}{\partial P_1} \\ q_2 &= -\frac{\partial F'}{\partial p_2}, \quad P_2 = -\frac{\partial F'}{\partial Q_2} \end{aligned} \tag{1.98}$$

with

$$\mathcal{H} = H + \frac{\partial F'}{\partial t} \tag{1.99}$$

1.9.1 *The Symplectic Approach to Canonical Transformations*

The matrix or symplectic formulation of the Hamilton's equations is another method for dealing with canonical transformations. Consider a canonical transformation given by the following equations of transformations:

$$\begin{aligned} Q_i &= Q_i(q, p) \\ P_i &= P_i(q, p) \end{aligned} \tag{1.100}$$

which does not change the Hamiltonian function. Consider calculating the time derivative of Q_i from Eq. (1.100):

$$\dot{Q}_i = \sum_{j=1}^{n} \left(\frac{\partial Q_i}{\partial q_j}\dot{q}_j + \frac{\partial Q_i}{\partial p_j}\dot{p}_j \right) = \sum_{j=1}^{n} \left(\frac{\partial Q_i}{\partial q_j}\frac{\partial H}{\partial p_j} - \frac{\partial Q_i}{\partial p_j}\frac{\partial H}{\partial q_j} \right) \tag{1.101}$$

From the inverses of Eq. (1.100), it can be written that

$$q_i = q_i(Q, P) \qquad p_i = p_i(Q, P) \tag{1.102}$$

Therefore, the Hamiltonian function $H(q, p, t)$ can be seen as a function of Q, P, and t. Then, the partial derivatives of H can be calculated as:

$$\dot{Q}_i = \frac{\partial H}{\partial P_i} = \sum_{j=1}^{n} \left(\frac{\partial H}{\partial p_j}\frac{\partial p_j}{\partial P_i} + \frac{\partial H}{\partial q_j}\frac{\partial q_j}{\partial P_i} \right) \tag{1.103}$$

Comparing Eqs. (1.101) and (1.103), it can be found that

$$\left(\frac{\partial Q_i}{\partial q_j}\right)_{q,p} = \left(\frac{\partial p_j}{\partial P_i}\right)_{Q,P}, \quad \left(\frac{\partial Q_i}{\partial p_j}\right)_{q,p} = -\left(\frac{\partial q_j}{\partial P_i}\right)_{Q,P} \tag{1.104}$$

The subscripts are used to indicate that Q_i function of q and p and q_j, p_j are functions of Q and P.

Similarly, the time derivative of P_i gives:

$$\dot{P}_i = \sum_{j=1}^{n} \left(\frac{\partial P_i}{\partial q_j}\dot{q}_j + \frac{\partial P_i}{\partial p_j}\dot{p}_j \right) = \sum_{j=1}^{n} \left(\frac{\partial P_i}{\partial q_j}\frac{\partial H}{\partial p_j} - \frac{\partial P_i}{\partial p_j}\frac{\partial H}{\partial q_j} \right) \tag{1.105}$$

Then again, the partial derivatives of H can be calculated as:

$$\dot{P}_i = -\frac{\partial H}{\partial Q_i} = -\sum_{j=1}^{n} \left(\frac{\partial H}{\partial p_j}\frac{\partial p_j}{\partial Q_i} + \frac{\partial H}{\partial q_j}\frac{\partial q_j}{\partial Q_i} \right) \tag{1.106}$$

Similarly, comparing Eqs. (1.105) and (1.106), it can be found that

$$\left(\frac{\partial P_i}{\partial q_j}\right)_{q,p} = -\left(\frac{\partial p_j}{\partial Q_i}\right)_{Q,P}, \quad \left(\frac{\partial P_i}{\partial p_j}\right)_{q,p} = \left(\frac{\partial q_j}{\partial Q_i}\right)_{Q,P} \tag{1.107}$$

The set of equations given by Eqs. (1.104) and (1.107) are also called the *direct conditions* for a canonical transformation. The derivations of these equations can be performed in a more elegant manner by introducing the symplectic terminology for the Hamiltonian formalism. For that, consider $\boldsymbol{\Phi}$ a column matrix with $2n$ elements (q_i, p_i), for $i = 1, 2, \cdots, n$:

$$\boldsymbol{\Phi} = \begin{bmatrix} q_1 \\ q_2 \\ \cdot \\ \cdot \\ q_n \\ p_1 \\ p_2 \\ \cdot \\ \cdot \\ p_n \end{bmatrix}$$

Then, the Hamilton's equation can be written as

$$\dot{\boldsymbol{\Phi}} = \mathbf{J}\frac{\partial H}{\partial \boldsymbol{\Phi}} \tag{1.108}$$

where $\mathbf{J}$ is a $2n \times 2n$ square matrix composed of four $n \times n$ zero and one matrices as the following

$$\mathbf{J} = \begin{bmatrix} \mathbf{0} & \mathbf{1} \\ -\mathbf{1} & \mathbf{0} \end{bmatrix} \tag{1.109}$$

where $\mathbf{0}$ is a $n \times n$ matrix with all elements zero, and $\mathbf{1}$ is a $n \times n$ matrix with all entries one. The transpose of the matrix $\mathbf{J}$ is

$$\mathbf{J}' = \begin{bmatrix} \mathbf{0} & -\mathbf{1} \\ \mathbf{1} & \mathbf{0} \end{bmatrix} \tag{1.110}$$

such that

$$\mathbf{J}\mathbf{J}' = \mathbf{J}'\mathbf{J} = \mathbf{I} = \begin{bmatrix} \mathbf{1} & \mathbf{0} \\ \mathbf{0} & \mathbf{1} \end{bmatrix} \tag{1.111}$$

Moreover, it can be seen that

$$\mathbf{J}' = -\mathbf{J} = \mathbf{J}^{-1}$$

and

$$\mathbf{J}^2 = -\mathbf{I}$$

and the determinant is

$$| \mathbf{J} | = +1$$

Now, consider $\boldsymbol{\xi}$ a column matrix with $2n$ elements in the new set of canonical variables (Q_i, P_i):

$$\boldsymbol{\xi} = \begin{bmatrix} Q_1 \\ Q_2 \\ \cdot \\ \cdot \\ Q_n \\ P_1 \\ P_2 \\ \cdot \\ \cdot \\ P_n \end{bmatrix}$$

Then, the canonical transformations given by Eq. (1.82) can be written as

$$\boldsymbol{\xi} = \boldsymbol{\xi}(\boldsymbol{\Phi}) \tag{1.112}$$

The time derivatives of the new canonical variables, which will give the equations of motion, can be found as

$$\dot{\xi}_i = \sum_{j=1}^{2n} \frac{\partial \xi_i}{\partial \Phi_j} \dot{\Phi}_j, \quad i = 1, 2, \cdots, 2n$$

or in matrix form as

$$\dot{\boldsymbol{\xi}} = \mathbf{M}\dot{\boldsymbol{\Phi}} \tag{1.113}$$

where $\mathbf{M}$ is the Jacobian matrix of the transformation with elements

$$M_{ij} = \frac{\partial \xi_i}{\partial \Phi_j}$$

Replacing the expression for $\dot{\boldsymbol{\Phi}}$, it can be found that

$$\dot{\boldsymbol{\xi}} = \mathbf{M}\boldsymbol{J}\frac{\partial H}{\partial \boldsymbol{\Phi}} \tag{1.114}$$

Now, consider the Hamiltonian function in terms of the new variable set, then the derivatives with respect to the old variable set are calculated using the chain rule as:

$$\frac{\partial H}{\partial \Phi_i} = \sum_{j=1}^{2n} \frac{\partial H}{\partial \xi_j}\frac{\partial \xi_j}{\partial \Phi_i}$$

Using the matrix notation, the last equation can be written as

$$\frac{\partial H}{\partial \boldsymbol{\Phi}} = \mathbf{M}'\frac{\partial H}{\partial \boldsymbol{\xi}} \tag{1.115}$$

where $\mathbf{M}'$ is the transpose matrix of $\mathbf{M}$. Replacing Eq. (1.115) into Eq. (1.114), we obtain

$$\dot{\boldsymbol{\xi}} = \mathbf{M}\boldsymbol{J}\mathbf{M}'\frac{\partial H}{\partial \boldsymbol{\xi}} \tag{1.116}$$

which are the equations of motion of the new set of variables $\boldsymbol{\xi}$ from the old canonical set $\boldsymbol{\Phi}$.

The transformation will be called canonical if $\mathbf{M}$ satisfies the condition

$$\mathbf{MJM}' = \mathbf{J} \tag{1.117}$$

In such case, the equations of motion of the new variables can be written as:

$$\dot{\boldsymbol{\xi}} = \boldsymbol{J}\frac{\partial H}{\partial \boldsymbol{\xi}} \tag{1.118}$$

Note that the condition Eq. (1.117) is also a necessary condition for restricted canonical transformation. Moreover, for an extended time-independent canonical transformation, with $\mathcal{H} = \lambda H$, the condition Eq. (1.117) would be replaced by

$$\mathbf{MJM}' = \lambda\mathbf{J} \tag{1.119}$$

Multiplying both sides of Eq. (1.117) by the inverse of $\mathbf{M}'$, we obtain

$$\mathbf{MJM}'(\mathbf{M}')^{-1} = \mathbf{MJ} = \mathbf{J}(\mathbf{M}')^{-1} \tag{1.120}$$

Multiplying the last equation by $\mathbf{J}$ from the left and $-\mathbf{J}$ from the right, we get

$$\mathbf{JM} = (\mathbf{M}')^{-1}\mathbf{J} \tag{1.121}$$

where the relation $\mathbf{J}^2 = -1$ is used. Or, by multiplying from the left with $\mathbf{M}'$ and using $\mathbf{M}'(\mathbf{M}')^{-1} = \mathbf{I}$, we obtain

$$\mathbf{M}'\mathbf{J}\mathbf{M} = \mathbf{J} \tag{1.122}$$

Both relations represented by Eq. (1.117) or Eq. (1.122) are known as the *symplectic condition* for a canonical transformation, and the matrix $\mathbf{M}$ satisfying this condition is called *symplectic matrix*.

It is important to note that for canonical transformations containing the time as a parameter, simple derivations given for the symplectic condition do not hold anymore. However, the symplectic condition still remains a necessary and sufficient condition for the canonical transformation. A canonical transformation involving time can be written in the form:

$$\boldsymbol{\xi} = \boldsymbol{\xi}(\boldsymbol{\Phi}, t) \tag{1.123}$$

which is assumed to involve continuously as the time increases from the initial value t_0. If the transformation

$$\boldsymbol{\Phi} \rightarrow \boldsymbol{\xi}(t) \tag{1.124}$$

is canonical, then also the transformation

$$\boldsymbol{\Phi} \rightarrow \boldsymbol{\xi}(t_0) \tag{1.125}$$

is canonical. Moreover, based on the definition of the canonical transformation,

$$\boldsymbol{\xi}(t_0) \rightarrow \boldsymbol{\xi}(t) \tag{1.126}$$

is also a canonical transformation. In Eq. (1.125) the initial time t_0 is a fixed constant, thus this condition satisfies the symplectic condition Eq. (1.122). Now, suppose that the transformation of Eq. (1.126) satisfies the symplectic condition, then it is easy to show that also the general transformation given by Eq. (1.124) will satisfy the symplectic condition.

In order to show that the symplectic condition holds for transformation of the type Eq. (1.126), we will start introducing the concept of the *infinitesimal canonical transformation*. In such concept, it is assumed that the new variables differ from the old only by infinitesimals. Only first-order terms in these infinitesimals are to be retained in all calculations. The transformation equations can be written as

$$Q_i = q_i + \delta q_i \tag{1.127}$$

$$P_i = p_i + \delta p_i \tag{1.128}$$

and in the matrix form as

$$\boldsymbol{\xi} = \boldsymbol{\Phi} + \delta\boldsymbol{\Phi} \tag{1.129}$$

Therefore, an infinitesimal canonical transformation differs only infinitesimally from the identity transformation. That is, in the generator formalism, a suitable generating function for an infinitesimal canonical transformation can be written as

$$F_2 = q_i P_i + \epsilon G(q, P, t) \tag{1.130}$$

where ϵ is a infinitesimal parameter of the transformation and G is a function of $2n + 1$ parameters. The transformation equation for the momenta is

$$p_j = \frac{\partial F_2}{\partial q_j} = P_j + \epsilon \frac{\partial G}{\partial q_j} \tag{1.131}$$

or

$$\delta p_j \equiv P_j - p_j = -\epsilon \frac{\partial G}{\partial q_j} \tag{1.132}$$

Similarly, transformation equation for Q_j is determined as

$$Q_j = \frac{\partial F_2}{\partial P_j} = q_j + \epsilon \frac{\partial G}{\partial P_j} \tag{1.133}$$

or

$$\delta q_j \equiv Q_j - q_j = \epsilon \frac{\partial G}{\partial P_j} \tag{1.134}$$

From Eq. (1.132) it can be seen that P and p differ from each other by only an infinitesimal, thus P_j can be replaced by p_j in the derivative function of Eq. (1.134). In such case, we may consider G as a function of q and p and possibly time t. Here, the function $G(q, p)$ will be referred as the *generating function of the infinitesimal canonical transformation*. Thus, Eq. (1.134) can be written as:

$$\delta q_j = \epsilon \frac{\partial G}{\partial p_j} \tag{1.135}$$

Both Eqs. (1.132) and (1.135) can be combined in a matrix form as:

$$\delta\boldsymbol{\Phi} = \epsilon \mathbf{J} \frac{\partial G}{\partial \boldsymbol{\Phi}} \tag{1.136}$$

A typical example of the infinitesimal canonical transformation is the transformation presented by Eq. (1.126) when t differs from t_0 by a small amount dt:

$$\boldsymbol{\xi}(t_0) \rightarrow \boldsymbol{\xi}(t_0 + dt) \tag{1.137}$$

where dt is the infinitesimal parameter ϵ. A continuous transformation $\boldsymbol{\xi}(t_0) \rightarrow \boldsymbol{\xi}(t)$ can be seen as a succession of infinitesimal canonical transformations of type Eq. (1.137) in steps of dt. Therefore, it is sufficient to show that Eq. (1.137) satisfies the symplectic condition Eq. (1.122). The Jacobian matrix for an infinitesimal transformation is

$$\mathbf{M} \equiv \frac{\partial \xi}{\partial \boldsymbol{\Phi}} = \mathbf{I} + \frac{\partial \delta \boldsymbol{\Phi}}{\partial \boldsymbol{\Phi}} \tag{1.138}$$

Or, by replacing Eq. (1.136) into Eq. (1.138), we get

$$\mathbf{M} = \mathbf{I} + \epsilon \mathbf{J} \frac{\partial^2 G}{\partial \boldsymbol{\Phi} \partial \boldsymbol{\Phi}} \tag{1.139}$$

with

$$\left(\frac{\partial^2 G}{\partial \boldsymbol{\Phi} \partial \boldsymbol{\Phi}} \right)_{ij} = \frac{\partial^2 G}{\partial \Phi_i \partial \Phi_j}$$

Using the property of the matrix $\mathbf{J}$ as anti-symmetrical matrix, the transpose of $\mathbf{M}$ can be written as

$$\mathbf{M}' = \mathbf{I} - \epsilon \frac{\partial^2 G}{\partial \boldsymbol{\Phi} \partial \boldsymbol{\Phi}} \mathbf{J} \tag{1.140}$$

The symplectic condition then can be written as

$$\mathbf{MJM}' = \left(\mathbf{I} + \epsilon \mathbf{J} \frac{\partial^2 G}{\partial \boldsymbol{\Phi} \partial \boldsymbol{\Phi}} \right) \mathbf{J} \left(\mathbf{I} - \epsilon \frac{\partial^2 G}{\partial \boldsymbol{\Phi} \partial \boldsymbol{\Phi}} \mathbf{J} \right) \tag{1.141}$$

Retaining only the first order terms, the expression in Eq. (1.141) can be further simplified as

$$\begin{aligned} \mathbf{MJM}' &= \mathbf{J} + \epsilon \mathbf{J} \frac{\partial^2 G}{\partial \boldsymbol{\Phi} \partial \boldsymbol{\Phi}} \mathbf{J} - \mathbf{J} \epsilon \frac{\partial^2 G}{\partial \boldsymbol{\Phi} \partial \boldsymbol{\Phi}} \mathbf{J} \\ &= \mathbf{J} \end{aligned} \tag{1.142}$$

indicating that the symplectic condition holds for any infinitesimal canonical transformation.

1.10 Poisson Brackets

The Poisson bracket of any two functions u and v with respect to canonical variables q and p is defined as

$$[u, v]_{q,p} = \sum_{i=1}^{n} \left(\frac{\partial u}{\partial q_i} \frac{\partial v}{\partial p_i} - \frac{\partial u}{\partial p_i} \frac{\partial v}{\partial q_i} \right) \tag{1.143}$$

This indicates the existence of a symplectic structure, similarly to the Hamilton's equations where q is coupled with p and p with $-q$. The Poisson bracket can be written as a matrix form

$$[u, v]_{\boldsymbol{\Phi}} = \left(\frac{\partial u}{\partial \boldsymbol{\Phi}} \right)' \mathbf{J} \frac{\partial v}{\partial \boldsymbol{\Phi}} \tag{1.144}$$

where $(\cdots)'$ stands for transpose. Consider that the functions u and v are from the set of canonical variables q and p. Then, we can write

$$\begin{aligned} \left[q_j, q_k\right]_{q,p} &= 0 \\ \left[p_j, p_k\right]_{q,p} &= 0 \\ \left[q_j, p_k\right]_{q,p} &= \delta_{jk} \\ \left[p_j, q_k\right]_{q,p} &= -\delta_{jk} \end{aligned} \tag{1.145}$$

which can be summarized into a matrix form by introducing the *square matrix Poisson bracket*, $[\boldsymbol{\Phi}, \boldsymbol{\Phi}]$, with elements

$$[\boldsymbol{\Phi}, \boldsymbol{\Phi}]_{ij} = \left[\Phi_i, \Phi_j\right]$$

as

$$[\boldsymbol{\Phi}, \boldsymbol{\Phi}]_{\boldsymbol{\Phi}} = \mathbf{J} \tag{1.146}$$

Consider now u and v are members of the new variables Q and P, or that $\boldsymbol{\xi}$ is defined in terms of old variables (q, p) from Eq. (1.123). Then, the Poisson brackets formed out of (Q, P) is defined as

$$[\boldsymbol{\xi}, \boldsymbol{\xi}]_{\boldsymbol{\Phi}} = \left(\frac{\partial \boldsymbol{\xi}}{\partial \boldsymbol{\Phi}} \right)' \mathbf{J} \frac{\partial \boldsymbol{\xi}}{\partial \boldsymbol{\Phi}} \tag{1.147}$$

Using the definition of the Jacobian matrix, we can re-write Eq. (1.147) as

$$[\boldsymbol{\xi}, \boldsymbol{\xi}]_{\boldsymbol{\Phi}} = \mathbf{M}'\mathbf{J}\mathbf{M} \tag{1.148}$$

If the transformation $\boldsymbol{\Phi} \rightarrow \boldsymbol{\xi}$ is canonical, then the symplectic condition holds, and Eq. (1.148) takes the form

$$[\boldsymbol{\xi}, \boldsymbol{\xi}]_{\boldsymbol{\Phi}} = \mathbf{J} \tag{1.149}$$

The inverse is also true, that is, if the condition given by Eq. (1.149) is satisfied, then the transformation is canonical.

The Poisson brackets of the canonical variables in Eq. (1.146) or Eq. (1.149) are called the *fundamental Poisson brackets*. Using Eq. (1.146), we have

$$[\boldsymbol{\xi}, \boldsymbol{\xi}]_{\boldsymbol{\xi}} = \mathbf{J} \tag{1.150}$$

Eq. (1.149) indicates that the fundamental Poisson brackets of the variable $\boldsymbol{\xi}$ have the same value when evaluated with respect to any canonical coordinate set. That is, the fundamental Poisson brackets are invariant under canonical transformation.

Equation (1.148) states that the invariance is a necessary and sufficient condition for the transformation matrix to be symplectic. Therefore, the invariance of the fundamental Poisson brackets is equivalent to the symplectic condition for a canonical transformation.

1.10.1 Equations of Motion in the Poisson Bracket Formulation

The framework of the Hamiltonian formalism can be reformulated using the Poisson brackets. Consider a function of the canonical variables and time, $u(q, p, t)$, then the total time derivative of u is

$$\frac{du}{dt} = \sum_{i=1}^{n} \left(\frac{\partial u}{\partial q_i} \dot{q}_i + \frac{\partial u}{\partial p_i} \dot{p}_i \right) + \frac{\partial u}{\partial t}$$

Using the Hamilton's equations of motion, this derivative can be written as

$$\frac{du}{dt} = \sum_{i=1}^{n} \left(\frac{\partial u}{\partial q_i} \frac{\partial H}{\partial p_i} - \frac{\partial u}{\partial p_i} \frac{\partial H}{\partial q_i} \right) + \frac{\partial u}{\partial t} \tag{1.151}$$

or

$$\frac{du}{dt} = [u, H] + \frac{\partial u}{\partial t} \tag{1.152}$$

Eq. (1.152) can also be written as the following using the symplectic notations:

$$\frac{du}{dt} = \left(\frac{\partial u}{\partial \boldsymbol{\Phi}} \right)' \dot{\boldsymbol{\Phi}} + \frac{\partial u}{\partial t} \tag{1.153}$$

Eq. (1.152) can be considered as the generalized equation of motion for an arbitrary function u in the Poisson bracket formulation. If the function u is one of the canonical variables, it reduces to the Hamilton's equations as a special case:

$$\dot{q}_i = [q_i, H], \quad \dot{p}_i = [p_i, H] \tag{1.154}$$

or using the symplectic notation:

$$\dot{\boldsymbol{\Phi}} = [\boldsymbol{\Phi}, H] \tag{1.155}$$

Using the definition of the Poisson bracket:

$$[\boldsymbol{\Phi}, H] = \mathbf{J}\frac{\partial H}{\partial \boldsymbol{\Phi}} \tag{1.156}$$

Replacing Eq. (1.156) into Eq. (1.155), we get

$$\dot{\boldsymbol{\Phi}} = \mathbf{J}\frac{\partial H}{\partial \boldsymbol{\Phi}} \tag{1.157}$$

which is the Hamilton's equation of motion.

If we take u as the Hamiltonian function H, then from Eq. (1.152) we get

$$\frac{dH}{dt} = \frac{\partial H}{\partial t}$$

The generalized equation of motion derived here is canonically invariant. That is, it is valued in any set of canonical variables (q, p) used to express the function u or to evaluate the Poisson bracket. Important here is that the Hamiltonian must be the one which is appropriate to the new set of canonical variables. Thus, upon transformation from one set of canonical variables to a new set of variables by a time-dependent canonical transformation, we have to change to the transformed Hamiltonian $\mathcal{H}$.

1.10.2 Conservation Laws in the Poisson Bracket Formulation

If u is a constant of motion ($du/dt = 0$), then from Eq. (1.152), we obtain

$$-[u, H] = \frac{\partial u}{\partial t} \tag{1.158}$$

or

$$[H, u] = \frac{\partial u}{\partial t} \tag{1.159}$$

In general, all the functions that satisfy Eq. (1.159) are constants of motion, and conversely, if the Poisson bracket of H with any function u, which is constant of motion, equals the explicit time derivative of u with time t. If the constant of motion does not depend on time explicitly, then $\partial u/\partial t = 0$, then Eq. (1.159) can be written as

$$[H, u] = 0 \tag{1.160}$$

Furthermore, if u and v are two constants of motion and not explicit function of time t, then

$$[H, [u, v]] = 0 \tag{1.161}$$

That is, the Poisson bracket of u and v is also a constant of motion. This is known as *Poisson's Law*:

> The Poisson bracket of any two constants of the motion represents a quantity that is a constant of the motion.

1.10.3 Infinitesimal Canonical Transformations

The Poisson bracket notation can also be used to derive the basic equations of an infinitesimal canonical transformation. Such transformation is a special case of a transformation that is a continuous function of a parameter. It can be started from the identity transformation at some initial value of the parameter, which often can be set to zero for convenience. If the parameter is small enough to be treated as a first-order infinitesimal, then the transformed canonical variables differ only infinitesimally from the initial conditions:

$$\boldsymbol{\xi} = \boldsymbol{\Phi} + \delta\boldsymbol{\Phi} \tag{1.162}$$

The change is given in terms of the generator function G as

$$\delta\boldsymbol{\Phi} = \epsilon\mathbf{J}\frac{\partial G(\boldsymbol{\Phi})}{\partial\boldsymbol{\Phi}} \tag{1.163}$$

Using the definition of the Poisson bracket given by Eq. (1.146), we obtain

$$[\boldsymbol{\Phi}, u] = \mathbf{J}\frac{\partial u}{\partial\boldsymbol{\Phi}} \tag{1.164}$$

where u can be any canonical variable. If u is taken to be G function, then

$$[\boldsymbol{\Phi}, G] = \mathbf{J}\frac{\partial G}{\partial\boldsymbol{\Phi}} \tag{1.165}$$

Combining Eqs. (1.163) and (1.165), we obtain

$$\delta\boldsymbol{\Phi} = \epsilon\,[\boldsymbol{\Phi}, G] \tag{1.166}$$

If it is considered an infinitesimal canonical transformation in which the continuous parameter is time t, such that, $\epsilon = dt$, and moreover, considering that G is the Hamiltonian function H, then we get

$$\delta\boldsymbol{\Phi} = dt\,[\boldsymbol{\Phi}, H] = \dot{\boldsymbol{\Phi}}dt = d\boldsymbol{\Phi} \tag{1.167}$$

These equations state that the transformation changes the coordinates and momenta at the time t to the values that have at the time $t + dt$. Therefore, the motion of the system in the time interval dt can be described by an infinitesimal contact transformation generated by the Hamiltonian. Furthermore, the motion of the system in a finite interval of time from t_0 to t is described by succession of infinitesimal contact transformations, which is equivalent to a single finite canonical transformation.

As a result, the values of canonical variables q and p at any time t can be determined from their initial values by a canonical transformation that is a continuous function of time. Thus, the motion of a mechanical system corresponds to the continuous time evolution of a canonical transformation. According to this view, the Hamiltonian is the generator of the system motion with time. Conversely, there exists a canonical transformation from the values of the coordinates and momenta at any time t to their constant initial values, which is equivalent to solving the equations of motion of the system.

After this discussion, we can have a better view of the canonical transformations. In the beginning, we started saying that the canonical transformations are a way of changing the coordinates that characterize the phase space of the mechanical system. That is, a change from the phase space vector $\boldsymbol{\Phi}$ with coordinates (q, p) to the vector $\boldsymbol{\xi}$ with coordinates (Q, P). Thus, if at some initial time the system is described by the point A in phase space, it could also be equally described by the transformed point B, as shown in Fig. 1.4. Therefore, any function of the coordinates of the system would have the same value for a system configuration whether it was at the point (q, p) or at the (Q, P). That is also called the *passive* view of a canonical transformation.

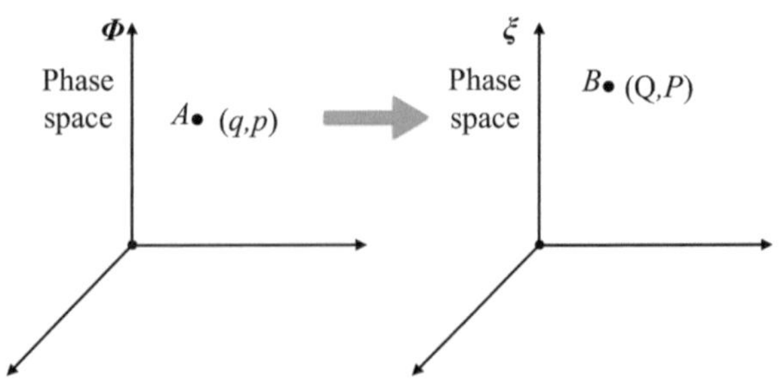

Fig. 1.4 The passive view of canonical transformations

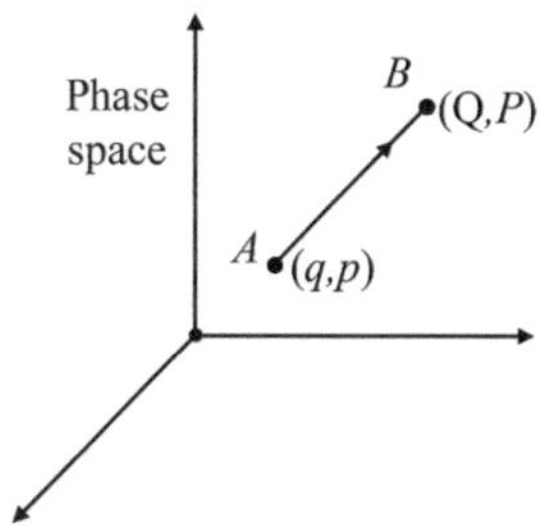

Fig. 1.5 The active view of canonical transformations

In contrast, in the Hamiltonian formalism, we described the canonical transformation as a relation between the coordinates of one point in phase space to those of another point in the same phase space. Using this point of view, the canonical transformation is a mapping of the points of phase space onto themselves. That is called the *active* view of the interpretation of the canonical transformation, as shown in Fig. 1.5. In effect, the active description considers the canonical transformation as moving the system configuration from one position with coordinates (q, p) to another point (Q, P) in phase space. However, the canonical transformation does not change the system configuration, but it expresses one configuration of the system in terms of another.

1.10.4 Liouville's Theorem

Another application of the Poisson brackets is the Liouville's theorem, which is a fundamental theorem of the statistical mechanics. As we described above, the exact motion of the classical mechanical system is entirely determined by the initial conditions. However, for complex systems, it is often impractical to calculate the exact solution. Besides, the initial conditions may also be unknown completely. Therefore, statistical mechanics does not attempt to determine the complete solution for complex systems containing many particles. The primary goal of the statistical mechanics is to predict the averages of the properties by examining the motion of a large number of identical systems, called *ensemble*.[2] Then, the values of some quantities of interest are calculated by performing the ensemble averages over all systems in the ensemble. Each system of the ensemble represents one single point in the phase space, thus, the ensemble corresponds to a swarm of a point in this phase space.

Liouville's theorem states:

> The systems density in the neighborhood of a given system in phase space remains constant in time.

[2]Note that principals of statistical mechanics will be described in more details in Chap. 3.

Denote with f the density of systems in the phase space, which will depend on the coordinates (q_i, p_i) that depend implicitly on the time t. There also may be an explicit dependence on the time t, that is the density may vary with time even when evaluated at a fixed point in phase space. The total time derivative of the density $f(q, p, t)$ due to both types of variation with time, can be calculated as

$$\begin{aligned}\frac{df}{dt} &= \sum_{i=1}^{n}\left(\frac{\partial f}{\partial q_i}\dot{q}_i + \frac{\partial f}{\partial p_i}\dot{p}_i\right) + \frac{\partial f}{\partial t} \\ &= \sum_{i=1}^{n}\left(\frac{\partial f}{\partial q_i}\frac{\partial H}{\partial p_i} - \frac{\partial f}{\partial p_i}\frac{\partial H}{\partial q_i}\right) + \frac{\partial f}{\partial t} \\ &= [f, H] + \frac{\partial f}{\partial t}\end{aligned} \tag{1.168}$$

where the definitions of the Hamilton's equations of motion are used, as well as, the definition of the Poisson brackets.

The motion of the ensemble of system points through the phase space corresponds to the motion of the fluid in a multidimensional space. The partial derivative $\partial f/\partial t$, called the *Eulerian* derivative, measures the variation at a fixed point in phase space, and the total derivative df/dt, called the *Lagrangian* derivative, gives the variation of f following the motion of a particular bit of the ensemble in time. These two derivatives correspond to two different views used to consider the fluid flow. The partial derivative at a fixed point (q, p) in the phase space is in line with the Eulerian viewpoint that looks on the coordinates as identifying a point in space. The total derivative corresponds to the Lagrangian point of view in which individual particles are followed in time.

Consider an infinitesimal volume in the phase space $d\Gamma = \prod_{i=1}^{n} dq_i dp_i$ surrounding a given system point (q, p), with the boundary of the volume formed by the surface of neighboring system points at the time $t = 0$. With time, the system points defining the volume move through the phase space, and the volume containing these points will change shapes as time processes, as illustrated in Fig. 1.6 for the two-dimensional case. However, the number of systems within the volume remains constant, that is, a system initially inside the volume can never get out of the volume. Similarly, a system initially outside the volume can never enter the volume.

In the active picture shown above of a canonical transformation, the motion of a system point in time is the evolution of a canonical transformation generated by the Hamiltonian. The canonical variables (q, p) at time t_2 are related to the variables at time t_1 by a particular canonical transformation (see Fig. 1.6.) Thus, the canonical transformation gives the change in the infinitesimal volume element about a system point in the time interval. Because the Poincaré's integral invariant says that a volume element in phase space is invariant under a canonical transformation, we can conclude that the size of volume element about the system point in the phase space can not vary with time.

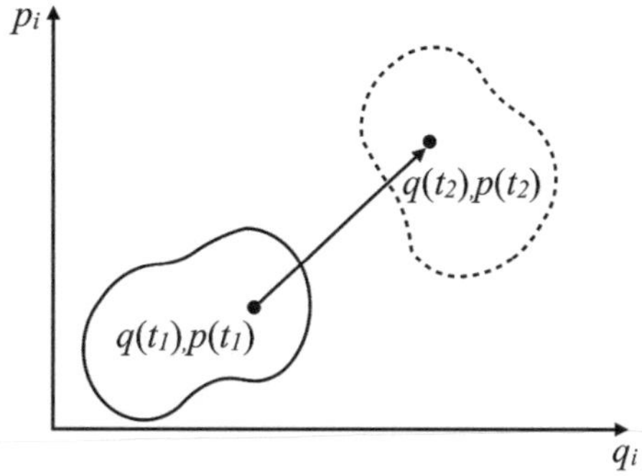

Fig. 1.6 The motion of a volume in two-dimensional phase space

Therefore, both the number of systems inside the infinitesimal volume, dN, and the volume $d\Gamma$ are constants. Hence, the density

$$f = \frac{dN}{d\Gamma}$$

is also constant in time, that means:

$$\frac{df}{dt} = 0$$

which is a prove of the Liouville's theorem. Thus, Eq. (1.168) can also be written as

$$\frac{\partial f}{\partial t} = -[f, H] \tag{1.169}$$

which is an alternative statement.

If the ensemble of systems is in statistical equilibrium, the number of systems in a given state must be constant in time, and hence the density of system points at a given point in phase space does not change with time. Since the variation of f with time at a fixed point corresponds to the partial derivative with respect to the time t, then $\partial f/\partial t$ must be zero in statistical equilibrium. Thus, in the equilibrium Eq. (1.169) reduces to

$$[f, H] = 0 \tag{1.170}$$

Therefore, in statistical equilibrium the density of states f is a function of those constants of the motion of the system that does not involve time explicitly, and then the Poisson bracket of f with H must vanish. For conservative systems, f can be any function of the energy, for which the equilibrium condition is automatically satisfied.

1.11 Invariant Measure

Consider again the dynamical system represented by Eq. (1.30) (Sect. 1.4), where $\boldsymbol{\eta} = (x_1, x_2, \cdots, x_n)$ is an n-dimensional generalized state vector and $\mathbf{V}(\boldsymbol{\eta}, t)$ is an n-dimensional generalized vector function of $\boldsymbol{\eta}$ and time t.

In general, knowing some initial conditions for $\boldsymbol{\eta}$ at $t = t_0$, $\boldsymbol{\eta}(t_0)$, the vector $\boldsymbol{\eta}(t)$ at any time $t > t_0$ can be determined. Therefore, we can write for $i = 1, 2, \cdots, n$

$$x_i(t) = x_i(t; \boldsymbol{\eta}(t_0)) \tag{1.171}$$

Following Tuckerman et al. (2001), Eq. (1.171) represents a coordinates transformation from their initial values at $t = t_0$ to the coordinates at $t > t_0$. The phase space volume transforms in time under the dynamical evolution according to

$$d\Gamma(t) = J\left(\boldsymbol{\eta}(t); \boldsymbol{\eta}(t_0)\right) d\Gamma(t_0) \tag{1.172}$$

where $d\Gamma = dx_1 dx_2 \cdots dx_n$ is the phase space volume and $J\left(\boldsymbol{\eta}(t); \boldsymbol{\eta}(t_0)\right)$ is the Jacobian of the transformation determined as determinant of the transformation matrix $\mathbf{M}$ (Tuckerman et al. 2001):

$$J\left(\boldsymbol{\eta}(t); \boldsymbol{\eta}(t_0)\right) = \det(\mathbf{M}) = \exp\left(\mathrm{Tr}\left(\ln \mathbf{M}\right)\right) \tag{1.173}$$

where

$$M_{ij} = \frac{\partial x_i(t)}{\partial x_j(t_0)} \tag{1.174}$$

In Eq. (1.173), Tr denotes the trace of a matrix and ln the natural logarithm function. According to Tuckerman et al. (2001), it can easily be shown that:

$$\begin{aligned} \frac{dJ\left(\boldsymbol{\eta}(t); \boldsymbol{\eta}(t_0)\right)}{dt} &= \exp\left(\mathrm{Tr}\left(\ln \mathbf{M}\right)\right) \mathrm{Tr}\left(\mathbf{M}^{-1}\frac{d\mathbf{M}}{dt}\right) \\ &= J\left(\boldsymbol{\eta}(t); \boldsymbol{\eta}(t_0)\right) \sum_{ij} M_{ij}^{-1}\frac{dM_{ji}}{dt} \\ &= J\left(\boldsymbol{\eta}(t); \boldsymbol{\eta}(t_0)\right) \sum_{ij} \frac{\partial x_i(t_0)}{\partial x_j(t)} \frac{d}{dt}\left(\frac{\partial x_j(t)}{\partial x_i(t_0)}\right) \\ &= J\left(\boldsymbol{\eta}(t); \boldsymbol{\eta}(t_0)\right) \sum_{ij} \frac{\partial x_i(t_0)}{\partial x_j(t)} \frac{\partial \dot{x}_j(t)}{\partial x_i(t_0)} \end{aligned} \tag{1.175}$$

$$= J\left(\eta(t); \eta(t_0)\right) \sum_{ijk} \frac{\partial x_i(t_0)}{\partial x_j(t)} \frac{\partial \dot{x}_j(t)}{\partial x_k(t)} \frac{\partial x_k(t)}{\partial x_i(t_0)}$$

$$= J\left(\eta(t); \eta(t_0)\right) \sum_{jk} \delta_{jk} \frac{\partial \dot{x}_j(t)}{\partial x_k(t)}$$

$$= J\left(\eta(t); \eta(t_0)\right) \sum_{k} \frac{\partial \dot{x}_k(t)}{\partial x_k(t)}$$

where

$$M_{ij}^{-1} = \frac{\partial x_i(t_0)}{\partial x_j(t)} \tag{1.176}$$

is the inverse matrix of $\mathbf{M}$. Denoting by

$$\kappa(\eta, t) = \nabla_\eta \cdot \dot{\eta} \tag{1.177}$$

the compressibility of the dynamical system, it can be written that (Tuckerman et al. 2001)

$$\frac{dJ\left(\eta(t); \eta(t_0)\right)}{dt} = J\left(\eta(t); \eta(t_0)\right) \kappa(\eta, t) \tag{1.178}$$

Similarly, the Jacobian of the inverse transformation is (Tuckerman et al. 2001):

$$J' = \det\left(\mathbf{M}^{-1}\right) = \exp\left(-\mathrm{Tr}\left(\ln \mathbf{M}\right)\right)$$

Therefore, it can be shown that

$$JJ' = 1 \tag{1.179}$$

Furthermore, it can be found that

$$\frac{dJ'\left(\eta(t); \eta(t_0)\right)}{dt} = -J'\left(\eta(t); \eta(t_0)\right) \kappa(\eta, t) \tag{1.180}$$

Multiplying both sides of Eq. (1.178) by J' and both sides of Eq. (1.180) by J, then using Eq. (1.179), we get

$$J' \frac{dJ}{dt} + J \frac{dJ'}{dt} = 0$$

In general, for a Hamiltonian system, the compressibility vanishes (Tuckerman et al. 2001), and thus

$$\frac{1}{J\left(\boldsymbol{\eta}(t);\,\boldsymbol{\eta}(t_0)\right)}\frac{dJ\left(\boldsymbol{\eta}(t);\,\boldsymbol{\eta}(t_0)\right)}{dt}=0$$

or (taking into account Eq. (1.179))

$$J\left(\boldsymbol{\eta}(t);\,\boldsymbol{\eta}(t_0)\right)=1$$

This can be shown for the simplest case of a Hamiltonian system with Hamiltonian function given as:

$$H(q,p)=\sum_{i=1}^{n}\frac{p_i^2}{2m_i}+U(q_1,q_2,\cdots,q_n)$$

where $U(q_1,q_2,\cdots,q_n)$ is the potential energy function of only coordinates and

$$\boldsymbol{\eta}=(q_1,q_2,\cdots,q_n,p_1,p_2,\cdots,p_n)$$

Using the Hamiltonian equations of motion for $i=1,2,\cdots,n$, we get

$$\dot{q}_i=\frac{p_i}{m_i} \tag{1.181}$$
$$\dot{p}_i=-\nabla_{q_i}U(q_1,q_2,\cdots,q_n)$$

Therefore, it can be found that the compressibility is:

$$\kappa(\boldsymbol{\eta},t)=\sum_{i=1}^{n}\left(\frac{\partial\dot{q}_i}{\partial q_i}+\frac{\partial\dot{p}_i}{\partial p_i}\right) \tag{1.182}$$
$$=\sum_{i=1}^{n}\left(\frac{1}{m_i}\frac{\partial p_i}{\partial q_i}-\frac{\partial\nabla_{q_i}U(q_1,q_2,\cdots,q_n)}{\partial p_i}\right)=0$$

because coordinates q_i (or conjugated momenta p_i) do not depend on conjugated momenta p_i (or coordinates q_i). Therefore, $J\left(\boldsymbol{\eta}(t);\,\boldsymbol{\eta}(t_0)\right)=1$, and

$$d\Gamma(t)=d\Gamma(t_0)$$

That indicates that the phase space of the Euclidean geometry is conserved, and hence the phase space is considered to be flat (Tuckerman et al. 2001).

In general, the Jacobian determines how the phase space volume transforms according to the dynamical system given by Eq. (1.171) and how the phase space metric transforms. For non-Hamiltonian systems $\kappa(\boldsymbol{\eta},t)$ does not vanish, and thus the volume of the phase space in not an invariant measure of dynamical systems (Tuckerman et al. 1999, 2001). In such case, by integrating the expression given by Eq. (1.178) from t_0 to t, we obtain (Tuckerman et al. 2001):

$$J\left(\eta(t);\,\eta(t_0)\right) = \exp\left(\int_{t_0}^{t} \kappa(\eta, t)\, dt\right) = \exp\left(W(\eta, t) - W(\eta, t_0)\right) \tag{1.183}$$

where

$$\kappa(\eta, t) = \frac{dW(\eta, t)}{dt}$$

Replacing expression given by Eq. (1.183) into Eq. (1.172), we obtain:

$$\exp\left(-W(\eta, t)\right) d\Gamma(t) = \exp\left(-W(\eta, t_0)\right) d\Gamma(t_0) \tag{1.184}$$

Comparing this expression with the one given in terms of the metric determinant factor:

$$\sqrt{g(\eta, t)}d\Gamma(t) = \sqrt{g(\eta, t_0)}d\Gamma(t_0)$$

where $\sqrt{g(\eta, t)}$ is the determinant of the metric tensor

$$\mathbf{G} = \left(\frac{\partial H}{\partial \mathbf{p}}, -\frac{\partial H}{\partial \mathbf{q}}\right)$$

we obtain that

$$\sqrt{g(\eta, t)} = \exp\left(-W(\eta, t)\right)$$

Chapter 2
Principles of Classical Thermodynamics

In the chapter, we discuss the principles of classical thermodynamics. In particular, we will describe the laws of thermodynamics, thermodynamic potential functions, Maxwell relations, and stability of thermodynamic systems.

For further information on the classical thermodynamics, the reader may consider the book by Callen (1985).

2.1 Introduction

The macroscopic quantities, such as the energy, temperature, and pressure are, in fact, statistical: i.e., in equilibrium, they exhibit random fluctuations around their mean values. Therefore, plotting out the probability distribution for the energy, for instance, for a system in thermal equilibrium with its surrounding, we would obtain a Gaussian with minimal fractional width. We expect

$$\frac{\Delta E}{\bar{E}} \sim \frac{1}{\sqrt{f}} \tag{2.1}$$

Here, the number of degrees of freedom f is about 10^{24} for laboratory scale systems. That means that the statistical fluctuations of macroscopic quantities around their mean values are typically only about 1 in 10^{-12}.

Since the statistical fluctuations of these equilibrium quantities are so small, we can neglect them to an excellent approximation, and replace macroscopic quantities, such as the energy, temperature, and pressure, and so on, by their mean values, such as $E \to \bar{E}$, $T \to \bar{T}$, and $p \to \bar{p}$, and so on.

H. Kamberaj, *Molecular Dynamics Simulations in Statistical Physics: Theory and Applications*, Scientific Computation,
https://doi.org/10.1007/978-3-030-35702-3_2

In the following discussion, we will drop the over bars altogether, for simplicity of notation. This prescription, which is the essence of *classical thermodynamics*, is equivalent to replacing all statistically varying quantities by their most probable values.

Although there are formally four laws of thermodynamics (from the zeroth to the third), the zeroth law is a result of the second law, and the third law is only relevant at temperatures close to absolute zero. Thus, for most purposes, the two rules which matter are the first law and the second law.

2.2 Microscopic and Macroscopic Views

There are two viewpoints for describing thermodynamics: *macroscopic* and *microscopic*. The macroscopic viewpoint is also known as *classical thermodynamics*, where the fact that matter is composed of molecules and that they are in movement does not count. The main focus is on the behavior of the entire system when the system is subject of the energy transfer or other thermodynamic processes. In this approximation, the mathematics is elementary, and hence it allows analyzing in a straightforward manner different complex systems that are of interests in industrial applications.

The microscopic approximation, also known as *statistical thermodynamics*, includes the motion of molecules or atoms of the system. That is done by using some mathematical models to describe the behavior of the constituent particles of the system, such as the molecules and atoms, and hence being able to derive conclusions in the response of the matter. However, this approximation is more complicated in terms of mathematical models used to predict the behaviour of the constituent particles.

In this chapter, we are focused on the macroscopic description of thermodynamics, or the so-called classical thermodynamics. While in the next chapter we will introduce statistical mechanics and describe statistical thermodynamics.

2.3 Some Definitions of Thermodynamics

The concept of the system is of particular importance. A *system* is everything that we want to analyze from a thermodynamic point of view.

We will draw the boundary between the content of a system from the environment (which is everything else outside the system) by constructing a *boundary wall*, which isolates the system from the surrounding. This boundary wall can be real, for example, walls of the container holding the system of interest, or it could also be an imagined boundary or a combination of both.

A system is called *isolated system* if its mass remains constant. For example, a water-like system in a container is considered to be an isolated system, if we assume that the vapors of water do not leave the boundary separating the system from the surrounding.

If the mass inside the system changes, for example, there is some mass that passes through the boundary (either enters or leaves the system or both), then it is called *open system*.

Everything that is outside the system boundary, but in close approximation with the system boundary, is called *surrounding*. In principle, everything that is too far from the system boundary may not be able to interact with the system (or the interactions with our system is so weak that may not be able to change its state), and hence it does not show any particular interest to be described in this context.

Combination of a system with its surrounding forms the so-called *universe*. Because of the universe includes only surrounding close to the system of interest, in this context, it is just an approximation.

The *thermodynamic property* is a characteristic or an attribute that can be used to describe the substance. It can be measured and quantified. In practice, to define a substance, those properties that naturally are easy to measure will be used. Moreover, we have to use enough number of properties to characterize the substance uniquely.

It is useful to determine two classes of properties: *extensive* and *intensive* properties. The extensive properties are those that change proportionally with the number of particles (or mass) of the substance. For example, the volume is an extensive property, from its definition as:

$$V = \frac{m}{\rho}$$

where m is the mass and ρ is the density, and hence V is proportional to m.

In contrast, the intensive properties are independent on the number of particles of the substance (or its mass), such as the pressure or temperature because they do not depend on the number of particles or the mass of a substance. For example, the pressure at an equal distance from the surface of either the ocean or pool is the same.[1]

If we merge two bottles of water at the same temperature in a big container, then this temperature of the water will be the same as the temperature of water in each bottle. But not the sum of the temperatures in each bottle![2]

[1] We have assumed here the same properties of the water, such as the density.

[2] Note that if the bottles are at different temperatures, let us say T_1 and T_2, if they mix in a bigger container, then because of the thermal equilibrium law (discussed in the following sections) the heat flows from the hot water into the cold water until the temperatures become equal. However, the final temperature of the water in the container will not be $T_1 + T_2$.

2.4 The First Law of Thermodynamics and Equilibrium

The first law is related to the *internal energy*. This quantity, denoted with E, determines the total energy of the system. It is postulated that the internal energy has two properties: First, the internal energy is extensive. Considering a composed system, as shown in Fig. 2.1, the total internal energy of the system, based on the fact that it is an extensive quantity, is

$$E = E_1 + E_2$$

Therefore, the total energy as an extensive property of the system is proportional to its size. If the size of the system doubles by keeping constant all the other features, then the energy of the systems doubles.

The second postulate says that the energy of an isolated system is constant. That is, the energy of system changes if we do something over the system; for example, we allow one form of energy to be given or taken from the system, such as doing work on the system or giving heat to the system. Thus, for an infinitesimal process, the *first law of thermodynamics* is written

$$dE = \delta Q + \delta W \tag{2.2}$$

Here, dE is the change in internal energy of the system, δQ is the heat absorbed by the system, and δW is the work done on the system by its surroundings. With δW we have denoted the differential of work done on the system, which is taken as $\delta W < 0$, and δQ is the differential of heat given to the system ($\delta Q > 0$).

Note that the first law of thermodynamics could equally be in terms of the heat emitted by the system or the work done by the system. Therefore, our definition in Eq. (2.2) is just conventional. Both descriptions are true as long as we are consistent in our definitions.

The work has the form:

$$\delta W = \mathbf{f} \cdot d\mathbf{X}$$

where $\mathbf{f}$ is the force exerted on the system, and $\mathbf{X}$ is an extensive mechanical parameter. A well known example for us is

$$\delta W = -p_{\text{ext}} dV$$

where V is the volume of the system and p_{ext} is the external pressure.

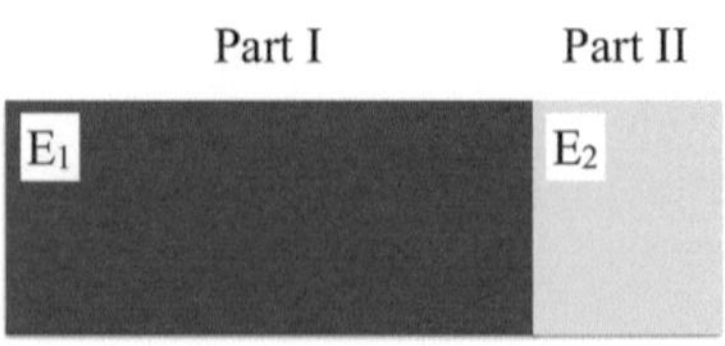

Fig. 2.1 A composed system

The experimental results indicate that isolated systems tend to go spontaneously towards simple final states, which are called equilibrium states. With simple states, we should understand states that are characterized by a small number of parameters from a macroscopic point of view. In particular, the equilibrium state from the macroscopic point of view is completely defined by E and $\mathbf{X}$. In the case of a system that can be characterized by extensive parameters, such as the volume and the number of particles (e.g., atoms, molecules), the entire set of parameters that characterize the system is

$$E, \underbrace{V, n_1, \cdots, n_j, \cdots, n_r}_{\mathbf{X}} \tag{2.3}$$

Here, V is the volume, n_j is the number of moles of the component j, and r is the total number of components. If on the system we apply an external electrical or magnetic field, then electric dipole momentum or magnetic dipole momentum of the system could enter on the list of parameters.

The complete list, which is characterizing the macroscopic equilibrium state, sometimes, it is difficult to be determined and it could be longer. However, this list is much smaller than the most significant number of degrees of freedom necessary to characterize the non-equilibrium macroscopic state of a system with many particles.

In general, there are no physical systems in complete equilibrium, but they are characterized by some metastable states, which can practically be a thermodynamic equilibrium. If the energy and the size of the system remain constant during the entire time of the observation, then the system can be considered in an equilibrium state, and its properties can be characterized by parameters given in expression of Eq. (2.3).

2.5 The Second Law of Thermodynamics

Figure 2.2 shows an isolated system, as an example.

We can impose these changes in the system: Allow the piston to move along the cylinder; Create a hole in the piston, so that a single molecule can pass through; Or remove the adiabatic wall and allow the system to exchange heat with the surrounding.

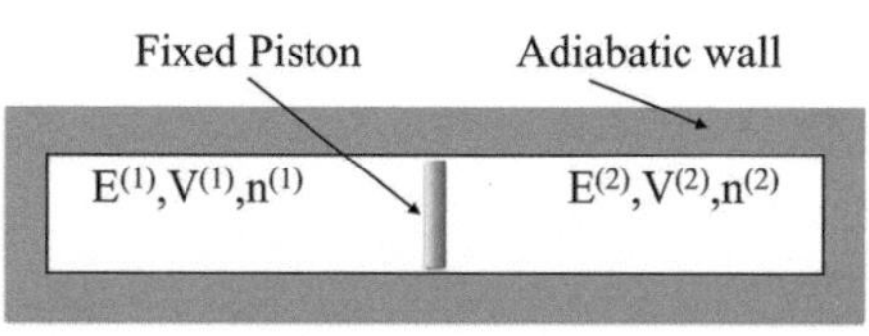

Fig. 2.2 An isolated system

Due to these possible changes, the system moves towards a final state, based on the so-called *second law of thermodynamics*, which is postulated as the following:

> Exists an extensive function of the state, $S(E, \mathbf{X})$, which is a monotonically increasing function of E and if the state B can be reached in an adiabatic way from the state A, then
>
> $$S_B \geq S_A$$

Note that if the state B is reachable from the state A in a reversible way, then the process $B \rightarrow A$ can adiabatically be realized as well. In this case, the postulate given above implies that $S_A \geq S_B$. Therefore, if the two states A and B are reachable in an adiabatic process and in a reversible way, then we can write

$$S_A = S_B$$

Differently, mathematically, for a reversible adiabatic process, this can be written as

$$\varDelta S = S_B - S_A = 0$$

For any other irreversible adiabatic process, it takes the following expression:

$$\varDelta S > 0$$

We can combine them for the general case of an adiabatic process, and write it as

$$(\varDelta S)_{adiabatic} \geq 0$$

Here, equality stands only for a reversible adiabatic process.

The extensive function $S(E, \mathbf{X})$ is called *entropy*. As we showed above, the change on the entropy is zero for any reversible adiabatic process. Moreover, entropy is a state function, which means that entropy is determined by these states that are functions of E and $\mathbf{X}$. These states are states of thermodynamic equilibrium. Let us consider first the differential of the entropy S:

$$dS = \left(\frac{\partial S}{\partial E}\right)_{\mathbf{X}} dE + \left(\frac{\partial S}{\partial \mathbf{X}}\right)_E \cdot d\mathbf{X} \tag{2.4}$$

The first law of thermodynamics for a reversible process can be written in the following form

$$dE = \delta Q + \mathbf{f} \cdot d\mathbf{X} \tag{2.5}$$

Substituting Eq. (2.5) into Eq. (2.4), we get

$$dS = \left(\frac{\partial S}{\partial E}\right)_{\mathbf{X}} \delta Q + \left[\left(\frac{\partial S}{\partial \mathbf{X}}\right)_E + \left(\frac{\partial S}{\partial E}\right)_{\mathbf{X}} \mathbf{f}\right] \cdot d\mathbf{X}$$

For an adiabatic and reversible process, $(\delta Q)_{\text{adiabatic}} = 0$ and $(dS)_{\text{rev}} = 0$, and hence

$$\left(\frac{\partial S}{\partial \mathbf{X}}\right)_E = -\left(\frac{\partial S}{\partial E}\right)_{\mathbf{X}} \mathbf{f} \tag{2.6}$$

We can define the temperature T as the derivative of the energy E with respect to the entropy S as

$$T \equiv \left(\frac{\partial E}{\partial S}\right)_{\mathbf{X}} \tag{2.7}$$

Since the entropy S is a monotonically increasing function of energy E, then $\left(\frac{\partial S}{\partial E}\right)_{\mathbf{X}} \geq 0$, or equivalently, $\left(\frac{\partial E}{\partial S}\right)_{\mathbf{X}} \geq 0$. Thus, $T \geq 0$, which is in agreement with our concept of temperature. Since both E and S are extensive quantities, then the temperature is an intensive property. That is, T is independent on the size of the system.

The combination of the above equations leads to

$$\left(\frac{\partial S}{\partial \mathbf{X}}\right)_E = -\frac{\mathbf{f}}{T} \tag{2.8}$$

From Eq. (2.4), we can write

$$dS = \left(\frac{1}{T}\right) dE - \left(\frac{\mathbf{f}}{T}\right) \cdot d\mathbf{X} \tag{2.9}$$

or we can re-arrange this equation into the following expression:

$$dE = TdS + \mathbf{f} \cdot d\mathbf{X} \tag{2.10}$$

This equation indicates that the energy, for a thermodynamic equilibrium, is characterized by S and $\mathbf{X}$, that is

$$E = E(S, \mathbf{X})$$

Comparing the first law of thermodynamic, Eq. (2.2), with expression given by Eq. (2.10), we get

$$dS = \frac{(\delta Q)_{rev}}{T}$$

and hence the second law of thermodynamics for a reversible process gives

$$(\delta Q)_{rev} = TdS, \tag{2.11}$$

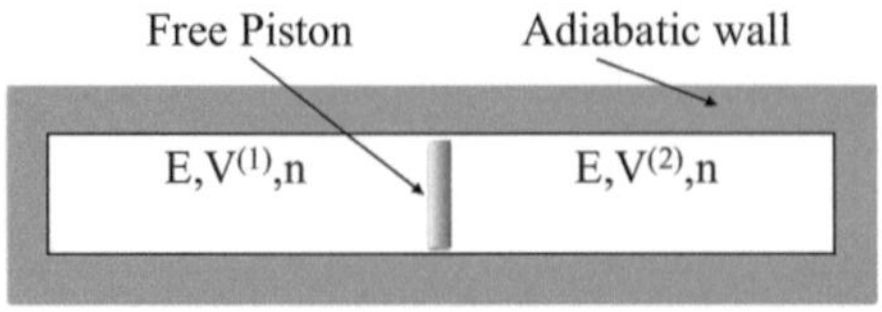

Fig. 2.3 Illustration of an internal constraint of a system. A piston can move freely along the cylinder by changing the volume $V^{(1)}$ and $V^{(2)}$ of each subsystem, but the total V is constant

for a quasi-static process, where T is now the thermodynamic temperature, and dS is the change in entropy of the system. While, for any process, we can write

$$dS \geq \frac{\delta Q}{T}$$

Here, the equality stands for a reversible process.

Consider a system that is partitioned into subsystems and the internal constraints related to the extensive parameters, such that a process applied to the system does not change the total energy E and $\mathbf{X}$ of the system, but the values of these quantities in each subsystem are allowed fluctuating. For example, Fig. 2.3 shows an example of internal constraints applied to a composed system of two subsystems, where the volume $V = V^{(1)} + V^{(2)}$ of the entire system is fixed, while the volume of each subsystem changes using, for example, an internal piston moving along the cylinder. To displace the internal piston along the cylinder we need to perform work, and hence the internal energy of the system will change.

The extremes are usually given in terms of variations far from the equilibrium states. Mathematically, we can express ΔS for these variations as a Taylor series:

$$\Delta S = S(E, \mathbf{X}; \delta\mathbf{Y}) - S(E, \mathbf{X}; 0) = (\delta S)_{E,\mathbf{X}} + (\delta^2 S)_{E,\mathbf{X}} + \cdots$$

where $\delta\mathbf{Y}$ is a variation due to the application of the internal constraint, such as

$$\mathbf{X} = \mathbf{X}^{(1)} + \mathbf{X}^{(2)} = \left[\mathbf{X}^{(1)} + \delta\mathbf{Y}\right] + \left[\mathbf{X}^{(2)} - \delta\mathbf{Y}\right],$$

and

$$\begin{aligned}(\delta S)_{E,\mathbf{X}} &= \left[\left(\frac{\partial S}{\partial \mathbf{Y}}\right)_{E,\mathbf{X}}\right]_{\mathbf{Y}=0} \delta\mathbf{Y} \\ (\delta^2 S)_{E,\mathbf{X}} &= \frac{1}{2}\left[\left(\frac{\partial^2 S}{\partial \mathbf{Y}^2}\right)_{E,\mathbf{X}}\right]_{\mathbf{Y}=0} (\delta\mathbf{Y})^2\end{aligned} \tag{2.12}$$

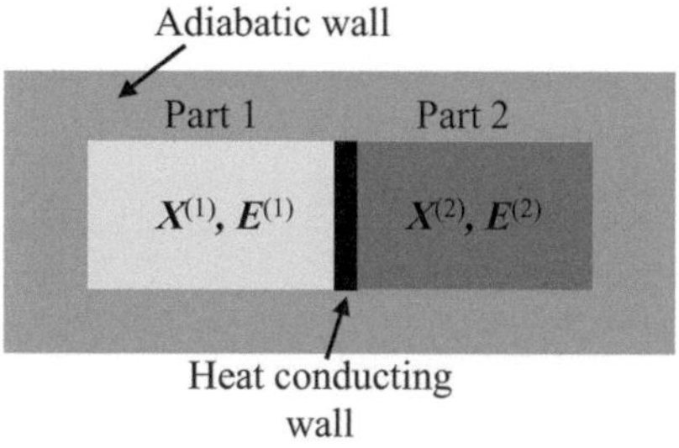

Fig. 2.4 A composite system isolated from surrounding with a heat conducting wall dividing two subsystems (1) and (2)

Based on the principle of entropy for all variations far from subspace of equilibrium states, we can write

$$(\delta S)_{E,\mathbf{X}} \leq 0$$

and for all variations far from stable equilibrium state, we can write

$$(\Delta S)_{E,\mathbf{X}} < 0$$

This implies that the entropy obtains its maximum value at equilibrium state, which is known as the principle of maximum entropy.

From the principle of the maximum entropy, we can obtain another principle, which is known as the principle of minimum energy. To derive this principle, we can consider a composed system, as shown in Fig. 2.4. Let us denote with $E^{(1)}$ and $E^{(2)}$ the equilibrium values of the energies. Using the principle of the maximum entropy, we can write

$$S(E^{(1)} - \Delta E, \mathbf{X}^{(1)}) + S(E^{(2)} + \Delta E, \mathbf{X}^{(2)}) < S(E^{(1)} + E^{(2)}, \mathbf{X}^{(1)} + \mathbf{X}^{(2)})$$

ΔE is the amount of energy reduced from the subsystem (1) and added to the subsystem (2). By the division of the energy, the value of entropy decreases as indicated by the sign of the inequality. For the composite system, we had used the extensive property of the entropy when we calculated the total energy of the composite system by adding the values of the entropy of each part.

The temperature is a positive quantity, therefore, the entropy is a monotonically increasing function of the energy. Thus,

$$E < E^{(1)} + E^{(2)}$$

for $\Delta E \neq 0$ such that

$$S(E^{(1)} - \Delta E, \mathbf{X}^{(1)}) + S(E^{(2)} + \Delta E, \mathbf{X}^{(2)}) = S(E^{(1)} + E^{(2)}, \mathbf{X}^{(1)} + \mathbf{X}^{(2)})$$

Therefore, we can suppose that the application of the internal constraints with S and $\mathbf{X}$ unchanged, will increase the total energy of the system. That is, $E(S, \mathbf{X})$ is the global minimum of

$$E(S, \mathbf{X}; \text{internal constraint})$$

This represents the so-called the principle of the minimum energy. Moreover, we can give the principle of the minimum energy in terms of the mathematical variations far from the equilibrium state. Thus, we can express ΔE for these variations as a Taylor series

$$\Delta E = E(S, \mathbf{X}; \delta Y) - E(S, \mathbf{X}; 0) = (\delta E)_{S,\mathbf{X}} + \left(\delta^2 E\right)_{S,\mathbf{X}} + \cdots$$

where δY is a variation or partition of internal extensive properties as the result of application of the internal constraint. Thus, we can write

$$(\delta E)_{S,\mathbf{X}} = \left[\left(\frac{\partial E}{\partial Y}\right)_{S,\mathbf{X}}\right]_{Y=0} \delta Y$$

$$\left(\delta^2 E\right)_{S,\mathbf{X}} = \left[\frac{1}{2}\left(\frac{\partial^2 E}{\partial Y^2}\right)_{S,\mathbf{X}}\right]_{Y=0} (\delta Y)^2$$

These principles are summarized for all variations far from the subspace of equilibrium states with $\delta Y = 0$ as the following:

$$(\delta E)_{S,\mathbf{X}} \geq 0$$

and for all variations far from the stable equilibrium states, we can write

$$(\Delta E)_{S,\mathbf{X}} > 0$$

2.6 Thermal Equilibrium and Temperature

The variation principle of the second law of thermodynamics can be applied to determine the criteria of the thermal equilibrium, which identifies T as the well-known temperature. First, let us consider the system shown in Fig. 2.5.

We are assuming that there are small displacements around the equilibrium state due to the internal constraints. Based on the variation principle for entropy, we can write

$$(\delta S)_{E,\mathbf{X}} \leq 0 \tag{2.13}$$

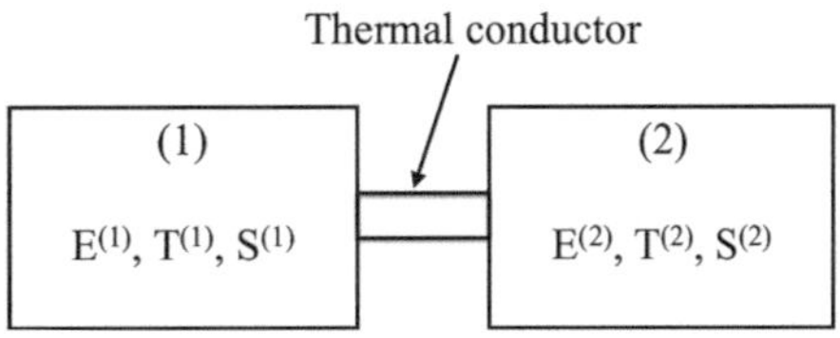

Fig. 2.5 A system used as heat conductor

where $E = E^{(1)} + E^{(2)}$ remains unchanged during the displacement. Thus, we get

$$\delta E = 0 = \delta E^{(1)} + \delta E^{(2)}$$

or

$$\delta E^{(1)} = -\delta E^{(2)} \tag{2.14}$$

Since the entropy is an extensive quantity, then

$$S = S^{(1)} + S^{(2)}$$

Thus,

$$\begin{aligned}(\delta S)_{E,\mathbf{X}} &= \delta S^{(1)} + \delta S^{(2)} \\ &= \left(\frac{\partial S^{(1)}}{\partial E^{(1)}}\right)_{\mathbf{X}} \delta E^{(1)} + \left(\frac{\partial S^{(2)}}{\partial E^{(2)}}\right)_{\mathbf{X}} \delta E^{(2)} \\ &= \left(\frac{1}{T^{(1)}} - \frac{1}{T^{(2)}}\right) \delta E^{(1)}\end{aligned} \tag{2.15}$$

where we have replaced

$$\left(\frac{\partial S}{\partial E}\right)_{\mathbf{X}} = \frac{1}{T}$$

and we have used Eq. (2.14). Using variation principle for entropy, Eq. (2.13), we get

$$\left(\frac{1}{T^{(1)}} - \frac{1}{T^{(2)}}\right) \delta E^{(1)} \leq 0$$

for all possible values of the small variations $\delta E^{(1)}$ (positive or negative). This implies that in equilibrium the following equality holds

$$T^{(1)} = T^{(2)}$$

Here, we initially assumed that system was at equilibrium, then by applying an internal constraint we analyzed the response of the system, and by comparing with the second law of thermodynamics we concluded about the initial equilibrium state. This is known as the *thermal equilibrium*, which states that:

> Two systems that are in contact with each other are in thermal equilibrium, if there is no change in their states.

To derive the condition of the thermal equilibrium, we assumed that the system is in the equilibrium state, and we learned about this state by applying an internal constraint on the system. Then, we observed the reaction of the system to this internal constraint, and using the second law of thermodynamics, we concluded about the equilibrium of the initial state. This procedure is often used to identify the equilibrium state of a system. In reality, this is also closely related to the nature of the experimental observations. In particular, any real experiment that describes the behavior of a system in equilibrium will observe the reaction of the system against a disturbance.

In this way, we only have proved using the second law of thermodynamics that the condition for thermal equilibrium is achieved when the two interacting subsystems have the same temperature. In this proof, we used the variation principle of the entropy. However, we can also apply the variation principle of the energy.

Now, we are going to consider what might happen when the system is not initially in thermal equilibrium, for example, initially $T^{(1)} \neq T^{(2)}$. In fact, in a stable thermal equilibrium, both temperatures equalize, as explained in the following.

For that, we are going to use again the second law, and this time we are going to emphasize that transition to equilibrium is a natural process, thus change in entropy, ΔS, is positive, $\Delta S > 0$. That is,

$$dS > \frac{\delta Q}{T}$$

and δQ for the entire system is zero. Therefore,

$$\Delta S = \Delta S^{(1)} + \Delta S^{(2)} > 0$$

Supposing that the differences are small, as a result of this we can write

$$\left(\frac{\partial S^{(1)}}{\partial E^{(1)}}\right)_{\mathbf{X}} \Delta E^{(1)} + \left(\frac{\partial S^{(2)}}{\partial E^{(2)}}\right)_{\mathbf{X}} \Delta E^{(2)} > 0$$

or, since $\Delta E^{(1)} = -\Delta E^{(2)}$ and

$$\left(\frac{\partial S}{\partial E}\right)_{\mathbf{X}} = \frac{1}{T}$$

then,

$$\left(\frac{1}{T^{(1)}} - \frac{1}{T^{(2)}}\right) \Delta E^{(1)} > 0$$

If we suppose that $T^{(1)} > T^{(2)}$, then $\left(\frac{1}{T^{(1)}} - \frac{1}{T^{(2)}}\right) < 0$, and in order that inequality to be satisfied, we must have that

$$\Delta E^{(1)} < 0$$

In similar way, if $T^{(1)} < T^{(2)}$, then $\left(\frac{1}{T^{(1)}} - \frac{1}{T^{(2)}}\right) > 0$, and the inequality is satisfied for $\Delta E^{(1)} > 0$.

Finally, the energy flows from the hotter system to the cold one. Besides, the heat is a form of the energy that transfers as a result of the gradient in the temperature, and the direction of this transfer is from the hotter system (the highest temperature) to the colder system (the lowest temperature).

2.7 Legendre Transformation

This section introduces the main mathematical method for analyzing the macroscopic thermodynamics of a system, known as *Legendre transformation*. Also, we will introduce a specific differential expression of the reversible work for the systems that are, in particular, studied in the field of biology and chemistry.

In this case for the reversible displacements we can write:

$$\mathbf{f} \cdot d\mathbf{X} = -pdV + \sum_{i=1}^{r} \mu_i dn_i \tag{2.16}$$

where p is the pressure of system, V is the volume, n_i is the number of moles of the component i, r is the total number of components in the system, and μ_i is the chemical potential of the component i. The chemical potential is determined from Eq. (2.16) and the first and second laws as:

$$dE = TdS + \mathbf{f} \cdot d\mathbf{X} = TdS - pdV + \sum_{i=1}^{r} \mu_i dn_i \tag{2.17}$$

then, we can determine μ_i as

$$\mu_i = \left(\frac{\partial E}{\partial n_i}\right)_{S,V,n_j \neq n_i}$$

That is, the chemical potential μ_i is the reversible change of the internal energy by changing the number of moles n_i and keeping constant S, V, and the number of moles n_j of other components in the system. Moreover, the chemical potential is an intensive property of the system, which controls the mass or particles equilibrium similarly to temperature T controls the thermal equilibrium. The gradient of the chemical potential determines the mass transfer or reorganization of atoms and molecules. In the absence of this gradient, an equilibrium of the mass establishes in the system. The process of the re-organization of atoms and molecules determines the equilibrium states between different phases of matter and between different chemical components.

Equation (2.17) indicates that the internal energy is a function of S, V and n_i (for $i = 1, 2, \cdots, r$):

$$E = E\,(S, V, n_1, \cdots, n_r)$$

The principles of variations are obtained in the form:

$$(\Delta E)_{S,V,n} > 0 \qquad (2.18)$$
$$(\delta E)_{S,V,n} \geq 0$$

which can give information about the equilibrium states that are characterized by S, V and n_i (for $i = 1, 2, \cdots, r$). Next, we will derive other thermodynamic functions of states, equivalent to the internal energy E or entropy S, which are functions of macroscopic properties of the system.

For this, let us consider a function $f = f(x_1, x_2, \cdots, x_n)$, then

$$df = \sum_{i=1}^{n} u_i dx_i \qquad (2.19)$$
$$u_i = \left(\frac{\partial f}{\partial x_i}\right)_{x_j \neq x_i}$$

In addition, we can determine another function g, such as

$$g = f - \sum_{i=r+1}^{n} u_i x_i \qquad (2.20)$$

Moreover, we can write that

$$dg = df - \sum_{i=r+1}^{n} (u_i dx_i + x_i du_i) \qquad (2.21)$$

$$= \sum_{i=1}^{n} u_i dx_i - \sum_{i=r+1}^{n} (u_i dx_i + x_i du_i)$$

$$= \sum_{i=1}^{r} u_i dx_i - \sum_{i=r+1}^{n} x_i du_i$$

This equation indicates that

$$g = g(x_1, x_2, \cdots, x_r, u_{r+1}, u_{r+2}, \cdots, u_n)$$

where $u_{r+1}, u_{r+2}, \cdots, u_n$ are the conjugated variables of $x_{r+1}, x_{r+2}, \cdots, x_n$. By definition, the function g is called *Legendre transformation* of f. The function g transforms the dependence from $x_{r+1}, x_{r+2}, \cdots, x_n$ of f to $u_{r+1}, u_{r+2}, \cdots, u_n$ of g.

Using the Legendre transformation and the expression given by Eq. (2.17), we can see that the conjugated variables of each other are:

$$T \leftrightarrow S, \quad -p \leftrightarrow V, \quad \mu_i \leftrightarrow n_i$$

Therefore, in order to define a new thermodynamic function of (T, V, n_i), we can use Eq. (2.20) with f being E, and write

$$F = E - TS = F(T, V, n_1, \cdots, n_r) \tag{2.22}$$

where F is called the *Helmholtz free energy*. The differential form of F is then given as

$$\begin{aligned} dF &= dE - TdS - SdT \\ &= TdS - pdV + \sum_{i=1}^{r} \mu_i dn_i - TdS - SdT \\ &= -SdT - pdV + \sum_{i=1}^{r} \mu_i dn_i \end{aligned} \tag{2.23}$$

Using Eq. (2.23), the macroscopic properties of the system can be determined as

$$\begin{aligned} S &= -\left(\frac{\partial F}{\partial T}\right)_{V,n_1,\cdots,n_r} \\ p &= -\left(\frac{\partial F}{\partial V}\right)_{T,n_1,\cdots,n_r} \\ \mu_i &= \left(\frac{\partial F}{\partial n_i}\right)_{T,V,n_j \neq n_i} \end{aligned} \tag{2.24}$$

Other Legendre transformations are also possible, for instance, we can determine a new thermodynamics function of S, p and $n_1, \cdots, n_r$ as

$$H = E + pV = H(S, p, n_1, \cdots, n_r) \tag{2.25}$$

which is called *enthalpy*. The differential form of H is given as

$$\begin{aligned} dH &= dE + pdV + Vdp \\ &= TdS - pdV + \sum_{i=1}^{r} \mu_i dn_i + pdV + Vdp \\ &= TdS + Vdp + \sum_{i=1}^{r} \mu_i dn_i \end{aligned} \tag{2.26}$$

Eq. (2.26) can determine the following macroscopic properties of the system:

$$\begin{aligned} T &= \left(\frac{\partial H}{\partial S}\right)_{p,n_1,\cdots,n_r} \\ V &= \left(\frac{\partial H}{\partial p}\right)_{S,n_1,\cdots,n_r} \\ \mu_i &= \left(\frac{\partial H}{\partial n_i}\right)_{S,p,n_j \neq n_i} \end{aligned} \tag{2.27}$$

Another thermodynamic function is the *Gibbs free energy*, G, given as

$$G = E - ST + pV = G(T, V, n_1, \cdots, n_r)$$

which is a natural function of T, p, and $n_1, \cdots, n_r$. We can derive the differential form of G as the following:

$$\begin{aligned} dG &= dE - SdT - TdS + pdV + Vdp \\ &= TdS - pdV + \sum_{i=1}^{r} \mu_i dn_i - SdT - TdS + pdV + Vdp \\ &= -SdT + Vdp + \sum_{i=1}^{r} \mu_i dn_i \end{aligned} \tag{2.28}$$

The macroscopic properties of the system can be determined as

$$S = -\left(\frac{\partial G}{\partial T}\right)_{p,n_1,\cdots,n_r} \tag{2.29}$$

$$V = \left(\frac{\partial G}{\partial p}\right)_{T,n_1,\cdots,n_r}$$

$$\mu_i = \left(\frac{\partial G}{\partial n_i}\right)_{T,p,n_j \neq n_i}$$

The principles of variations of the thermodynamic functions F, H, and G are given as

$$(\delta F)_{T,V,n_1,\cdots,n_r} \geq 0, \quad (\Delta F)_{T,V,n_1,\cdots,n_r} > 0 \tag{2.30}$$

$$(\delta H)_{S,p,n_1,\cdots,n_r} \geq 0, \quad (\Delta H)_{S,p,n_1,\cdots,n_r} > 0$$

$$(\delta G)_{T,p,n_1,\cdots,n_r} \geq 0, \quad (\Delta G)_{T,p,n_1,\cdots,n_r} > 0$$

2.8 Maxwell Relations

When changing the order of taking mixed derivatives of a thermodynamic potential creates a class of identities known as *Maxwell relations*. Using Maxwell relations, different other quantities can be found to be related to each other.

Let us consider first the Helmholtz free energy for a system with one component in its differential form:

$$dF = -SdT - pdV + \mu dn$$

Then, we can write that

$$S = -\left(\frac{\partial F}{\partial T}\right)_{V,n}$$

$$p = -\left(\frac{\partial F}{\partial V}\right)_{T,n}$$

Using these two relations, we can determine the second partial derivatives of S and p as the following:

$$\left(\frac{\partial S}{\partial V}\right)_{T,n} = -\left(\frac{\partial}{\partial V}\left(\frac{\partial F}{\partial T}\right)_{V,n}\right)_{T,n} \tag{2.31}$$

$$\left(\frac{\partial p}{\partial T}\right)_{V,n} = -\left(\frac{\partial}{\partial T}\left(\frac{\partial F}{\partial V}\right)_{T,n}\right)_{V,n} = -\left(\frac{\partial}{\partial V}\left(\frac{\partial F}{\partial T}\right)_{V,n}\right)_{T,n}$$

From Eq. (2.31), since the right-hand-side are equal, then we obtain that

$$\left(\frac{\partial S}{\partial V}\right)_{T,n} = \left(\frac{\partial p}{\partial T}\right)_{V,n} \tag{2.32}$$

which is known as the *first Maxwell relation*. The left-hand-side indicates that the entropy S is a function of V, T, and n. Equation (2.32) is a very useful relation, because practically it is very difficult to measure entropy S; on the other hand measuring the pressure p as a function of T is much easier, thus Eq. (2.32) can be used to determine S.

Using the following differential form of the Gibbs free energy:

$$dG = -SdT + Vdp + \mu dn$$

another Maxwell relation can be derived. In particular, we can obtain

$$S = -\left(\frac{\partial G}{\partial T}\right)_{p,n}$$

$$V = \left(\frac{\partial G}{\partial p}\right)_{T,n}$$

The second order partial derivatives of S and V can now be derived as the following:

$$\left(\frac{\partial S}{\partial p}\right)_{T,n} = -\left(\frac{\partial}{\partial p}\left(\frac{\partial G}{\partial T}\right)_{p,n}\right)_{T,n} \tag{2.33}$$

$$\left(\frac{\partial V}{\partial T}\right)_{p,n} = \left(\frac{\partial}{\partial T}\left(\frac{\partial G}{\partial p}\right)_{T,n}\right)_{p,n} = \left(\frac{\partial}{\partial p}\left(\frac{\partial G}{\partial T}\right)_{p,n}\right)_{T,n}$$

From Eq. (2.33), we obtain:

$$\left(\frac{\partial S}{\partial p}\right)_{T,n} = -\left(\frac{\partial V}{\partial T}\right)_{p,n} \tag{2.34}$$

which is known as the *second Maxwell relation*.

The first and the second Maxwell relations can be used to obtain also other thermodynamic relations of practical interests. Taking the entropy S a function of T, V and n

$$S = S(T, V, n)$$

then the differential form of S for fixed n is given as

$$(dS)_n = \left(\frac{\partial S}{\partial T}\right)_{V,n} (dT)_n + \left(\frac{\partial S}{\partial V}\right)_{T,n} (dV)_n \tag{2.35}$$

Dividing both sides of Eq. (2.35) by dT, keeping the pressure p constant, we obtain

$$\left(\frac{\partial S}{\partial T}\right)_{p,n} = \left(\frac{\partial S}{\partial T}\right)_{V,n} + \left(\frac{\partial S}{\partial V}\right)_{T,n} \left(\frac{\partial V}{\partial T}\right)_{p,n}$$

Using the definitions of the heat capacities

$$C_p = T\left(\frac{\partial S}{\partial T}\right)_{p,n}, \quad C_V = T\left(\frac{\partial S}{\partial T}\right)_{V,n}$$

and the first Maxwell relation, we obtain

$$\frac{1}{T}C_p = \frac{1}{T}C_V + \left(\frac{\partial p}{\partial T}\right)_{V,n} \left(\frac{\partial V}{\partial T}\right)_{p,n} \tag{2.36}$$

Now, consider the volume V to be a function of p, T, and n:

$$V = V(p, T, n)$$

then, the differential form of V for fixed n is given as

$$(dV)_n = \left(\frac{\partial V}{\partial p}\right)_{T,n} (dp)_n + \left(\frac{\partial V}{\partial T}\right)_{p,n} (dT)_n$$

If we divide both sides by dT and keep fixed V and n, we obtain

$$0 = \left(\frac{\partial V}{\partial p}\right)_{T,n} \left(\frac{\partial p}{\partial T}\right)_{V,n} + \left(\frac{\partial V}{\partial T}\right)_{p,n}$$

or

$$\left(\frac{\partial p}{\partial T}\right)_{V,n} = -\left(\frac{\partial p}{\partial V}\right)_{T,n} \left(\frac{\partial V}{\partial T}\right)_{p,n} \tag{2.37}$$

Substituting Eq. (2.37) into Eq. (2.36), we obtain

$$\frac{1}{T}C_p = \frac{1}{T}C_V - \left(\frac{\partial p}{\partial V}\right)_{T,n}\left(\frac{\partial V}{\partial T}\right)_{p,n}\left(\frac{\partial V}{\partial T}\right)_{p,n} \tag{2.38}$$

or

$$C_p - C_V = -T\left(\frac{\partial p}{\partial V}\right)_{T,n}\left[\left(\frac{\partial V}{\partial T}\right)_{p,n}\right]^2 \tag{2.39}$$

Introducing the coefficient of thermal expansion as

$$\alpha = \frac{1}{V}\left(\frac{\partial V}{\partial T}\right)_{p,n}$$

and the isothermal compressibility as

$$\kappa_T = -\frac{1}{V}\left(\frac{\partial V}{\partial p}\right)_{T,n}$$

Eq. (2.39) can also be written as

$$C_p - C_V = TV\frac{\alpha^2}{\kappa_T} \tag{2.40}$$

This expression is very popular in relating the heat capacity with the coefficient of thermal expansion and the isothermal compressibility.

Equation (2.40) can also be written in terms of the specific heat capacities as

$$c_p - c_V = \frac{TV}{n}\frac{\alpha^2}{\kappa_T}$$

where n is the number of moles. For an ideal gas, $pV = nRT$, therefore

$$\alpha = 1/T$$

and

$$\kappa_T = 1/p$$

Thus, we obtain

$$c_p - c_V = \frac{TV}{n}\frac{1/T^2}{1/p} = R$$

which is a well-known relation that has been derived for the ideal gas.

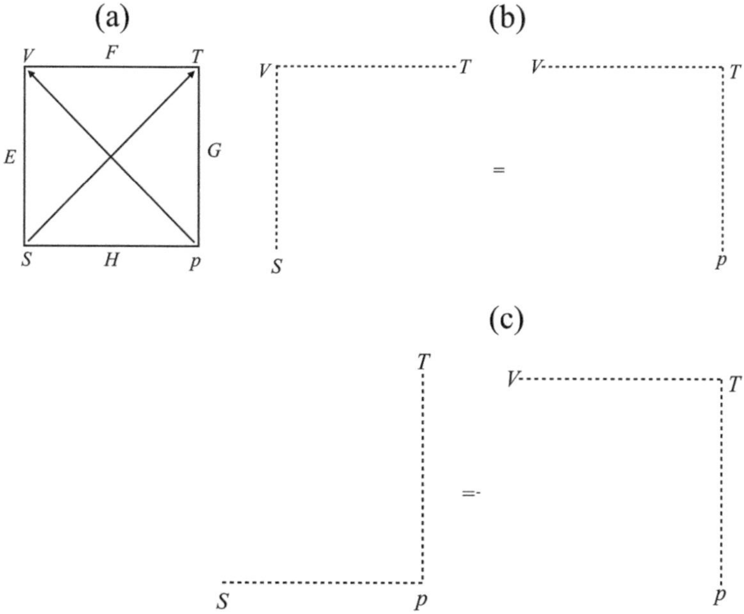

Fig. 2.6 (**a**) A mnemonic diagram; (**b**) the diagram for the first Maxwell relation; (**c**) the diagram for the second Maxwell relation

Note that the Maxwell relations can also be derived using the mnemonic diagram shown in Fig. 2.6a. This diagram consists of sides labeled by four thermodynamic potential functions flanked by their respective natural independent variables. Using Fig. 2.6a, in the differential expression for each potential function in terms of the physical variables the arrow pointing away from the variable implies of positive sign, while an arrow pointing towards the variable involves negative sign. The vertices of the diagram give Maxwell relations. For example, the graph shown in Fig. 2.6b represents the first Maxwell relation, and a diagram of Fig. 2.6c represents the second Maxwell relation.

Other Maxwell relations can also be derived, as also shown from the diagram in Fig. 2.7. In particular, the following Maxwell relation can be derived using the diagram in Fig. 2.7:

$$\left(\frac{\partial V}{\partial S}\right)_{p,n} = \left(\frac{\partial T}{\partial p}\right)_{S,n} \tag{2.41}$$

It is important to note that this relation can be derived using the differential form of the enthalpy:

$$dH = Vdp + TdS + \mu dn$$

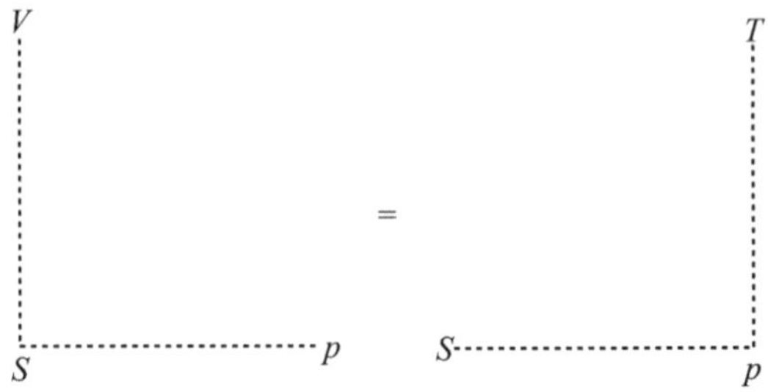

Fig. 2.7 A mnemonic diagram the Maxwell relation $\left(\frac{\partial V}{\partial S}\right)_{p,n} = \left(\frac{\partial T}{\partial p}\right)_{S,n}$

From this equation, we can write

$$V = \left(\frac{\partial H}{\partial p}\right)_{S,n}$$

$$T = \left(\frac{\partial H}{\partial S}\right)_{p,n}$$

Taking the partial derivatives of both sides in each equation, we obtain

$$\left(\frac{\partial V}{\partial S}\right)_{p,n} = \left(\frac{\partial}{\partial S}\left(\frac{\partial H}{\partial p}\right)_{S,n}\right)_{p,n}$$

$$\left(\frac{\partial T}{\partial p}\right)_{S,n} = \left(\frac{\partial}{\partial p}\left(\frac{\partial H}{\partial S}\right)_{p,n}\right)_{S,n} = \left(\frac{\partial}{\partial S}\left(\frac{\partial H}{\partial p}\right)_{S,n}\right)_{p,n}$$

Comparing these two expressions, it is easy to find that Eq. (2.41) holds. We call this equation here as the *third Maxwell relation*. It should be mentioned that this relation, Eq. (2.41), has less practical applications because it is very difficult to control entropy during an experiment.

Another relation can be obtained using the mnemonic diagram shown in Fig. 2.8, called here the *fourth Maxwell relation*:

$$\left(\frac{\partial T}{\partial V}\right)_{S,n} = -\left(\frac{\partial p}{\partial S}\right)_{V,n} \tag{2.42}$$

To derive this relation, the differential form of the internal energy E can be used:

$$dE = TdS - pdV + \mu dn$$

From here, we obtain

$$T = \left(\frac{\partial E}{\partial S}\right)_{V,n}$$

$$p = -\left(\frac{\partial E}{\partial V}\right)_{S,n}$$

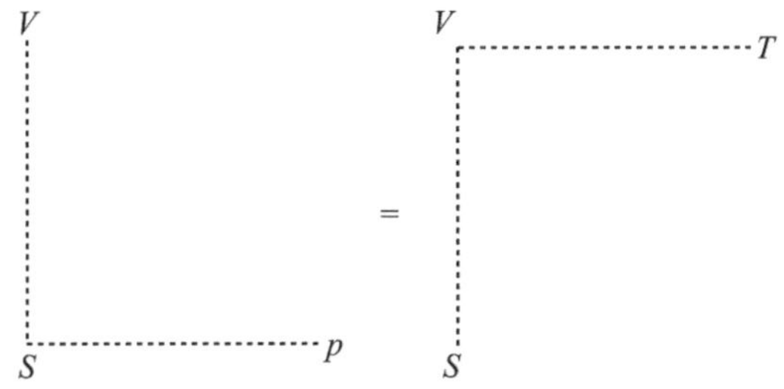

Fig. 2.8 A mnemonic diagram the Maxwell relation $\left(\frac{\partial T}{\partial V}\right)_{S,n} = -\left(\frac{\partial p}{\partial S}\right)_{V,n}$

Taking the partial derivatives of the both sides of each equation, we can obtain

$$\left(\frac{\partial T}{\partial V}\right)_{S,n} = \left(\frac{\partial}{\partial V}\left(\frac{\partial E}{\partial S}\right)_{V,n}\right)_{S,n}$$

$$\left(\frac{\partial p}{\partial S}\right)_{V,n} = -\left(\frac{\partial}{\partial S}\left(\frac{\partial E}{\partial V}\right)_{S,n}\right)_{V,n} = -\left(\frac{\partial}{\partial V}\left(\frac{\partial E}{\partial S}\right)_{V,n}\right)_{S,n}$$

It can be seen that the right-hand-sides of the both equations are equal, thus equalizing the left-hand-sides, we obtain Eq. (2.42).

2.9 Extensive Functions

Consider the internal energy, which is an extensive quantity, and it depends on entropy S and mechanical parameters $\mathbf{X}$, which also are extensive. Therefore,

$$E(\lambda S, \lambda \mathbf{X}) = \lambda E(S, \mathbf{X}) \tag{2.43}$$

where λ is a real constant. This indicates that from mathematical point of view $E(S, \mathbf{X})$ is a homogeneous function of the first order of S and $\mathbf{X}$. Let us denote

$$S' = \lambda S, \quad \mathbf{X}' = \lambda \mathbf{X}$$

Then,

$$\left(\frac{\partial E(S', \mathbf{X}')}{\partial \lambda}\right)_{S,\mathbf{X}} = E(S, \mathbf{X}) \tag{2.44}$$

Moreover, the differential form of E is

$$dE(S', \mathbf{X}') = \left(\frac{\partial E}{\partial \mathbf{X}'}\right)_{S'} \cdot d\mathbf{X}' + \left(\frac{\partial E}{\partial S'}\right)_{\mathbf{X}'} dS'$$

Dividing both sides with $d\lambda$ and keeping constant $\mathbf{X}$ and S, we get

$$\begin{aligned}\left(\frac{\partial E}{\partial \lambda}\right)_{S,\mathbf{X}} &= \left(\frac{\partial E}{\partial \mathbf{X}'}\right)_{S'} \cdot \left(\frac{\partial \mathbf{X}'}{\partial \lambda}\right)_{S,\mathbf{X}} + \left(\frac{\partial E}{\partial S'}\right)_{\mathbf{X}'} \left(\frac{\partial S'}{\partial \lambda}\right)_{S,\mathbf{X}} \\ &= \left(\frac{\partial E}{\partial \mathbf{X}'}\right)_{S'} \cdot \mathbf{X} + \left(\frac{\partial E}{\partial S'}\right)_{\mathbf{X}'} S\end{aligned} \tag{2.45}$$

Combining Eq. (2.44) with Eq. (2.45), we obtain

$$E(S, \mathbf{X}) = \left(\frac{\partial E}{\partial \mathbf{X}'}\right)_{S'} \cdot \mathbf{X} + \left(\frac{\partial E}{\partial S'}\right)_{\mathbf{X}'} S \tag{2.46}$$

If we take the constant $\lambda = 1$, then

$$d\mathbf{X}' = d\mathbf{X}, \quad dS' = dS$$

Thus, Eq. (2.46) can be written as

$$E(S, \mathbf{X}) = \left(\frac{\partial E}{\partial \mathbf{X}}\right)_{S} \cdot \mathbf{X} + \left(\frac{\partial E}{\partial S}\right)_{\mathbf{X}} S \tag{2.47}$$

This is also known as the *Euler Theorem*. From Eq. (2.47) we can further write that

$$E(S, \mathbf{X}) = \mathbf{f} \cdot \mathbf{X} + TS \tag{2.48}$$

The differential form of Eq. (2.48) gives

$$dE(S, \mathbf{X}) = d\mathbf{f} \cdot \mathbf{X} + \mathbf{f} \cdot d\mathbf{X} + TdS + SdT \tag{2.49}$$

On the other hand, we know that

$$dE(S, \mathbf{X}) = TdS + \mathbf{f} \cdot d\mathbf{X}$$

Thus, we obtain

$$d\mathbf{f} \cdot \mathbf{X} + SdT = 0 \tag{2.50}$$

Replacing

$$d\mathbf{f} \cdot \mathbf{X} = -Vdp + \sum_{i=1}^{r} n_i d\mu_i$$

into Eq. (2.50), we get

$$SdT - Vdp + \sum_{i=1}^{r} n_i d\mu_i = 0 \tag{2.51}$$

which is also known as the *equation of the Gibbs-Duhem*.

In addition, using the Euler Theorem for the Gibbs free energy, we can obtain the following:

$$\begin{aligned} G &= E - TS + pV \\ &= \left(TS - pV + \sum_{i=1}^{r} \mu_i n_i\right) - TS + pV \\ &= \sum_{i=1}^{r} \mu_i n_i \end{aligned} \tag{2.52}$$

If the system has just one component, then

$$G = \mu n$$

or

$$\mu = \frac{G}{n}$$

That is, for a system with one component, the chemical potential μ is the Gibbs free energy per mol.

Equation (2.51) is the special case of the most general case

$$\mathbf{X}(T, p, n_1, \cdots, n_r) = \sum_{i=1}^{r} x_i n_i \tag{2.53}$$

where $\mathbf{X}$ is an extensive function of the intensive parameters T, p and x_i partial moles of the component i and n_i is the number of moles of the component i. We can write:

$$x_i = \left(\frac{\partial \mathbf{X}}{\partial n_i}\right)_{T,p,n_{j\neq i}} = x_i(T, p, n_1, \cdots, n_r) \tag{2.54}$$

Moreover, if T and p are constants, then using the Euler Theorem, $\mathbf{X}(T, p, n_1, \cdots, n_r)$ is a homogeneous function of the first order of the number of moles, thus:

$$\mathbf{X}(T, p, \lambda n_1, \cdots, \lambda n_r) = \lambda \mathbf{X}(T, p, n_1, \cdots, n_r) \tag{2.55}$$

2.10 Intensive Functions

The intensive functions are homogeneous of the zeroth order of the extensive parameters. For example, the pressure is an intensive quantity, if we consider it as a function of S, $V, n_1, \cdots, n_r$, we can write

$$p(\lambda S,\ \lambda V,\ \lambda n_1,\ \cdots,\ \lambda n_r) = p(S,\ V,\ n_1,\ \cdots,\ n_r)$$

which holds for every λ. Choosing λ to be

$$\lambda = \frac{1}{\sum_{i=1}^{r} n_i} = \frac{1}{n}$$

where n denotes the total number of moles, then

$$p = p(S/n, V/n, n_1/n, \cdots, n_r/n) \tag{2.56}$$

Denoting

$$x_i = \frac{n_i}{n}$$

the fraction of moles of the component i, then

$$p = p(S/n, V/n, x_1, \cdots, x_r) \tag{2.57}$$

In addition, the fraction of moles are not independent, that is

$$\sum_{i=1}^{r} x_i = 1$$

Thus,

$$p = p(S/n, V/n, x_1, \cdots, x_{r-1}, 1 - x_1 - x_2 - \cdots - x_{r-1}) \tag{2.58}$$

While Eq. (2.55) indicates that $2 + r$ extensive parameters are necessary to determine the value of an extensive parameter for a system in equilibrium, Eq. (2.58) suggests that only $1 + r$ parameters are essential to deciding on the values of other intensive parameters. Because the intensive properties are independent of the size of the system implies this reduction of the degrees of freedom from $2 + r$ to $1 + r$.

2.11 Stability of Thermodynamic Systems

In the following, we will consider a heterogeneous system consisting of multiple phases with multiple components in each phase. Each phase contains another different subsystem. Also, a new partition of the extensive parameters is considered between different phases. For example, since E is an extensive quantity, the internal energy is

$$E = \sum_{\alpha=1}^{\nu} E^{(\alpha)} \tag{2.59}$$

Here, ν is the total number of the phases. If we make a new partition of the energy between the phases, then changes on E^{α} are such that the total internal energy of the system E remains constant. Note that here we have neglected the energy term related to the interface surface, which is just an approximation because the surface internal energy term is of the order $N^{2/3}$, where N is the total number of particles in the phase.

First, we will observe the entropy representation, then the total entropy is

$$S = \sum_{\alpha=1}^{\nu} S^{(\alpha)} \tag{2.60}$$

Similarly, the volume is

$$V = \sum_{\alpha=1}^{\nu} V^{(\alpha)} \tag{2.61}$$

and the number of moles n_i of the ith component is

$$n_i = \sum_{\alpha=1}^{\nu} n_i^{(\alpha)} \tag{2.62}$$

From the principles of the thermodynamic equilibrium state, the stationarity of the equilibrium state requires that

$$(\delta E)_{S,V,n} = 0$$

and stability

$$(\delta^2 E)_{S,V,n} > 0$$

From Eq. (2.60), we can calculate δE as a variational displacement of the first order with respect to E:

$$\begin{aligned}\delta E &= \sum_{\alpha=1}^{\nu} \delta E^{(\alpha)} \\ &= \sum_{\alpha=1}^{\nu} \left[T^{(\alpha)} \delta S^{(\alpha)} - p^{(\alpha)} \delta V^{(\alpha)} + \sum_{r=1}^{r} \mu_i^{(\alpha)} \delta n_i^{(\alpha)} \right]\end{aligned} \tag{2.63}$$

From the condition of the equilibrium state

$$(\delta E)_{S,V,n} \geq 0$$

implies that from all the processes, we consider those processes that partition $S^{(\alpha)}$, $V^{(\alpha)}$ and $n_i^{(\alpha)}$ by keeping the total S, V, and n_i constants, which means that

$$\begin{aligned}&\sum_{\alpha=1}^{\nu} \delta S^{(\alpha)} = 0 \\ &\sum_{\alpha=1}^{\nu} \delta V^{(\alpha)} = 0 \\ &\sum_{\alpha=1}^{\nu} \delta n_i^{(\alpha)} = 0, \quad (i = 1, 2, \cdots, r)\end{aligned} \tag{2.64}$$

For a system with two phases, $\nu = 2$, relations in Eq. (2.64) reduce to

$$\begin{aligned}\delta S^{(1)} &= -\delta S^{(2)} \\ \delta V^{(1)} &= -\delta V^{(2)} \\ \delta n_i^{(1)} &= -\delta n_i^{(2)}, \quad (i = 1, 2, \cdots, r)\end{aligned} \tag{2.65}$$

Combining Eq. (2.63) with Eq. (2.65), we get

$$\begin{aligned}(\delta E)_{S,V,n} = &\left(T^{(1)} - T^{(2)}\right) \delta S^{(1)} - \left(p^{(1)} - p^{(2)}\right) \delta V^{(1)} \\ &+ \sum_{r=1}^{r} \left(\mu_i^{(1)} - \mu_i^{(2)}\right) \delta n_i^{(1)}\end{aligned} \tag{2.66}$$

For the condition of the equilibrium state $(\delta E)_{S,V,n} \geq 0$ to be satisfied for every infinitesimal variations $\delta S^{(1)}$, $\delta V^{(1)}$, and $\delta n_i^{(1)}$, which could be either positive or negative, we must have:

$$
\begin{aligned}
T^{(1)} &= T^{(2)} \\
p^{(1)} &= p^{(2)} \\
\mu_i^{(1)} &= \mu_i^{(2)}, \quad (i = 1, 2, \cdots, r)
\end{aligned}
\tag{2.67}
$$

which guarantees that $(\delta E)_{S,V,n} = 0$ for any infinitesimal displacement from the equilibrium state. The first is called the *thermal equilibrium*, the second one is the *mechanical equilibrium*, and the last one is the *mass equilibrium*.

Let us consider the case when all the infinitesimal displacements from the equilibrium state have only one sign, such as positive. Then, the conditions of the equilibrium will be expressed as inequalities instead of equality:

$$
\begin{aligned}
T^{(1)} &\geq T^{(2)} \\
p^{(1)} &\leq p^{(2)} \\
\mu_i^{(1)} &\geq \mu_i^{(2)}, \quad (i = 1, 2, \cdots, r)
\end{aligned}
\tag{2.68}
$$

We can also write the conditions of the equilibrium for the fluctuations about any number of the phases:

$$
\begin{aligned}
T^{(1)} &= T^{(2)} = T^{(3)} = \cdots \\
p^{(1)} &= p^{(2)} = p^{(3)} = \cdots \\
\mu_i^{(1)} &= \mu_i^{(2)} = \mu_i^{(3)} = \cdots, \quad (i = 1, 2, \cdots, r)
\end{aligned}
\tag{2.69}
$$

Eq. (2.69) gives an ensemble of equilibrium criteria, which are at the same time necessary and sufficient conditions for equilibrium.

Consider a system composed of two subsystems as shown in Fig. 2.9. Let us furthermore suppose that the system initially is in the state with $\mu^{(1)} > \mu^{(2)}$. Transfer of the mass will bring the system in an equilibrium final state, such that:

$$\mu_{\text{final}}^{(1)} = \mu_{\text{final}}^{(2)}$$

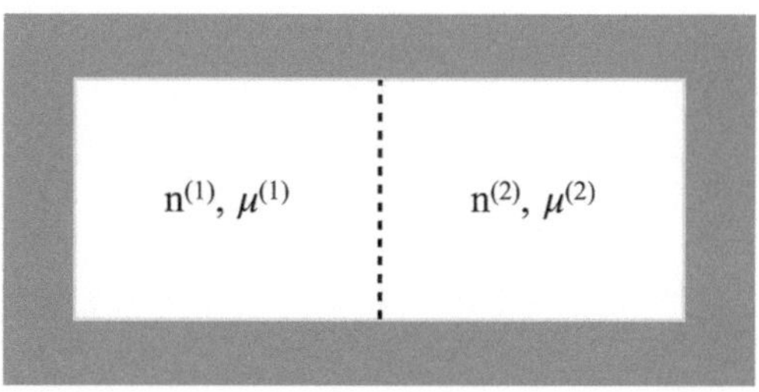

Fig. 2.9 Illustration of a composed system

If we assume that there is no work applied on the system and there is no transfer of the heat on the system from the surroundings, then for an equilibrium process:

$$\Delta S > 0$$

Suppose that the deviations from the equilibrium are small, then for a system with one component we can write

$$0 = \Delta E = T\Delta S - p\Delta V + \mu\Delta n$$

If we assume that the total volume does not change, then $\Delta V = 0$, and we obtain

$$\begin{aligned} \Delta S &= -\frac{\mu^{(1)}}{T}\Delta n^{(1)} - \frac{\mu^{(2)}}{T}\Delta n^{(2)} \\ &= -\left(\frac{\mu^{(1)}}{T} - \frac{\mu^{(2)}}{T}\right)\Delta n^{(1)} \end{aligned} \tag{2.70}$$

where $\Delta n^{(1)} = -\Delta n^{(2)}$ is the change in the number of moles in the subsystem (1) during this process, assuming that the total number of moles remains constant.

Therefore, since $\mu^{(1)} > \mu^{(2)}$, and $\Delta S > 0$ implies that $\Delta n^{(1)} < 0$. That is the mass transfers from the subsystem with the largest μ to the subsystem with smaller μ. Thus, we can say that the gradient in μ/T creates a transfer of mass. As a result, $-\nabla(\mu/T)$ is a general force. Similarly, $-\nabla(1/T)$ is a general force causing transfer of the heat. In general, the gradient of the intensive parameters causing the transfer of their respective conjugate parameters, are called the *thermodynamic fields*.

As we mentioned, the condition for a stable equilibrium state is given as the following for all variations far from the subspace of equilibrium states:

$$(\Delta E)_{S,V,n} > 0$$

Thus, for infinitesimal deviations

$$(\delta E)_{S,V,n} \geq 0$$

Above we found out that for unconstrained systems, for which the internal extensive parameters can change in both directions, either positive or negative,

$$(\delta E)_{S,V,n} = 0 \tag{2.71}$$

Therefore, the variation close to the equilibrium state can be written as

$$(\Delta E)_{S,V,n} = \left(\delta^2 E\right)_{S,V,n} + \left(\delta^3 E\right)_{S,V,n} + \cdots \tag{2.72}$$

where it is supposed that Eq. (2.71) is satisfied.

Since the terms of the second order will dominate for infinitesimal deviations, we can write

$$\left(\delta^2 E\right)_{S,V,n} \geq 0 \tag{2.73}$$

which is called the *stability condition of the thermodynamic equilibrium.*

If the inequality given by Eq. (2.73) is satisfied, the system is stable concerning small fluctuations from the equilibrium state. In other words, after the small fluctuations apply to the system, it will return to the previous equilibrium state. If the following equality is satisfied:

$$\left(\delta^2 E\right)_{S,V,n} = 0 \tag{2.74}$$

the thermodynamic stability is unpredictable, and the variations of higher order are needed to predict it. If

$$\left(\delta^2 E\right)_{S,V,n} < 0 \tag{2.75}$$

we can say that the system is not stable and the smallest fluctuation or disturbance will cause the system to change from the macroscopic point of view.

Consider the example of a composed system of two subsystems, namely (1) and (2), as shown in Fig. 2.10. Since the total entropy of the system is constant, then fluctuations for this system are:

$$\delta S = 0 = \delta S^{(1)} + \delta S^{(2)}$$

or

$$\delta S^{(1)} = -\delta S^{(2)} \tag{2.76}$$

In addition, we assume that the fluctuations of the volume and number of moles in each subsystem are zero, that is

$$\delta V^{(1)} = \delta V^{(2)} = \delta n^{(1)} = \delta n^{(2)} = 0$$

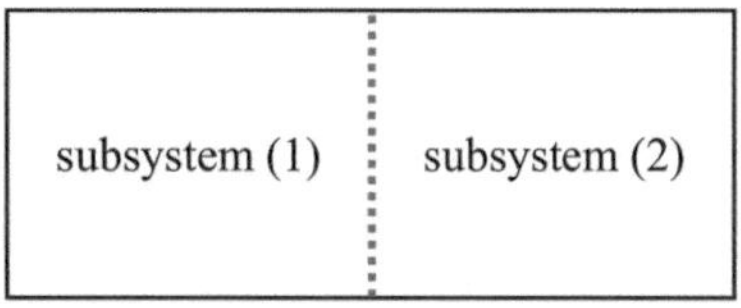

Fig. 2.10 Illustration of a composed system

Then,

$$
\begin{aligned}
\left(\delta^2 E\right)_{S,V,n} &= \left(\delta^2 E\right)^{(1)}_{V,n} + \left(\delta^2 E\right)^{(2)}_{V,n} \\
&= \frac{1}{2}\left(\frac{\partial^2 E}{\partial S^2}\right)^{(1)}_{V,n}\left(\delta S^{(1)}\right)^2 \\
&+ \frac{1}{2}\left(\frac{\partial^2 E}{\partial S^2}\right)^{(2)}_{V,n}\left(\delta S^{(2)}\right)^2
\end{aligned}
\tag{2.77}
$$

In Eq. (2.77), the subscripts (1) and (2) represent the derivatives calculated in equilibrium for the subsystems (1) and (2). Using Eq. (2.76) and the definition of the temperature

$$
\left(\frac{\partial^2 E}{\partial S^2}\right)_{V,n} = \left(\frac{\partial T}{\partial S}\right)_{V,n} = \frac{T}{C_V}
$$

we obtain

$$
\begin{aligned}
\left(\delta^2 E\right)_{S,V,n} &= \frac{1}{2}\left[\frac{T^{(1)}}{C_V^{(1)}} + \frac{T^{(2)}}{C_V^{(2)}}\right]\left(\delta S^{(1)}\right)^2 \\
&= \frac{1}{2}T\left[\frac{1}{C_V^{(1)}} + \frac{1}{C_V^{(2)}}\right]\left(\delta S^{(1)}\right)^2
\end{aligned}
\tag{2.78}
$$

where we have used the thermal equilibrium condition: $T^{(1)} = T^{(2)} = T$. Since, $\left(\delta^2 E\right)_{S,V,n} \geq 0$, then we get

$$
T\left[\frac{1}{C_V^{(1)}} + \frac{1}{C_V^{(2)}}\right] \geq 0 \tag{2.79}
$$

For any division of the entire system into two subsystems (1) and (2), Eq. (2.79) implies that

$$
T/C_V \geq 0
$$

or

$$
C_V \geq 0 \tag{2.80}
$$

Therefore, if the condition given by Eq. (2.80) is satisfied, then the system is in stable thermodynamic equilibrium. Otherwise, if the two subsystems (1) and (2) are in thermal contact with each-other, but not yet in equilibrium (i.e., $T^{(1)} \neq T^{(2)}$), then the gradient in T will cause a transfer of heat from the subsystem with higher T to the subsystem with lower T. If $C_V < 0$, then the direction of the heat transfer will cause an increase in the gradient of T, and the system will not be able to find the equilibrium state, which also gives a physical explanation of the criteria of stability. If this criterion is satisfied, then the spontaneous processes created as a result of a deviation from the equilibrium will be in that direction that will establish the equilibrium state.

Let us consider in the following the Helmholtz free energy principles of thermodynamic equilibrium:

$$\begin{aligned}(\Delta F)_{T,V,n} &> 0 \\ (\delta F)_{T,V,n} &= 0 \\ \left(\delta^2 F\right)_{T,V,n} &\geq 0\end{aligned} \tag{2.81}$$

These principles of variations apply to the experiments in which internal constraints imply changes on the internal extensive parameters, but the total internal extensive parameters remain constants. Since T is an intensive parameter, then it is not allowed to consider fluctuations for T.

Consider again the system shown in Fig. 2.10. The principles of variations can be applied for these variations:

$$\begin{aligned}\delta V &= 0 = \delta V^{(1)} + \delta V^{(2)} \\ \delta n^{(1)} &= \delta n^{(2)} = 0\end{aligned} \tag{2.82}$$

The second order variation of the Helmholtz free energy F is

$$\begin{aligned}\left(\delta^2 F\right)_{T,V,n} &= \frac{1}{2}\left[\left(\frac{\partial^2 F}{\partial V^2}\right)_{T,n}^{(1)} \left(\delta V^{(1)}\right)^2 + \left(\frac{\partial^2 F}{\partial V^2}\right)_{T,n}^{(2)} \left(\delta V^{(2)}\right)^2\right] \\ &= \frac{1}{2}\left(\delta V^{(1)}\right)^2 \left[\left(\frac{\partial^2 F}{\partial V^2}\right)_{T,n}^{(1)} + \left(\frac{\partial^2 F}{\partial V^2}\right)_{T,n}^{(2)}\right]\end{aligned} \tag{2.83}$$

where $\delta V^{(1)} = -\delta V^{(2)}$ is used derived by Eq. (2.82). From the differential form of F (see Eq. (2.23)) and Eq. (2.24), we obtain

$$\left(\frac{\partial p}{\partial V}\right)_{T,n} = -\left(\frac{\partial^2 F}{\partial V^2}\right)_{T,n}$$

After replacing this expression into Eq. (2.83) and considering that $\left(\delta^2 F\right)_{T,V,n} \geq 0$, we obtain

$$\left(\delta^2 F\right)_{T,V,n} = -\frac{1}{2}\left(\delta V^{(1)}\right)^2 \left[\left(\frac{\partial p}{\partial V}\right)_{T,n}^{(1)} + \left(\frac{\partial p}{\partial V}\right)_{T,n}^{(2)}\right] \geq 0 \tag{2.84}$$

Or, since the partition into two subsystems is arbitrary, every term inside the parentheses of Eq. (2.84) has the same sign, thus

$$-\left(\frac{\partial p}{\partial V}\right)_{T,n} \geq 0 \tag{2.85}$$

Note that the coefficient of the isothermal compressibility is given as

$$\kappa_T = -\frac{1}{V}\left(\frac{\partial V}{\partial p}\right)_{T,n}$$

Combining this equation with Eq. (2.85), we obtain that

$$\kappa_T \geq 0 \tag{2.86}$$

This is considered as another condition for a state of system to be in the stable thermodynamic equilibrium. Therefore, increasing thermally the pressure of a system in stable equilibrium state, implies that its volume decreases.

Similarly, if the second order variation vanishes (i.e., $\left(\delta^2 F\right)_{T,V,n} = 0$), then we have to look at higher order variation terms, that is

$$\left(\delta^3 F\right)_{T,V,n} \geq 0$$

and consider infinitesimal deviations from equilibrium state.

Let us look now at a general principle of the criteria of thermodynamic stability. For that, consider Ψ being one of the thermodynamic potential functions

$$\Psi\left(X_1,\ X_2,\ \cdots,\ X_r,\ I_{r+1},\ I_{r+2},\ \cdots,\ I_n\right)$$

where $X_1,\ X_2,\ \cdots,\ X_r$ are extensive parameters and $I_{r+1},\ I_{r+2},\ \cdots,\ I_n$ intensive parameters. The differential form of Ψ is given as

$$d\Psi = \sum_{i=1}^{r} I_i dX_i - \sum_{i=r+1}^{n} X_i dI_i$$

Then, the criteria of the stable thermodynamic equilibrium are given as

$$\left(\frac{\partial I_i}{\partial X_i}\right)_{X_{j\neq i}, I_{r+1}, \cdots, I_n} \geq 0 \tag{2.87}$$

That is, the criteria of the stable thermodynamic equilibrium are formulated in terms of the derivatives of the internal parameters with respect to their conjugated extensive parameters.

For example, If Ψ is the internal energy of the system, from the differential form of the internal energy of a system of r components:

$$dE = TdS - pdV + \sum_{i=1}^{r} \mu_i dn_i \tag{2.88}$$

we can see that the extensive parameters are: S, V, and n_i (for $i = 1, 2, \cdots, r$) and their respective conjugated intensive parameters are: T, $-p$, and μ_i (for $i = 1, 2, \cdots, r$). Therefore, the criteria of the stable thermodynamic equilibrium states can be written as the following:

$$\begin{aligned} \left(\frac{\partial T}{\partial S}\right)_{V,n} &\geq 0 \\ -\left(\frac{\partial p}{\partial V}\right)_{S,n} &\geq 0 \\ \left(\frac{\partial \mu_i}{\partial n_i}\right)_{S,V,n_{j\neq i}} &\geq 0, \quad i = 1, 2, \cdots, r \end{aligned} \tag{2.89}$$

We can also consider other thermodynamic potential functions, for example the free energy of Helmholtz F. From Eq. (2.23), we can see that

$$F = F(T, V, n_1, \cdots, n_r)$$

In this case, the criteria of the stable thermodynamic equilibrium states can be formulated as the following:

$$\begin{aligned} -\left(\frac{\partial p}{\partial V}\right)_{T,n} &\geq 0 \\ \left(\frac{\partial \mu_i}{\partial n_i}\right)_{T,V,n_{j\neq i}} &\geq 0, \quad i = 1, 2, \cdots, r \end{aligned} \tag{2.90}$$

Similarly, the criteria for the stable thermodynamic equilibrium for enthalpy $H(S, p, n)$, with a differential form given by Eq. (2.26), can be formulated as

$$\begin{aligned} \left(\frac{\partial T}{\partial S}\right)_{p,n} &\geq 0 \\ \left(\frac{\partial \mu_i}{\partial n_i}\right)_{S,p,n_{j\neq i}} &\geq 0, \quad i = 1, 2, \cdots, r \end{aligned} \tag{2.91}$$

Moreover, the same criteria for the stable thermodynamic equilibrium for Gibbs free energy $G(T, p, n)$, with a differential form given by Eq. (2.28), can be formulated as

$$\left(\frac{\partial \mu_i}{\partial n_i}\right)_{T,p,n_{j\neq i}} \geq 0, \quad i = 1,\ 2,\ \cdots,\ r \tag{2.92}$$

It is worth noting that from the second law, the stability of the equilibrium state does not imply anything regarding the sign of, for example,

$$\left(\frac{\partial p}{\partial T}\right)_{V,n},\ \left(\frac{\partial \mu_i}{\partial n_j}\right)_{T,V,n_{i\neq j}}$$

and so on, since they do not represent the derivatives of the intensive parameters related to respective conjugated extensive parameters.

Chapter 3
Principles of Statistical Mechanics

In this chapter, we will describe some fundamental topics of thermodynamics and statistical mechanics. Furthermore, we will discuss the energy or particle number fluctuations in different statistical ensembles and their differences.

For further reading on the statistical mechanics, one should consider the books by Gibbs (1902), Hansen and McDonald (1986), and McQuarrie (1976, 2000).

3.1 Systems

In statistical mechanics systems play the same role as particles in kinetic theory. The system has a very general concept in statistical mechanics, and it may include any physical object.

For example, we can mention the galaxy, a planet, crystal and its fundamental mode of vibration, an atom in a crystal, an electron of the atom, and a quantum state in which that electron could reside.

Statistical mechanics pays special attention to systems that couple only weakly to the rest of the universe. With other words, in statistical mechanics, the focus is the systems whose relevant internal evolution timescales, τ_{int}, are short compared with the external timescales, τ_{ext}, on which they exchange energy, entropy, particles, and so on, with their surrounding environments. These systems are also called *semi-closed*. In contrast, a system for which in the idealized limit external interactions are completely ignored, is called *closed* system.

The statistical mechanics formalism for dealing with closed systems relies on the assumption $\tau_{int}/\tau_{ext} \ll 1$. Therefore, it depends on the two length scale expansions, τ_{int} and τ_{ext}.

If a semi-closed classical system does not interact with the external universe, then it is considered a closed system, and Hamiltonian dynamics (Poole 2001) describe its time evolution. In this textbook we are discussing the classical systems, therefore,

H. Kamberaj, *Molecular Dynamics Simulations in Statistical Physics: Theory and Applications*, Scientific Computation,
https://doi.org/10.1007/978-3-030-35702-3_3

within this context, we can determine the *phase space* as the $2f$-dimensional space such that the coordinates determine every point in this space:

$$(\mathbf{q}_1, \cdots, \mathbf{q}_N, \mathbf{p}_1, \cdots, \mathbf{p}_N)$$

where $\mathbf{q} = \{q_i\}_{i=1}^{f}$ are the generalized coordinates and $\mathbf{p} = \{p_i\}_{i=1}^{f}$ are generalized conjugated momenta. Here, f denotes the number of degrees of freedom related to the number of particles N of a system, $f = 3N$, assuming that each particle is in a three-dimensional space. Each point in this phase space corresponds to one of the *microscopic states* of the system.

The time evolution of $\mathbf{p}$ and $\mathbf{q}$ is governed by Hamilton's equations of motion (as discussed in Chap. 1):

$$\begin{aligned} \frac{dq_i}{dt} &= \frac{\partial H}{\partial p_i} \\ \frac{dp_i}{dt} &= -\frac{\partial H}{\partial q_i}, \qquad i = 1, \cdots, f \end{aligned} \tag{3.1}$$

where $H(\mathbf{q}, \mathbf{p})$ is the Hamiltonian of system. Here, we have considered that the Hamiltonian does not depend explicitly on time t, because it is assumed dealing with closed systems. In general, some physical systems, such as those with strong internal dissipation, are not described by Hamiltonian dynamics (Poole 2001).

The differential equations given by Eq. (3.1) can be solved by determining first the initial conditions. These conditions define a point in the phase space of the system. The motion of this phase point will determine the microscopic state of the system at any time t. The path of phase point in microscopic state space is often called *trajectory*, as mentioned in Chap. 1.

Note that for a closed system the energy is constant and equal to Hamiltonian:

$$H(\mathbf{q}, \mathbf{p}) = E$$

Thus, the orbit of this motion corresponds to a constant energy surface. Other parameters that we can fix include the volume V, temperature T, pressure p, the chemical potential μ, and the number of particles N.

3.2 Ensembles

While the kinetic theory aims to study a system of a vast number of particles statistically, the statistical mechanics aims to study statistically an *ensemble* of a vast number of systems. Note that the concept of an ensemble in statistical mechanics is a theoretical concept. That is, it forms a statistical argument for describing a *thought* experiment. Hence, there can be different ways of thinking about an ensemble.

Often, it is required that an ensemble is formed by systems which are *closed* and *identical*, in the sense that systems have identically the same number of degrees of freedom and described by Hamiltonian with identically the same functional forms $H(\mathbf{q}, \mathbf{p})$, and have the same volume V and total internal energy E. However, the generalized coordinates $\mathbf{q}$ and conjugated momenta $\mathbf{p}$ do not need to be the same at any time t, and hence, the systems are not all at the same state at time t. According to the Boltzmann, such a theoretically conceptual ensemble of identical closed systems evolves until reaches the so-called *statistical equilibrium*, and it is called *microcanonical* ensemble.

In practice, we often deal with ensembles that exchange energy in the form of heat with the surrounding environment such that the internal energy of each ensemble's system can fluctuate. Such a surrounding environment is also called *heat bath*. The heat bath is considered to have a much larger number of degrees of freedom, and so a far higher heat capacity than each system of the ensemble. If the statistical equilibrium is obtained, then the ensemble is called *canonical* ensemble.

If the systems of the ensemble can also exchange volume with surrounding as well as energy, then if the statistical equilibrium is reached, this ensemble is called *Gibbs* or *isothermal-isobaric* ensemble. Another often described ensemble is *grand canonical* ensemble in which each ensemble's system exchange energy and particles, but not volume, with its surrounding environment. In this chapter, we will also describe another ensemble, where the system exchanges energy, particles, and volume with the surrounding environment, keeping in this way constant the chemical potential μ, pressure p and temperature T, which is called *grand isothermal-isobaric* ensemble.

Like in kinetic theory, where we describe the statistical properties of a system by the distribution function $\mathcal{N}(t, \mathbf{q}, \mathbf{p})$, which gives the number of particles per unit volume of $6N$-dimensional phase space, in the statistical mechanics, the properties of the ensemble are statistically described by a *distribution function* which equals the number of systems per unit volume in $2f$-dimensional phase space. Here, we will introduce all these distribution functions mathematically applied to each ensemble.

Statistical mechanics assumes that if we follow one trajectory for a very long time, the system will go through all possible microscopic states and it will eventually come very close to the initial point imposed on the system. In such case, if we perform, as the system moves, a set of T measurements on the system, then an observable property $\mathcal{A}$ will have a value given as the following:

$$\mathcal{A}_T = \frac{1}{T}\sum_{t=1}^{T} \mathcal{A}(t)$$

Here, $\mathcal{A}(t)$ is the value of $\mathcal{A}$ at the t measurement in a short time interval during which system remains in the same microscopic state. If we denote n the total number of states visited during a measurement time interval, then the sum can also be written as

$$\mathcal{A}_T = \sum_{i=1}^{n} \frac{T_i}{T} \mathcal{A}_i$$

T_i is the number of observations of the system in the state i during T measurements, and $\mathcal{A}_i$ is the value of quantity $\mathcal{A}$ in this state. By definition, $\frac{T_i}{T}$ gives the probability, p_i, of visiting the state i during the measurement time interval. Therefore, we can write

$$\mathcal{A}_T = \sum_{i=1}^{n} p_i \mathcal{A}_i \equiv \langle \mathcal{A} \rangle \tag{3.2}$$

Here, $\langle \mathcal{A} \rangle$ represents the so-called *ensemble average*, and it characterizes an operation for calculation of the average value of any physical property of a system. That also gives a new meaning for the concept of an ensemble as a set of all possible microscopic states which characterize the macroscopic system under conditions for which it is determined. For example, the microcanonical ensemble is the set of all microscopic states with constant energy E, number of particles N and volume V; the canonical ensemble is the set of all microscopic states with constant temperature T, number of particles N and volume V. Thus, the microcanonical ensemble is suitable for describing an isolated system, and the canonical ensemble is suitable for describing a closed system in thermal contact with a heat bath.

Conceptually, the time average $\mathcal{A}_T$ uses the classical mechanics description of the system as a collection of particles moving under physics laws, for example using Hamilton's equations of motion to obtain a trajectory. If the observation time of the measurements is very long, then eventually this trajectory represents all possible microscopic states of a system in phase space. Therefore, the time average is equal to the ensemble average. Thus, the central assumption of statistical mechanics is that the value of an observed quantity corresponds to the ensemble average.

Equivalence between the ensemble and time average is not such apparent as it looks, because its main assumption is that during the observation time interval system has visited all possible microscopic states, which is not that simple to verify. Dynamical systems obeying to this equivalence are called *ergodic*. Practically, we assume that the ergodic condition is satisfied for all systems in nature. For small size molecular systems, the ergodic condition is considered to be satisfied based on their dynamical nature of interactions between molecules from molecular kinetic theories.

However, this equivalence can be satisfied not only when the observation time is too long, but also when the value is an average over a vast number of independent observations. Both these approaches are equivalent if with "long time" we understand the time that is longer than the *relaxation time*, τ, of the system. For a molecular system, the relaxation time corresponds to the time that the system has lost all correlations with its initial conditions. In such case, if a measurement is performed for some time $\mathcal{T}$ such that $\mathcal{T} = T\tau$, then it corresponds to T independent measurements.

Often for macroscopic systems, the measurements are performed for relatively short timescales. The ensemble average is also applicable to these cases. However, for these cases, this could be understood as the partition of the macroscopic system into a set of smaller subsystems which are macroscopic as well, in the sense that molecular behavior in each subsystem is independent on that of neighboring subsystems. In other words, each of these subsystems is large enough such that its distance from all other subsystems is larger than the correlation length, and hence, the subsystems can be considered as macroscopic. Then, the set of subsystems can characterize an ensemble, and a measurement of the entire macroscopic system is equivalent to the set of independent measurements in each subsystem, which corresponds to an ensemble average.

3.3 Microcanonical Partition Function

As mentioned above, the main assumption governing the statistical mechanics is that during a measurement all possible microscopic states are observed, and the value of some observing quantity is an average over all these microscopic states. Therefore, it is necessary to characterize the distribution probability of these microscopic states.

For a *microcanonical ensemble* in which each isolated system of the ensemble has a constant energy E, volume V and the number of particles N, the assumption made can be postulated as:

> All microscopic states are equally possible in a thermodynamic equilibrium macroscopic state.

In other words, the thermodynamic equilibrium state corresponds to the most disordered situation, i.e., the distribution of microscopic states with the same E, V, and N is completely uniform.

We denote with $\Omega(E, V, N)$ the number of microscopic states characterized by N particles, volume V and energy in the interval $(E,\ E - \delta E)$. The value of δE characterizes the uncertainty of determining E of the macroscopic system for some of E values. If $\delta E = 0$, then $\Omega(E, V, N)$ would be a discontinuous function of E, and for $\delta E \neq 0$, $\Omega(E, V, N)$ will be the degeneracy of energy level E. For a macroscopic system, the energy levels often are distributed very close to each other, such that they can be considered continuously distributed. In the limit of continuous distribution for $\Omega(E, V, N)$, we can denote with

$$\Omega(E, V, N)\, dE$$

the number of energy states with E in the interval between E and $E + dE$. In this case, $\Omega(E, V, N)$ determines the so-called *density of states*.

Based on the above statistical assumption, the probability of observing a microscopic state n of the ensemble for a system in equilibrium is given by

$$P_n = \begin{cases} \dfrac{1}{\Omega(E, V, N)}, & \text{for } E_n = E \\ 0, & \text{for } E_n \neq E \end{cases} \tag{3.3}$$

which indicates that P_n is equal for all microscopic states.

By definition, the *entropy* is defined as the following quantity:

$$S = k_B \ln \Omega(E, V, N) \tag{3.4}$$

where k_B is an arbitrary constant known as the Boltzmann's constant with value:

$$k_B = 1.38 \times 10^{-16} \text{ erg/deg}$$

The quantity S, determined by Eq. (3.4), satisfies the extensive property. For instance, imagine we divide the system into two independent subsystems, let's say I and II, with, respectively, number of states Ω_I and Ω_{II}. Then, the joint density of states will be given by

$$\Omega = \Omega_I \Omega_{II}$$

and the entropy of the entire systems is

$$S_{I+II} = k_B \ln(\Omega_I \Omega_{II}) = k_B \ln \Omega_I + k_B \ln \Omega_{II} = S_I + S_{II}$$

which indicates that S is additive.

Now, let us show that this definition of the entropy is also in agreement with the variation principle of the second law of thermodynamics. For that, we are going to assume that the system with constant E, N and V partitions into two subsystems with, respectively, N_I, N_{II}; V_I, V_{II}; and E_I, E_{II}, such that

$$\begin{aligned} N &= N_I + N_{II} \\ V &= V_I + V_{II} \\ E &= E_I + E_{II} \end{aligned} \tag{3.5}$$

Each particular partition of the system in this way is a subset of all possible states. Therefore, the number of states in this partition, $\Omega(E, V, N; \text{internal condition})$ is smaller than the total number of states, $\Omega(E, V, N)$:

$$\Omega(E, V, N) > \Omega(E, V, N; \text{internal condition})$$

Since the logarithm function is a monotonically increasing function, we can write

$$S(E, V, N) > S(E, V, N; \text{internal condition})$$

This inequality is the second law of thermodynamics seen in Chap. 2. From the statistical mechanics' point of view, this inequality indicates that the maximum of entropy corresponds to the maximum disorder, and hence, more microscopic disorder, larger the entropy.

The temperature can also be determined as

$$\frac{1}{T} = \left(\frac{\partial S}{\partial E}\right)_{N,V} \tag{3.6}$$

From here, we can get

$$\beta = \frac{1}{k_B T} = \left(\frac{\partial \ln \Omega}{\partial E}\right)_{N,V} \tag{3.7}$$

It can be seen that since Ω is a monotonically increasing function of E, then $\ln \Omega$ also is a monotonically increasing function of E. This satisfies the thermodynamic condition that the temperature is a positive quantity, $T > 0$.

3.4 Canonical Partition Function

The partition function describes the statistical properties of a system in thermodynamic equilibrium, which contains all of the essential information about the system under consideration. It is a function of the macroscopic properties of the system, such as the temperature, T, number of particles, N, and volume, V. The canonical partition function of a classical system has the following general form:

$$Q = \sum_n e^{-E_n/k_B T}, \tag{3.8}$$

where n ($n = 1, 2, 3, \cdots$) labels exact (microscopic) states occupied by the system, and E_n is the total energy of the system in the state n. The term $e^{-E_n/k_B T}$ is known as the Boltzmann's factor.

If the system has multiple quantum states n sharing the same E_n, then the energy levels of the system are degenerate. The partition function is a sum of the contributions from energy levels as the following:

$$Q = \sum_\nu g_\nu e^{-E_\nu/k_B T}, \tag{3.9}$$

where g_ν is the degeneracy factor characterizing the number of quantum states ν with the same energy level: $E_\nu = E_n$. This is the case of a quantum statistical mechanics system, such as a physical system inside a finite-sized box that is characterized by a discrete set of energy eigenstates defining the states n. In classical statistical mechanics, however, it is not exactly correct to express the partition function in terms of a summation of the discrete terms. Instead, in classical mechanics, the position (**r**) and momentum (**p**) variables of a particle can vary continuously. Hence, the microscopic states are actually uncountable. In this case, the integral replaced the sum in the definition of the partition function. Thus, the partition function of a gas of N identical classical particles is

$$Q = \frac{1}{N!h^{3N}} \times \int \exp\left[-\beta H(\mathbf{p}_1, \cdots, \mathbf{p}_N, \mathbf{r}_1, \cdots, \mathbf{r}_N)\right] d^3\mathbf{p}_1 \cdots d^3\mathbf{p}_N d^3\mathbf{r}_1 \cdots d^3\mathbf{r}_N, \tag{3.10}$$

where $\beta = 1/k_B T$, $\mathbf{r}_i$ is i-th particle position vector, $\mathbf{p}_i$ is its conjugated momentum vector, and H is the classical Hamiltonian, which depends on the positions and momenta. The constant factor (h^{3N}) in the denominator makes Q to be a quantity without dimensions. Here, h denotes the Planck's constant:

$$h = 6.622477 \times 10^{-34}\ \mathrm{kg} \cdot \mathrm{m}^2/\mathrm{s}$$

The other factor, $N!$, takes into account that the particles are actually identical particles. This ensures that no "over-count" of the number of the microscopic states occurs. While this may seem like an unusual requirement, it is actually necessary to preserve the existence of a thermodynamic limit for such systems, known also as the *Gibbs paradox* (Gibbs 1902).

There are a few examples where it is possible to calculate the partition function analytically for some vast systems of interacting particles. In general, it can not be evaluated precisely. Even enumerating the terms in partition function on a computer can be an impossible task. For instance, consider a system of only 10,000 interacting particles, which is a very small fraction of the Avogadro's number (6.022×10^{23} particles/mol), with only two possible states per particle, the partition function would contain $2^{10,000}$ terms, which can be quite impossible to store even in the most powerful computers in nowadays.

The partition function depends on the temperature T and the microscopic state energies E_n ($n = 1, 2, \cdots$). Other thermodynamic variables determine the microscopic state energies, such as the number of particles N, the volume V, and the microscopic quantities, such as the mass of constituent particles. This dependence on microscopic variables is the main goal of statistical mechanics. Using a particular model of the microscopic constituents of a system, such as molecules and atoms, we can calculate the microscopic state energies, and thus the partition function. Knowing the partition function allows us to calculate all the other thermodynamic properties of the system. In particular, the partition function can relate to the thermodynamic properties because it is related to a fundamental statistical property, such as the probability P_n that the system occupies microscopic state n given as

$$P_n = \frac{1}{Q} \exp\left(-\beta E_n\right), \tag{3.11}$$

Here, the partition function, Q, represents a normalization constant.

3.5 Entropy, Free Energy and Internal Energy

The partition function can be related to different thermodynamic quantities, such as the entropy, free energy, and internal energy.

The entropy of the canonical ensemble is defined in statistical mechanics by

$$S = -k_B \sum_n P_n \ln P_n \tag{3.12}$$

where P_n is given by Eq. (3.11). Substituting Eq. (3.11) into Eq. (3.12) we obtain

$$\begin{aligned} TS &= -k_B T \sum_n \frac{1}{Q} e^{-\beta E_n} \left(-\ln Q - \beta E_n\right) \\ &= k_B T \sum_n \frac{\ln Q}{Q} e^{-\beta E_n} + \frac{1}{Q} \sum_n E_n e^{-\beta E_n} \\ &= \frac{1}{\beta} \ln Q - \frac{\partial \ln Q}{\partial \beta} \end{aligned} \tag{3.13}$$

The Helmholtz free energy for a canonical ensemble is determined by

$$F = -k_B T \ln Q \tag{3.14}$$

This relation is introduced by Callen (1985) and it provides a relationship between the statistical mechanics and thermodynamics. Indeed, the expression in Eq. (3.14) can also be obtained from the free energy using

$$S = -\left(\frac{\partial F}{\partial T}\right)_{V,N} \tag{3.15}$$

The free energy of the system takes us to the internal energy using the following relation:

$$U = -T^2 \frac{\partial}{\partial T} \left(\frac{F}{T}\right) \tag{3.16}$$

This expression also implies that knowing the internal energy of a system, the free energy can be obtained by the integration, if we assume that we know already the free energy at a reference temperature. This is particularly important for computer simulations where the free energy is not directly calculated but rather the internal energy. Then, the free energy can be calculated by integration:

$$\Delta\left(\frac{F}{T}\right) = \frac{F(T)}{T} - \frac{F(T_0)}{T_0} = \int_{T_0}^{T} U(T)\, d(\frac{1}{T}).$$

Using Eq. (3.16), we can easily obtain an expression for the internal energy in terms of the partition function:

$$U = -\frac{\partial \ln Q}{\partial \beta} \tag{3.17}$$

where the following relation

$$dT = -k_B T^2 d\beta$$

is used, knowing that

$$\beta = \frac{1}{k_B T}$$

If the microscopic state energies depend on a parameter λ according to

$$E_n = E_n^{(0)} + \lambda A_n, \tag{3.18}$$

where A_n is the value of A at microscopic state n and $E_n^{(0)}$ is the value of reference energy at microscopic state n.

Then, the expected value of A is calculated as the following:

$$\begin{aligned}
\langle A \rangle &= \sum_n A_n P_n \\
&= \sum_n A_n \frac{1}{Q_\lambda} \exp(-\beta E_n) \\
&= -\frac{1}{Q_\lambda \beta} \sum_n \frac{\partial}{\partial \lambda} \exp(-\beta E_n) \\
&= -\frac{1}{\beta}\left[\frac{1}{Q_\lambda}\frac{\partial}{\partial \lambda}\sum_n \exp(-\beta E_n)\right]
\end{aligned} \tag{3.19}$$

$$= -\frac{1}{\beta}\left[\frac{1}{Q_\lambda}\frac{\partial}{\partial\lambda}Q_\lambda\right]$$
$$= -\frac{1}{\beta}\frac{\partial \ln Q_\lambda}{\partial\lambda}.$$

This procedure provides a method for calculating the expected values of many microscopic quantities. We add the quantity artificially to the microscopic state energies (or to the Hamiltonian for the isolated systems), calculate the new partition function and expected value. For this, the expression given by Eq. (3.19) can be rewritten as:

$$\begin{aligned}\langle A\rangle &= \sum_n A_n P_n \\ &= \sum_n A_n \frac{1}{Q_\lambda}\exp(-\beta E_n) \\ &= \frac{1}{Q_\lambda}\sum_n\left(\frac{\partial E_n}{\partial\lambda}\right)\exp(-\beta E_n) \\ &= \langle\frac{\partial E_n}{\partial\lambda}\rangle_\lambda\end{aligned} \tag{3.20}$$

where $\langle\frac{\partial E_n}{\partial\lambda}\rangle_\lambda$ represents the expected value of derivative of E_n (or Hamiltonian) with respect to λ for the new partition function.

3.6 Thermodynamic Potentials

As we have seen, the internal energy depends on the extensive variables, such as entropy S, volume V, number of particles N, and so on. In some cases, it is appropriate to replace some of these variables with their conjugated intensive variables. For this purpose, additional thermodynamic potentials can be defined using the Legendre transformations of the internal energy as described in Chap. 2:

$$F = U - TS\,, \tag{3.21}$$

$$H = U + pV\,, \tag{3.22}$$

$$G = U - TS + pV\,, \tag{3.23}$$

where F is the Helmholtz free energy, H is the enthalpy, and G is the Gibbs free energy. The Helmholtz free energy F is particularly important because it has a minimum value in equilibrium for N, T and V held fixed. On the other hand, G has a minimum value in the case when N, T and p are held fixed. In addition, the difference in free energies between two states does not depend on the path connecting these states. This means, two different paths connecting points 1 and 2, have the same difference on the free energy:

$$F_2 - F_1 = \int_{\text{path I}} dF = \int_{\text{path II}} dF. \tag{3.24}$$

3.7 Generalized Ensembles

Consider a system with $\mathbf{x} = \{x_1, x_2, \cdots\}$ a vector of mechanical extensive variables and $\mathbf{f}$ the corresponding conjugated intensive variables vector. Now imagine a system in thermodynamic equilibrium in which both E and $\mathbf{x}$ can fluctuate. This system can be considered as a part of an isolated composite system in which the other part can be viewed as a big reservoir for both E and $\mathbf{x}$ (see also Fig. 3.1). Both, the energy E_n and $\mathbf{x}_n$ fluctuate because the system is in contact with the bath, but $E = E_b + E_n$ and $\mathbf{x} = \mathbf{x}_b + \mathbf{x}_n$ are constant. If the system is in one definite state n, then the density of states of the microscopic states accessible from the system and the bath is:

$$\Omega(E_b, \mathbf{x}_b) = \Omega(E - E_n)\Omega(\mathbf{x} - \mathbf{x}_n) ,$$

where we have assumed that fluctuations of the energy and extensive parameter $\mathbf{x}$ are independent.

The probability observing the system in a microscopic state n is given by

$$\begin{aligned} P_n &\propto \Omega(E - E_n, \mathbf{x} - \mathbf{x}_n) \\ &= \Omega(E - E_n)\Omega(\mathbf{x} - \mathbf{x}_n) \\ &= \exp\left(\ln\left(\Omega(E - E_n)\Omega(\mathbf{x} - \mathbf{x}_n)\right)\right) \\ &= \exp\left(\ln\left(\Omega(E - E_n)\right)\right)\exp\left(\ln\left(\Omega(\mathbf{x} - \mathbf{x}_n)\right)\right) \end{aligned}$$

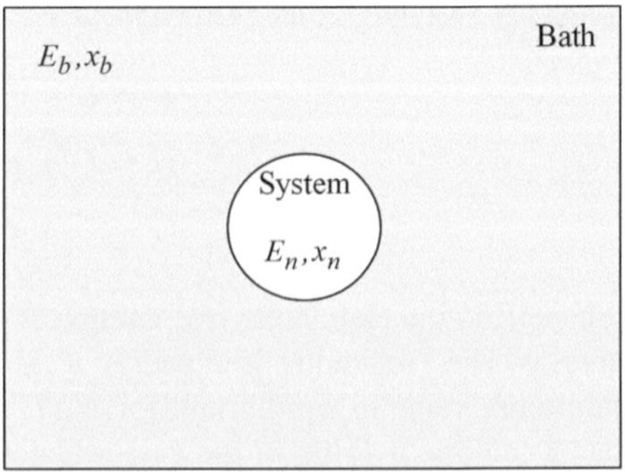

Fig. 3.1 Illustration of a system immersed in a bath

Now we can express $\ln(\Omega(E - E_n))$ and $\ln(\Omega(\mathbf{x} - \mathbf{x}_n))$ according to Taylor series for $E_n \ll E$ and $\mathbf{x}_n \ll \mathbf{x}$:

$$\ln(\Omega(E - E_n)) = \ln \Omega(E) - E_n \frac{d \ln \Omega}{dE} + \cdots \approx \ln \Omega(E) - \beta E_n \ , \tag{3.25}$$

$$\ln(\Omega(x - x_n)) = \ln \Omega(\mathbf{x}) - \mathbf{x}_n \cdot \frac{d \ln \Omega}{d\mathbf{x}} + \cdots \approx \ln \Omega(\mathbf{x}) - \mathbf{f} \cdot \mathbf{x}_n \ , \tag{3.26}$$

where

$$\beta = \left(\frac{\partial \ln \Omega}{\partial E} \right)_{\mathbf{x}} , \tag{3.27}$$

$$f_i = \left(\frac{\partial \ln \Omega}{\partial x_i} \right)_{E, x_{j \neq i}} . \tag{3.28}$$

Finally, we can write:

$$P_n \propto \exp(-\beta E_n - \mathbf{f} \cdot \mathbf{x}_n) \ , \tag{3.29}$$

where the proportionality constant is independent on the particular state of the system and it is determined from the normalization condition:

$$\sum_n P_n = 1 \ .$$

Hence,

$$P_n = \Xi^{-1} \exp(-\beta E_n - \mathbf{f} \cdot \mathbf{x}_n) \ , \tag{3.30}$$

where

$$\Xi = \sum_n \exp(-\beta E_n - \mathbf{f} \cdot \mathbf{x}_n) \ , \tag{3.31}$$

are the probability distribution and partition function of the generalized ensemble, respectively. Thermodynamic values of E and x_i are given through ensemble averages:

$$\langle E \rangle = \sum_n P_n E_n = - \left(\frac{\partial \ln \Xi}{\partial \beta} \right)_{\mathbf{f}, \mathbf{x}} , \tag{3.32}$$

$$\langle x_i \rangle = \sum_n P_n x_{i,n} = - \left(\frac{\partial \ln \Xi}{\partial f_i} \right)_{T, E, x_{j \neq i}} \tag{3.33}$$

3.8 Isothermal-Isobaric Ensemble

An example of the generalized canonical ensemble is the so-called *isothermal-isobaric ensemble* or *Gibbs ensemble*. In the isothermal-isobaric ensemble, the temperature T, pressure p, and the number of particles N are held fixed, but both, the energy E and volume V are allowed fluctuating. The conjugated fields that control these fluctuations are T (or β) and βp, respectively.

We denote with n the microscopic state of the system at volume V_n and energy E_n, then from Eqs. (3.30) and (3.31) we get

$$P_n = \Xi^{-1} \exp\left(-\beta E_n - \beta p V_n\right) , \tag{3.34}$$

where

$$\Xi = \sum_n \exp\left(-\beta E_n - \beta p V_n\right) . \tag{3.35}$$

Using the Gibbs formula for the entropy

$$\begin{aligned} S &= -k_B \sum_n P_n \ln P_n \\ &= -k_B \sum_n P_n \left[-\ln \Xi - \beta E_n - \beta p V_n\right] \\ &= -k_B \left[-\ln \Xi - \beta\langle E\rangle - \beta p\langle V\rangle\right] . \end{aligned}$$

This expression can also be written as:

$$- k_B T \ln \Xi = \langle E\rangle - TS + p\langle V\rangle , \tag{3.36}$$

which can be compared with the formula for Gibbs free energy: $G = E - TS + pV$. From this comparison we obtain

$$G = -k_B T \ln \Xi = G(N, p, T) \tag{3.37}$$

Thus, it can be seen that Gibbs free energy G is a natural function of macroscopic variables of the system, namely N, p, and T.

3.9 Grand Canonical Ensemble

Another very useful application of generalized ensemble is *grand canonical ensemble*. This ensemble includes the set of all microscopic states of an open system at constant volume V, chemical potential μ, and temperature T. Both, the energy and number of particles are allowed fluctuating, and the conjugated fields that control the magnitude of these fluctuations are β and $-\beta\mu$, respectively. Thus, if we denote with n the microscopic state with N_n particles and energy E_n, from Eqs. (3.30) and (3.31), we get

$$P_n = \Xi^{-1} \exp\left(-\beta E_n + \beta\mu N_n\right) , \tag{3.38}$$

where Ξ denotes the grand canonical partition function given by

$$\Xi = \sum_n \exp\left(-\beta E_n + \beta\mu N_n\right) \tag{3.39}$$

which ensures that P_n is normalized to one.

Using the Gibbs formula for the entropy, we get

$$\begin{aligned} S &= -k_B \sum_n P_n \ln P_n \\ &= -k_B \sum_n P_n \left[-\ln \Xi - \beta E_n + \beta\mu N_n\right] \\ &= k_B \ln \Xi + \frac{1}{T}\langle E\rangle - \frac{1}{T}\mu\langle N\rangle . \end{aligned} \tag{3.40}$$

Or

$$\langle E\rangle = ST - k_B T \ln \Xi + \mu\langle N\rangle$$

Comparing this expression with the one for internal energy: $E = TS - pV + \mu N$, we obtain

$$pV = k_B T \ln \Xi , \tag{3.41}$$

which gives the free energy for the open system. Since the energy, E, depends on the volume, we can say that the free energy pV of an open system depends on T, V and μ, which characterize the nature macroscopic variables for a grand canonical ensemble.

3.10 Grand Isothermal-Isobaric Ensemble

Grand isothermal-isobaric ensemble is defined as an ensemble with a fixed external pressure p, temperature T, and chemical potential μ. Thus, in this ensemble, the energy E, volume V and number of particles N are allowed fluctuating. The conjugated fields that control these fluctuations are β, βp and $-\beta\mu$, respectively.

Therefore, the probability observing the system in a microscopic state n characterized by the volume V_n, energy E_n and number of particles N_n is obtained from Eqs. (3.30) and (3.31) as:

$$P_n = \Xi^{-1} \exp\left(-\beta E_n - \beta p V_n + \beta \mu N_n\right) , \tag{3.42}$$

where the normalization factor Ξ represents the grand isothermal-isobaric partition function given by

$$\Xi = \sum_n \exp\left(-\beta E_n - \beta p V_n + \beta \mu N_n\right) . \tag{3.43}$$

Using the Gibbs formula for the entropy, we have

$$\begin{aligned} S &= -k_B \sum_n P_n \ln P_n \\ &= -k_B \sum_n P_n \left[-\ln \Xi - \beta E_n - \beta p V_n + \beta \mu N_n\right] \\ &= k_B \ln \Xi + \frac{1}{T}\langle E\rangle + \frac{1}{T} p\langle V\rangle - \frac{1}{T}\mu\langle N\rangle , \end{aligned} \tag{3.44}$$

or

$$TS = k_B T \ln \Xi + \langle E\rangle + p\langle V\rangle - \mu\langle N\rangle .$$

This expression indicates that

$$k_B T \ln \Xi = 0 ,$$

or

$$\Xi = 1 ,$$

since $T > 0$.

3.11 Fluctuations

3.11.1 Canonical Ensemble

We showed that each microscopic state n in canonical ensemble has a certain probability P_n given by expression Eq. (3.11). Since the number of different microscopic states is very large, we are not only interested on the probability of microscopic states (P_n), but also on the probability of the macroscopic state variables. For instance, the internal energy U. We first calculate the average internal energy $\langle E\rangle$ (or U):

$$\begin{aligned} U \equiv \langle E\rangle &= \sum_n E_n P_n \\ &= \frac{1}{Q}\sum_n E_n e^{-\beta E_n} \\ &= -\left(\frac{\partial \ln Q}{\partial \beta}\right)_{N,V}. \end{aligned} \tag{3.45}$$

Similarly, the second moment of U is

$$\begin{aligned} \langle E^2\rangle &= \sum_n E_n^2 P_n \\ &= \frac{1}{Q}\sum_n E_n^2 e^{-\beta E_n} \\ &= \frac{1}{Q}\sum_n \frac{\partial^2}{\partial \beta^2} e^{-\beta E_n} \\ &= \frac{1}{Q}\left(\frac{\partial^2 Q}{\partial \beta^2}\right)_{N,V}. \end{aligned} \tag{3.46}$$

Then, the variance of energy fluctuations can be obtained as:

$$\begin{aligned} \langle(\delta E)^2\rangle &= \langle(E-\langle E\rangle)^2\rangle \\ &= \langle E^2\rangle - \langle E\rangle^2 \\ &= \frac{1}{Q}\left(\frac{\partial^2 Q}{\partial \beta^2}\right)_{N,V} - \left[\left(\frac{\partial \ln Q}{\partial \beta}\right)_{N,V}\right]^2 \\ &= -\left(\frac{\partial \langle E\rangle}{\partial \beta}\right)_{N,V} \end{aligned} \tag{3.47}$$

Since $(\partial\langle E\rangle/\partial T)_{N,V} = C_V$ is the specific heat, then we can further write

$$\langle(\delta E)^2\rangle = k_B T^2 C_V \tag{3.48}$$

Practically, this is a very important result, because it relates the amount of spontaneous fluctuations, $\langle(\delta E)^2\rangle$, with the degree of energy change with temperature, C_V. Since the specific heat and the energy are extensive quantities, then they are of order N (the number of particles in the system). Thus, the ratio of the standard deviation of the energy fluctuations to its average value is of order $N^{-1/2}$, i.e:

$$\frac{\sqrt{\langle (E - \langle E \rangle)^2 \rangle}}{\langle E \rangle} = \frac{\sqrt{k_B T^2 C_V}}{\langle E \rangle} \sim O\left(\frac{1}{\sqrt{N}}\right) .$$

For large systems, such as $N \sim 10^{23}$, $N^{-1/2}$ becomes a very small number and we can consider the average value of the energy as an accurate prediction of the experimental internal energy.

For example, consider an ideal gas of structureless particles, we know that

$$C_V = \frac{3}{2} N k_B$$

and

$$\langle E \rangle = \frac{3}{2} N k_B T$$

If we consider $N \sim 10^{22}$, then the above ration is numerically $\sim 10^{-11}$, which is a very small number.

3.11.2 Generalized Ensemble

For general statistical ensemble, from statistical mechanics we have

$$\langle (\delta x)^2 \rangle = -\left(\frac{\partial \langle x \rangle}{\partial f}\right)_{T,E} . \tag{3.49}$$

The left hand side is always positive, and the right hand side determines the curve or convexity of the thermodynamic free energy.

3.11.3 Isothermal-Isobaric Ensemble

For isothermal-isobaric ensemble, $x \equiv V$ and $f \equiv \beta p$, thus

$$\begin{aligned} \langle (\delta V)^2 \rangle &= -\left(\frac{\partial \langle V \rangle}{\partial (\beta p)}\right)_{T,E} \qquad (3.50) \\ &= -k_B T \left(\frac{\partial \langle V \rangle}{\partial p}\right)_{T,E} \geq 0 . \end{aligned}$$

The last inequality is true since from the condition of the thermodynamic equilibrium stability, discussed in Chap. 2, we obtained that

$$\left(\frac{\partial \langle V\rangle}{\partial p}\right)_{T,E} \leq 0\,.$$

For this ensemble, the fluctuations of energy are given as:

$$\langle(\delta E)^2\rangle = -\left(\frac{\partial \langle E\rangle}{\partial \beta}\right)_{N,p}\,, \tag{3.51}$$

which can further be written as

$$\begin{aligned}\langle(\delta E)^2\rangle &= -\left(\frac{\partial \langle H\rangle - p\langle V\rangle}{\partial \beta}\right)_{N,p} \\ &= -\left(\frac{\partial \langle H\rangle}{\partial \beta}\right)_{N,p} + p\left(\frac{\partial \langle V\rangle}{\partial \beta}\right)_{N,p}\,,\end{aligned} \tag{3.52}$$

where $\langle H\rangle$ is the average value of the enthalpy:

$$\langle H\rangle = \langle E\rangle + p\langle V\rangle$$

Using the relation of the specific heat at constant pressure:

$$C_p = \left(\frac{\partial \langle H\rangle}{\partial T}\right)_{N,p}$$

we can write

$$\langle(\delta E)^2\rangle = k_B T^2 \left[C_p - p\left(\frac{\partial \langle V\rangle}{\partial T}\right)_{N,p}\right]\,. \tag{3.53}$$

This expression can be simplified using the relation:

$$C_p = C_V + T\left(\frac{\partial p}{\partial T}\right)_{N,V}\left(\frac{\partial V}{\partial T}\right)_{N,p}\,.$$

Substituting this expression into Eq. (3.53), we get

$$\langle(\delta E)^2\rangle = k_B T^2 C_V + k_B T^2 \left[T\left(\frac{\partial p}{\partial T}\right)_{N,V} - p\right]\left(\frac{\partial \langle V\rangle}{\partial T}\right)_{N,p}\,. \tag{3.54}$$

Substituting the expression

$$\left(\frac{\partial \langle V\rangle}{\partial T}\right)_{N,p} = -\left(\frac{\partial p}{\partial T}\right)_{N,V}\left(\frac{\partial V}{\partial p}\right)_{N,T}$$

into Eq. (3.54), we obtain

$$\begin{aligned}\langle(\delta E)^2\rangle &= k_B T^2 C_V \\ &\quad - k_B T\left[T\left(\frac{\partial p}{\partial T}\right)_{N,V} - p\right]\left[T\left(\frac{\partial p}{\partial T}\right)_{N,V}\right]\left(\frac{\partial V}{\partial p}\right)_{N,T} \\ &= k_B T^2 C_V \\ &\quad - k_B T\left[T\left(\frac{\partial p}{\partial T}\right)_{N,V} - p\right]\left[\left(\frac{\partial (pT)}{\partial T}\right)_{N,V} - p\right]\left(\frac{\partial V}{\partial p}\right)_{N,T}\end{aligned} \tag{3.55}$$

where the following relation was used:

$$\left(\frac{\partial (pT)}{\partial T}\right)_{N,V} = p + T\left(\frac{\partial p}{\partial T}\right)_{N,V}$$

By direct comparison of the expression given by Eq. (3.48) and the one given by Eq. (3.55), it can be seen that the energy fluctuations in a canonical ensemble (with N, V and T constant) differ from the fluctuations in the isothermal-isobaric ensemble (with N, p and T constant), namely by the term

$$-k_B T\left[T\left(\frac{\partial p}{\partial T}\right)_{N,V} - p\right]\left[\left(\frac{\partial (pT)}{\partial T}\right)_{N,V} - p\right]\left(\frac{\partial V}{\partial p}\right)_{N,T}$$

which includes the changes of both the pressure (p) and (pT) with temperature (at constant N and V) and changes of the volume with pressure (at constant N and T).

3.11.4 Grand Canonical Ensemble

Formulas of the fluctuations for the grand canonical ensemble can be defined similarly to the canonical ensemble. For instance, the fluctuations in the number of the particles are given by:

$$\langle(\delta N)^2\rangle = \langle(N - \langle N\rangle)^2\rangle = \langle N^2\rangle - \langle N\rangle^2 \, .$$

It can easily be shown that

$$\langle N \rangle = \left(\frac{\partial \ln \Xi}{\partial (\beta\mu)} \right)_V , \tag{3.56}$$

$$\langle N^2 \rangle = \left(\frac{1}{\Xi} \frac{\partial^2 \Xi}{\partial (\beta\mu)^2} \right)_V . \tag{3.57}$$

Substituting these two expressions into the expression for $\langle (\delta N)^2 \rangle$, we finally get

$$\langle (\delta N)^2 \rangle = \left(\frac{\partial \langle N \rangle}{\partial (\beta\mu)} \right)_V . \tag{3.58}$$

Note that $\langle (\delta N)^2 \rangle \geq 0$, hence,

$$\left(\frac{\partial \langle N \rangle}{\partial \mu} \right)_{T,V} \geq 0 .$$

Since $\langle N \rangle = nN_0$, where n is the number of moles and N_0 is the Avogadro's number, we get

$$\left(\frac{\partial n}{\partial \mu} \right)_{T,V} \geq 0 ,$$

which is one of the thermodynamic conditions of the stability for the thermodynamic equilibrium of the system, obtained in Chap. 2, derived here in the context of statistical thermodynamics.

Fluctuations of the energy are:

$$\langle (\delta E)^2 \rangle = - \left(\frac{\partial \langle E \rangle}{\partial \beta} \right)_{V,\mu} . \tag{3.59}$$

The above relation is similar to the relationship found for the canonical ensemble, but it is different because the derivative is evaluated at a constant N for the canonical ensemble and a constant chemical potential μ for the grand canonical ensemble. Hence, the energy fluctuations are different in the canonical and grand canonical ensembles.

The reason for this difference between the fluctuations in the grand canonical and canonical ensembles is not easy to determine from the above expression since the chemical potential is difficult to measure. The difference can be made explicit by using thermodynamics, in which case we identify the average energy E in the grand canonical ensemble with U. That is, since on holding V fixed

$$\langle N \rangle = \langle N(T, \mu) \rangle , \tag{3.60}$$

one has

$$U = U(T, \langle N(T, \mu)\rangle) \,. \tag{3.61}$$

Thus, the differential of U can be written as

$$\begin{aligned} dU &= \left(\frac{\partial U}{\partial T}\right)_{\langle N\rangle} dT + \left(\frac{\partial U}{\partial \langle N\rangle}\right)_T d\langle N\rangle \\ &= \left(\frac{\partial U}{\partial T}\right)_{\langle N\rangle} dT + \left(\frac{\partial U}{\partial \langle N\rangle}\right)_T \\ &\times \left[\left(\frac{\partial \langle N\rangle}{\partial T}\right)_\mu dT + \left(\frac{\partial \langle N\rangle}{\partial \mu}\right)_T d\mu\right] . \end{aligned} \tag{3.62}$$

Hence, the derivative of U with respect to T with μ held fixed is given by

$$\begin{aligned} \left(\frac{\partial U}{\partial T}\right)_{\mu,V} &= \left(\frac{\partial U}{\partial T}\right)_{\langle N\rangle} + \left(\frac{\partial U}{\partial \langle N\rangle}\right)_T \left(\frac{\partial \langle N\rangle}{\partial T}\right)_\mu \\ &= C_{N,V} + \left(\frac{\partial U}{\partial \langle N\rangle}\right)_T \left(\frac{\partial \langle N\rangle}{\partial T}\right)_\mu . \end{aligned} \tag{3.63}$$

Therefore, the fluctuations of energy are given by

$$\langle(\delta E)^2\rangle = k_B T^2 C_{N,V} + k_B T^2 \left(\frac{\partial U}{\partial \langle N\rangle}\right)_T \left(\frac{\partial \langle N\rangle}{\partial T}\right)_\mu . \tag{3.64}$$

As one can see, the first term of the energy fluctuation in the grand canonical ensemble is the same as energy fluctuations in the canonical ensemble where the number of particles and the volume are fixed and the other contribution (see second term in Eq. (3.64)) originates from the temperature dependence on the number of particles.

We can further simplify the above expression to obtain more insights into the origin of the energy fluctuations in grand canonical ensemble. First, we can consider the differential of the internal energy U for one component system

$$dU = TdS - pdV + \mu d\langle N\rangle \,. \tag{3.65}$$

Dividing both sides by $d\langle N\rangle$ keeping T and V constant gives

$$\left(\frac{\partial U}{\partial \langle N\rangle}\right)_{T,V} = T\left(\frac{\partial S}{\partial \langle N\rangle}\right)_{T,V} + \mu \,. \tag{3.66}$$

Substituting the Maxwell equation

$$\left(\frac{\partial S}{\partial \langle N\rangle}\right)_{T,V} = -\left(\frac{\partial \mu}{\partial T}\right)_{\langle N\rangle,V}$$

into Eq. (3.66) we obtain the first thermodynamic relation

$$\left(\frac{\partial U}{\partial \langle N\rangle}\right)_{T,V} = -T\left(\frac{\partial \mu}{\partial T}\right)_{\langle N\rangle,V} + \mu \,. \tag{3.67}$$

Dividing both sides of Eq. (3.62) by $d\mu$ keeping T and V constant, we obtain

$$\left(\frac{\partial U}{\partial \mu}\right)_{T,V} = \left(\frac{\partial U}{\partial \langle N\rangle}\right)_{T,V}\left(\frac{\partial \langle N\rangle}{\partial \mu}\right)_{T,V} = T\left(\frac{\partial \langle N\rangle}{\partial T}\right)_{\mu,V} ,$$

which holds assuming that the internal energy U show minimal changes with fluctuations in the number of particles, that is $\left(\frac{\partial U}{\partial \langle N\rangle}\right)_{T,V} \approx 0$, such that the following relation holds using Eq. (3.67):

$$\left(\frac{\partial \mu}{\partial T}\right)_{\langle N\rangle,V} = \frac{\mu}{T}$$

Hence,

$$\left(\frac{\partial \langle N\rangle}{\partial T}\right)_{\mu,V} = \frac{1}{T}\left(\frac{\partial U}{\partial \langle N\rangle}\right)_{T,V}\left(\frac{\partial \langle N\rangle}{\partial \mu}\right)_{T,V} \tag{3.68}$$

$$= \frac{1}{T}\left(\frac{\partial \langle N\rangle}{\partial \mu}\right)_{T,V}\left[\mu - T\left(\frac{\partial \mu}{\partial T}\right)_{\langle N\rangle,V}\right] . \tag{3.69}$$

Then, we obtain

$$\langle(\delta E)^2\rangle = k_B T^2 C_{N,V} + k_B T\left(\frac{\partial \langle N\rangle}{\partial \mu}\right)_{T,V}\left[\mu - T\left(\frac{\partial \mu}{\partial T}\right)_{\langle N\rangle,V}\right]^2 \tag{3.70}$$

Using the relation

$$\langle(\delta N)^2\rangle = k_B T\left(\frac{\partial \langle N\rangle}{\partial \mu}\right)_{T,V}$$

and expressions in Eqs. (3.67) and (3.70), we obtain

$$\langle(\delta E)^2\rangle = k_B T^2 C_{N,V} + \langle(\delta N)^2\rangle\left[\left(\frac{\partial U}{\partial \langle N\rangle}\right)_{T,V}\right]^2 \tag{3.71}$$

This expression indicates that the mean squared energy fluctuations have two contributions, one originating from the mean squared energy fluctuations with a fixed number of particles and the second contribution comes from the mean squared fluctuations of the number of particles where each particle that is exchanged with the reservoir carries with it the energy.

Chapter 4
Thermodynamics of Biological Phenomena

This chapter aims to discuss the application of the statistical mechanics (or the so-called statistical thermodynamics) in understanding biological phenomena, based on the theoretical framework introduced by Lazaridis and Karplus (2003).

4.1 Introduction

The biological systems, similarly to other mechanical systems, live under the universal laws of physics and chemistry. Currently, the studies of understanding the function of the leaving organisms and high complex interactions involved in many critical biological processes support this too.

Experimental results provide a vast amount of information regarding the biological phenomena in living cells. However, this information is only qualitatively and intuitively explained, due in part, that a theoretical framework is missing, which will be able to interpret the phenomena based on the laws of nature, even when the structures of the components involved, such as protein and nucleic acids, are well known. Therefore, the laws of physics, in particular, statistical mechanics and classical mechanics, can be useful in understanding the phenomena of biological interests.

For simple isolated systems, the movement towards the equilibrium state is fast, which is characterized by a maximum of the entropy function, or a minimum of other thermodynamic potential functions based on the principle of the thermodynamic equilibrium discussed in Chap. 2. However, for complex biological systems, the use of thermodynamics is limited due to the complexity of the processes involved. Moreover, identification of the equilibrium state of the complex biological systems in living systems is in many cases impossible, since they never reach this thermodynamic equilibrium. Therefore, the complete description of thermodynamics of such complex biological systems is often tricky (Zotin 1972).

H. Kamberaj, *Molecular Dynamics Simulations in Statistical Physics: Theory and Applications*, Scientific Computation,
https://doi.org/10.1007/978-3-030-35702-3_4

From the thermodynamic point of view, biological systems may be considered as open systems far from the equilibrium state. However, some orders can be observed in the biological system with the entropy of the entire universe still increasing, although why it occurs is not explained.

From the microscopic point of view, it is more likely that the biological systems through their processes develop gradually towards the equilibrium states, but due to the change in the external conditions, these equilibrium states are shifted to other microscopic states and thus making these equilibrium states unreachable. A typical example is the folded conformation of a protein, which may represent the lowest free energy state for some environment conditions (such as, pH), and thus the equilibrium state of the system. However, if the environmental conditions change, then the system may gradually develop towards a new equilibrium state, which may be the unfolded conformation of a protein. In spite of that the biological processes may not evolve towards thermodynamic stable equilibrium states, equilibrium thermodynamics has been extensively used to characterize those systems. It is common, that specific isolated biological processes (such as protein folding, proton and electron transfer, protein complex interactions, and so on) to have been studied under equilibrium conditions either experimentally or using computational methods.

Equilibrium thermodynamic measurements can be instrumental and relevant to the situations in vivo because, despite the lack of overall thermodynamic equilibrium in the cell, there is a partial equilibrium, either within a specific timescale or within a particular region. Therefore, the limitations of such experimental measurements should be considered when comparing experimental and theoretical models results. For example, in protein-ligand binding measurements, equilibrium binding constant will be measured accurately only if the kinetics of binding is fast relative to the rate of transport of the protein and ligand.

Since the biological processes depend significantly on the nature of the molecules involved in the process, statistical thermodynamics is the method of choice in biology because it deals with the microscopic description of the phenomena, in contrast to classical thermodynamics is concerned with the macroscopic nature of the matter. It is worth noting that the origin of interactions between molecules modeled using the molecular models is critical.

4.2 Stability of Macromolecular Conformations

Macromolecules, such as proteins and nucleic acids, may be in different unique conformations depending on the physiological conditions. If the conformation adopted by the system is the most stable, then it is under thermodynamic control. Otherwise, it corresponds to the kinetically most accessible, and it is said to be under kinetic control (Wetlaufer and Ristow 1973; Anfinsen and Scheraga 1975). There could also exist transitions from one conformation to another, depending if the barriers are small enough to be traversed within experimental timescales. In such a case, thermodynamic equilibrium establishes, and the Boltzmann's distribution will give the probability distribution of each conformation in the ensemble. In contrast, if the barriers between the configurations would have been too high, then the system will not behave ergotically, and the macromolecule will occupy only the lowest energetically local minimum that will be reached within the available timescale. With the ergodic system, we will understand a system that has sampled all possible conformations.

There is experimental evidence of both hypotheses, either of thermodynamic control (Anfinsen and Scheraga 1975) or kinetic control as recently suggested for large complex protein systems, both experimentally (Goldberg 1985; Baker and Agard 1994; Baker 1998) and from computer simulation models of protein folding (Dinner and Karplus 1998). The kinetic control is more likely to happen in complex cellular processes (Lazaridis and Karplus 2003), where non-equilibrium states in biological systems during the lifetime of a cell are maintained, for example, the composition of biological membranes (Jain 1988).

All concepts of the thermodynamic equilibrium apply if the thermodynamic control is valid. However, they can also be in use, if the thermodynamic equilibrium is achieved for some degrees of freedom, although the system has not reached the overall thermodynamic equilibrium, yet. For example, one concern is about solvent degrees of freedom. It is known that for most of the macromolecular changes in conformation, the solvent equilibrates in picoseconds time scales (Halle et al. 1981). Sometimes, equilibration can also take longer (Otting et al. 1991; Ernst et al. 1995), for example, equilibration of possible configurations of water inside protein cavities during the transition from bulk water. Since the equilibration of the solvent is fast most of the time compare to equilibration of the macromolecule, its degrees of freedom can be integrated out to give the so-called *equilibrium solvation free energy*, which is added to the internal macromolecular energy to provide the so-called *effective energy* or *potential of mean force* for each macromolecular conformation. This potential of mean force defines a hypersurface in the conformational space of the molecule, known as the *energy landscape* (Frauenfelder et al. 1991), with a shape that determines the conformational properties of the macromolecule under either thermodynamic or kinetic control regarding the macromolecular degrees of freedom.

The formalism in Lazaridis and Karplus (1999) using standard mechanical methods (McQuarrie 1976) is the basics of the following discussion. For this, consider a macromolecule consisting of M atoms with Cartesian coordinates

$$\mathbf{R}_i = (X_i,\ Y_i,\ Z_i)\,, \quad i = 1,\ 2,\ \cdots,\ M$$

of the ith atom, and internal coordinates

$$\mathbf{q} = (q_1,\ q_2,\ \cdots,\ q_{3M-6})$$

Moreover, we suppose that this macromolecule is immersed in a solution of N solvent molecules, considered to be rigid, with coordinates

$$r_i = (x_i,\ y_i,\ z_i,\ \alpha_i,\ \beta_i,\ \gamma_i)\,, \quad i = 1,\ 2,\ \cdots,\ N$$

where (x, y, z) are the Cartesian coordinates of the centre of mass, and (α, β, γ) are the Euler angles specifying the orientation. Furthermore, we can consider $(N,\ V,\ T)$ constant conditions corresponding to the canonical ensemble. Then, the partition function is

$$Q = \frac{\prod_{i=1}^{N} (2\pi m_i k_B T)^{3/2} \prod_{i=1}^{M} (2\pi M_i k_B T)^{3/2}}{N!} Z \tag{4.1}$$

where Z is the classical configurational integral

$$Z = \int d\mathbf{r}^N \int d\mathbf{R}^M \exp(-\beta H) \tag{4.2}$$

with H being the Hamiltonian and $\beta = 1/k_B T$. We can denote

$$\begin{aligned} \Lambda^{3N} &\equiv \frac{1}{\prod_{i=1}^{N} (2\pi m_i k_B T)^{3/2}} \\ \Lambda^{3M} &\equiv \frac{1}{\prod_{i=1}^{M} (2\pi M_i k_B T)^{3/2}} \end{aligned} \tag{4.3}$$

to obtain

$$Q = \frac{Z}{N! \Lambda^{3M} \Lambda^{3N}} \tag{4.4}$$

The Helmholtz free energy is given by

$$\begin{aligned} F &= -k_B T \ln Q \\ &= -k_B T \ln Z + k_B T \ln \left(N! \Lambda^{3M} \Lambda^{3N} \right) \end{aligned} \tag{4.5}$$

The Hamiltonian can be formally split into different terms based on the interactions in the system:

$$H = H_{ww} + H_{wm} + H_{mm} \tag{4.6}$$

where H_{ww} is the Hamiltonian of the solvent-solvent interactions, H_{wm} is the Hamiltonian of the solvent-macromolecule interactions, and H_{mm} is the Hamiltonian of the intra-macromolecule interactions. Here, we have assumed that the

Hamiltonian is additive, which is a fair assumption since most of the molecular mechanics force fields are additive (MacKerell et al. 1998).

Performing the integration over the solvent degrees of freedom in Eq. (4.2), we get

$$Z_{ww} = \int d\mathbf{r}^N \exp\left(-\beta H_{ww}\right) \tag{4.7}$$

which is the so-called *pure solvent configurational integral.*

Then, we can define the potential of the mean force as

$$\exp\left(-\beta W\right) = Z_{ww}^{-1} \int d\mathbf{r}^N \exp\left(-\beta H\right) \tag{4.8}$$

Thus, Eq. (4.2) can also be written in terms of W as

$$Z = Z_{ww} \int d\mathbf{R}^M \exp\left(-\beta W\right) \tag{4.9}$$

The integral in Eq. (4.9) depends only on the macromolecular degrees of freedom R. Combining Eq. (4.8) with Eq. (4.6), we can write

$$\begin{aligned} \exp\left(-\beta W\right) &= \exp\left(-\beta H_{mm}\right) Z_{ww}^{-1} \int d\mathbf{r}^N \exp\left(-\beta\left(H_{mw} + H_{ww}\right)\right) \\ &= \exp\left(-\beta\left(H_{mm} + X\right)\right) \end{aligned} \tag{4.10}$$

where

$$\exp\left(-\beta X\right) = Z_{ww}^{-1} \int d\mathbf{r}^N \exp\left(-\beta\left(H_{mw} + H_{ww}\right)\right) \tag{4.11}$$

or

$$\begin{aligned} \exp\left(-\beta X\right) &= Z_{ww}^{-1} \int d\mathbf{r}^N \left[\exp\left(-\beta H_{mw}\right)\right] \exp\left(-\beta H_{ww}\right) \\ &= \langle \exp\left(-\beta H_{mw}\right) \rangle_w \end{aligned} \tag{4.12}$$

where $\langle \cdots \rangle_w$ denotes an ensemble average over all pure solvent degrees of freedom. From Eq. (4.12), we can write that

$$X = -k_B T \ln\left(\langle \exp\left(-\beta H_{mw}\right) \rangle_w\right) \equiv \Delta G^{\text{solv}} \tag{4.13}$$

where ΔG^{solv} is the solvation free energy of the macromolecule or excess chemical potential (Ben-Naim 1978). It also gives the work done to bring M atoms of the macromolecule from the infinity to the solvent plus the work done to create a cavity in solvent with size big enough to accommodate the macromolecule.

The potential of the mean force can then be:

$$W = H_{mm} + X$$

W contains two terms: the intra-molecular interactions energy and the solvation free energy. It defines a hypersurface in the conformation space of the macromolecule in the presence of the equilibrated solvent, and hence it also includes solvation entropy. This hypersurface is also called *energy landscape* and it determines the thermodynamics and kinetics of macromolecular conformation transitions.

This separation in potential of the mean force W can also be obtained in the case when the Hamiltonian is not additive, such as in the case of three-body interactions molecular mechanics force fields. In this case, the Hamiltonian is as the following:

$$H = (H_{mm} + H_{mmm}) + (H_{mw} + H_{ww} + H_{www} + H_{wwm} + H_{mmw}) \tag{4.14}$$

then, the potential of the mean force W is

$$\begin{aligned} W = {} & (H_{mm} + H_{mmm}) \\ & - k_B T \ln \langle \exp\left(-\beta \left(H_{mw} + H_{mmw} + H_{mww}\right)\right) \rangle_w \end{aligned} \tag{4.15}$$

It is more convenient to use the internal coordinates $\mathbf{q}$ for a description of the macromolecule degrees of freedom (Lazaridis and Karplus 2003). The Jacobian of this transformation will depend on the bond lengths and bond angles. Therefore, the Jacobian is constant for all conformations, and it is out of the integral in Eq. (4.9). The integral over six external coordinates of the overall rotational and translational motions can be performed, since the system is homogeneous, and it gives

$$I_{rot+trans} = 8\pi^2 V$$

where V is the volume. Therefore,

$$Z = Z_{ww} I_{rot+trans} \mid J \mid \int d\mathbf{q} \exp\left(-\beta W\right) \tag{4.16}$$

or

$$Z = 8\pi^2 V \mid J \mid Z_{ww} \int d\mathbf{q} \exp\left(-\beta W\right) \tag{4.17}$$

The probability of finding the system at some configuration with internal coordinates $\mathbf{q}$ is (Lazaridis and Karplus 2003) (and the reference therein Lazaridis and Karplus 1999)

$$p\left(\mathbf{q}\right) = \frac{\exp\left(-\beta W(\mathbf{q})\right)}{\int d\mathbf{q} \exp\left(-\beta W(\mathbf{q})\right)} \tag{4.18}$$

From Eq. (4.5), we obtain

$$F = -k_B T \ln Z + k_B T \ln \left(N! \Lambda^{3N} \right) + k_B T \ln \left(\Lambda^{3M} \right) \tag{4.19}$$

$$= -k_B T \ln \left(8\pi^2 V \mid J \mid \right) + k_B T \ln \left(\frac{N! \Lambda^{3N}}{Z_{ww}} \right)$$

$$- k_B T \ln \left(\int d\mathbf{q} \exp\left(-\beta W(\mathbf{q})\right) \right) + k_B T \ln \left(\Lambda^{3M} \right)$$

$$= -k_B T \ln \left(8\pi^2 V \mid J \mid \right) + k_B T \ln \left(\frac{N! \Lambda^{3N}}{Z_{ww}} \right)$$

$$- k_B T \ln \left(\frac{\exp\left(-\beta W(\mathbf{q})\right)}{p(\mathbf{q})} \right) + k_B T \ln \left(\Lambda^{3M} \right)$$

$$= F^0 + k_B T \ln \left(\frac{\Lambda^{3M}}{8\pi^2 V \mid J \mid} \right) + W(\mathbf{q}) + k_B T \ln p(\mathbf{q})$$

Thus, the average value is

$$\langle F \rangle = F^0 + k_B T \ln \left(\frac{\Lambda^{3M}}{8\pi^2 V \mid J \mid} \right) + \int d\mathbf{q} p(\mathbf{q}) W(\mathbf{q}) \tag{4.20}$$

$$+ k_B T \int d\mathbf{q} p(\mathbf{q}) \ln p(\mathbf{q})$$

or

$$\langle F \rangle = F^0 + k_B T \ln \left(\frac{\Lambda^{3M}}{8\pi^2 V \mid J \mid} \right) + \langle W \rangle - T S^{conf} \tag{4.21}$$

where F^0 is the free energy of the pure solvent, the second term is the free energy of an ideal gas from macromolecular translational and rotational degrees of freedom, the third term is the average of the potential of mean force, which equals the average intramolecular energy plus the average solvation energy:

$$\langle W \rangle = \int d\mathbf{q} \left(H_{mm}(\mathbf{q}) + \Delta G^{solv}(\mathbf{q}) \right) p(\mathbf{q})$$

The fourth term is the contribution from the configurational entropy of the macromolecule to the free energy.

The solvent contribution to the entropy is included in the term $\Delta G^{solv}(\mathbf{q})$ and in $p(\mathbf{q})$. Note that the Gibbs free energy is

$$\langle G \rangle = \langle F \rangle + \langle pV \rangle$$

where the term pV under the standard conditions may be negligible.

The entropy of the system is given as

$$S = k_B \ln Q + k_B T \left(\frac{\partial \ln Q}{\partial T} \right)_{V,N} \tag{4.22}$$

Using the expression for Q given by Eq. (4.4), we get

$$\begin{aligned} S &= k_B \left[\ln Z - \ln \left(N! \Lambda^{3M} \Lambda^{3N} \right) \right] + k_B T \left(\frac{\partial \ln Z}{\partial T} \right)_{V,N} \\ &\quad - k_B T \left(\frac{\partial \ln \Lambda^{3N}}{\partial T} \right)_{V,N} - k_B T \left(\frac{\partial \ln \Lambda^{3M}}{\partial T} \right)_{V,N} \\ &= k_B \left[\ln Z + T \left(\frac{\partial \ln Z}{\partial T} \right)_{V,N} \right] - k_B \ln \left(N! \Lambda^{3M} \Lambda^{3N} \right) \\ &\quad + \frac{3}{2} k_B (N + M) \\ &= -k_B \ln p(\mathbf{q}, \mathbf{r}^N) - k_B \ln \left(N! \Lambda^{3M} \Lambda^{3N} \right) + \frac{3}{2} k_B (N + M) \end{aligned} \tag{4.23}$$

where

$$p(\mathbf{q}, \mathbf{r}^N) = \frac{\exp\left(-\beta H(\mathbf{q}, \mathbf{r}^N)\right)}{\int d\mathbf{q} d\mathbf{r}^N \exp\left(-\beta H(\mathbf{q}, \mathbf{r}^N)\right)} = Z^{-1} \exp\left(-\beta H(\mathbf{q}, \mathbf{r}^N)\right) \tag{4.24}$$

Taking the averages of both sides with probability $p(\mathbf{q}, \mathbf{r}^N)$, we obtain

$$\begin{aligned} \langle S \rangle &= -k_B \int d\mathbf{q} d\mathbf{r}^N p(\mathbf{q}, \mathbf{r}^N) \ln p(\mathbf{q}, \mathbf{r}^N) \\ &\quad - k_B \ln \left(N! \Lambda^{3M} \Lambda^{3N} \right) + \frac{3}{2} k_B (N + M) \end{aligned} \tag{4.25}$$

We can determine the conditional probability $p(\mathbf{r}^N|\mathbf{q})$ of finding the solvent at a configuration $\mathbf{r}^N$, given that the macromolecule is in a conformation $\mathbf{q}$ as

$$p(\mathbf{r}^N|\mathbf{q}) = \frac{p(\mathbf{q}, \mathbf{r}^N)}{p(\mathbf{q})} \tag{4.26}$$

or,

$$p(\mathbf{q}, \mathbf{r}^N) = p(\mathbf{q}) p(\mathbf{r}^N|\mathbf{q}) \tag{4.27}$$

Replacing Eq. (4.27) into Eq. (4.25), we obtain

$$\langle S\rangle = -k_B \int d\mathbf{q}d\mathbf{r}^N p(\mathbf{q})p(\mathbf{r}^N|\mathbf{q})\left[\ln p(\mathbf{q}|\mathbf{r}^N) + \ln p(\mathbf{q})\right] \\ - k_B \ln\left(N!\Lambda^{3M}\Lambda^{3N}\right) + \frac{3}{2}k_B(N+M) \tag{4.28}$$

or,

$$\langle S\rangle = -k_B \int d\mathbf{q}p(\mathbf{q}) \ln p(\mathbf{q}) \\ - k_B \int d\mathbf{q}p(\mathbf{q}) \int d\mathbf{r}^N p(\mathbf{r}^N|\mathbf{q}) \ln p(\mathbf{q}|\mathbf{r}^N) \\ - k_B \ln\left(N!\Lambda^{3M}\Lambda^{3N}\right) + \frac{3}{2}k_B(N+M) \tag{4.29}$$

In Eq. (4.29), the first term is the configurational entropy of the macromolecule; the second term is the average solvent entropy for all possible conformations of a macromolecule, which arises from the solute-solvent and solvent-solvent correlations.

The internal energy can also be calculated similarly as

$$E = k_B T^2 \left(\frac{\partial \ln Q}{\partial T}\right)_{V,N} \tag{4.30}$$

Or,

$$E = k_B T^2 \left(\frac{\partial \ln Z}{\partial T}\right)_{V,N} \\ - k_B T^2 \left(\frac{\partial \ln \Lambda^{3N}}{\partial T}\right)_{V,N} - k_B T^2 \left(\frac{\partial \ln \Lambda^{3M}}{\partial T}\right)_{V,N} \\ = k_B T^2 \left(\frac{\partial \ln Z}{\partial T}\right)_{V,N} - k_B T \ln\left(N!\Lambda^{3M}\Lambda^{3N}\right) \\ + \frac{3}{2}k_B T(N+M) \\ = H(\mathbf{q},\mathbf{r}^N) - k_B T \ln\left(N!\Lambda^{3M}\Lambda^{3N}\right) + \frac{3}{2}k_B T(N+M) \tag{4.31}$$

Taking the average of both sides, we obtain

$$\langle E \rangle = \int d\mathbf{q} d\mathbf{r}^N p(\mathbf{q}, \mathbf{r}^N) H(\mathbf{q}, \mathbf{r}^N) \\ - k_B T \ln \left(N! \Lambda^{3M} \Lambda^{3N} \right) + \frac{3}{2} k_B T (N + M) \tag{4.32}$$

where the first term gives the average potential energy and the last two terms give the average kinetic energy. Using the definition of the condition probability (see Eq. (4.27)) and Eq. (4.6), we can further write that

$$\langle E \rangle = \int d\mathbf{q} p(\mathbf{q}) H_{mm}(\mathbf{q}) \\ + \int d\mathbf{q} p(\mathbf{q}) \int d\mathbf{r}^N p(\mathbf{r}^N | \mathbf{q}) \left[H_{mw}(\mathbf{q}, \mathbf{r}^N) + H_{ww}(\mathbf{r}^N) \right] \\ - k_B T \ln \left(N! \Lambda^{3M} \Lambda^{3N} \right) + \frac{3}{2} k_B T (N + M) \tag{4.33}$$

where the first term is the average potential energy of intra-macromolecular interactions, and the second term gives the average potential energy of solute-solvent and solvent-solvent interactions for all possible conformations of the macromolecule.

The main assumption of the following analysis is that the probability distribution $p(\mathbf{q})$ completely specifies the conformational states of the macromolecule. In Eq. (4.21), there are two competing terms: $\langle W \rangle$ and $-TS^{conf}$. The first term tends to localize the macromolecule in the most minimum value of the energy landscape, on the other hand, the configurational entropy (second term) tends to make $p(\mathbf{q})$ as uniform as possible. Therefore, the most probable state, the one with the lowest value of the free energy, may not be the one with the lowest value of the potential of the mean force, since some of these states may be so narrow that the vibrational entropy of the macromolecule could be minimal, and thus the term $-TS^{conf}$ is unfavorable. That is a common problem on deterring the native state of a protein (Lazaridis and Karplus 2003).

4.3 Gibbs Free Energy of the Transition

The thermodynamics stability of a macromolecular state A can be expressed in terms of the Gibbs free energy, ΔG of the state A. Suppose that we have a transition state reaction

$$A \rightleftharpoons B$$

where B can be any other state of the macromolecule. Let us denote with K the equilibrium constant of this reaction. Then, the free energy can be written as

$$\Delta G = -RT \ln K = -RT \ln \frac{c_A}{c_B} \tag{4.34}$$

where c_A fraction of the state A and c_B is the fraction of the state B of macromolecule. Under standard physiological conditions the Gibbs free energy is equal to the Helmholtz free energy given that $p\Delta V$ term is negligible. Often the experimental results are obtained under constant pressure conditions, and the quantity ΔG is used. Our aim is to determine an expression for the free energy of the transition from the state B to the state A in terms of the interactions and the distributions of microscopic states using the statistical mechanics. Practically, we can divide the configurational space of macromolecule into subsets consisting of the configurations of type A and B. Then, the free energy of the conformations from the set A (or B) is

$$F_{A/B} = -k_B T \ln Z_{A/B} + k_B T \ln \left(N! \Lambda^{3N} \Lambda^{3M} \right) \tag{4.35}$$

where

$$Z_{A/B} = Z_{ww} 8\pi^2 V \mid J \mid \int_{A/B} \exp\left(-\beta W\right) d\mathbf{q}$$

Here, the integral is over all the configurations either in set A or set B. By definition, we assume

$$\sum_{A,B} Z_{A/B} = Z$$

The average free energy of the conformations from the set A or B can be found as

$$\begin{aligned} \langle F_{A/B} \rangle &= F^0 + k_B T \ln \left(\frac{\Lambda^{3M}}{8\pi^2 V \mid J \mid} \right) \\ &+ \int_{A/B} d\mathbf{q} p_{A/B}(\mathbf{q}) W(\mathbf{q}) \\ &+ k_B T \int_{A/B} d\mathbf{q} p_{A/B}(\mathbf{q}) \ln p_{A/B}(\mathbf{q}) \\ &= F^0 + F^{id} + \langle W \rangle_{A/B} - T S^{conf}_{A/B} \end{aligned} \tag{4.36}$$

where $p_{A/B}(\mathbf{q})$ is the probability distribution normalized within either the set A or B, such that

$$p_{A/B}(\mathbf{q}) = \frac{\exp\left(-\beta W(\mathbf{q})\right)}{\int_{A/B} d\mathbf{q} \exp\left(-\beta W(\mathbf{q})\right)} = Z_{A/B}^{-1} \exp\left(-\beta W(\mathbf{q})\right)$$

The first two terms in Eq. (4.36) are the free energy of pure solvent and the ideal gas translational and rotational free energy of the macromolecule, which are constant for any configurational either A or B. The last two terms in Eq. (4.36) are the average potential of the mean force and configurational entropy of either state A of state B, which are different. Thus, the free energy difference between the state A and B will be

$$\begin{aligned} \Delta\langle F\rangle &= \langle F_B\rangle - \langle F_A\rangle \\ &= \left[\langle W\rangle_B - \langle W\rangle_A\right] - T\left[S_B^{conf} - S_A^{conf}\right] \\ &= \Delta\langle W\rangle - T\Delta S^{conf} \\ &= \Delta\langle H_{mm}\rangle + \Delta\langle\Delta G^{solv}\rangle - T\Delta S^{conf} \end{aligned} \tag{4.37}$$

where it is assumed the free energy is a function of the distribution function. As one can see, the evaluation of the difference in Eq. (4.37) requires an arbitrary separation of the conformational space into state A and B regions. In the last expression of Eq. (4.37), the first term $\Delta\langle H_{mm}\rangle$ gives the difference in the average intra-macromolecular interactions, the second term gives the difference in solvation free energy term, which contains the difference in solute-solvent and solvent-solvent interactions plus the change in the solvent entropy, and the last term includes the change in the configurational entropy of the macromolecule.

The Gibbs free energy difference between the states A and B is then calculated by adding the term $p\Delta V$:

$$\Delta G = \Delta\langle H_{mm}\rangle + \Delta\langle\Delta G^{solv}\rangle - T\Delta S^{conf} + p\Delta V \tag{4.38}$$

4.4 The Binding Free Energy

Consider the following macromolecular reaction

$$A + B \rightleftharpoons AB$$

where A represents the macromolecule of type A in free state, B represents a macromolecule of type B in free state, and AB represents the two macromolecules in bound state. The thermodynamic stability of the bound state (AB) can be expressed in terms of the binding free energy ΔG, given the equilibrium constant K of this reaction:

$$\Delta G = -RT \ln K = -RT \ln \frac{c_{AB}}{c_A c_B} \tag{4.39}$$

where c_{AB}, c_A and c_B are the concentrations of the macromolecules of the type AB, A, and B, respectively, in solution. The Eq. (4.39) can also be written in terms of the probabilities as

$$\Delta G = -RT \ln K = -RT \ln \frac{p_{AB}}{p_A p_B} \tag{4.40}$$

where p_{AB} denotes the probability of observing the complex AB in solution, and p_A and p_B characterize the probabilities of observing A and B in solution after the equilibrium of the reaction is reached. This is a macroscopic picture of determination of the binding free energy. The statistical mechanics is the approach used to derive an expression for the binding free energy in terms of the interactions and the distributions of microscopic states.

We can divide the solution into three type of species: A, B and AB, where each of this macromolecules is characterized by set of conformations. The free energy of the conformation set of the type A free in solution is:

$$F_A = -k_B T \ln Z_A + k_B T \ln \left(N! \Lambda^{3N} \Lambda^{3M} \right) \tag{4.41}$$

where

$$Z_A = Z_{ww} 8\pi^2 V \mid J \mid \int_A d\mathbf{q} \exp(-\beta W)$$

with the integral over all possible configurations of macromolecule A in free state. The average free energy of the conformations from the set A can be found as

$$\begin{aligned} \langle F_A \rangle [p_A(\mathbf{q})] = & F^0 + k_B T \ln \left(\frac{\Lambda^{3M}}{8\pi^2 V \mid J \mid} \right) \\ & + \int_A d\mathbf{q} p_A(\mathbf{q}) W(\mathbf{q}) \\ & + k_B T \int_A d\mathbf{q} p_A(\mathbf{q}) \ln p_A(\mathbf{q}) \\ & = F^0 + F^{id} + \langle W \rangle_A - T S_A^{conf} \end{aligned} \tag{4.42}$$

where $p_A(\mathbf{q})$ is the probability distribution normalized within the set of configurations of macromolecule A, such that

$$p_A(\mathbf{q}) = \frac{\exp(-\beta W(\mathbf{q}))}{\int_A d\mathbf{q} \exp(-\beta W(\mathbf{q}))} = Z_A^{-1} \exp(-\beta W(\mathbf{q}))$$

Similarly, we can write the expressions of the average free energy of the conformations from the set B:

$$\langle F_B \rangle [p_B(\mathbf{q})] = F^0 + F^{id} + \langle W \rangle_B - T S_B^{conf} \tag{4.43}$$

and AB:

$$\langle F_{AB} \rangle [p_{AB}(\mathbf{q})] = F^0 + F^{id} + \langle W \rangle_{AB} - T S_{AB}^{conf} \tag{4.44}$$

The notation $\langle F \rangle [p(\mathbf{q})]$ indicates that the free energy is a function of the probability distribution function $p(\mathbf{q})$. The following difference gives the free energy difference between the bound and unbound states:

$$\begin{aligned} \Delta\langle F \rangle &= \langle F_{AB} \rangle - (\langle F_A \rangle + \langle F_B \rangle) \\ &= [\langle W \rangle_{AB} - (\langle W \rangle_A + \langle W \rangle_B)] - T\left[S_{AB}^{conf} - \left(S_A^{conf} + S_B^{conf}\right)\right] \\ &= \Delta\langle W \rangle - T\Delta S^{conf} \\ &= \Delta\langle H_{mm} \rangle + \Delta\langle \Delta G^{solv} \rangle - T\Delta S^{conf} \end{aligned} \tag{4.45}$$

where

$$\Delta\langle H_{mm} \rangle = \langle H_{mm} \rangle_{AB} - (\langle H_{mm} \rangle_A + \langle H_{mm} \rangle_B) \tag{4.46}$$

is the change on the average intra-macromolecular interactions upon binding;

$$\Delta\langle \Delta G^{solv} \rangle = \langle \Delta G^{solv} \rangle_{AB} - \left(\langle \Delta G^{solv} \rangle_A + \langle \Delta G^{solv} \rangle_B\right) \tag{4.47}$$

is the change on the solvation free energy upon binding (which contains solute-solvent interactions, solvent-solvent interactions, and solvent reorganization entropy term); and

$$\Delta S^{conf} = S_{AB}^{conf} - \left(S_A^{conf} + S_B^{conf}\right) \tag{4.48}$$

is the change on the macromolecule configuration entropy upon binding.

By adding the term $p\Delta V$ (where $\Delta V = V_{AB} - V_A - V_B$), the binding Gibbs free energy becomes:

$$\Delta G = \Delta\langle H_{mm}\rangle + \Delta\langle \Delta G^{solv}\rangle - T\Delta S^{conf} + p\Delta V \tag{4.49}$$

Here, we assume that the measurements performed under constant N, p, T conditions. The above expressions for the binding free energy, either Eq. (4.45) or Eq. (4.49), indicate that the macromolecular binding stability, such as protein-protein, is a result of the balance between the change in the average potential of the mean force and the change in macromolecular configurational entropy upon binding. The change in the average potential of the mean force is related to the change in depth of the bound state well on the energy landscape upon binding. The change in the macromolecular configurational entropy is related to the change in the width of the bound state well on the energy landscape upon binding.

4.5 Theoretical Models

In molecular dynamics simulations, solvation is treated explicitly by surrounding the macromolecule with a sufficiently large number of water model molecules. There exist two limitations of this approach (Lazaridis and Karplus 2003). First, the procedure is computationally expensive. Most of the Computer Processing Unit (CPU) time is spent on calculating the motions of the solvent molecules with no direct interest in most of the cases. The second limitation is the lack of knowledge of the potential of mean force in explicit solvent simulations for a macromolecular conformation, whereas the intramolecular energy is known. Note that the calculation of the solvent-solute and solvent-solvent energies is also possible, but they are not directly related to the solvation free energy. An alternative way to explicit solvation is to include in the potential energy function a model for the solvation free energy, such as performing simulations with a potential of the mean force. Often this approach is called *implicit solvation* model and it has resulted in being about two orders of magnitude faster than the simulations using the explicit solvent model (Lazaridis and Karplus 2003).

The solvation free energy given by Eq. (4.13) is also known as the *excess chemical potential*, which is the part of the chemical potential that depends on the interactions between the solute and solvent, and it is zero for an ideal gas of particles. The solvation free energy, ΔG^{solv} represents the free energy for transferring the macromolecule from the gas phase to the solution, which include the electrostatic work done for bringing the set of partial atomic charges of macromolecule from the infinity to a configuration in solution, plus the work done for creating a cavity in solution of size of the solute. Besides, it also includes the re-organization entropy of the solvent molecules displaced from the formed cavity into the solution.

There are many theoretical studies of the conformational properties of macromolecules that depend significantly on the development of quantitative models for the solvation free energy of these systems. Statistical thermodynamics has been used to develop different solvation models. For example, analytical theories for predicting thermodynamic properties of hard sphere fluids (Reiss 1965) and integral equation theories, such as Percus-Yevick approximation (Hansen and McDonald 1986). For solutions, we can mention the X-reference interaction site model (XRISM) (Hirata and Rossky 1981; Pettitt and Rossky 1986; Yu and Karplus 1988), which has been used to calculate the thermodynamic solvation properties of small solutes.

More accurate methods are also employed, such as the free energy perturbation theory and thermodynamic integration methods (Beveridge and DiCapua 1989; Kollman 1993), which are exact within the limit of given intermolecular potentials. The only limitation is the computational demands; they require very long simulations for reaching the convergence of the configuration sampling.

Another approach is to consider the system solute-solvent as inhomogeneous system. Based on this approach the solvation free energy contributions of energetics and entropy are considered separately (Matubayasi et al. 1994; Lazaridis 1998b):

$$\Delta G^{solv} = \Delta E^{solv} - T\Delta S^{solv} + p\Delta V^{solv} \tag{4.50}$$

where the first term is solvation energy, the second is solvation entropy and the third term includes changes on the volume at constant pressure. First two terms are expressed as sum of the solute-solvent and solvent-solvent contributions:

$$\begin{aligned} \Delta E^{solv} &= E_{mw} + \Delta E_{ww} \\ \Delta S^{solv} &= S_{mw} + \Delta S_{ww} \end{aligned} \tag{4.51}$$

For inserting the macromolecule (solute) in a fixed point in the solvent solution, we can write (Bartels and Karplus 1998; Lazaridis 1998b):

$$E_{mw} = \rho \int d\mathbf{r} g^{(1)}(\mathbf{r}) u_{mw}(\mathbf{r}) \tag{4.52}$$

$$\Delta E_{ww} = \frac{1}{2}\rho^2 \int d\mathbf{r} d\mathbf{r}' g^{(1)}(\mathbf{r}) \left[g^{(1)}(\mathbf{r}') - 1 \right] g^{(2)}(\mathbf{r}, \mathbf{r}') u_{ww}(\mathbf{r}, \mathbf{r}') \tag{4.53}$$

$$S_{mw} = -k_B \rho \int d\mathbf{r} g^{(1)}(\mathbf{r}) \ln g^{(1)}(\mathbf{r}) \tag{4.54}$$

$$\begin{aligned} \Delta S_{ww} = &-\frac{1}{2} k_B \rho^2 \int d\mathbf{r} d\mathbf{r}' g^{(1)}(\mathbf{r}) \left[g^{(1)}(\mathbf{r}') - 1 \right] g^{(2)}(\mathbf{r}, \mathbf{r}') \\ &\times \left[g^{(2)}(\mathbf{r}, \mathbf{r}') \ln g^{(2)}(\mathbf{r}, \mathbf{r}') - g^{(2)}(\mathbf{r}, \mathbf{r}') + 1 \right] \end{aligned} \tag{4.55}$$

where $\rho = N/V$ is the solvent number density, $\rho g^{(1)}(\mathbf{r})$ is the local density of the solvent located at $\mathbf{r}$ (where $g^{(1)}(\mathbf{r})$ is the pair correlation function between the macromolecule and the solvent), $u_{mw}(\mathbf{r})$ is the potential interaction function between the macromolecule and a solvent molecule, $g^{(2)}(\mathbf{r}, \mathbf{r}')$ is the pair correlation function between the solvent molecules, and $u_{ww}(\mathbf{r}, \mathbf{r}')$ is the interaction potential function between two solvent molecules at positions $\mathbf{r}$ and $\mathbf{r}'$, respectively. The main approximation assumed from Eq. (4.52) to Eq. (4.55) includes neglecting the correlations between more than two particles (Lazaridis 1998b). Here, ΔV^{solv} is the excess partial molar volume of the macromolecule (Lazaridis and Karplus 2003)

$$\Delta V^{solv} = \int \left[1 - g^{(1)}(\mathbf{r})\right] d\mathbf{r} \tag{4.56}$$

In general, the term $p\Delta V^{solv}$ is too small under standard conditions, and thus it can be neglected.

Here, E_{mw} representing the macromolecule-solvent interaction energy derives from the statistical thermodynamics. The macromolecule-solvent entropy S_{mw} include correlations between the macromolecule and solvent, such as, positional correlations (which include the fluctuations of the water around macromolecule) and orientation correlations (which include preferential orientations of the solvent around macromolecule). ΔE_{ww} and ΔS_{ww} characterize the solvent reorganization energy and entropy, respectively, due to changes in solvent-solvent interactions and correlations upon insertion of the macromolecule in the solvent.

The components of the solvation free energy can be written as integrals over the space around the macromolecule as

$$\Delta G^{solv} = \int f(\mathbf{r}) d\mathbf{r} \tag{4.57}$$

where $f(\mathbf{r})$ is the solvation free energy density. Here, we will neglect the term $p\Delta V^{solv}$, and write

$$\begin{aligned} f(\mathbf{r}) &= \rho g^{(1)}(\mathbf{r}) u_{mw}(\mathbf{r}) \\ &+ \frac{1}{2}\rho^2 g^{(1)}(\mathbf{r}) \int d\mathbf{r}' \left[g^{(1)}(\mathbf{r}') - 1\right] g^{(2)}(\mathbf{r}, \mathbf{r}') u_{ww}(\mathbf{r}, \mathbf{r}') \\ &+ k_B T \rho g^{(1)}(\mathbf{r}) \ln g^{(1)}(\mathbf{r}) \\ &+ \frac{1}{2} k_B \rho^2 g^{(1)}(\mathbf{r}) \int d\mathbf{r}' \left[g^{(1)}(\mathbf{r}') - 1\right] \\ &\times \left[g^{(2)}(\mathbf{r}, \mathbf{r}') \ln g^{(2)}(\mathbf{r}, \mathbf{r}') - g^{(2)}(\mathbf{r}, \mathbf{r}') + 1\right] \end{aligned} \tag{4.58}$$

The main advantage of the definitions in Eqs. (4.57) and (4.58) is that they give a direct relation between the solvation free energy and structure of the solvent around the macromolecule immersed on it, and a detailed decomposition of the solvation free energy. This definition of the solvation free energy has already been applied to determine the thermodynamics of hydrophobic hydration (Lazaridis 2000; Lazaridis and Paulaitis 1992, 1994) and solvation in simple fluids (Lazaridis 1998a, 2001), and it has proven to be useful for understanding of the solvation process.

4.6 Energy Function

As we discussed above, one of the main characteristics on the thermodynamics of the biological processes is the potential energy function, which will lead in the calculation of the Hamiltonian function given by Eq. (4.6) in terms of the atomic coordinates.

The potential energy function of macromolecules must be accurate, but at the same time must be fast in the evaluation. Thus simple models are used in practice to express the potential energy function mathematically. In general, it is defined by a calibration procedure, which consists of fitting it to either experimental or quantum mechanical data.

The potential energy function $U(\mathbf{R}^M)$ of the macromolecule as a function of the atomic coordinates, $\mathbf{R}^M$, has the form

$$\begin{aligned} U(\mathbf{R}^M) = & \sum_{\text{bonds}} k_b(b-b_0)^2 + \sum_{\text{angles}} k_\theta(\theta-\theta_0)^2 \\ & + \sum_{\text{dihedrals}} k_\phi(1+\cos(n\phi-\delta)) + \sum_{\text{impropers}} k_\eta(\eta-\eta_0)^2 \\ & + \sum_{i>j} \varepsilon_{ij}\left[\left(\frac{R_{ij}^{(min)}}{r_{ij}}\right)^{12} - 2\left(\frac{R_{ij}^{(min)}}{r_{ij}}\right)^{6}\right] + \frac{q_i q_j}{4\pi\varepsilon\varepsilon_0 r_{ij}} \end{aligned} \tag{4.59}$$

where k_b, k_θ, k_ϕ, and k_η are the bond stretching, angle bending, dihedral angle, and improper dihedral angle force constants, respectively. b_0, θ_0, δ and η_0 are the equilibrium values of the bond length, angle bending, dihedral angle, and improper dihedral angle, respectively; while b, θ, ϕ and η are the values of the bond length, angle bending, dihedral angle, and improper dihedral angle, respectively, which are all functions of the coordinates $\mathbf{R}^M$.

Note that dihedral term depends on the parameters n and δ, which are, respectively, the multiplicity and the phase, The non-bounded interactions include the Lennard-Jones potential 12-6 and the electrostatic interactions. Here, ε_{ij} is the Lennard-Jones well-depth and $R_{ij}^{(min)}$ is the distance at the Lennard-Jones minimum, as illustrated graphically in Fig. 4.1.

The electrostatic interactions are characterized by the Coulomb interaction between two partial atomic charges q_i and q_j separated by a distance r_{ij}, ε_0 is the vacuum permitability and ε is the dielectric constant. The Lennard-Jones

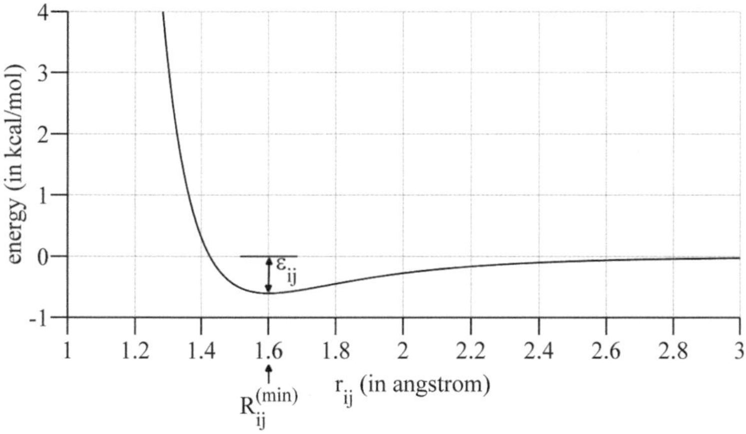

Fig. 4.1 A plot of the Lennard-Jones 12-6 potential; ε_{ij} indicates the well-depth and $R_{ij}^{(min)}$ is the distance of the potential energy function minimum

parameters between pairs of different atoms are obtained using the Lorentz-Berthelot combination rules:

$$\varepsilon_{ij} = \sqrt{\varepsilon_i \varepsilon_j}, \quad R_{ij}^{(min)} = \frac{1}{2}\left(R_i^{(min)} + R_j^{(min)}\right)$$

which are known as the geometric and arithmetic rules, respectively. The form of the potential energy function given by Eq. (4.59) is simple and easy to calculate, in particular, the derivatives for the Cartesian coordinates. Moreover, simplicity still guarantees the accuracy of the macromolecular properties representation. Use of the harmonic terms for the internal motions is sufficient for main condensed phase simulations, which run around the room temperatures. The parameters are determined by a calibration procedure, which tries to reproduce the experimental pure liquid, solution and crystal data, and sometimes also the quantum mechanical results.

Note that there are different forms of the potential energy function given by Eq. (4.59), which differ slightly from each other; mainly from additional of other cross terms for bonded interactions and the protocol used during the calibration for determination of the force constants. Equation (4.59) determines the form of the potential energy function defined in CHARMM program (Brooks et al. 2009) with parameters developed as in Dinner and Karplus (1998). Next chapters discuss the details of parametrization and explanation of every term.

Chapter 5
Free Energy Calculation Methods Used in Computer Simulations

In this chapter, we will present the most advanced methods used in the calculation of free energy from the computer simulations. First, in this chapter, we will discuss the methods employed in molecular dynamics simulations using explicit solvent models, such as the thermodynamic free energy perturbation method, thermodynamic integration method, and slow growth method. Then, the implicit solvation models will be discussed using either Poisson-Boltzmann or Generalized Born approximation for treating the electrostatic interactions. Besides, in this chapter, we will discuss rigorous methods used on the free energy decomposition for predicting the contributions from different physical terms into the total free energy value.

5.1 The Free Energy Calculations

There exist several methods for calculation of absolute free energy of (bio)molecules immersed in an explicit solvent solution, discussed in the following.

5.1.1 Thermodynamic Perturbation Method

Let us consider two systems characterized by states 1 and 2, respectively, with Hamiltonian $H_1(\mathbf{r}^N, \mathbf{p}^N)$ and $H_2(\mathbf{r}^N, \mathbf{p}^N)$, where N is the number of particles in the system. Then, the free energy difference between the two systems is given

$$\Delta F = F_2 - F_1 \tag{5.1}$$

H. Kamberaj, *Molecular Dynamics Simulations in Statistical Physics: Theory and Applications*, Scientific Computation,
https://doi.org/10.1007/978-3-030-35702-3_5

Using the thermodynamic relationship between the Helmholtz free energy and partition function:

$$F = -k_B T \ln Q(T)$$

we obtain

$$\Delta F = -k_B T \ln\left(\frac{Q_2}{Q_1}\right) \tag{5.2}$$

or

$$\begin{aligned}
\Delta F &= -k_B T \ln\left(\frac{\int d^N\mathbf{r} \int d^N\mathbf{p} \exp(-H_2/k_B T)}{\int d^N\mathbf{r} \int d^N\mathbf{p} \exp(-H_1/k_B T)}\right) \\
&= -k_B T \ln\left(\frac{\int d\boldsymbol{\Gamma} \exp\left(-\frac{H_2}{k_B T}\right) \exp\left(-\frac{H_1}{k_B T}\right) \exp\left(+\frac{H_1}{k_B T}\right)}{\int d\boldsymbol{\Gamma} \exp\left(-\frac{H_1}{k_B T}\right)}\right) \\
&= -k_B T \ln\left(\frac{\int d\boldsymbol{\Gamma} \exp\left(-\frac{H_2 - H_1}{k_B T}\right) \exp\left(-\frac{H_1}{k_B T}\right)}{\int d\boldsymbol{\Gamma} \exp\left(-\frac{H_1}{k_B T}\right)}\right) \\
&= -k_B T \ln\langle \exp\left(-\frac{H_2 - H_1}{k_B T}\right)\rangle_1
\end{aligned} \tag{5.3}$$

where $\langle \cdots \rangle_1$ indicates an ensemble average over the configurations with initial state 1 and $d\boldsymbol{\Gamma} = d^N\mathbf{r}\, d^N\mathbf{p}$. This is known as forward thermodynamic perturbation method developed by Zwanzig (1954).

Similarly, we can perform the averaging over configurations corresponding to the state 2 as:

$$\begin{aligned}
\Delta F &= +k_B T \ln\left(\frac{\int d^N\mathbf{r} \int d^N\mathbf{p} \exp(-H_1)/k_B T)}{\int d^N\mathbf{r} \int d^N\mathbf{p} \exp(-H_2/k_B T)}\right) \\
&= +k_B T \ln\left(\frac{\int d\boldsymbol{\Gamma} \exp\left(-\frac{H_1}{k_B T}\right) \exp\left(-\frac{H_2}{k_B T}\right) \exp\left(+\frac{H_2}{k_B T}\right)}{\int d\boldsymbol{\Gamma} \exp\left(-\frac{H_2}{k_B T}\right)}\right) \\
&= +k_B T \ln\left(\frac{\int d\boldsymbol{\Gamma} \exp\left(-\frac{H_1 - H_2}{k_B T}\right) \exp\left(-\frac{H_2}{k_B T}\right)}{\int d\boldsymbol{\Gamma} \exp\left(-\frac{H_2}{k_B T}\right)}\right) \\
&= +k_B T \ln\langle \exp\left(-\frac{H_1 - H_2}{k_B T}\right)\rangle_2
\end{aligned} \tag{5.4}$$

which simulates the reverse process. In general, to perform the thermodynamic perturbation simulation one has first to define H_1 and H_2 and then run the simulation at state 1 and calculate the average given by Eq. (5.3). Similarly, the simulations could start from state 2 and perform the average given by Eq. (5.4).

Note that if the probability distributions of the two states 1 and 2 do not overlap the free energy calculations by either using formula given by Eq. (5.3) or Eq. (5.4), the value of free energy difference ΔF will not be evaluated accurately. This is because the state 2 will not be sampled efficiently when starting the simulations from state 1, and vice-versa. This happens, in particular, when the energy difference between the two states is much larger than $k_B T$, i.e.,

$$\mid H_2 - H_1 \mid \gg k_B T$$

In order to obtain accurate calculations of the free energy differences, an intermediate state between 1 and 2 is introduced, called here *reference state*, with Hamiltonian H_r and free energy A_r. Then, the free energy difference can be written as:

$$\begin{aligned} \Delta F &= F_2 - F_1 \\ &= (F_2 - F_r + F_r - F_1) \\ &= -k_B T \ln\left(\frac{Q_2}{Q_r} \cdot \frac{Q_r}{Q_1}\right) \\ &= -k_B T \ln\langle \exp\left(-\beta(H_2 - H_r)\right)\rangle_r \\ &+ k_B T \ln\langle \exp\left(-\beta(H_1 - H_r)\right)\rangle_r \end{aligned} \tag{5.5}$$

where $\langle \cdots \rangle_r$ denotes an ensemble average over the configurations generates starting from state r. Using this method, the sampling can be improved, and so more reliable value can be obtained, if the reference state is such that its distribution overlaps with distributions of the states 1 and 2.

In principle, we can determine more than one reference state, for example, N such states can be defined with Hamiltonian $H_r^{(1)}, \cdots, H_r^{(N)}$, then the free energy difference is given by

$$\begin{aligned} \Delta F &= F_2 - F_1 \\ &= \left(F_2 - \sum_{i=1}^{N} F_r^{(i)} + \sum_{i=1}^{N} F_r^{(i)} - F_1\right) \\ &= -k_B T \ln\left(\frac{Q_2}{Q_r^{(N)}} \cdot \frac{Q_r^{(N)}}{Q_r^{(N-1)}} \cdot \cdots \cdot \frac{Q_r^{(2)}}{Q_r^{(1)}} \frac{Q_r^{(1)}}{Q_1}\right) \\ &= -k_B T \ln\langle \exp\left(-\beta(H_2 - H_r^{(N)})\right)\rangle_{(N)} \end{aligned} \tag{5.6}$$

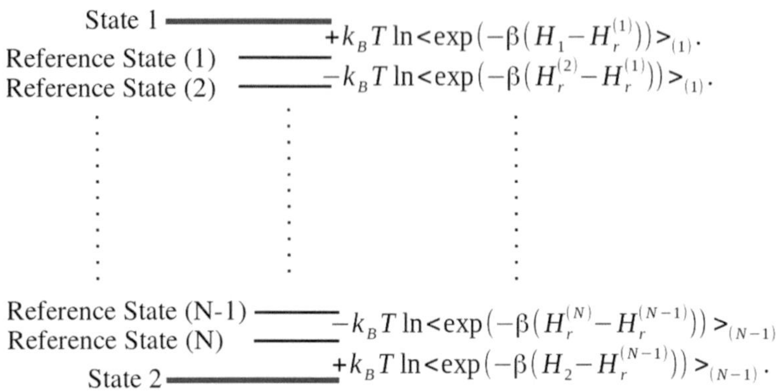

Fig. 5.1 Free energy difference calculation using multiple simulation run from the reference states (1), (2), $\cdots$, (N)

$$+ k_B T \ln\langle\exp\left(-\beta(H_r^{(N-1)} - H_r^{(N)})\right)\rangle_{(N)}$$

$$- \cdots$$

$$- k_B T \ln\langle\exp\left(-\beta(H_r^{(2)} - H_r^{(1)})\right)\rangle_{(1)}$$

$$+ k_B T \ln\langle\exp\left(-\beta(H_1 - H_r^{(1)})\right)\rangle_{(1)}$$

This simulation can significantly improve the sampling of phase space and in this way it can also improve the accuracy of free energy difference calculations, especially, when the two states 1 and 2 are separated by many intermediate metastable states. Practically, we can start multiple simulation runs using different initial conditions, each having as a starting configuration one of the intermediate reference states, 1 to N, as illustrated graphically in Fig. 5.1.

Note that the efficiency of this method depends on the overlap of distributions between neighboring reference states, and it also depends on the overlap between distributions of state 1 and reference state (1) and between state 2 and the reference state (N). However, in practice, the distribution of phase space for a system is not known a priory. Therefore, the choice of reference states is crucial in improving the accuracy of the method.

To overcome the above-mentioned problems, different simulation techniques can be suggested. For instance, it can be suggested to run multiple copies of the same system starting with different initial conditions and configurations from the reference states (1), (2), $\cdots$, (N), called *replica*. Then, we can frequently at regular time intervals swap the configurations of neighboring runs using energy criteria to ensure the detailed balance. This way, we will have the configurations from different simulation runs *traveling* along the replicas increasing overlap of energy distribution between the neighboring replicas. Acceptance probability of an attempting configuration swap between two neighboring replicas, let us say i and j, can be calculated as

$$p_{acc} = \min\{1, \exp\left(-\beta(E_i - E_j)\right)\} \tag{5.7}$$

where E_i and E_j are the configuration energies of replica i and j, respectively. To create a *pathway* of multiple reference states, an order parameter λ has often been assigned to the transformation path, which is gradually changed from 0 to 1. This parameter is used to describe progress in the transformation relative to the end states by parameterizing the Hamiltonian of the system at some intermediate reference state as the following

$$H_\lambda(\mathbf{r}^N, \mathbf{p}^N) = \lambda H_2(\mathbf{r}^N, \mathbf{p}^N) + (1-\lambda)H_1(\mathbf{r}^N, \mathbf{p}^N) \tag{5.8}$$

It can be seen that for $\lambda = 0$, $H_\lambda(\mathbf{r}^N, \mathbf{p}^N) = H_1(\mathbf{r}^N, \mathbf{p}^N)$, and for $\lambda = 1$, we get $H_\lambda(\mathbf{r}^N, \mathbf{p}^N) = H_2(\mathbf{r}^N, \mathbf{p}^N)$. Then, the simulations are performed at different values of λ between 0 and 1 generating in this way a series of reference states to determine ensemble averages from which free energy differences are calculated using Eq. (5.6).

In general, the empirical force field potential energy function used to describe the inter- and intramolecular interactions of system is given by Eq. (4.59) (in Chap. 4). Each of these terms in the potential energy function is modified at an intermediate reference state λ and written as a linear combination of the values for states 1 and 2:

$$\begin{aligned}
k_{b,i}(\lambda) &= \lambda k_{b,i}(2) + (1-\lambda)k_{b,i}(1) \\
b_{i,0}(\lambda) &= \lambda b_{i,0}(2) + (1-\lambda)b_{i,0}(1) \\
k_{\theta,i}(\lambda) &= \lambda k_{\theta,i}(2) + (1-\lambda)k_{\theta,i}(1) \\
\theta_{i,0}(\lambda) &= \lambda \theta_{i,0}(2) + (1-\lambda)\theta_{i,0}(1) \\
k_{\phi,i}(\lambda) &= \lambda k_{\phi,i}(2) + (1-\lambda)k_{\phi,i}(1) \\
\phi_i(\lambda) &= \lambda \phi_i(2) + (1-\lambda)\phi_i(1) \\
\delta_i(\lambda) &= \lambda \delta_i(2) + (1-\lambda)\delta_i(1) \\
k_{\eta,i}(\lambda) &= \lambda k_{\eta,i}(2) + (1-\lambda)k_{\eta,i}(1) \\
\eta_{i,0}(\lambda) &= \lambda \eta_{i,0}(2) + (1-\lambda)\eta_{i,0}(1) \\
\varepsilon_{ij}(\lambda) &= \lambda \varepsilon_{ij}(2) + (1-\lambda)\varepsilon_{ij}(1) \\
R_{ij}^{(min)}(\lambda) &= \lambda R_{ij}^{(min)}(2) + (1-\lambda)R_{ij}^{(min)}(1) \\
q_i(\lambda) &= \lambda q_i(2) + (1-\lambda)q_i(1)
\end{aligned} \tag{5.9}$$

where the subscript i runs overall elements of each potential energy term and the subscript ij runs overall pairs of the atoms i and j.

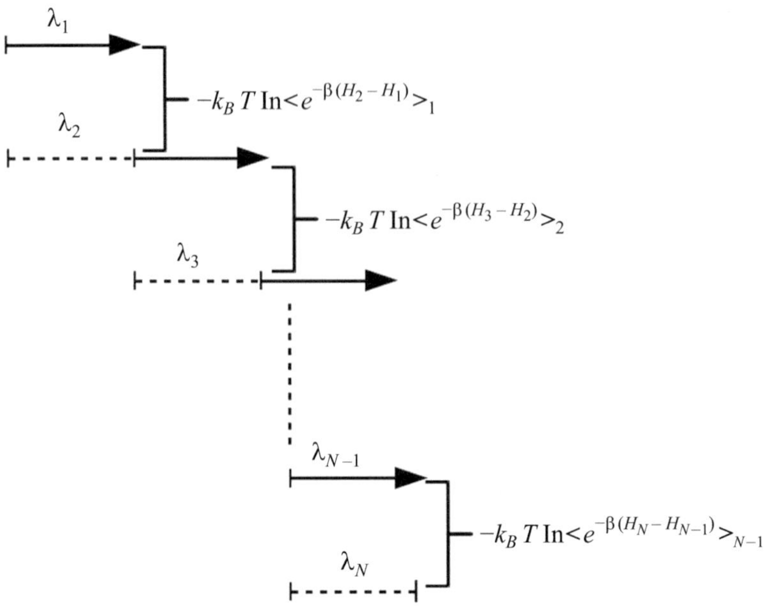

Fig. 5.2 Forward free energy difference calculation using λ pathway from state 1 to 2

For each value λ_i we can perform a molecular dynamics simulation with corresponding force field parameters scaled according to an appropriate value of λ_i as in Eq. (5.9). Figure 5.2 shows a setup of the free energy difference calculated using molecular dynamics simulations. For every simulation at a value of λ_i, first the system is equilibrated, then a production run is generated during which the free energy difference $\Delta F(\lambda_i \rightarrow \lambda_{i+1})$ is calculated according to the ensemble average:

$$\Delta F(\lambda_i \rightarrow \lambda_{i+1}) = -\frac{1}{\beta} \ln \langle \exp\left(-\beta(H_{i+1} - H_i)\right) \rangle_i \tag{5.10}$$

where $\langle \cdots \rangle_i$ denotes an ensemble average with initial distribution from the simulation run with λ_i. For $\lambda_1 = 0$, the initial configuration is random and the initial velocities randomly chosen from the Maxwell-Boltzmann distribution. For every other λ_i, the initial state is chosen from the last previous run for λ_{i-1}. Then the free energy difference between the states 1 and 2 is given by

$$\Delta F = \sum_{i=1}^{N-1} \Delta F(\lambda_i \rightarrow \lambda_{i+1}) \tag{5.11}$$

The approach described above is also called the *forward sampling*, i.e., the free energy difference is calculated based on transition $\lambda_i \rightarrow \lambda_{i+1}$. Similarly,

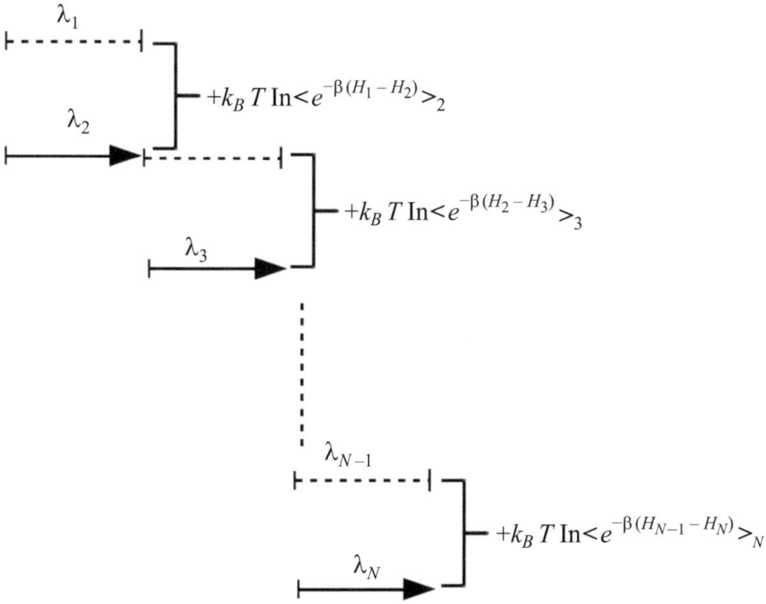

Fig. 5.3 Backward free energy difference calculation using λ pathway from state 1 to 2

the *backward sampling* approach can also be introduced, in which the free energy difference between λ_i and λ_{i-1} is calculated. Also in this approach, λ changes from 0 to 1, and the total free energy difference is

$$\Delta F = \sum_{i=2}^{N} \Delta F(\lambda_i \rightarrow \lambda_{i-1})$$

where

$$\Delta F(\lambda_i \rightarrow \lambda_{i-1}) = +k_B T \ln\langle \exp(-\beta(H_{i-1} - H_i))\rangle_i$$

The simulation diagram is shown in Fig. 5.3.

Note that although the formula for forward and backward analysis are formally equivalent, their convergences may be different (Widom 1963). Therefore, there could be a preferred direction to carry out the required transformation between the states.

In *double-wide sampling* approach, the free energy difference is obtained for transitions $\lambda_i + \Delta\lambda/2 \rightarrow \lambda_i$ and $\lambda_i + \Delta\lambda/2 \rightarrow \lambda_{i+1}$ using the simulations shown in Fig. 5.4, where $\Delta\lambda$ is the step on change in λ. This approach produces for each λ_i two free energy differences, $\Delta F(\lambda_{i-1/2} \rightarrow \lambda_i)$ and $\Delta F(\lambda_i \rightarrow \lambda_{i+1/2})$, which is more efficient approach of calculating the desired free energy since two free energy

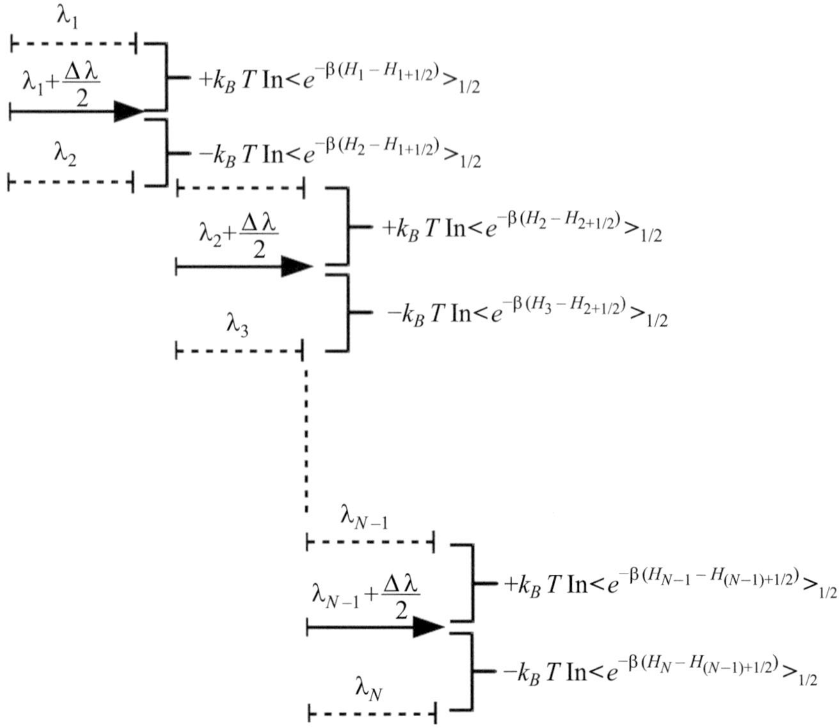

Fig. 5.4 Double-wide sampling free energy difference calculation using λ pathway from state 1 to 2

difference calculations are produced from a single simulation. The total free energy difference is then calculated as

$$
\begin{aligned}
\Delta F &= \Delta F(\lambda_{i-1/2} \to \lambda_i) + \Delta F(\lambda_i \to \lambda_{i+1/2}) \qquad (5.12)\\
&= \sum_{i=1}^{N-1} \left[+k_B T \ln\langle \exp\left(-\beta(H_i - H_{i+1/2})\right)\rangle_{i+1/2} \right.\\
&\left. -k_B T \ln\langle \exp\left(H_{i+1} - H_{i+1/2}\right)\rangle_{i+1/2} \right]\\
&= -k_B T \sum_{i=1}^{N-1} \ln \frac{\langle \exp\left(H_{i+1} - H_{i+1/2}\right)\rangle_{i+1/2}}{\langle \exp\left(-\beta(H_i - H_{i+1/2})\right)\rangle_{i+1/2}}
\end{aligned}
$$

It can be seen that in all calculations, ΔF depends on the average of a quantity which is a function of $\Delta H = H_{i+1} - H_i$. In general, we can take this average as an integral of $\exp(-\beta \Delta H)$ weighted with probability distribution of ΔH $P_i(\Delta H)$:

$$\Delta F_i = -k_B T \ln \int \exp(-\beta \Delta H) P_i(\Delta H)\, d(\Delta H) \tag{5.13}$$

If H_i and H_{i+1} would be functions of some number of identically distributed random variables, then ΔH would have a Gaussian distribution:

$$P_i(\Delta H) = \frac{1}{\sqrt{2\pi\sigma^2}} \exp\left(-\frac{(\Delta H - \langle \Delta H \rangle_i)^2}{2\sigma^2}\right) \tag{5.14}$$

where

$$\sigma^2 = \langle \Delta H^2 \rangle_i - (\langle \Delta H \rangle_i)^2$$

However, $P_i(\Delta H)$, in general, may have slightly different shape from the Gaussian distribution, although it is close to a Gaussian-like shape. Replacing Eq. (5.14) into Eq. (5.13), we get

$$\Delta F_i = \langle \Delta H \rangle_i - \frac{\beta}{2}\sigma^2 \tag{5.15}$$

where the first term is the average of ΔH measured in the reference state i and the second term depends on the fluctuations of ΔH. While the first term can be positive or negative, the second one is always negative. Therefore, the accuracy in measuring ΔF_i depends on the balance between two terms in Eq. (5.15). We can write Eq. (5.15) as

$$\Delta F_i = \langle \Delta H \rangle_i - \gamma\sigma \tag{5.16}$$

where

$$\gamma = \frac{\sigma}{2k_B T}$$

Thus, for $\sigma = nk_B T$, where n integer number, we get

$$\Delta F_i = \langle \Delta H \rangle_i - \frac{n}{2}\sigma \tag{5.17}$$

Simple calculations show that for $n = 1$ (i.e., $\sigma = k_B T$), then 95% of the values of ΔH fall in the region $\langle \Delta H \rangle_i \pm 2\sigma$, and from Eq. (5.17) we can see that $\Delta F_i = \langle \Delta H \rangle_i - \sigma/2$, which falls inside the region where most of the ΔH are sampled. That is, the simulations will result in accurate measure of ΔF_i. However, for $n > 4$ (i.e., $\sigma > 4k_B T$), more than 97% of the ΔH values fall in the region $\langle \Delta H \rangle_i \pm 2\sigma$, and from Eq. (5.17) we can see that $\Delta F_i < \langle \Delta H \rangle_i - 2\sigma$, which falls outside the region where most of the ΔH are sampled, hence, in this case inaccurate measure of ΔF_i will be produced due to the sampling inefficiency.

Thus, the use of free energy perturbation method implies that $P_i(\Delta H)$ should have small fluctuations around the mean value of ΔH, although this does not imply that the free energy difference between the reference states i and $i+1$ must be small. Depending also on the problem of interest, choosing intermediate reference states separated by a constant $\Delta\lambda$ may not be the best possible way. Therefore, in practice, both N (number of reference states) and $\Delta\lambda_i$ have to be optimized. For example, one can start with a large value of N, and then optimize N and $\Delta\lambda_i$ from initial runs such that $P(\Delta H_{i,i+1})$ are sufficiently small and approximately equal (Pearlman and Kollman 1989).

In order to improve the sampling, so-called *enveloping distribution sampling* (EDS), can be used. In this method, instead of performing the simulations using Hamiltonian of any two reference states 1 and 2, we can simulate using an effective Hamiltonian given as

$$H_{eff}(\mathbf{r}^N, \mathbf{p}^N) = -\frac{1}{\beta} \ln \left(\exp\left(-\beta H_1(\mathbf{r}^N, \mathbf{p}^N)\right) + \exp\left(-\beta H_2(\mathbf{r}^N, \mathbf{p}^N)\right)\right) \tag{5.18}$$

As an illustration, we considered a two states system, namely the state 1 with a potential energy function

$$U_1(x) = k_1(x+15)^2$$

and state 2 with the potential energy function

$$U_2(x) = k_2(x-15)^2$$

with $k_1 = k_2 = 0.1$ kJ/mol. These two states can represent two Hamiltonian systems with probability distributions centered around $x = -15$ and $x = 15$, respectively, in one-dimensional space. We performed two different molecular dynamics simulation runs. The first one run corresponds to standard molecular dynamics simulation starting from state 1 or 2, and the second one corresponds to molecular dynamics simulation run with EDS starting either from state 1 or 2 with effective potential energy function given by

$$U_{eff}(x) = -\frac{1}{\beta} \ln\left(\exp\left(-\beta U_1(x)\right) + \exp\left(-\beta U_2(x)\right)\right)$$

In Fig. 5.5 we present the results of molecular dynamics simulation runs at constant temperature $T = 600$ K. In Fig. 5.5a for both simulations, standard and EDS, the runs started from the state 1, i.e., initially $x = -15$, and in Fig. 5.5b the results are presented for simulations with initial position $x = +15$. The initial velocities sampled from the Maxwell-Boltzmann distribution at $T = 600$ K. The results show

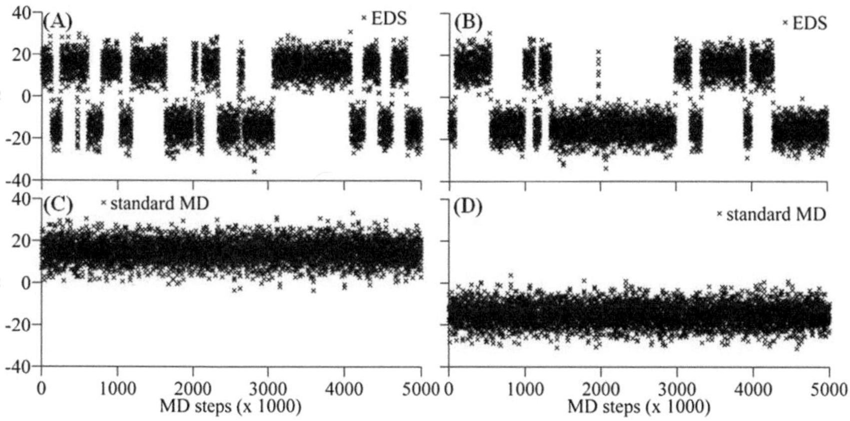

Fig. 5.5 Position distribution of simulation run for standard and EDS molecular dynamics simulations for a two states system with potential energy functions $U_1(x) = k_1(x + 15)^2$ and $U_2(x) = k_2(x - 15)^2$, respectively. EDS molecular dynamics (**a**) started from the same configuration $x = +15$ and (**b**) from $x = -15$. Standard molecular dynamics (**c**) started from the same configuration $x = +15$ and (**d**) from $x = -15$. Velocity according to the Maxwell-Boltzmann distribution at temperature $T = 600\,\text{K}$, and $k_1 = k_2 = 0.1\,\text{kJ/mol}$

that regardless of which state the simulations started both states (1 and 2) are visited during the EDS molecular dynamics simulations. In contrast, in the case of standard molecular dynamics simulations runs, the system was not able to visit the other state (see also Fig. 5.5). These results indicate that molecular dynamics simulations with EDS provide a better sampling of the configuration space and can significantly improve the accuracy of free energy difference calculations using the thermodynamic perturbation method.

5.1.2 *Thermodynamic Integration Method*

Another used method for calculation of free energy differences is the so-called *thermodynamic integration* approach. Using this method, the free energy difference between the two states 1 and 2 is given by

$$\Delta F = \int_0^1 \left(\frac{\partial F_\lambda}{\partial \lambda} \right)_T d\lambda \qquad (5.19)$$

where λ is the coupling parameter of the Hamiltonian

$$H_\lambda(\mathbf{r}^N, \mathbf{p}^N) = (1 - \lambda) H_1(\mathbf{r}^N, \mathbf{p}^N) + \lambda H_2(\mathbf{r}^N, \mathbf{p}^N) \qquad (5.20)$$

The Helmholtz free energy is related to the partition function $Q(\beta)$ as

$$F_\lambda = -\frac{1}{\beta}\ln Q_\lambda(\beta)$$

where

$$Q_\lambda(\beta) = \frac{1}{N!h^{3N}}\int d^N\mathbf{r}\int d^N\mathbf{p}\exp\left(-\beta H_\lambda(\mathbf{r}^N,\mathbf{p}^N)\right)$$

Then, we can first calculate the derivative of Helmholtz free energy with respect to λ as

$$\begin{aligned}\left(\frac{\partial F_\lambda}{\partial\lambda}\right)_T &= -\frac{1}{\beta}\frac{1}{Q_\lambda}\left(\frac{\partial Q_\lambda(\beta)}{\partial\lambda}\right)_T \\ &= \frac{1}{Q_\lambda}\frac{1}{N!h^{3N}}\int d\boldsymbol{\Gamma}\exp\left(-\beta H_\lambda(\mathbf{r}^N,\mathbf{p}^N)\right)\left(\frac{\partial H_\lambda}{\partial\lambda}\right) \\ &= \langle\frac{\partial H_\lambda}{\partial\lambda}\rangle_\lambda\end{aligned} \tag{5.21}$$

Replacing Eq. (5.21) into Eq. (5.19), we obtain

$$\Delta F = \int_0^1\langle\frac{\partial H_\lambda}{\partial\lambda}\rangle_\lambda\, d\lambda \tag{5.22}$$

where $\langle\cdots\rangle_\lambda$ denotes the ensemble average at the reference state with a particular λ.

Equation (5.22) indicates that to calculate free energy differences we need to perform numerically the integral with respect to λ. This practically is done by performing several simulation runs at different λ_i from 0 to 1: $\lambda_1, \lambda_2, \cdots, \lambda_K$. For each λ_i, the ensemble average of the following quantity is calculated:

$$\langle\frac{\partial H_\lambda}{\partial\lambda}\rangle_{\lambda_i}$$

Then, using simple numerical integration, Eq. (5.22) can be approximated as

$$\Delta F \approx \sum_{i=1}^{K}\langle\frac{\partial H_\lambda}{\partial\lambda}\rangle_{\lambda_i}\Delta\lambda_i \tag{5.23}$$

Other numerical integration approaches can also be used, such as, the trapezoidal method

$$\Delta F \approx \sum_{i=1}^{K-1}\left[\langle\frac{\partial H_\lambda}{\partial\lambda}\rangle_{\lambda_i} + \langle\frac{\partial H_\lambda}{\partial\lambda}\rangle_{\lambda_{i+1}}\right]\frac{\Delta\lambda_i}{2} \tag{5.24}$$

or Gaussian quadrature approach:

$$\Delta F \approx \sum_{i=1}^{K} \langle \frac{\partial H_\lambda}{\partial \lambda} \rangle_{\lambda_i} w_i \tag{5.25}$$

where w_i are the Gaussian's weights, which depend on K.

Note that in principle thermodynamic integration can be performed for any expression of H_λ as a function of λ, including nonlinear coupling between H_1 and H_2, as long as the function is differentiable and satisfies the boundary conditions:

$$H_{\lambda=0} = H_1, \;\; H_{\lambda=1} = H_2$$

However, the linearity of λ dependence of H_λ has a property that makes this expression practically very convenient, that is the sign of second derivative $\partial^2 F/\partial\lambda^2$ is a known non-positive value. In particular, it is as:

$$\left(\frac{\partial^2 F}{\partial \lambda^2}\right)_{N,V,T} = -\beta \left[\langle (H_2 - H_1)^2 \rangle_\lambda - \langle H_2 - H_1 \rangle_\lambda^2 \right] \leq 0$$

That means, $\partial F/\partial\lambda$ can never increase with increasing λ. That is also known as the Gibbs-Bogoliubov inequality, and it can be used to validate the accuracy of free energy difference calculation results.

It is interesting to note that performing numerical integration requires that the integrand in Eq. (5.22) must be a continuous function of λ. However, this property could not always be satisfied with a linear parameterization of H_λ. In particular, as it is most of the time the case, λ parametrization is applied to the potential energy function as

$$U_\lambda = (1 - \lambda)U_1 + \lambda U_2$$

It could happen that for $\lambda = 0$, U_λ exhibit a singularity in Eq. (5.22) for $\lambda = 0$, which can be avoided by using special techniques (Frenkel and Smit 2001).

5.1.3 The Slow Growth Method

The other method for calculation of free energy differences by means of computer simulations is *slow growth* approach. In this method, the Hamiltonian changes by a very small constant amount at each step of the calculation. In other words, the changes $H_{\lambda_{i+1}} - H_{\lambda_i}$ are very small and constant. The free energy difference can be derived using the free energy perturbation expression:

$$\Delta F = -k_B T \sum_{i=1}^{K-1} \ln\langle \exp\left(-\beta(H_{i+1} - H_i)\right)\rangle_{\lambda_i,NVT} \tag{5.26}$$
$$\approx -k_B T \sum_{i=1}^{K-1} \ln\langle 1 - \beta(H_{i+1} - H_i) + \cdots\rangle_{\lambda_i,NVT}$$
$$\approx -k_B T \sum_{i=1}^{K-1} \ln\left[1 - \beta\langle(H_{i+1} - H_i)\rangle_{\lambda_i,NVT} + \cdots\right]$$
$$\approx \sum_{i=1}^{K-1} \langle(H_{\lambda_{i+1}} - H_{\lambda_i})\rangle_{\lambda_i,NVT}$$

where $\langle\cdots\rangle_{\lambda_i,NVT}$ denotes an ensemble average for constant NVT simulations with Hamiltonian of the system determined by $H(\lambda = \lambda_i)$.

5.2 Free Energy Decomposition

In this section, we will describe a method for a decomposition of the free energy into different terms, such as the terms belonging to the diverse group of atoms, or the other kind of interactions, as developed in Bren et al. (2006, 2007) for the perturbation method.

In particular, if the Hamiltonian difference in Eq. (5.10) can be decomposed into the terms that are either from different group of atoms or different types of interactions, then we can write:

$$\Delta H_i \equiv H_{i+1} - H_i = \sum_{g=1}^{N_g} \left(H_{i+1}^{(g)} - H_i^{(g)}\right) \equiv \sum_{g=1}^{N_g} \Delta H_i^{(g)} \tag{5.27}$$

where N_g denotes the number of decomposition terms and $\Delta H_i^{(g)}$ gives the difference of the Hamiltonian between the nearest reference states at the i window. This, of course, is valid if the force field does not include cross terms, which is the case of many biomolecular force fields, but for the polarized force fields and quantum-chemical calculations, it may not be the case.

If we can write Eq. (5.27), then expression given by Eq. (5.10) can be simplified as the following:

$$\Delta F(\lambda_i \to \lambda_{i+1}) = -\frac{1}{\beta} \ln\langle \exp\left(-\beta \sum_{g=1}^{N_g} \Delta H_i^{(g)}\right)\rangle_i \tag{5.28}$$
$$= -\frac{1}{\beta} \ln\langle \prod_{g=1}^{N_g} \exp\left(-\beta \Delta H_i^{(g)}\right)\rangle_i$$

If we assume that different terms of the Hamiltonian decomposition are additionally independent, which is not the case even for the pair-wise additive force fields given by Eq. (4.59) (in Chap. 4), because different energy terms include in their definition atoms that belong to other energy terms, then we can write that

$$\langle\prod_{g=1}^{N_g} \exp\left(-\beta \Delta H_i^{(g)}\right)\rangle_i = \prod_{g=1}^{N_g} \langle\exp\left(-\beta \Delta H_i^{(g)}\right)\rangle_i \tag{5.29}$$

Then, Eq. (5.28) can take the following form:

$$\begin{aligned}
\Delta F(\lambda_i \to \lambda_{i+1}) &= -\frac{1}{\beta} \ln \prod_{g=1}^{N_g} \langle\exp\left(-\beta \Delta H_i^{(g)}\right)\rangle_i \\
&= \sum_{g=1}^{N_g} -\frac{1}{\beta} \ln\langle\exp\left(-\beta \Delta H_i^{(g)}\right)\rangle_i \\
&= \sum_{g=1}^{N_g} \Delta F^{(g)}(\lambda_i \to \lambda_{i+1})
\end{aligned} \tag{5.30}$$

where $\Delta F^{(g)}(\lambda_i \to \lambda_{i+1})$ is the contribution to the total free energy from the term g of the Hamiltonian decomposition, which is calculated as:

$$\Delta F^{(g)}(\lambda_i \to \lambda_{i+1}) = -\frac{1}{\beta} \ln\langle\exp\left(-\beta \Delta H_i^{(g)}\right)\rangle_i \tag{5.31}$$

To estimate the error made by the two assumptions given by Eqs. (5.27) and (5.29), in Bren et al. (2007), the Thiele cumulants theory is used, which is briefly discussed in the following.

Any function $f(x)$ can be expressed in power series as:

$$f(x) = 1 + \sum_{n=1}^{\infty} \frac{\mu_n}{n!} x^n \tag{5.32}$$

where (!) denotes the factorial of an integer number, such as $n! = 1 \cdot 2 \cdot \cdots \cdot n$. Furthermore, the function $\ln f(x)$ can be expressed in power series as:

$$\ln f(x) = \sum_{n=1}^{\infty} \frac{\kappa_n}{n!} x^n \tag{5.33}$$

where κ_n are the so-called the *Thiele cumulants*:

$$\kappa_n = n! \sum_{s} S(n_1^{(s)}, \cdots, n_{K_s}^{(s)}) \tag{5.34}$$

where the sum runs over all sequences

$$S(n_1^{(s)}, \cdots, n_{K_s}^{(s)}) = (-1)^{-1+\sum_{k=1}^{K_s} n_k^{(s)}} \left(-1 + \sum_{k=1}^{K_s} n_k^{(s)}\right)! \times \prod_{k=1}^{K_s} \frac{\left(\frac{\mu_k}{k!}\right)^{n_k^{(s)}}}{n_k^{(s)}!} \tag{5.35}$$

and

$$\sum_{k=1}^{K_s} \left(k \cdot n_k^{(s)}\right) = n$$

Using Eq. (5.32), then Eq. (5.31) can be written as

$$\begin{aligned} \Delta F^{(g)}(\lambda_i \rightarrow \lambda_{i+1}) &= -\frac{1}{\beta} \ln\langle \sum_{n=0}^{\infty} \frac{(-\beta)^n}{n!} (\Delta H_i^{(g)})^n \rangle_i \\ &= -\frac{1}{\beta} \ln \sum_{n=0}^{\infty} \frac{(-\beta)^n}{n!} \langle (\Delta H_i^{(g)})^n \rangle_i \\ &= -\frac{1}{\beta} \ln \left(1 + \sum_{n=1}^{\infty} \frac{(-\beta)^n}{n!} \langle (\Delta H_i^{(g)})^n \rangle_i \right) \\ &= -\frac{1}{\beta} \sum_{n=1}^{\infty} \frac{(-\beta)^n}{n!} \kappa_n^{(g)} \end{aligned} \tag{5.36}$$

where the linearity of the expectation value is assumed, that is the expected value of the sum of terms is equal to the sum of the expectations of each term of the sum. Here, $\kappa_n^{(g)}$ is given by Eqs. (5.34) and (5.35) with $\mu_k = \langle (\Delta H_i^{(g)})^k \rangle_i$. Similarly, we can obtain in terms of the Thiele cumulants, the total free energy for the transition $\lambda_i \rightarrow \lambda_{i+1}$ as:

$$\Delta F(\lambda_i \rightarrow \lambda_{i+1}) = -\frac{1}{\beta} \sum_{n=1}^{\infty} \frac{(-\beta)^n}{n!} \kappa_n \tag{5.37}$$

κ_n is given by Eqs. (5.34) and (5.35) with $\mu_k = \langle (\Delta H_i)^k \rangle_i$.

Then, the error of order n, which gives an estimate of the error due to assumption of additive property of the Hamiltonian and independence between the terms of the decomposition, is given as

$$\begin{aligned}\epsilon_n(\lambda_i \to \lambda_{i+1}) &= \Delta F(\lambda_i \to \lambda_{i+1}) - \sum_{g=1}^{N_g} \Delta F^{(g)}(\lambda_i \to \lambda_{i+1}) \\ &= -\frac{1}{\beta}\sum_{n=1}^{\infty}\frac{(-\beta)^n}{n!}\kappa_n - \sum_{g=1}^{N_g} -\frac{1}{\beta}\sum_{n=1}^{\infty}\frac{(-\beta)^n}{n!}\kappa_n^{(g)} \\ &= -\frac{1}{\beta}\sum_{n=1}^{\infty}\frac{(-\beta)^n}{n!}\left(\kappa_n - \sum_{g=1}^{N_g}\kappa_n^{(g)}\right)\end{aligned} \tag{5.38}$$

We can calculate κ_n and $\kappa_n^{(g)}$ for $n = 1$ and $n = 2$. For $n = 1$, we have the following sequence of $\{n_k^s\}$:

$$\{n_1^{(1)}\} = \{1\}$$

Thus, there is just one sequence. Then, from Eq. (5.35) we obtain:

$$\kappa_1 = 1!\left((-1)^{-1+1}(-1+1)!\frac{\mu_1}{1!}\right) = \mu_1 = \langle \Delta H_i \rangle_i \tag{5.39}$$

and

$$\begin{aligned}\sum_{g=1}^{N_g}\kappa_1^{(g)} &= 1!\left((-1)^{-1+1}(-1+1)!\frac{\mu_1}{1!}\right) = \mu_1 = \sum_{g=1}^{N_g}\langle \Delta H_i^{(g)}\rangle_i \\ &= \langle\sum_{g=1}^{N_g}\Delta H_i^{(g)}\rangle_i = \langle \Delta H_i\rangle_i\end{aligned} \tag{5.40}$$

where the linearity of the expectation value is used. Combining, Eqs. (5.39) and (5.40), we obtain the first order error as:

$$\epsilon_1(\lambda_i \to \lambda_{i+1}) = 0$$

For $n = 2$, we have the following sequence of integers $\{n_k^{(s)}\}$:

$$\begin{aligned}\{n_1^{(1)}, n_2^{(1)}\} &= \{2, 0\} \\ \{n_1^{(2)}, n_2^{(2)}\} &= \{0, 1\}\end{aligned} \tag{5.41}$$

Then, the Thiele cumulants are given as:

$$\kappa_2 = 2!\left((-1)^{-1+2+0}(-1+2+0)!\frac{(\mu_1/1!)^2}{2!}\frac{(\mu_2/2!)^0}{0!} + (-1)^{-1+0+1}(-1+0+1)!\frac{(\mu_1/1!)^0}{0!}\frac{(\mu_2/2!)^1}{1!}\right) = -\mu_1^2 + \mu_2 \tag{5.42}$$

Substituting Eq. (5.42) into Eq. (5.38), we can obtain the second order error as:

$$\begin{aligned}
\epsilon_2(\lambda_i \to \lambda_{i+1}) &= -\frac{1}{\beta}\sum_{n=1}^{2}\frac{(-\beta)^n}{n!}\left(\kappa_n - \sum_{g=1}^{N_g}\kappa_n^{(g)}\right) \\
&= \epsilon_1(\lambda_i \to \lambda_{i+1}) - \frac{1}{\beta}\frac{(-\beta)^2}{2!}\left(\kappa_2 - \sum_{g=1}^{N_g}\kappa_2^{(g)}\right) \\
&= 0 - \frac{\beta}{2}\left(-(\langle\Delta H_i\rangle_i)^2 + \langle(\Delta H_i)^2\rangle_i - \sum_{g=1}^{N_g}\left(-\left(\langle\Delta H_i^{(g)}\rangle_i\right)^2 + \langle\left(\Delta H_i^{(g)}\right)^2\rangle_i\right)\right) \\
&= \frac{\beta}{2}\left(\left(\sum_{g=1}^{N_g}\langle\Delta H_i^{(g)}\rangle_i\right)^2 - \langle\left(\sum_{g=1}^{N_g}\Delta H_i^{(g)}\right)^2\rangle_i - \sum_{g=1}^{N_g}\left(\langle\Delta H_i^{(g)}\rangle_i\right)^2 + \sum_{g=1}^{N_g}\langle\left(\left(\Delta H_i^{(g)}\right)^2\right)\rangle_i\right) \\
&= \beta\sum_{g=1}^{N_g}\sum_{g'>g}^{N_g}\left(\langle\Delta H_i^{(g)}\rangle_i\langle\Delta H_i^{(g')}\rangle_i - \langle\Delta H_i^{(g)}\Delta H_i^{(g')}\rangle_i\right)
\end{aligned} \tag{5.43}$$

where the following binomial formulas were used:

$$\left(\sum_{g=1}^{N_g}\langle\Delta H_i^{(g)}\rangle_i\right)^2 = \sum_{g=1}^{N_g}\left(\langle\Delta H_i^{(g)}\rangle_i\right)^2 + 2\sum_{g=1}^{N_g}\sum_{g'>g}^{N_g}\langle\Delta H_i^{(g)}\rangle_i\langle\Delta H_i^{(g')}\rangle_i \tag{5.44}$$

$$
\begin{aligned}
\langle\left(\sum_{g=1}^{N_g}\Delta H_i^{(g)}\right)^2\rangle_i &= \sum_{g=1}^{N_g}\langle\left(\Delta H_i^{(g)}\right)^2\rangle_i \\
&+ 2\sum_{g=1}^{N_g}\sum_{g'>g}^{N_g}\langle\Delta H_i^{(g)}\Delta H_i^{(g')}\rangle_i
\end{aligned}
$$

and the linearity of the expectation value.

It can be seen that if the decomposition terms are independent, then

$$
\langle\Delta H_i^{(g)}\rangle_i\langle\Delta H_i^{(g')}\rangle_i = \langle\Delta H_i^{(g)}\Delta H_i^{(g')}\rangle_i \tag{5.45}
$$

and the second order error is zero, $\epsilon_2(\lambda_i \to \lambda_{i+1}) = 0$. It is worth noting that higher orders of the error include cross terms of higher order, such as

$$
\left(\langle\Delta H_i^{(g)}\rangle_i\right)^{\alpha}\left(\langle\Delta H_i^{(g')}\rangle_i\right)^{\nu} - \langle\left(\Delta H_i^{(g)}\right)^{\alpha}\left(\Delta H_i^{(g')}\right)^{\nu}\rangle_i \tag{5.46}
$$

where α and ν are integers, which are different from zero, if there are cross terms in the force fields, or if the decomposition terms are dependent. Therefore, the error increases with N_g and the number of the cross terms or dependent terms included in the decomposition.

An alternative use of the free energy decomposition is to use the thermodynamic integration method with the following Hamiltonian decomposition:

$$
\begin{aligned}
H_\lambda(\mathbf{r}^N,\mathbf{p}^N) &= (1-\lambda)\sum_{g=1}^{N_g} H_1^{(g)}(\mathbf{r}^N,\mathbf{p}^N) + \lambda\sum_{g=1}^{N_g} H_2^{(g)}(\mathbf{r}^N,\mathbf{p}^N) \\
&= \sum_{g=1}^{N_g}\left((1-\lambda)H_1^{(g)}(\mathbf{r}^N,\mathbf{p}^N) + \lambda H_2^{(g)}(\mathbf{r}^N,\mathbf{p}^N)\right) \\
&= \sum_{g=1}^{N_g} H_\lambda^{(g)}(\mathbf{r}^N,\mathbf{p}^N)
\end{aligned} \tag{5.47}
$$

where

$$
H_\lambda^{(g)}(\mathbf{r}^N,\mathbf{p}^N) = (1-\lambda)H_1^{(g)}(\mathbf{r}^N,\mathbf{p}^N) + \lambda H_2^{(g)}(\mathbf{r}^N,\mathbf{p}^N) \tag{5.48}
$$

Here, the assumption is that the same terms of the Hamiltonian decomposition exist in both $H_1(\mathbf{r}^N,\mathbf{p}^N)$ and $H_2(\mathbf{r}^N,\mathbf{p}^N)$. Then, using Eq. (5.19), we obtain

$$
\Delta F^{(g)} = \int_0^1 \langle\frac{\partial H_\lambda^{(g)}}{\partial\lambda}\rangle_\lambda\, d\lambda \tag{5.49}
$$

and

$$\Delta F = \sum_{g=1}^{N_g} \Delta F^{(g)} \tag{5.50}$$

5.3 Implicit Models for Free Energy Calculations

Computer simulations combined with implicit solvation models, such as Poisson-Boltzmann and Generalized Born methods, are widely used in free energy calculations, known in the literature, respectively, as Molecular Mechanics Poisson-Boltzmann Surface Area (MM/PBSA) and Molecular Mechanics Generalized Born Surface Area (MM/GBSA) (Kollman et al. 2000; Wang et al. 2001). An advantage of these two methods is the rigorous decomposition of free energy into contributions coming from the different group of atoms or even type of interactions (Gohlke et al. 2003).

The binding free energy between two (bio)molecules (e.g., proteins, protein and ligand, protein and DNA, etc.) A and B in water is evaluated based on the thermodynamic cycle depicted in Fig. 5.6.

In this approach binding free energy is estimated as a sum of some reference phase (often gas phase) energies, solvation free energies, and entropic contributions averaged over a series of snapshots from molecular dynamics simulations:

$$\Delta G_b^w = \langle \Delta G_b^g \rangle + \langle \Delta G_{AB}^{solv} \rangle - \langle \Delta G_A^{solv} \rangle - \langle \Delta G_B^{solv} \rangle \tag{5.51}$$

where $\langle \Delta G_b^g$ is the binding energy in the reference (e.g. gas) phase and $\langle \Delta G_{AB}^{solv} \rangle$, $\langle \Delta G_A^{solv} \rangle$, $\langle \Delta G_B^{solv} \rangle$ are the solvation free energies of (bio)molecules A, B and their

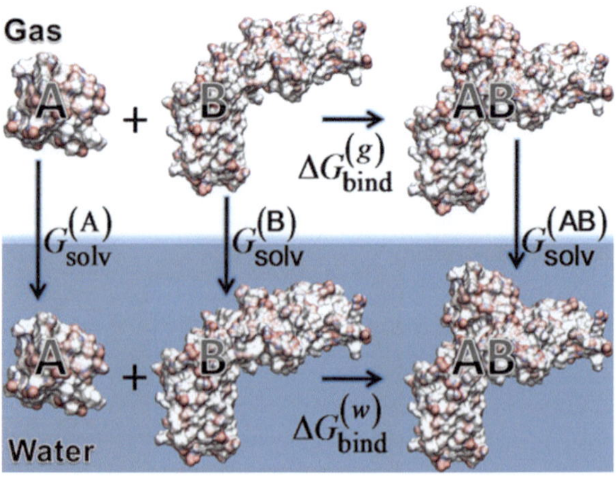

Fig. 5.6 Thermodynamic cycle for absolute binding free energy calculation of protein-protein association. ΔG_A^{solv}, ΔG_B^{solv}, and ΔG_{AB}^{solv} are solvation free energies of A, B and AB, respectively. ΔG_b^g and ΔG_b^w are binding free energies in gas phase and condensed phase (e.g., water)

complex AB, respectively. The binding energy in the reference phase is a sum of molecular mechanics force field used and entropic terms as:

$$\langle \Delta G_b^g \rangle = \langle \Delta E_{elec}^{MM} \rangle + \langle \Delta E_{vdw}^{MM} \rangle + \langle \Delta E_{int}^{MM} \rangle + T \langle \Delta S \rangle \tag{5.52}$$

Here, the first three terms characterize the change on electrostatic, van der Waals and internal molecular mechanics energy, respectively, upon binding calculated in the reference phase (e.g., gas phase). The last term is the change in configuration entropy associated with (bio)molecular motions, and it is the sum of conformational, translational and rotational entropy terms. The solvation energy is calculated using the continuum solvent models as discussed in the following sections. In our notations, $\langle \cdots \rangle$ denotes an ensemble average.

Combinations of computer simulations with implicit solvent continuum models for free energy calculations have been proposed previously (Gilson and Honig 1988). The methods are applied in estimating the binding free energy of protein-protein binding (Gohlke et al. 2003) and protein-ligand binding (Kollman et al. 2000; Massova and Kollman 2000).

These approaches can also be implemented in calculation of average relative binding free energy change upon the mutation of a residue i to alanine, j, $\langle \Delta\Delta G_b^w(i \to j) \rangle$:

$$\langle \Delta\Delta G_b^w(i \to j) \rangle = \langle \Delta G_b^w(j) \rangle - \langle \Delta G_b^w(i) \rangle \tag{5.53}$$

where $\langle \Delta G_b^w(i) \rangle$ and $\langle \Delta G_b^w(j) \rangle$ are the binding free energies of the wild type and mutated system, respectively.

5.3.1 Empirical Solvation Models

The statistical mechanics models of the macromolecular solvation have shown to be difficult, and thus several different empirical models have also been developed. In the following, we will discuss these approaches, used in the free energy decomposition.

In general, the total solvation free energy is calculated as a sum of two contributions, namely the non-polar solvation free energy and polar (electrostatic) solvation free energy:

$$\Delta G^{solv} = \Delta G_{nonpolar}^{solv} + \Delta G_{elec}^{solv} \tag{5.54}$$

Here, $\Delta G_{nonpolar}^{solv}$ is the non-polar contribution and ΔG_{elec}^{solv} is the electrostatic contribution to solvation free energy, discussed in the following sections. Note that ΔG^{solv} is just the Helmholtz free energy. To obtain the Gibbs free energy, the term $p\Delta V$ must be added, which usually is a small term because the changes on the volume are small.

5.3.1.1 Implicit Nonpolar Solvation Free Energy

The nonpolar contribution of the solvation free energy associates with attractive short-range dispersion interactions for creating a cavity in a solvent where the solute resize, and solute-solvent dispersion interactions:

$$\Delta G_{nonpolar}^{solv} = \Delta G_{cav}^{solv} + \Delta G_{vdW}^{solv} \tag{5.55}$$

ΔG_{cav}^{solv} is the free energy cost for creation of a cavity in solvent with size of macromolecule, plus the reorganization energy (change on solvent-solvent energy) and entropic cost of solvent around the cavity, and ΔG_{vdW}^{solv} is the free energy cost of compensating for solute-solvent van der Waals dispersion interactions. Note that both terms have opposite signs and are anti-correlated with each other. Besides, both these interactions include only the first solvent shell. Moreover, the non-polar contribution of the solvation free energy includes the entropic contribution due to the changes in solvent structures near the solute. The length-scale dependence of the cavity creation and solvent screening of solute-solvent dispersion interactions have to be properly described to have an accurate representation of the implicit modeling of non-polar solvation free energy.

The *Atomic Solvation Parameter* (ASP) model is the simplest one, in which the cavity non-polar contribution of the solvation free energy is the sum of atomic contributions (Eisenberg and McLachlan 1986). In ASP model, the non-polar term of the solvation free energy of an atom (or group of atoms) is assumed to be proportional to its molecular surface area, A_i, and it is given by

$$\Delta G_{cav}^{solv} = \sum_i \gamma_i A_i \tag{5.56}$$

where the proportionality constant γ_i depends on the type of atom, calibrated using the experimental data by a fitting procedure. In most of Generalized Born Molecular Surface Area (GB/SA) and Poisson-Boltzmann Molecular Surface Area (PB/SA) models, further approximations apply by assuming a universal Γ value for all atom types, and Eq. (5.56) is as:

$$\Delta G_{cav}^{solv} = \Gamma A \tag{5.57}$$

A is the total molecular surface area of the macromolecule. This relationship showed to be fairly accurate by investigating the experimental solvation free energies of different molecular systems, such as linear alkanes and neutral organic compounds (Eisenberg and McLachlan 1986; Ooi et al. 1987; Simonson and Brunger 1994; Chothia 1974), or by theoretical and computational studies of non polar solvation and hydrophobic interactions (Nemethy and Scheraga 1962; Pierotti 1976; Ashbaugh et al. 1999; Raschke et al. 2001). Often, the coefficient Γ and parametrization of the surface effects in Eq. (5.57) are carefully designed also to include short-range effects, dominated by the first solvation shell, such as nonlinear response of solvent to the local electric field near solute and charge transfer to or from solvent (Chen and Brooks III 2008) (and the reference therein Cramer and Truhlar 1999). Equation (5.57) is also used to model other effects rather than non-polar contributions to solvation free energy, for instance, as a measure of thermodynamic parameters of peptide hydration in Ooi et al. (1987). Thus, based on the empirical nature of the factor Γ, different values have been used for its parametrization, different underlying effects, including macromolecular force field, electrostatic solvation model, and choice of the macromolecule-solvent boundary (Chen and Brooks III 2008). Indeed, the values of Γ ranging from 5 to 7 cal mol^{-1} Å^{-2} (Still et al. 1990; Sitkoff et al. 1994; Simonson and Brunger 1994), or from 40 to 70 cal mol^{-1} Å^{-2} (Tanford 1979; Sharp et al. 1991). This wide range of the values of Γ is because of different models used to represent the molecular surface (Chen and Brooks III 2008). However, small values of Γ indicate that molecular surface contribution has only a small impact on the free energy changes due to large-scale conformation changes.

Note that a precise definition of the solute-solvent interface is a valuable physical property for both non-polar and electrostatic (as shown in the next section) solvation free energies. Although, molecular surface non-polar term contains empirical corrections to compensate for various effects related to the first hydration shell and force field terms. Moreover, the optimal choice of the solute-solvent interface boundary for the non-polar contribution to the solvation free energy may not necessarily coincide with that of the electrostatic contribution to solvation free energy.

Different methods have been introduced to calculate the molecular surfaces (Lee and Richards 1971; Richards 1977, 1985; Connolly 1983a,b, 1985; Vorobjev and Hermans 1997). In general, the molecular surface can be seen as a set of overlapping spheres, each having the van der Waals radius of its constituent atom, which is the so-called *van der Waals surface* (see Fig. 5.7a). The so-called *solvent-accessible surface* of a molecule is defined as the van der Waals envelope of a molecule expanded by the radius of the solvent sphere about each atom center, as shown in Fig. 5.7b (Lee and Richards 1971; Richards 1977). Then, the so-called *molecular surface* which is defined as the contour drawn by a sphere of radius r_p, representing the solvent molecule, rolling over a set of the van der Waals beads centered at the atomic positions, as shown in Fig. 5.7c (Connolly 1983a,b, 1985). More recently (Vorobjev and Hermans 1997), a so-called *smooth invariant molecular surface* is also defined using a smoothing sphere with radius r_s rolling over the generated molecular surface. The molecular surface comprises the convex spherical patches, saddle-shaped toroidal patches, and the concave patches, which then is partitioned in a set of triangles (Connolly 1983a,b, 1985).

Note that by definition, the solvent-accessible surface has no reentrant sections, and hence the molecule comprises only convex spherical patches. That may yield a loss of information since the ratio of the contact-to-reentrant surface could be a measure of molecular surface roughness (Richards 1985). Besides, the choice of the probe radius will also influence the value of the area of the molecular surface

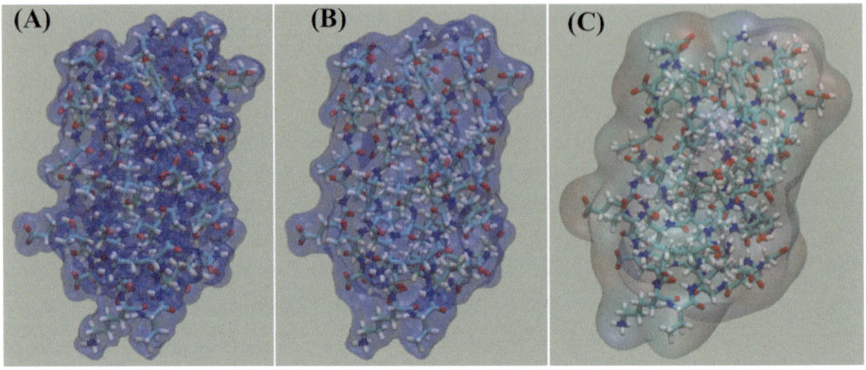

Fig. 5.7 Different molecular surface representations: (**a**) the van der Waals surface; (**b**) the solvent-accessible surface; and (**c**) the molecular surface

in each case. Use of a small probe radius reveals a large number of features from the molecular surface, and for $r_p = 0$, the solvent-accessible surface is equal to the molecular surface. Since the probe sphere mimics the solvent molecule, i.e., water molecule, the smallest physically accepted radius is 1.4 Å (the radius of a water molecule). As the probe radius increases, the molecular surface and solvent-accessible surface become smoother, and for $r_p = \infty$, the molecular surface is the convex hull of the set of atomic spheres. From the geometrical point of view, the molecular surface tends towards a finite limiting value as the probe radius increases, whereas the solvent-accessible surface tends towards infinite. Therefore, the molecular surface may be a better representation of the surface of a molecule. The calculation of the surface of a molecule is a geometrical problem, and hence the accuracy of this calculation will depend on a geometrical representation of the molecular surface. To accurately represent the surface of a molecule, several conditions are required to be satisfied (Vorobjev and Hermans 1997): maximum homogeneity of dot distribution, a smoothing surface near the singular points, small changed on the dot density, independence on the rotation of molecule and stability with changing the molecular conformation. The integration of functions over the molecular surface, as required here, require a numerical representation of the molecular surface as a set of

$$\{(x_i, y_i, z_i), \mathbf{n}_i, \Delta\sigma_i\}$$

where (x_i, y_i, z_i), $\mathbf{n}_i$, $\Delta\sigma_i$ are, respectively, the coordinates, normal vector and area of a small element of the molecular surface.

Then, Eq. (5.57) can be written as

$$\Delta G_{cav}^{solv} = \Gamma \sum_{t=1}^{\mathcal{T}} \Delta\sigma_t \tag{5.58}$$

where the sum runs over all triangle molecular surface patches $\mathcal{T}$. This term will represent the non-polar contribution to the cavity creation, and the reorganization solvent energy and entropy.

5.3.1.2 Implicit Dielectric Surface Boundary Free Energy Term

Here, we are going to consider another term to the free energy decomposition due to the forces acting on the dielectric surface boundary between the solvent and solute, namely ΔG^{solv}_{diel}.

In order to evaluate ΔG^{solv}_{diel}, we are going to calculate the forces acting on the dielectric surface boundary between the solvent and solute using the relationship of forces on the dielectric medium and the Maxwell stress tensor $\boldsymbol{\sigma}$. The following expression is going to be used to evaluate the forces acting on an elementary volume of dielectric due to the electrostatic forces of charge distribution in macromolecule:

$$\mathbf{f} = \rho^f \mathbf{E} + \frac{\varepsilon_0}{2}\nabla\left(\mathbf{E}\cdot\mathbf{E}g\frac{d\varepsilon}{dg}\right) - \frac{\varepsilon_0}{2}(\mathbf{E}\cdot\mathbf{E})\nabla\varepsilon \tag{5.59}$$

where ρ^f is the fixed charge density in dielectric and g is the mass density. $\mathbf{D}$ and $\mathbf{E}$ are the electric displacement and field vectors, respectively, related as

$$\mathbf{D} = \varepsilon_0\varepsilon\mathbf{E}$$

In Eq. (5.59), the first term gives the force acting on the fixed charges in dielectric, the second term is related to inhomogeneity of the field, and the third term is related to the inhomogeneity of dielectric media. Ignoring the second term and using the Maxwell equation $\nabla\cdot\mathbf{D} = \rho^f$, then Eq. (5.59) can be written as

$$\begin{aligned}
\mathbf{f} &= (\nabla\cdot\mathbf{D})\mathbf{E} - \frac{\varepsilon_0}{2}(\mathbf{E}\cdot\mathbf{E})\nabla\varepsilon \\
&= \nabla\cdot(\mathbf{D}\otimes\mathbf{E}) - (\mathbf{D}\cdot\nabla)\mathbf{E} - \frac{\varepsilon_0}{2}\nabla(\varepsilon\mathbf{E}\cdot\mathbf{E}) + \varepsilon_0(\varepsilon\mathbf{E}\cdot\nabla)\mathbf{E} \\
&= \nabla\cdot(\mathbf{D}\otimes\mathbf{E}) - (\mathbf{D}\cdot\nabla)\mathbf{E} - \frac{1}{2}\nabla(\mathbf{D}\cdot\mathbf{E}) + (\mathbf{D}\cdot\nabla)\mathbf{E} \\
&= \nabla\cdot(\mathbf{D}\otimes\mathbf{E}) - \frac{1}{2}\nabla(\mathbf{D}\cdot\mathbf{E}) \\
&= \nabla\cdot\left[\mathbf{D}\otimes\mathbf{E} - \frac{1}{2}(\mathbf{D}\cdot\mathbf{E})\mathbf{I}\right]
\end{aligned} \tag{5.60}$$

where $\mathbf{D}\otimes\mathbf{E}$ is a tensor of second rank with elements $(\mathbf{D}\otimes\mathbf{E})_{ij} = D_i E_j$, for $i, j = 1, 2, 3$, and $\mathbf{I}$ is the unitary diagonal tensor:

$$I_{ij} = \begin{cases} 0 \text{ if } i \neq j \\ 1 \text{ if } i = j \end{cases}$$

Note that Eq. (5.60) is also valid for the dielectric regions where $\rho^f = 0$. In such case,

$$\nabla \cdot (\mathbf{D} \otimes \mathbf{E}) - (\mathbf{D} \cdot \nabla)\mathbf{E} = 0$$

or

$$\nabla \cdot (\mathbf{D} \otimes \mathbf{E}) = (\mathbf{D} \cdot \nabla)\mathbf{E}$$

Therefore, for the regions where $\rho^f = 0$ (e.g., at the dielectric boundary between solvent-solute), the density of force acting on dielectric media is

$$\begin{aligned}\mathbf{f} &= -\frac{\varepsilon_0}{2}(\mathbf{E} \cdot \mathbf{E})\nabla\varepsilon \\ &= -\frac{\varepsilon_0}{2}\nabla(\varepsilon\mathbf{E} \cdot \mathbf{E}) + \varepsilon_0(\varepsilon\mathbf{E} \cdot \nabla)\mathbf{E} \\ &= -\frac{1}{2}\nabla(\mathbf{D} \cdot \mathbf{E}) + (\mathbf{D} \cdot \nabla)\mathbf{E} \\ &= \nabla \cdot (\mathbf{D} \otimes \mathbf{E}) - \frac{1}{2}\nabla(\mathbf{D} \cdot \mathbf{E}) \\ &= \nabla \cdot \left[\mathbf{D} \otimes \mathbf{E} - \frac{1}{2}(\mathbf{D} \cdot \mathbf{E})\mathbf{I}\right]\end{aligned} \tag{5.61}$$

Defining the Maxwell stress tensor $\boldsymbol{\sigma}$ as

$$\boldsymbol{\sigma} = \varepsilon_0\varepsilon\mathbf{E} \otimes \mathbf{E} - \frac{1}{2}\varepsilon_0\varepsilon(\mathbf{E} \cdot \mathbf{E})\mathbf{I} \tag{5.62}$$

Eq. (5.60) can then be written as

$$\mathbf{f} = \nabla \cdot \boldsymbol{\sigma} \tag{5.63}$$

Physically, the Maxwell stress tensor is the force acting on the unit area (or *stress*), that is, σ_{ij} is the force per unit area in the i-th direction acting on the surface element oriented in the j-th direction. Thus, the diagonal elements (σ_{xx}, σ_{yy}, σ_{zz}) represent the *stresses* and off-diagonal elements are the *shear stresses*. Time-average force acting on the dielectric body is

$$\mathbf{F} = \int_V \mathbf{f} dV = \int_V \nabla \cdot \boldsymbol{\sigma} dV$$

The volume integral can be expressed as the surface integral surrounding the dielectric body by applying the Gauss's theorem,

$$\mathbf{F} = \oint_S \boldsymbol{\sigma} \cdot \mathbf{n} dS \tag{5.64}$$

where $\mathbf{n}$ is the unit vector outward the surface element dS. Since the normal component of the electrical field $\mathbf{E}$ is discontinuous at the boundary between solvent and solute media, the product $\boldsymbol{\sigma} \cdot \mathbf{n}$ in Eq. (5.64) exhibits a jump. Therefore, the difference between $\boldsymbol{\sigma} \cdot \mathbf{n}$ in the two media will give the surface force $\mathbf{F}_S$, which is a measure of the force per unit of area, which can be written as

$$\mathbf{F}_S = P\mathbf{n} + S_t\mathbf{t} = \left(\boldsymbol{\sigma}^{(2)} - \boldsymbol{\sigma}^{(1)}\right) \cdot \mathbf{n} \tag{5.65}$$

where $\boldsymbol{\sigma}^{(2)}$ and $\boldsymbol{\sigma}^{(1)}$ are the Maxwell stress tensors defined in the solute and solvent media, respectively. P is the normal stress to the molecular solute surface and S_t is the shear stress, which is the tangent component of $\mathbf{F}_S$. $\mathbf{t}$ is the tangential unit vector to the element surface dS. The normal stress and the shear stress are defined as

$$P = \mathbf{F}_S \cdot \mathbf{n} = \sum_{i=1}^{3} F_{S,i} n_i \tag{5.66}$$

$$S_t = \mathbf{F}_S \cdot \mathbf{t} = \sum_{i=1}^{3} F_{S,i} t_i$$

Using Eq. (5.65), we can determine

$$F_{S,i} = \sum_{j=1}^{3} \left(\sigma_{ij}^{(2)} - \sigma_{ij}^{(1)}\right) n_j \tag{5.67}$$

Combining expression for P from Eq. (5.66) with $F_{S,i}$ from Eq. (5.67), we obtain

$$\begin{aligned} P &= \sum_{i=1}^{3} \left(\sum_{j=1}^{3} \left(\sigma_{ij}^{(2)} - \sigma_{ij}^{(1)}\right) n_j \right) n_i \\ &= \sum_{i=1}^{3} \sum_{j=1}^{3} \left(D_i^{(2)} E_j^{(2)} - \frac{1}{2} D^{(2)} E^{(2)} \delta_{ij} - D_i^{(1)} E_j^{(1)} \right. \\ &\left. + \frac{1}{2} D^{(1)} E^{(1)} \delta_{ij} \right) n_j n_i \\ &= D_n^{(2)} E_n^{(2)} - \frac{1}{2} D^{(2)} E^{(2)} - D_n^{(1)} E_n^{(1)} + \frac{1}{2} D^{(1)} E^{(1)} \end{aligned} \tag{5.68}$$

Using the dielectric boundary conditions

$$E_t^{(1)} = E_t^{(2)},\ D_n^{(1)} = D_n^{(2)},\ \varepsilon_w E_n^{(1)} = \varepsilon_m E_n^{(2)}$$

under the assumption that there is no free charge at the dielectric boundary surface, Eq. (5.68) reduces to

$$\begin{aligned} P &= \varepsilon_0 \varepsilon_m E_n^{(2)} E_n^{(2)} - \frac{\varepsilon_0 \varepsilon_m}{2} \left(E_n^{(2)} E_n^{(2)} + E_t^{(2)} E_t^{(2)} \right) \\ &\quad - \varepsilon_0 \varepsilon_m E_n^{(2)} \frac{\varepsilon_m}{\varepsilon_w} E_n^{(2)} + \frac{1}{2} \varepsilon_0 \varepsilon_w \left(E_n^{(1)} E_n^{(1)} + E_t^{(1)} E_t^{(1)} \right) \\ &= \frac{\varepsilon_0 \varepsilon_m}{2} E_n^{(2)} E_n^{(2)} - \frac{\varepsilon_0 \varepsilon_m}{2} E_t^{(2)} E_t^{(2)} \\ &\quad - \varepsilon_0 \frac{\varepsilon_m^2}{\varepsilon_w} E_n^{(2)} E_n^{(2)} + \frac{\varepsilon_0 \varepsilon_w}{2} \left(\frac{\varepsilon_m}{\varepsilon_w} \right)^2 E_n^{(2)} E_n^{(2)} + \frac{\varepsilon_0 \varepsilon_w}{2} E_t^{(2)} E_t^{(2)} \\ &= \frac{\varepsilon_0}{2} (\varepsilon_w - \varepsilon_m) \left[\left(\frac{\varepsilon_m}{\varepsilon_w} - 1 \right) E_n^2 + E^2 \right] \end{aligned} \tag{5.69}$$

where E_n and E are the normal component and total electric field on the second medium, i.e., solute.

Similarly, the tangential component S_t is determined as

$$\begin{aligned} S_t &= \sum_{i=1}^{3} \left(\sum_{j=1}^{3} \left(\sigma_{ij}^{(2)} - \sigma_{ij}^{(1)} \right) n_j \right) t_i \\ &= \sum_{i=1}^{3} \sum_{j=1}^{3} \left(D_i^{(2)} E_j^{(2)} - \frac{1}{2} D^{(2)} E^{(2)} \delta_{ij} - D_i^{(1)} E_j^{(1)} \right. \\ &\quad \left. + \frac{1}{2} D^{(1)} E^{(1)} \delta_{ij} \right) n_j t_i \\ &= D_t^{(2)} E_n^{(2)} - D_t^{(1)} E_n^{(1)} \\ &= \varepsilon_0 \varepsilon_m E_t^{(2)} E_n^{(2)} - \varepsilon_0 \varepsilon_w E_t^{(1)} E_n^{(1)} \\ &= \varepsilon_0 \varepsilon_m E_t^{(2)} E_n^{(2)} - \varepsilon_0 \varepsilon_m E_t^{(2)} E_n^{(2)} \\ &= 0 \end{aligned} \tag{5.70}$$

where ε_m and ε_w are the solute and solvent dielectric constants, respectively. Here, E_n and E_t are the components of electric field in solute medium. These results indicate that only the normal component (normal stress) of the surface force exist. On the other hand, the tangential component (shear stress) vanishes because fixed charge density is zero close to the dielectric boundary surface, and only the surface polarization induced charges exist in this boundary.

Based on the new formalism, the electrostatic contribution to the cavity creating in solvent can be calculated as the following:

$$\begin{aligned}\Delta G_{diel}^{solv} &= P\Delta V = \int \left(\left[\oint_S (\boldsymbol{\sigma}^{(m)} - \boldsymbol{\sigma}^{(w)}) \cdot \mathbf{n}\, dS \right] \cdot \mathbf{n} \right) d\mathbf{r} \\ &= \sum_{i=1}^{\mathcal{T}} (\mathbf{F}_{S,i} \cdot \mathbf{n}_i)\delta V_i \\ &= \sum_{i=1}^{\mathcal{T}} P_i \delta V_i \end{aligned} \tag{5.71}$$

where δV_i is an element volume of the solute, calculated as

$$\delta V_i = \left[\left(\mathbf{r}_{s,i} - \mathbf{r}_c \right) \cdot \mathbf{n}_i \right] \Delta\sigma_i$$

where $\mathbf{r}_c$ is the solute centre of gravity defined as

$$\mathbf{r}_c = \frac{\sum_{i=1}^{\mathcal{T}} \mathbf{r}_{s,i} \Delta\sigma_i}{\sum_{i=1}^{\mathcal{T}} \Delta\sigma_i}$$

Note that P_i is the normal stress at surface point i given by expression in Eq. (5.69), where E_n and E are the normal electric field component and total electrical field in the surface solute-solvent interface at the surface point i calculated in the solute side.

5.3.1.3 Implicit van der Waals Free Energy Term

An approach for evaluating the van der Waals contributions is introduced to non-polar solvation free energy using a continuum model. That is valid for interpreting the solvent screening of solute-solvent dispersion interactions (Levy et al. 2003). According to this model, the solvent (e.g., water) is a simple single site (e.g., oxygen) with a constant density in the region outside the solute volume, and the dispersion interaction energy between each atomic site of the solute and solvent is evaluated as a volume integral:

$$\Delta G_{vdW}^{solv} = \rho_w \sum_{i=1}^{M} \int_{solvent} u_{i,w}^{(vdW)} (|\, \mathbf{r} - \mathbf{r}_i \,|)\, d^3\mathbf{r} \tag{5.72}$$

where ρ_w is the solvent bulk number density at standard conditions, for example for water $\rho_w = 0.033428$ Å^{-3}, $u_{i,w}^{(vdW)}$ is the dispersion interaction potential between the atom i of the solute and solvent, which is defined as the attractive part of the following decomposition of the Lennard-Jones potential (Levy et al. 2003; Gallicchio and Levy 2004):

$$u_{i,j}^{(vdw)}(r) = \begin{cases} -\varepsilon_{ij}, & r \leq 2^{1/6}\sigma_{ij} \\ 4\varepsilon_{ij}\left[\left(\frac{\sigma_{ij}}{r}\right)^{12} - \left(\frac{\sigma_{ij}}{r}\right)^{6}\right], & r > 2^{1/6}\sigma_{ij} \end{cases} \tag{5.73}$$

and

$$u_{i,j}^{(rep)}(r) = \begin{cases} 4\varepsilon_{ij}\left[\left(\frac{\sigma_{ij}}{r}\right)^{12} - \left(\frac{\sigma_{ij}}{r}\right)^{6}\right] + \varepsilon_{ij}, & r \leq 2^{1/6}\sigma_{ij} \\ 0, & r > 2^{1/6}\sigma_{ij} \end{cases} \tag{5.74}$$

Thus, Eq. (5.72) can be written as

$$\Delta G_{vdW}^{solv} = \rho_w \sum_{i=1}^{M} \int_{\Omega_w} \left(\frac{A_i}{|\mathbf{r} - \mathbf{r}_i|^{12}} - \frac{B_i}{|\mathbf{r} - \mathbf{r}_i|^{6}} \right) d^3\mathbf{r} \tag{5.75}$$

for $|\mathbf{r} - \mathbf{r}_i| \geq 2^{1/6}\sigma_{iw}$, and

$$\Delta G_{vdW}^{solv} = -\rho_w \sum_{i=1}^{M} \int_{\Omega_w} \varepsilon_{iw} d^3\mathbf{r} \tag{5.76}$$

for $|\mathbf{r} - \mathbf{r}_i| < 2^{1/6}\sigma_{iw}$. Here, $A_i = 4\varepsilon_{iw}\sigma_{iw}^{12}$ and $B_i = 4\varepsilon_{iw}\sigma_{iw}^{6}$ are the force constants, which depend on the atom i and solvent site of molecule (e.g., oxygen of water molecule) determined using the arithmetic rules for the Lennard-Jones type of potential:

$$\varepsilon_{iw} = \sqrt{\varepsilon_i \varepsilon_w}, \ \sigma_{iw} = \frac{\sigma_i + \sigma_w}{2}$$

Ω_w denotes the entire solvent volume.

The integral, in principle, can be evaluated using different approaches, such as surface integral (Zacharias 2003), by a pair-wise descreening approximation (Gallicchio and Levy 2004), or directly using the numerical quadrature techniques (Lee et al. 2002).

Another approach is introduced, converting the above solvent volume integrals into surface ones, where this characterizes the closed surface enclosing the solute volume. It can be seen that calculation of ΔG_{vdw} includes estimation of the integrals of the following form:

$$I = \int_{\Omega_w} \frac{1}{\mid \mathbf{r} - \mathbf{r}_i \mid^n} d^3\mathbf{r} \tag{5.77}$$

where integration is performed over solvent volume. This integral can be split into two terms, where the first term is the integral over entire space and the second one over solute volume as:

$$I = \int_{\Omega} \frac{1}{\mid \mathbf{r} - \mathbf{r}_i \mid^n} d^3\mathbf{r} - \int_{\Omega_s} \frac{1}{\mid \mathbf{r} - \mathbf{r}_i \mid^n} d^3\mathbf{r} \tag{5.78}$$

where Ω is the all space volume and Ω_s is the solute volume. The first term of integral can be further split into two other terms. The first is the integral over volume of the i-th solute atom and the second term is over all space excluding the volume of that atom. Thus, we obtain:

$$\begin{aligned} I = & \int_{V_0} \frac{1}{\mid \mathbf{r} - \mathbf{r}_i \mid^n} d^3\mathbf{r} + \int_{\Omega - V_0} \frac{1}{\mid \mathbf{r} - \mathbf{r}_i \mid^n} d^3\mathbf{r} \\ & - \int_{\Omega_s} \frac{1}{\mid \mathbf{r} - \mathbf{r}_i \mid^n} d^3\mathbf{r} \end{aligned} \tag{5.79}$$

We can now use the Gauss's theorem:

$$\int_{\Omega} \nabla \cdot \mathbf{A} d^3\mathbf{r} = \oint_S \mathbf{A} \cdot \mathbf{n} \, dS$$

where Ω is the integration volume and S is surface enclosing that volume. $\mathbf{n}$ is the outward unit vector to the surface element dS. In our case, when

$$\nabla \cdot \mathbf{A} = \frac{1}{\mid \mathbf{r} - \mathbf{r}_i \mid^n}$$

it can be found that

$$\mathbf{A} = \frac{1}{3-n} \cdot \frac{\mathbf{r} - \mathbf{r}_i}{\mid \mathbf{r} - \mathbf{r}_i \mid^n}$$

Therefore, we obtain that

$$\begin{aligned} I = & \frac{1}{3-n} \oint_{S_0} \frac{(\mathbf{r} - \mathbf{r}_i) \cdot \mathbf{n}_0}{\mid \mathbf{r} - \mathbf{r}_i \mid^n} dS \\ & + \frac{1}{3-n} \oint_{S_0} \frac{(\mathbf{r} - \mathbf{r}_i) \cdot \tilde{\mathbf{n}}_0}{\mid \mathbf{r} - \mathbf{r}_i \mid^n} dS + \frac{1}{3-n} \oint_{S_\infty} \frac{(\mathbf{r} - \mathbf{r}_i) \cdot \mathbf{n}_\infty}{\mid \mathbf{r} - \mathbf{r}_i \mid^n} dS \\ & - \frac{1}{3-n} \oint_{S} \frac{(\mathbf{r} - \mathbf{r}_i) \cdot \mathbf{n}}{\mid \mathbf{r} - \mathbf{r}_i \mid^n} dS \end{aligned} \tag{5.80}$$

where S is the molecular surface of solute, S_0 is spherical surface of i-th solute atom, S_∞ is the surface at the infinity of solvent, $\mathbf{n}_0$ is the outward unit vector normal to dS element of S_0 surface and $\tilde{\mathbf{n}}_0$ is the inward unit vector normal to dS element of S_0 surface, which is given as

$$\tilde{\mathbf{n}}_0 = -\mathbf{n}_0$$

and hence, the first to integral cancel out. Here, $\mathbf{n}_\infty$ is the outward unit vector normal to dS element of S_∞ surface and $\mathbf{n}$ is the outward unit vector normal to dS element of S surface. At the infinity we have

$$\frac{1}{\mid \mathbf{r} - \mathbf{r}_i \mid^n} \to 0$$

and hence the third integral is zero. Therefore, we can finally write that

$$I = \frac{1}{n-3} \oint_S \frac{(\mathbf{r} - \mathbf{r}_i) \cdot \mathbf{n}}{\mid \mathbf{r} - \mathbf{r}_i \mid^n} dS \tag{5.81}$$

Thus, ΔG_{vdw}^{solv} from Eqs. (5.75) and (5.76) takes the following form for $\mid \mathbf{r} - \mathbf{r}_i \mid \geq 2^{1/6}\sigma_{iw}$:

$$\Delta G_{vdw}^{solv} = \sum_{i=1}^{M} \frac{\rho_w A_i}{9} \oint_S \frac{(\mathbf{r} - \mathbf{r}_i) \cdot \mathbf{n}}{\mid \mathbf{r} - \mathbf{r}_i \mid^{12}} dS - \sum_{i=1}^{M} \frac{\rho_w B_i}{3} \oint_S \frac{(\mathbf{r} - \mathbf{r}_i) \cdot \mathbf{n}}{\mid \mathbf{r} - \mathbf{r}_i \mid^{6}} dS \tag{5.82}$$

and for $\mid \mathbf{r} - \mathbf{r}_i \mid < 2^{1/6}\sigma_{iw}$:

$$\Delta G_{vdw}^{solv} = \sum_{i=1}^{M} \frac{\rho_w \varepsilon_{iw}}{3} \oint_S (\mathbf{r} - \mathbf{r}_i) \cdot \mathbf{n} \, dS \tag{5.83}$$

Since the molecular surface of solute can be represented by a set discrete surface elements, we can rewrite the above surface integrals as summations as shown in the following. For $\mid \mathbf{r}_t - \mathbf{r}_i \mid \geq 2^{1/6}\sigma_{iw}$:

$$\begin{aligned} \Delta G_{vdw}^{solv} = & \sum_{i=1}^{M} \frac{\rho_w A_i}{9} \sum_{t=1}^{T} \frac{(\mathbf{r}_t - \mathbf{r}_i) \cdot \mathbf{n}_t}{\mid \mathbf{r}_t - \mathbf{r}_i \mid^{12}} \Delta\sigma_t \\ & - \sum_{i=1}^{M} \frac{\rho_w B_i}{3} \sum_{t=1}^{T} \frac{(\mathbf{r}_t - \mathbf{r}_i) \cdot \mathbf{n}_t}{\mid \mathbf{r}_t - \mathbf{r}_i \mid^{6}} \Delta\sigma_t \end{aligned} \tag{5.84}$$

and for $| \mathbf{r}_t - \mathbf{r}_i |< 2^{1/6}\sigma_{iw}$:

$$\Delta G_{vdw}^{solv} = \sum_{i=1}^{M} \frac{\rho_w \varepsilon_{iw}}{3} \sum_{t=1}^{\mathcal{T}} (\mathbf{r}_t - \mathbf{r}_i) \cdot \mathbf{n}_t \Delta\sigma_t \tag{5.85}$$

Eqs. (5.58), (5.84) and (5.85) can be used to calculate the non-polar contributions to the cavity creation in the solvent to resize the macromolecule, and Eq. (5.71) can be used to calculate the electrostatic contribution to cavity creation.

In the following, we are going to discuss the methods used to calculate the electrostatic contribution to the solvation free energy, namely the so-called Poisson-Boltzmann model and the Generalized Born approximation.

5.3.1.4 The Poisson-Boltzmann Model

The Poisson-Boltzmann equation was described in these references Gouy (1910) and Chapman (1913). They derived a relation between the chemical potential and the force acting on small adjacent volumes in an ionic solution between two plates at a different voltage. Later, the approach was generalized by Debye and Hückel (1923), applying their work to the theory of ionic solutions.

Gronwall et al. (1928) provided the solutions to the nonlinearized Poisson-Boltzmann equation as a function of the powers of the inverse of the dielectric constant as coefficients. Onsager (1933) formulated the statistical mechanics basis of the Poisson-Boltzmann and Fowler and Guggenheimer (1939) in terms of the potential of mean force.

Kirkwood (1934a) formulated some approximations from the formalism, pointing out that the Poisson-Boltzmann method used the assumption that it is possible to replace the potential of mean force with the mean electrostatic potential. Due to the success of the theory on explaining the behavior of ionic solutions, the Poisson-Boltzmann approach started to apply in other fields such as colloid chemistry, and later the method was used in calculating the free energy of interacting particles based on the framework of Derjaguin and Landau (1941), and Verwey and Overbeek (1948).

Kirkwood (1934b), Linderstrom-Lang (1924), and Nozaki and Tanford (1967) formulated the first, simple electrostatic models of globular proteins, and for DNA and other linear polyelectrolytes the cylindrical symmetric models were used by Lifson and Katchalski (1954), Alfrey et al. (1951), Katchalski (1971) and Manning (1978). All these models use either the Poisson-Boltzmann equation or its linear approximation leading to high accuracy results. Initially, the models used simple shape molecular models, such as spheres for proteins and rods for DNA. During the 1980s the models improved by developing methods for solving the Poisson-Boltzmann equation for any arbitrary shape using the finite difference algorithm in software such as DelPhi, Grasp, and UHBD, allowing the study of electrostatic models to the atomic level.

Later, also fast approaches for dealing with electrostatic interactions were developed by Lazaridis and Karplus (1999), Simonson (2001), and Roux and Simonson (1999).

In the Poisson-Boltzmann approach, all the macromolecular atoms are considered explicitly as particles with partial point charges at the atomic positions, and the dielectric constant of the macromolecule itself is often considered to be low, typically in the range 2–4. The solvent environment surrounding the macromolecule is taken implicitly into account as a dielectric medium with the dielectric constant of about 80. Sometimes, a solvent-macromolecule interaction model used a surface model with a surface tension typically 5–70 cal Å^{-2} (Fogolari et al. 2002) using the models described in the previous section. The macromolecular dielectric value does not take into account the rearrangement of polar and charged amino acids with external electric fields, which could result into a larger dielectric constants (Gilson and Honig 1986; Simonson 1998). For example, Schutz and Warshel (2001) suggested that the increase of the dielectric can compensate for the need for group re-orientations.

Let us consider first a homogeneous medium with dielectric constant and with no external charges (i.e., $\rho = 0$). The Maxwell equation for the medium will give

$$\nabla \cdot \mathbf{D} = 0 \tag{5.86}$$

where $\mathbf{D}$ is the electric displacement vector. Knowing the relation between the vector $\mathbf{D}$ and the electrical field vector $\mathbf{E}$ for the approximation of linear polarised medium,

$$\mathbf{D} = \varepsilon_0 \varepsilon \mathbf{E}$$

and the relation between the electrostatic potential ϕ and the vector $\mathbf{E}$

$$\mathbf{E} = -\nabla \phi$$

we get

$$\nabla \cdot (\nabla \phi(\mathbf{r})) = 0 \tag{5.87}$$

or

$$\Delta \phi(\mathbf{r}) = 0 \tag{5.88}$$

which is known as Laplace equation.

If the external charges are present (i.e., $\rho \neq 0$), for example a macromolecule is immersed in the solvent medium, Eq. (5.86) is written as

$$\nabla \cdot \mathbf{D} = \rho(\mathbf{r}) \tag{5.89}$$

which leads to the so-called Poisson equation

$$\varepsilon \nabla \cdot (\nabla \phi(\mathbf{r})) = -\frac{\rho(\mathbf{r})}{\varepsilon_0} \tag{5.90}$$

In the general case of the nonhomogeneous medium, the polarised charges created at the dielectric boundaries must be taken into account as well (Jackson 1962), and Eq. (5.90) is modified as the following:

$$\nabla \cdot (\varepsilon(\mathbf{r}) \nabla \phi(\mathbf{r})) = -\frac{\rho(\mathbf{r})}{\varepsilon_0} \tag{5.91}$$

where ρ is the sum of the distribution of the macromolecule fixed charge density $\rho_m(\mathbf{r})$ and ionic charge density $\rho_I(\mathbf{r})$:

$$\rho(\mathbf{r}) = \rho_m(\mathbf{r}) + \rho_I(\mathbf{r})$$

In non homogeneous interacting particles system, density of a particle at any point $\mathbf{r}$ can be written as

$$\sigma_{I,i}(\mathbf{r}) = g_i(\mathbf{r}) \sigma_{I,i}^0(\mathbf{r}) \tag{5.92}$$

where $\sigma_{I,i}^0(\mathbf{r})$ is the particle density of the same system considered as ideal gas (i.e., non-interacting particle system), and $g_i(\mathbf{r})$ is the i-th particle distribution, which is taken to follow the Boltzmann distribution

$$g_i(\mathbf{r}) = \exp(-\beta W_i(\mathbf{r})) \tag{5.93}$$

In Eq. (5.93), $W_i(\mathbf{r})$ is the potential of mean force for the particle i. The assumption made here is that the potential of mean force is equal to the average electrostatic potential at the point of the charge multiplied by the charge of particle:

$$W_i(\mathbf{r}) = q_i \phi(\mathbf{r})$$

where $q_i = z_i e$ with z_i being its valency and e being the charge of proton.

Thus, Eq. (5.92) can be written as

$$\sigma_{I,i}(\mathbf{r}) = \sigma_{I,i}^0(\mathbf{r}) \exp(-\beta q_i \phi(\mathbf{r})) \tag{5.94}$$

Then, the charge density is given as

$$\rho_I(\mathbf{r}) = \sum_i q_i \sigma_{I,i}(\mathbf{r}) = \sum_i q_i \sigma_{I,i}^0(\mathbf{r}) \exp(-\beta q_i \phi(\mathbf{r})) \tag{5.95}$$

where

$$\sigma^0_{I,i}(\mathbf{r}) = c_i^\infty \lambda(\mathbf{r})$$

where c_i^∞ is the bulk constant concentration of the ith ionic species, satisfying the condition of the electrostatic neutrality:

$$\sum_i q_i c_i^\infty = 0$$

$\lambda(\mathbf{r})$ is the accessibility of ions at point $\mathbf{r}$ (i.e., $\lambda(\mathbf{r}) = 0$ in the region inside the macromolecule and $\lambda(\mathbf{r}) = 1$ in the solvent region). Therefore, we can write

$$\rho_I(\mathbf{r}) = \lambda(\mathbf{r}) \sum_i q_i c_i^\infty \exp\left(-\beta q_i \phi(\mathbf{r})\right) \tag{5.96}$$

Using Eq. (5.96), the Poisson equation (see Eq. (5.91)) takes the form of the so-called nonlinear Poisson-Boltzmann equation

$$\nabla \cdot (\varepsilon(\mathbf{r})\nabla\phi(\mathbf{r})) + \frac{\lambda(\mathbf{r})}{\varepsilon_0} \sum_i q_i c_i^\infty \exp\left(-\beta q_i \phi(\mathbf{r})\right) = -\frac{\rho_m(\mathbf{r})}{\varepsilon_0} \tag{5.97}$$

For an electrostatic neutral solvent, we can write

$$\sum_{i=1}^{N_+} q_i^{(+)} c_i^{+,\infty} = \sum_{i=1}^{N_-} q_i^{(-)} c_i^{-,\infty}$$

where two kind of ionic species are assumed to exist in the solution, positive and negative with N_+ and N_- being the number of positive and negative ions, respectively. Assuming that $N_+ = N_- = N_I$, and since $q_i^{(+)} = -q_i^{(-)} \equiv q_i$ and $c_i^{+,\infty} = c_i^{-,\infty} = c_i^\infty/2$, we get from Eq. (5.97) that

$$\nabla \cdot (\varepsilon(\mathbf{r})\nabla\phi(\mathbf{r})) - \frac{\lambda(\mathbf{r})}{\varepsilon_0} \sum_{i=1}^{N_I} q_i c_i^\infty \sinh\left(\beta q_i \phi(\mathbf{r})\right) = -\frac{\rho_m(\mathbf{r})}{\varepsilon_0} \tag{5.98}$$

which is a form often found in the literature and it represents a nonlinear partial differential equation. In Eq. (5.98), sinh represents the function: $\sinh(x) = \left(e^x - e^{-x}\right)/2$.

Assuming that the potential is small, the linear form of the equation can be obtained as

$$\nabla \cdot (\varepsilon(\mathbf{r})\nabla\phi(\mathbf{r})) = -\frac{\rho_m(\mathbf{r})}{\varepsilon_0} + \varepsilon \left[\frac{\beta}{\varepsilon\varepsilon_0} \sum_i q_i^2 c_i^\infty \right] \lambda(\mathbf{r})\phi(\mathbf{r}) \tag{5.99}$$

We can determine the so-called Debye screening constant κ as

$$\kappa^2 = \frac{\beta}{\varepsilon\varepsilon_0}\sum_i q_i^2 c_i^\infty = \frac{\beta}{\varepsilon\varepsilon_0} I \equiv \frac{1}{l_D^2} \tag{5.100}$$

which also describes the exponential decay of the potential in the solvent, with l_D being the Debye length, and I

$$I = \sum_i q_i^2 c_i^\infty$$

being the ionic strength. Note that $\kappa = 0$ in the macromolecule region because the mobile ions are present only in the solvent region.

Equation (5.99) can then be written as

$$\nabla \cdot (\varepsilon(\mathbf{r})\nabla\phi(\mathbf{r})) - \varepsilon(\mathbf{r})\kappa^2\lambda(\mathbf{r})\phi(\mathbf{r}) = -\frac{\rho_m(\mathbf{r})}{\varepsilon_0} \tag{5.101}$$

Although for biological systems ϕ is not small, and therefore the linearisation condition does not hold, comparisons between the linear and nonlinear forms of the Poisson-Boltzmann equation (Fogolari et al. 1999) show that both forms are in good agreement with each other. Moreover, these comparisons have shown that small differences are related to the charge density, and hence to the electric field magnitude, at the interface solvent-solute.

From solving either the linear Poisson-Boltzmann equation (see Eq. (5.101)) or the nonlinear Poisson-Boltzmann equation (see Eq. (5.97)), the electrostatic potential, $\phi(\mathbf{r})$, will be obtained at any point $\mathbf{r}$ in space. It can be seen, that knowing ϕ, we may calculate the local concentration of ions through the formula

$$c_i(\mathbf{r}) = c_i^\infty \exp(-\beta q_i \phi(\mathbf{r})) \tag{5.102}$$

which involves the Boltzmann distribution. Moreover, the gradient of the electrostatic potential can give the electric field, $\mathbf{E}(\mathbf{r}) = -\nabla\phi(\mathbf{r})$.

Another quantity of interest calculated using the electrostatic potential is the electrostatic component of the solvation free energy. The electrostatic term of solvation free energy gives the work done for a possible process of charging the macromolecule and ions in an ionic discharged atmosphere. Using these processes in thermodynamic cycles, we can compute the electrostatic component of free energies for real processes such as solvation. The free energy for charging the solute (e.g., a macromolecule) in an ionic environment can be calculated using different approaches, for example, by direct integration of the charge (Zhou 1994), by considering a variation principle (Reiner and Radke 1990; Sharp and Honig 1990; Fogolari and Briggs 1997), or using thermodynamic arguments (Marcus 1955).

Based on Marcus theory (Marcus 1955), the electrostatic energy G_{elec}^{solv} contains three different terms. The first term is the classical electrostatic energy, G_{elec}^{cl}

$$G_{elec}^{cl} = \frac{1}{2} \int d^3\mathbf{r}\rho_m(\mathbf{r})\phi(\mathbf{r}) \tag{5.103}$$

The second term is arising from mixing the mobile species G_{elec}^{mob}:

$$G_{elec}^{mob} = k_B T \int d^3\mathbf{r} \sum_i c_i(\mathbf{r}) \ln \frac{c_i(\mathbf{r})}{c_i^{\infty}} \tag{5.104}$$

Combining Eqs. (5.102) and (5.104), we obtain

$$G_{elec}^{mob} = -\int d^3\mathbf{r} \left(\sum_i c_i(\mathbf{r}) q_i \right) \phi(\mathbf{r}) \tag{5.105}$$

The third term is the so-called *osmotic* term, which due to nonuniform ionic concentration, and it is calculated as a volume integral:

$$\begin{aligned} G_{elec}^{solvent} &= k_B T \int d^3\mathbf{r} \sum_i \left(c_i^{\infty} - c_i(\mathbf{r})\right) \\ &= k_B T \int d^3\mathbf{r} \sum_i c_i(\mathbf{r}) \left[\exp\left(\beta q_i \phi(\mathbf{r})\right) - 1\right] \\ &= \int d^3\mathbf{r} \left(\sum_i c_i(\mathbf{r}) q_i \right) \phi(\mathbf{r}) \end{aligned} \tag{5.106}$$

where the linearity of the exponential term is applied for ϕ small.

Therefore, the total electrostatic energy is

$$G_{elec} = G_{elec}^{cl} + G_{elec}^{mob} + G_{elec}^{solvent}$$

or

$$G_{elec} = G_{elec}^{cl} = \frac{1}{2} \int d^3\mathbf{r}\rho_m(\mathbf{r})\phi(\mathbf{r}) \tag{5.107}$$

where ρ_m is the charge density of fixed charges (i.e., nonionic charges such as partial atomic charges of macromolecule). It is possible to also calculate the free energy by knowing the electrostatic potential, for example, for the possible charging process of a macromolecule in an ionic environment. That is done by combining different

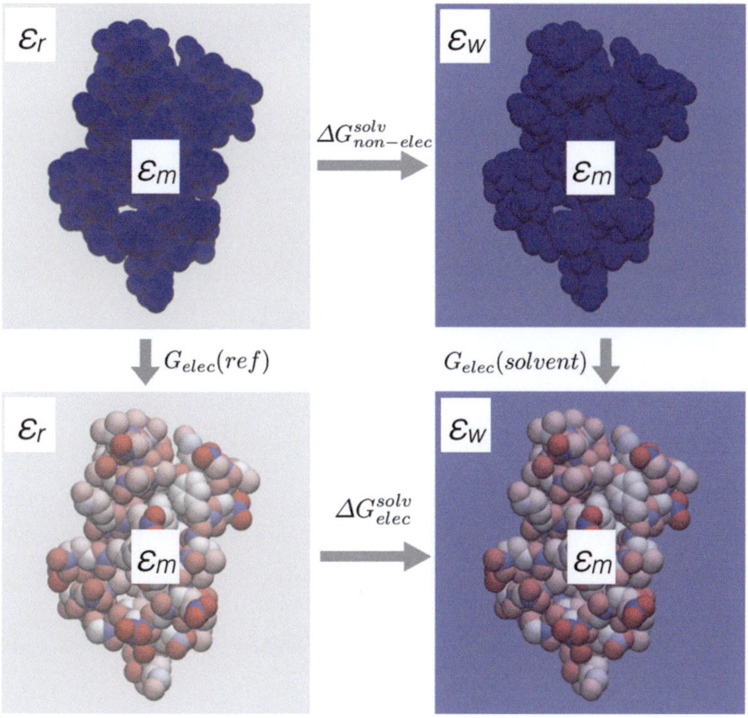

Fig. 5.8 Thermodynamic cycle for calculation of electrostatic solvation free energy. The difference in the charging energy, ΔG_{elec} in reference surrounding phase and in solvent is the electrostatic solvation free energy. Colors in bottom plots indicate the partial atom charges. In the top plots, the blue colour indicates no partial charges on atoms

possible processes in a thermodynamic cycle that can lead to the computation of the theoretical free energy of some real process, as discussed in Misra et al. (1994).

For instance, the thermodynamic cycle shown in Fig. 5.8 can be used to calculate the electrostatic component of the solvation free energy. This thermodynamic cycle indicates that the electrostatic component of the solvation free energy is the difference in the free energies related to two purely hypothetical charging processes; one in some reference surrounding environment phase (e.g., with dielectric constant equal to that of the macromolecule) and the other the solvent surrounding environment with dielectric constant ε_w: Thus, using Eq. (5.107), we can write

$$\begin{aligned}\Delta G^{solv}_{elec} &= G_{elec}(solvent) - G_{elec}(ref) \\ &= \frac{1}{2}\int d^3\mathbf{r}\rho_m(\mathbf{r})\,[\phi_w(\mathbf{r}) - \phi_r(\mathbf{r})]\end{aligned} \tag{5.108}$$

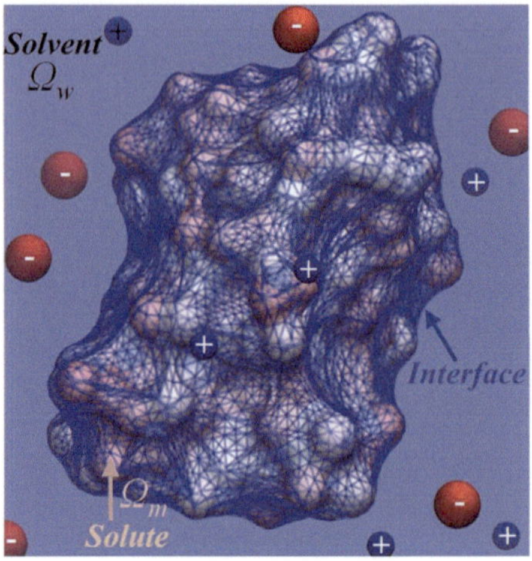

Fig. 5.9 Different computational regions of interest: The solute (macromolecule) region, Ω_m, with dielectric constant ε_m; solvent region, Ω_w, with dielectric constant ε_w where different mobile ions are denoted; dielectric interface (wireframes surface)

Often, $\phi_w(\mathbf{r}) - \phi_r(\mathbf{r})$ is called *reaction* potential, $\phi_{reac}(\mathbf{r})$, and thus Eq. (5.108) can be written as

$$\Delta G_{elec}^{solv} = \frac{1}{2}\int d^3\mathbf{r}\rho_m(\mathbf{r})\phi_{reac}(\mathbf{r}) \tag{5.109}$$

In Eq. (5.109), ΔG_{elec}^{solv} represents the work done by electrostatic forces for transferring a set of partial atomic charges of macromolecule from a fixed point in some reference surrounding environment with dielectric constant ε_r to a fixed point in solvent surrounding environment with dielectric constant ε_w.

Depending on the shape and charge distribution of macromolecule, numerical solutions of the Poisson-Boltzmann equation could be difficult. Solvated macromolecular systems are in general modeled by regions with different dielectric constants. Figure 5.9 illustrates a solvated macromolecule occupying the region Ω, where the macromolecule region is represented by Ω_m as a solid surface; the solvent region is represented by Ω_w. The dielectric interface σ is defined by the molecular surface and represents the region not penetrated by mobile ions, and $\mathbf{n}$ will represent an unit vector normal to σ pointing from Ω_m to Ω_w. The transition from solute (with low-dielectric constant) to solvent (with high-dielectric constant) is modelled to be abrupt, giving rise to the dielectric interface σ. There are two conditions on σ that are usually satisfied:

$$\begin{aligned}(\phi(\mathbf{r}))_{\Omega_m} &= (\phi(\mathbf{r}))_{\Omega_w} \\ \left(\varepsilon\frac{\partial\phi(\mathbf{r})}{\partial\mathbf{n}}\right)_{\Omega_m} &= \left(\varepsilon\frac{\partial\phi(\mathbf{r})}{\partial\mathbf{n}}\right)_{\Omega_w}\end{aligned} \tag{5.110}$$

These conditions are used in the methods based on *boundary integral equations*, but may not apply to *finite difference methods*. Usually, the boundary of the entire computational domain is also defined, Γ. In addition, approximated Dirichlet boundary condition is imposed in the boundary Γ.

The widely used numerical methods include finite difference method (FDM), the boundary element method (BEM), and finite element method (FEM).

5.3.1.5 Generalized Born Model

The *generalized Born* (GB) model is another approach to describe the electrostatic interactions in a multiple-dielectric environment in fewer computation efforts. To obtain the electrostatic potential ϕ in such a model, we have to solve the following Poisson equation:

$$\nabla\left[\varepsilon(\mathbf{r})\nabla\phi(\mathbf{r})\right] = -\frac{1}{\varepsilon_0}\rho(\mathbf{r}) \tag{5.111}$$

where ρ is the charge distribution and ε is the dielectric constant, which is equal to ε_m in the macromolecular interior, and it is equal to ε_w elsewhere. For a macromolecule immersed in a reference environment (e.g., gas phase), the dielectric constant of the exterior region is ε_r (e.g., for the gas phase $\varepsilon_r = 1$), and hence $\varepsilon = \varepsilon_r$. Then, we can solve Eq. (5.111) under these two conditions:

$$\nabla\left[\varepsilon(\mathbf{r})\nabla\phi_w(\mathbf{r})\right] = -\frac{1}{\varepsilon_0}\rho(\mathbf{r}), \tag{5.112}$$

$$\varepsilon(interior) = \varepsilon_m,\ \varepsilon(exterior) = \varepsilon_w$$

$$\nabla\left[\varepsilon(\mathbf{r})\nabla\phi_r(\mathbf{r})\right] = -\frac{1}{\varepsilon_0}\rho(\mathbf{r}),$$

$$\varepsilon(interior) = \varepsilon_m,\ \varepsilon(exterior) = \varepsilon_r \tag{5.113}$$

with solutions, respectively, ϕ_w and ϕ_r. The difference between these two potentials gives the reaction field

$$\phi_{reac} = \phi_w - \phi_r \tag{5.114}$$

The electrostatic contribution in solvation free energy is then given by

$$\Delta G_{elec}^{solv} = \frac{1}{2}\int d^3\mathbf{r}\phi_{reac}(\mathbf{r})\rho(\mathbf{r}) \tag{5.115}$$

Approximating the macromolecule charge distribution by a set of partial atomic point charges q_i for $i = 1, 2, \cdots, M$, then

$$\Delta G_{elec}^{solv} = \frac{1}{2}\sum_{i=1}^{M} q_i\phi_{reac}(\mathbf{r}_i) \tag{5.116}$$

In the case of a single ion of radius a and charge q, the potentials can be found analytically from Eq. (5.112), as

$$\phi_w = \frac{q}{4\pi\varepsilon_0\varepsilon_w a} \tag{5.117}$$
$$\phi_r = \frac{q}{4\pi\varepsilon_0 a}$$

where $\varepsilon_r = 1$.

Then, the electrostatic part of solvation free energy is given by the well known *Born formula* (Born 1920)

$$\Delta G_{elec}^{solv} = \frac{q^2}{8\pi\varepsilon_0 a}\left(\frac{1}{\varepsilon_w} - 1\right) \tag{5.118}$$

If then the macromolecule would have been considered as a set of charges $q_1, \cdots, q_M$ embedded in spheres of radii $a_1, \cdots, a_M$, and assuming that the separation between any of these two spheres i and j is r_{ij} is sufficiently large in comparison to their radii, the solvation free energy can be written as

$$\Delta G_{elec}^{solv} = \frac{1}{2}\sum_{i=1}^{M}\frac{q_i^2}{4\pi\varepsilon_0 a_i}\left(\frac{1}{\varepsilon_w} - 1\right) + \frac{1}{2}\sum_{i\neq j}^{M}\frac{q_j q_j}{4\pi\varepsilon_0 r_{ij}}\left(\frac{1}{\varepsilon_w} - 1\right) \tag{5.119}$$

where the first term is the sum of individual Born terms and the second term is the sum of the pair-wise Coulomb interactions.

In the GB theory, we attempt to find the same relatively simple analytical formula as in Eq. (5.119) by solving the Poisson equation directly. First, we will assume no salt effects in the description, then the Poisson equation is linear, which provides a ΔG_{elec}^{solv} quadratic in the charges. Besides, as we will see, the effect of the dielectric constants of solvent and macromolecule is different.

From classical electrostatics, assuming a linearly polarized medium (i.e., isotropic medium), the work needed to assemble a charge distribution can also be formulated in terms of the scalar product of the electric field $\mathbf{E}$ and electric displacement $\mathbf{D}$:

$$\mathbf{D} = \varepsilon_0\varepsilon\mathbf{E}$$

where ε is the dielectric constant. The work is

$$W = \frac{1}{2}\int \mathbf{E}\cdot\mathbf{D}d^3\mathbf{r} = \frac{1}{2}\int\left(\frac{\mathbf{D}}{\varepsilon_0\varepsilon}\right)\cdot\mathbf{D}d^3\mathbf{r} \tag{5.120}$$

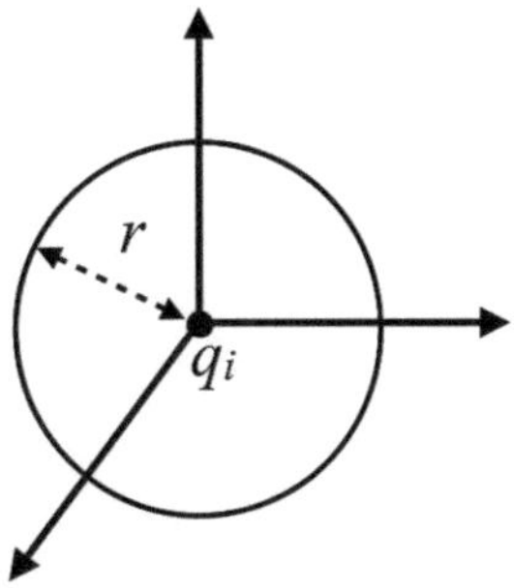

Fig. 5.10 The electrostatic field of a point charge at the origin

Let us consider the charge q_i which is assumed to be placed at the origin of some coordinative systems (as shown in Fig. 5.10) representing the partial charge of atom i. Using the Maxwell equation, we can write

$$\nabla \cdot \mathbf{D} = \rho$$

Integration according to the volume and using the Gaussian theorem, we get

$$\int \nabla \cdot \mathbf{D} \, d^3\mathbf{r} = \int \rho \, d^3\mathbf{r} = q_i$$

or

$$\int \mathbf{D} \cdot d\mathbf{S} = q_i$$

since the electric displacement vector **D** has a radial symmetry, then we can write

$$\int D(r) dS = q_i$$

where $dS = d(4\pi r^2) = 8\pi r dr$, then we obtain the solution as

$$D(r) = \frac{q_i}{4\pi r^2}$$

As the vector, it can be written as

$$\mathbf{D} = \frac{q_i \mathbf{r}}{4\pi r^3}$$

The electric field vector is

$$\mathbf{E} = \frac{q_i \mathbf{r}}{4\pi \varepsilon_0 \varepsilon r^3}$$

The work was done to place a charge q_i at the origin within some macromolecule whose interior part has a dielectric constant ε_m, surrounded by a reference medium with dielectric constant ε_r with no other charges placed yet, is

$$
\begin{aligned}
W_i &= \frac{1}{2}\int d^3\mathbf{r}\frac{\mathbf{D}}{\varepsilon_0\varepsilon}\cdot\mathbf{D} \\
&= \frac{1}{8\pi\varepsilon_0}\int_{interior} d^3\mathbf{r}\frac{q_i\mathbf{r}}{\varepsilon_m r^3}\cdot\frac{q_i\mathbf{r}}{r^3} \\
&+ \frac{1}{8\pi\varepsilon_0}\int_{exterior} d^3\mathbf{r}\frac{q_i\mathbf{r}}{\varepsilon_r r^3}\cdot\frac{q_i\mathbf{r}}{r^3} \\
&= \frac{q_i^2}{8\pi\varepsilon_0}\left[\int_{interior} d^3\mathbf{r}\frac{1}{\varepsilon_m r^4} + \int_{exterior} d^3\mathbf{r}\frac{1}{\varepsilon_r r^4}\right]
\end{aligned}
\tag{5.121}
$$

If the reference medium has the same dielectric constant as the macromolecule, $\varepsilon_r = \varepsilon_m$, then

$$
W_i(ref) = \frac{q_i^2}{8\pi\varepsilon_0}\left[\int_{interior} d^3\mathbf{r}\frac{1}{\varepsilon_m r^4} + \int_{exterior} d^3\mathbf{r}\frac{1}{\varepsilon_m r^4}\right] \tag{5.122}
$$

If the reference medium is the solvent, $\varepsilon_r = \varepsilon_w$, then

$$
W_i(wat) = \frac{q_i^2}{8\pi\varepsilon_0}\left[\int_{interior} d^3\mathbf{r}\frac{1}{\varepsilon_m r^4} + \int_{exterior} d^3\mathbf{r}\frac{1}{\varepsilon_w r^4}\right] \tag{5.123}
$$

Then, the electrostatic contribution to solvation free energy, which is the work done for moving a partial charge of a macromolecule placed at some fixed point origin in a reference medium to another fixed point in the macromolecule in a solvent medium, is

$$
\Delta G_{elec}^{solv} = W_i(wat) - W_i(ref) = \frac{q_i^2}{8\pi\varepsilon_0}\int_{exterior}\left[\frac{1}{\varepsilon_w} - \frac{1}{\varepsilon_m}\right]\frac{d^3\mathbf{r}}{r^4} \tag{5.124}
$$

or in the form

$$
\Delta G_{elec}^{solv} = \frac{1}{2}\frac{q_i^2}{R_i\varepsilon_0}\left[\frac{1}{\varepsilon_w} - \frac{1}{\varepsilon_m}\right] \tag{5.125}
$$

where R_i is the effective Born radius,

$$
\frac{1}{R_i} = \frac{1}{4\pi}\int_{exterior}\frac{d^3\mathbf{r}}{r^4} \tag{5.126}
$$

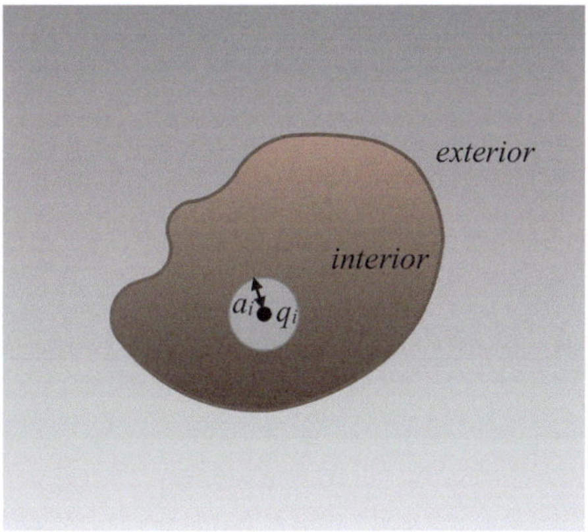

Fig. 5.11 The illustration of the Born sphere with radius a_i, interior and exterior part of the macromolecule

In general, the effective Born radius is an integral over the interior region of the macromolecule, excluding a radius a_i around the point of charge q_i, as illustrated in Fig. 5.11,

$$\begin{aligned}\frac{1}{R_i} &= \frac{1}{4\pi}\int_{exterior}\frac{d^3\mathbf{r}}{r^4} \\ &= \frac{1}{4\pi}\int_{all\,space}\frac{d^3\mathbf{r}}{r^4} \\ &-\left[\frac{1}{4\pi}\int_{interior(r>a_i)}\frac{d^3\mathbf{r}}{r^4}+\frac{1}{4\pi}\int_{r\leq a_i}\frac{d^3\mathbf{r}}{r^4}\right] \\ &= \frac{1}{a_i}-\frac{1}{4\pi}\int_{interior(r>a_i)}\frac{d^3\mathbf{r}}{r^4}\end{aligned} \tag{5.127}$$

where the integration over all space gives zero because $\lim_{r\to\infty} 1/r^4 = 0$. The second term in Eq. (5.127) is the so-called *excluded volume integral*. It can be seen that if the molecular boundary is simply the sphere of radius a_i, then $R_i = a_i$.

For a macromolecular system of M partial atomic charges, the electrostatic term of the solvation free energy for bringing the system of charges for a fixed point in surrounding medium with dielectric constant $\varepsilon_r = \varepsilon_m$ to a fixed point in surrounding medium being the solvent with dielectric constant ε_w is

$$\Delta G_{elec}^{solv} = \frac{1}{2}\sum_{i=1}^{M}\sum_{j=1}^{M}\frac{q_i q_j}{4\pi\varepsilon_0 f_{GB}(r_{ij})}\left[\frac{1}{\varepsilon_w}-\frac{1}{\varepsilon_m}\right] \tag{5.128}$$

where $f_{GB}(r_{ij})$ is a function, which if $i = j$ becomes the effective Born radius R_i and if $i \neq j$ (i.e., in pair-wise terms) becomes the effective interaction distance. The form chosen for this functions is (Still et al. 1990):

$$f_{GB}(r_{ij}) = \left(r_{ij}^2 + R_i R_j \exp\left(-\frac{r_{ij}^2}{4R_i R_j} \right) \right)^{1/2} \tag{5.129}$$

The electrostatic term of the potential of mean force is then given as

$$W_{elec} = H_{mm}^{elec} + \Delta G_{elec}^{solv}$$

where H_{mm}^{elec} is the intra-macromolecular electrostatic interaction energy

$$H_{mm}^{elec} = \frac{1}{2} \sum_{i=1}^{M} \sum_{j=1 \neq i}^{M} \frac{q_i q_j}{4\pi \varepsilon_0 \varepsilon_m r_{ij}}$$

Thus, we obtain

$$\begin{aligned} W_{elec} &= \frac{1}{2} \sum_{i=1}^{M} \sum_{j=1 \neq i}^{M} \frac{q_i q_j}{4\pi \varepsilon_0 \varepsilon_m r_{ij}} \\ &+ \frac{1}{2} \sum_{i=1}^{M} \sum_{j=1}^{M} \frac{q_i q_j}{4\pi \varepsilon_0 f_{GB}(r_{ij})} \left[\frac{1}{\varepsilon_w} - \frac{1}{\varepsilon_m} \right] \end{aligned} \tag{5.130}$$

It can be seen that both dielectric constants, of the solvent and macromolecule, appear in the formula (see Eq. (5.130)), where ε_m of macromolecule appears in the second term of Eq. (5.130) since ε_m is considered as the dielectric constant of the reference environment system.

The GB model can be extended to consider low salt concentrations at the Debye-Hückel approximation level as discussed in Srinivasan et al. (1999). In this approximation Eq. (5.128) is written as (Tjong and Zhou 2007b)

$$\Delta G_{elec}^{solv} = \frac{1}{2} \sum_{i=1}^{M} \sum_{j=1}^{M} \frac{q_i q_j}{4\pi \varepsilon_0 f_{GB}(r_{ij})} \left[\frac{\exp\left(-\alpha \kappa f_{GB}(r_{ij})\right)}{\varepsilon_w} - \frac{1}{\varepsilon_m} \right] \tag{5.131}$$

where κ is the Debye-Hückel screening parameter, given

$$\kappa = \left(\frac{I}{\varepsilon_w \varepsilon_0 k_B T} \right)^{1/2} \tag{5.132}$$

where I is the ionic strength. In Eq. (5.132), α depends on the ionic strength according to (Tjong and Zhou 2007b)

$$\alpha = \frac{1 + 0.0169\sqrt{I}}{1 + 0.075\sqrt{I}} \tag{5.133}$$

The GB methods discussed in the literature do not have any other dependence on the macromolecule and solvent dielectric constants beyond the scaling factor shown in Eq. (5.128). In fact, the linear Poisson-Boltzmann solvation energy method has more complicated dependencies on ε_m and ε_w. In Tjong and Zhou (2007b), it has been found that these dependencies can be modeled accurately by multiplying ΔG_{elec}^{solv} in Eq. (5.131) by a scaling factor $S(\varepsilon_m, \varepsilon_w)$:

$$S(\varepsilon_m, \varepsilon_w) = \frac{A + 2B\varepsilon_m/\varepsilon_w}{1 + 2\varepsilon_m/\varepsilon_w} \tag{5.134}$$

with

$$A = -1.63 \times 10^{-3} \mid Q \mid^{0.65} + 2.18 \times 10^{-6} M + 1.016$$

$$B = 3.31 \times 10^{-2} \mid Q \mid^{0.65} - 4.77 \times 10^{-5} M + 0.683$$

where Q denotes the net charge of the macromolecule. Then, Eq. (5.131) is modified as the following

$$\Delta G_{elec}^{solv} = S(\varepsilon_m, \varepsilon_w)\frac{1}{2}\sum_{i=1}^{M}\sum_{j=1}^{M}\frac{q_i q_j}{4\pi\varepsilon_0 f_{GB}(r_{ij})} \times \left[\frac{\exp\left(-\alpha\kappa f_{GB}(r_{ij})\right)}{\varepsilon_w} - \frac{1}{\varepsilon_m}\right] \tag{5.135}$$

Thus, we can rewrite Eq. (5.130) as

$$W_{elec} = \frac{1}{2}\sum_{i=1}^{M}\sum_{j=1\neq i}^{M}\frac{q_i q_j}{4\pi\varepsilon_0\varepsilon_m r_{ij}} + S(\varepsilon_m, \varepsilon_w)\frac{1}{2}\sum_{i=1}^{M}\sum_{j=1}^{M}\frac{q_i q_j}{4\pi\varepsilon_0 f_{GB}(r_{ij})} \times \left[\frac{\exp\left(-\alpha\kappa f_{GB}(r_{ij})\right)}{\varepsilon_w} - \frac{1}{\varepsilon_m}\right] \tag{5.136}$$

The effective Born radii in GB models can be determined using the continuum dielectric models and Coulomb field approximation. Other approximations are also introduced to reduce the computation efforts, such as those that attempt to adjust these parameters using the fitting to the experimental data or the numerical continuum dielectric data. For relatively small molecules, the extension of the standard GB models was introduced by carrying out integrals over the macromolecule dielectric regions numerically (Luo et al. 1997; Majeux et al. 1999). In particular, we can mention the SMx models of Cramer and Truhlar (1999). In SMx models, the electrostatic contribution of GB models combines with the surface-area dependent solvation term and various quantum mechanical treatments of the solute charge distribution. These models, although accurate, have generally not used for macromolecules, since they are not fast enough.

Instead, the pair-wise methods have shown to be fast in applications to macromolecules. In this approach, the integral in Eq. (5.127) is approximated as a summation over the contributions of each atom. Thus, based on this model, if the molecule is consisting of a set of spheres with non-vanishing radius a_i. Consider two spheres, i and j, respectively, where sphere i is assumed to be placed at the origin and sphere j at position r_{ij} with respect to i, then Eq. (5.127) can be written as summation of integrals over spherical volumes:

$$\frac{1}{R_i} = \frac{1}{a_i} - \frac{1}{4\pi}\sum_{j=1}^{M}\int_{a_i}^{\infty}\frac{1}{r^4}d^3\mathbf{r} \tag{5.137}$$

The analytical solution of the integral in Eq. (5.137) is given in (Schaefer and Froemmel 1990):

$$\int_{a_i}^{\infty}\frac{d^3\mathbf{r}}{r^4} = \tag{5.138}$$

$$\begin{cases} \dfrac{a_j}{2(r_{ij}^2 - a_j^2)} + \dfrac{1}{4r_{ij}}\log\dfrac{r_{ij} - a_j}{r_{ij} + a_j}, \\ \left(r_{ij} \geq a_i + a_j\right) \\ \dfrac{2-\theta}{4a_i} - \dfrac{1}{4(r_{ij} + a_j)} + \dfrac{1}{4r_{ij}}\log\dfrac{a_i}{r_{ij} + a_j}, \\ \left(\mid a_i - a_j \mid \leq r_{ij} \leq a_i + a_j\right) \\ \dfrac{a_j}{2(r_{ij}^2 - a_j^2)} + \dfrac{1}{a_i} + \dfrac{1}{4r_{ij}}\log\dfrac{a_j - r_{ij}}{a_j + r_{ij}}, \\ \left(r_{ij} \leq \mid a_i - a_j \mid \wedge a_i \leq a_j\right) \\ \dfrac{1}{a_i} - \dfrac{1}{a_j}, \\ \left(r_{ij} = 0 \wedge a_i \leq a_j\right) \\ 0, \\ \left(r_{ij} \leq \mid a_i - a_j \mid \wedge a_i \geq a_j\right) \end{cases}$$

where

$$\theta = \frac{r_{ij}^2 + a_i^2 - a_j^2}{2r_{ij}a_i}$$

Other approached have also been introduced to compensate for overlapping of spheres. For example, Hawkins et al. (1995, 1996) have proposed the following expression:

$$\frac{1}{R_i} = \frac{1}{a_i} - \frac{1}{4\pi} \sum_{j=1}^{M} \int_{a_i}^{\infty} dr \frac{1}{r^2} H_{ij}(r_{ij}, S_{ij}, a_j) \tag{5.139}$$

where $H_{ij}(r_{ij}, S_{ij}, a_j)$ represents the fraction of the area of a sphere of radius r centred at atom i that is shielded by a sphere of scaled radius $S_{ij}a_j$ at a distance r_{ij}. By using the scaled radius $S_{ij}a_j$ instead of a_j in Eq. (5.139), the overlapping of van der Waals spheres is calculated. The scaling factors are estimated by a fitting procedure to the experimental data of the free energies of solvation, with values initially restricted in the range between 0.5 and 1.0. $H_{ij}(r_{ij}, S_{ij}, a_j)$ can be calculated analytically (Hawkins et al. 1995):

$$\begin{aligned} \frac{1}{4\pi} \sum_{j=1}^{M} \int_{a_i}^{\infty} dr \frac{1}{r^2} H_{ij}(r_{ij}, S_{ij}, a_j) &= \frac{1}{2} \sum_{j=1}^{M} \left[\frac{1}{L_{ij}} - \frac{1}{U_{ij}} \right. \\ &+ \frac{r_{ij}}{4} \left(\frac{1}{U_{ij}^2} - \frac{1}{L_{ij}^2} \right) + \frac{1}{2r_{ij}} \ln \frac{L_{ij}}{U_{ij}} \\ &\left. + \frac{S_{ij}^2 a_j^2}{4r_{ij}} \left(\frac{1}{L_{ij}^2} - \frac{1}{U_{ij}^2} \right) \right] \end{aligned} \tag{5.140}$$

where

$$L_{ij} = \begin{cases} 1 & r_{ij} + S_{ij}a_j \le a_i \\ a_i & r_{ij} - S_{ij}a_j \le a_i < r_{ij} + S_{ij}a_j \\ r_{ij} - S_{ij}a_j & a_i \le r_{ij} - S_{ij}a_j \end{cases} \tag{5.141}$$

and

$$U_{ij} = \begin{cases} 1 & r_{ij} + S_{ij}a_j \le a_i \\ r_{ij} + S_{ij}a_j & a_i < r_{ij} + S_{ij}a_j \end{cases} \tag{5.142}$$

The fitting parameters are considered the set of $\{a_i, S_{ij}\}$, which are optimised by a minimisation procedure which minimises the sums of the squares of the errors:

$$\begin{aligned} U = \frac{1}{r+I} &\left[\sum_{i=1}^{r} \mid \Delta G^{solv}(exp_i) - \Delta G^{solv}(calc_i) \mid \right. \\ &\left. + \frac{1}{6} \sum_{i=1}^{I} \mid \Delta G^{solv}(exp_i) - \Delta G^{solv}(calc_i) \mid \right] \end{aligned}$$

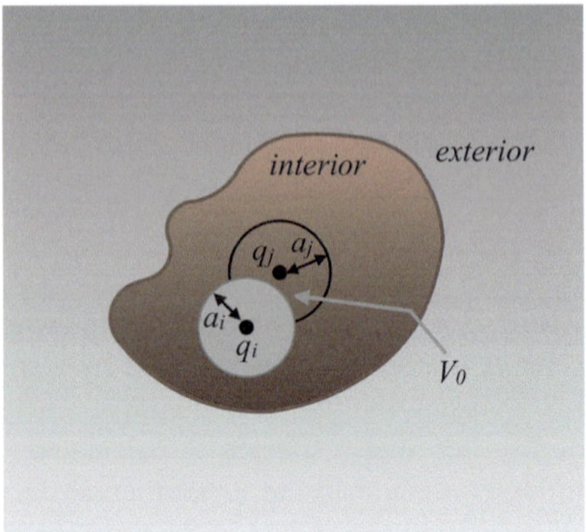

Fig. 5.12 The illustration of the contribution from the atom j to the integral Z_{ij}

where r is the total number of neutral molecules and I is the total number of ionic compounds in the training set. $\Delta G^{solv}(exp)$ is the experimental free energy of solvation and $\Delta G^{solv}(calc)$ is the calculated free energy of solvation for the given set of parameters.

These approximations are known as the generalized Born model with the Coulomb-field approximation. Recently, Tjong and Zhou (2007a,b) introduced a parametrization-free and accurate method for calculation of the effective Born radius in GB model based on a previously proposed approach (Grycuk 2003). In this approximation, known as generalized Born model with the Grycuk-Kirkwood approximation, the integral of Eq. (5.127) is replaced by

$$\frac{1}{R_i^3} = \frac{1}{a_i^3} - \frac{3}{4\pi} \int_{interior(r>a_i)} \frac{d^3\mathbf{r}}{r^6} \tag{5.143}$$

To a zeroth order approximation, the volume integral of Eq. (5.143) can be written as a sum of contributions from individual atoms. As such, the contribution from the j $(j \neq i)$ atom is the integral over the region of its Born sphere laying outside atom i:

$$Z_{ji} = \frac{3}{4\pi} \int_{V_0} \frac{d^3\mathbf{r}}{r^6} \tag{5.144}$$

where V_0 is the volume illustrated in Fig. 5.12.

We can distinguish four different cases:

1. The atoms i and j do not intersect, i.e., $r_{ij} > a_i + a_j$. In this case,

$$Z_{ji} = \frac{a_j^3}{\left(r_{ij}^2 - a_j^2\right)^3} \tag{5.145}$$

2. The atoms i and j intersect, but neither is completely laying inside the other, i.e., $|a_i - a_j| \leq r_{ij} < a_i + a_j$. The integral is

$$Z_{ji} = \frac{1}{16r_{ij}}\left[-6\left(\frac{1}{a_i^2} - \frac{1}{\left(r_{ij}+a_j\right)^2}\right)\right.$$
$$+ 8r_{ij}\left(\frac{1}{a_i^3} - \frac{1}{\left(r_{ij}+a_j\right)^3}\right)$$
$$\left. - 3\left(r_{ij}^2 - a_j^2\right)\left(\frac{1}{a_i^4} - \frac{1}{\left(r_{ij}+a_j\right)^4}\right)\right] \tag{5.146}$$

3. The atom i is completely inside atom j, i.e., $r_{ij} \leq |a_i - a_j|$ and $a_i \leq a_j$. In this case

$$Z_{ji} = \frac{1}{a_i^3} - \frac{a_j^3}{\left(a_j^2 - r_{ij}^2\right)^3} \tag{5.147}$$

4. The atom j is completely inside atom i, i.e. $r_{ij} \leq |a_i - a_j|$ and $a_i \geq a_j$. In this case the atom j does not contribute, thus

$$Z_{ji} = 0 \tag{5.148}$$

Then, the effective Born radius is calculated as

$$\frac{1}{R_i} = \left(\frac{1}{a_i^3} - \sum_{j=1\neq i}^{M} Z_{ji}\right)^{1/3} \tag{5.149}$$

Comparison of the approximation given by Eq. (5.149) with Coulomb-field approximation (Eq. (5.138)) shows that the latter results involve a logarithmic function, which makes the calculations computationally more expensive, about 10% additional CPU time as reported elsewhere (Tjong and Zhou 2007b).

Following the above discussion, Z_{ji} overestimates the contribution of atom j if its region outside the atom i intersects with another atom k. In such case, one can use the scaling Born radius strategy ($S_{ji}a_j$), which is originally provided as a solution to the problem (Hawkins et al. 1995; Gallicchio and Levy 2004). The volume of overlapping spheres according to the Poincaré exclusion principle is (Petitjean 1994)

$$V = \sum_{j=1}^{M} V_j - \sum_{j=1}^{M}\sum_{k>j}^{M} V_{jk} + \sum_{j=1}^{M}\sum_{k>j}^{M}\sum_{l>k>j}^{M} V_{jkl} - \cdots \tag{5.150}$$

where V_j is the volume of atom j, V_{jk} is the intersection volume of atoms j and k, V_{jkl} is the intersection volume of atoms j, k, and l, and so on. The self-volume V_{jj} is found as

$$V_{jj} = V_j - \frac{1}{2}\sum_{k=1}^{M} V_{jk} + \frac{1}{3}\sum_{k=1}^{M}\sum_{l=1}^{M} V_{jkl} - \cdots \tag{5.151}$$

The scaling factor is then calculated as

$$S_{ji} = \frac{V_{jj} + \frac{1}{2}V_{ji}}{V_j} \tag{5.152}$$

If no other atoms intersect with atom j, then $S_{ji} = 1$. The intersection volumes can also be well approximated using the Gaussian integrals as in Gallicchio and Levy (2004), which works well if the intersecting atoms are heavy atoms, because the hydrogen atoms are buried inside the heavy atoms in molecules and the Gaussian approximation may give large errors. Then, the effective Born radius given by Eq. (5.149) can be modified as the following:

$$\frac{1}{R_i} = \left(\frac{1}{a_i^3} - \sum_{j=1\neq i}^{M} S_{ji} Z_{ji}\right)^{1/3} \tag{5.153}$$

Chapter 6
Molecular Dynamics Methods in Simulations of Macromolecules

Molecular dynamics simulations at atomic level have widely been used in studying macromolecular systems, such as protein, DNA and their complexes, mainly because the laws of classical statistical mechanics can largely govern the processes involved at the experimental conditions. Macromolecules, such as proteins, are characterized by dynamics with time scales ranging from nanoseconds to milliseconds. In this chapter, we discuss the molecular dynamics method as one of the most common computer simulation approach used to study molecular systems. In particular, we will present the equations of motion in the most relevant statistical ensembles used in the molecular dynamics simulations of molecular systems.

6.1 Introduction

Molecular dynamics (MD) method is playing a major role in studying macromolecular systems (Karplus and McCammon 2002) in part because the laws of classical statistical mechanics can mainly govern the processes involved at the experimental conditions (van Gunsteren et al. 2006). MD is a computer simulation approach used to numerically integrate the equations of motion of atoms and molecules by approximations of known physics (Allen and Tildesley 1989; Hoover 1991; Frenkel and Smit 2001). Ciccotti and Vanden-Eijnden (2015) argue that MD is an engine that is used to sample both time-independent and time-dependent statistical mechanical properties of molecular systems. In general, molecular systems may consist of a large number of particles. Thus it is impossible to find the properties of such complex systems analytically (Abraham 1986). When the number of bodies is more than two no analytical solutions can be seen and result in chaotic motion (Posch et al. 1986). MD simulation circumvents this problem by using numerical methods (Allen and Tildesley 1989; Hoover 1991). The first large-scale

H. Kamberaj, *Molecular Dynamics Simulations in Statistical Physics: Theory and Applications*, Scientific Computation,
https://doi.org/10.1007/978-3-030-35702-3_6

atomistic molecular dynamics simulations include the work of Abraham and co-workers (Abraham et al. 1984).

MD represents an interface between the experiment and theory and can be understood as a virtual experiment (Allen and Tildesley 1989; Frenkel and Smit 2001). MD probes the relationship between molecular structure, movement, and function (Karplus and McCammon 2002; Karplus and Kuriyan 2005). Molecular dynamics is a multidisciplinary method, and its laws and theories originate from different fields, such as mathematics, physics, and chemistry. Also, it implements algorithms from computer science and information theory (van Gunsteren et al. 2006). It was initially conceived within theoretical physics (Alder and Wainwright 1959), but today applies to other fields too, such as the computer simulations of the materials science and biomolecular systems. In the beginning, before the computers were used to perform molecular dynamics simulations, simple, but that required a lot of hard work, physical models were used, such as macroscopic spheres to prove the concepts (Bernal 1964; Hoover and Ree 1968; Alder et al. 1968).

Molecular dynamics as a discipline includes molecular modeling and computer simulations based on statistical mechanics laws. The main justification of MD method, as a specialized discipline of molecular modeling and computer simulation based on statistical mechanics, is that statistical ensemble averages are equal to time averages of the system, known as the ergodic hypothesis (Allen and Tildesley 1989; Leach 2001):

$$\langle \mathcal{A} \rangle = \lim_{T \to \infty} \int_0^T \mathrm{d}t \, \mathcal{A}(t) \tag{6.1}$$

where $\langle \cdots \rangle$ is an ensemble average and $\mathcal{A}$ can be any physical property of the system.

MD has also been termed *statistical mechanics by numbers* and *Laplace's vision of Newtonian mechanics* of predicting the future by animating nature's forces (Bernal 1964; Schlick 1996) and allowing insight into molecular motion on an atomic scale. However, long MD simulations are mathematically limited, due to cumulative errors in the numerical integration of equations of motion. However, these numerical errors can be minimized using more sophisticated algorithms and parameter selections (Martyna et al. 1994; Minary et al. 2004), but not eliminated (Piana et al. 2014). Furthermore, current potential functions may not, in all cases, be sufficiently accurate to reproduce the dynamics of molecular systems, so the much more computationally demanding Ab-Initio Molecular Dynamics method must be used combined with recent advances in parallel supercomputing (Hardy et al. 2011; Stone et al. 2011, 2013; Scarpazza et al. 2013; Phillips et al. 2014). Nevertheless, molecular dynamics method is successfully used to provide a detailed time and space resolution into representative behavior in phase space when combined with appropriate algorithms for optimizing the conformation search (Andricioaei and Straub 1996b).

MD approach has been used for a very long time to study macromolecular systems. These studies include many applications, such as the internal atomic motion (Amadei et al. 1993; Karplus and McCammon 2002), protein folding (Rogal and Bolhuis 2008), transition path sampling (Bolhuis et al. 2002), free energy calculations (Seyler and Beckstein 2014) and protein structure prediction (Perez et al. 2016). In all these studies, an observed limitation of standard MD method is the time and size scales covered in the simulations. For instance, in studying slow conformation motions of macromolecular systems (Palmer 1982; Clarage et al. 1995). To date there exist MD simulations that are far longer than standard MD simulation timescales, for instance, in Arkhipov et al. (2008) there are reported up to ten-microsecond MD simulation of a fast folding protein that was made possible through an improved scaling and parallel performance of the MD engine, or even longer, up to milliseconds time scale by computer engineering (Friedrichs et al. 2009; Dror et al. 2011). Long MD simulations are also possible using multiple time step integrator algorithms (Tuckerman et al. 1992; Tuckerman and Martyna 2000; Minary et al. 2004). Different advanced methods have been introduced to enhance conformation sampling of MD simulations (see the review Kamberaj 2019).

6.2 Equations of Motion

In this section, we discuss the equations of motion used in molecular dynamics simulations of different statistical ensembles.

6.2.1 Microcanonical Ensemble

The simplest form of equations of motion used in molecular dynamics corresponds to the microcanonical ensemble (namely NVE ensemble), in which the system is isolated and does not allow changes in number of particles (N), volume (V) and energy (E) (Allen and Tildesley 1989; Frenkel and Smit 2001). It corresponds to an adiabatic process with no heat exchange. During generation of a molecular dynamics trajectory in microcanonical ensemble, the energy is exchanged between potential and kinetic energy, but the total energy is constant. For an N particles system with coordinates $\mathbf{r} = \{\mathbf{r}_1, \cdots, \mathbf{r}_N\}$ and momenta $\mathbf{p} = \{\mathbf{p}_1, \cdots, \mathbf{p}_N\}$, the first order differential equations for each atom i with mass m_i are in Newton's notation as follows (Goldstein 2002):

$$\begin{aligned} \dot{\mathbf{p}}_i &= -\nabla U(\mathbf{r}), \\ \dot{\mathbf{r}}_i &= \mathbf{p}_i / m_i, \quad i = 1, \cdots, N \end{aligned} \tag{6.2}$$

The potential energy function $U(\mathbf{r})$ of the system is a function of the particle coordinates $\mathbf{r}$. In Physics, it is simply called the potential and in Chemistry as the force field (Leach 2001). The first equation comes from Newton's second law, where the force $\mathbf{F}_i$ acting on each particle i in the system is the negative gradient of the potential.

In a MD simulation run, every time step the positions and momenta of particles may be determined numerically using a numerical integration algorithm, such as Verlet or Leap-frog (Allen and Tildesley 1989). The time evolution of $\mathbf{r}$ and $\mathbf{p}$ defines a trajectory in the phase space. Given the initial positions (e.g. from theoretical knowledge) and momenta (e.g. according to Maxwell-Boltzmann distribution), the positions and velocities of the particles advance in time using Eq. (6.2).

It is interesting to explain the meaning of temperature in molecular dynamics simulations. From the macroscopic point of view, the temperature involves a huge number of particles. From the statistical point of view, the temperature is a statistical quantity, and as such only if there is a large enough number of atoms, statistical temperature can be estimated from the instantaneous temperature $\mathcal{T}$, which is found by equating the kinetic energy, E_k, of the system to $gk_B\mathcal{T}/2$ where g denotes the number of degrees of freedom of the system (Allen and Tildesley 1989):

$$\begin{aligned} T &= \langle \mathcal{T} \rangle, \\ \mathcal{T} &= 2E_k/gk_B \end{aligned} \tag{6.3}$$

If the number of atoms used in MD simulations is small, then the instantaneous temperature may show high fluctuations and may not converge to the thermodynamic temperature, known as the *temperature-related phenomenon* (Allen and Tildesley 1989). Something similar happens in biophysical simulations. The temperature of the system in NVE increases naturally when macromolecules, such as proteins, undergo exothermic conformation changes and binding.

6.2.2 Canonical Ensemble

The canonical (NVT) ensemble corresponds to the ensemble with the number of the atoms (N), volume (V) and temperature (T) constant. Often, it is called constant temperature molecular dynamics. In NVT, the energy of the system exchanges with the external bath also called as the *thermostat*.

There exist different thermostat methods introduced to absorb or dissipate energy at the boundaries of an MD system with a surrounding thermostat using the laws of physics, approximating in this way the canonical ensemble. In the MD simulations of macromolecular systems, the most popular methods to control temperature are the Nosé-Hoover thermostat, the Berendsen thermostat, and Langevin dynamics.

In the following, we will introduce all the methods that can be used to control the temperature in a molecular dynamics simulation.

6.2.2.1 Isokinetic Dynamics Method

The isokinetic dynamics method is the earliest technique used in the molecular dynamics simulations to re-scale the velocities sufficiently frequently for controlling the temperature of the system. In this approach, after each time step, Δt, of solving the microcanonical equations of motion (as given by Eqs. (6.2)), the particles momenta $\mathbf{p}(t)$ are updated to $\mathbf{p}(t + \Delta t)$, then each of the particle momentum is multiplied by a re-scaling factor giving in this way a new momentum $\mathbf{p}'(t + \Delta t)$ with the same kinetic energy as the previous time step (t):

$$\mathbf{p}_i'(t + \Delta t) = \sqrt{\frac{T_0}{\mathcal{T}}}\mathbf{p}_i(t + \Delta t), \quad i = 1, \cdots, N \tag{6.4}$$

where T_0 is the target temperature of the system and $\mathcal{T}$ is the instantaneous temperature calculated according to Eq. (6.3). The isokinetic dynamics method represents the simplest deterministic and time-reversible thermostat for sufficiently small integration time step Δt used in the molecular dynamics method.

6.2.2.2 Berendsen Thermostat

Another approach for controlling the temperature is the *Berendsen thermostat*, which represents a weak coupling of the system to an external heat bath at a fixed temperature T_0 (Berendsen et al. 1984). To control the temperature of system, each particle velocity is scaled at every timestep such that the rate change of temperature is proportional to the difference in temperature given as:

$$\frac{dT(t)}{dt} = \frac{1}{\tau}(T_0 - T(t)) \tag{6.5}$$

Here, τ represents a coupling parameter, which determines how strongly the bath and the system are coupled to each other. This method yields an exponential decay of the system's temperature towards the desired temperature with a change in the temperature at every time step given as:

$$\Delta T = \frac{\Delta t}{\tau}(T_0 - T(t)) \tag{6.6}$$

The scaling factor of the velocities is (Berendsen et al. 1984)

$$\lambda = \left[1 + \frac{\Delta t}{\tau}\left(\frac{T_0}{T(t - \Delta t/2)} - 1\right)\right]^{1/2} \tag{6.7}$$

The $T(t - \Delta t/2)$ indicates the use of the so-called leap-frog algorithm for the numerical integration. Alternatively, if the velocity-Verlet algorithm is used, then the scaling factor is given as:

$$\lambda = \left[1 + \frac{\Delta t}{\tau}\left(\frac{T_0}{T(t + \Delta t)} - 1\right)\right]^{1/2} \tag{6.8}$$

The velocities are scaled using the following scheme:

$$\mathbf{v}'_i(t + \Delta t) = \lambda \mathbf{v}_i(t + \Delta t), \quad i = 1, \cdots, N \tag{6.9}$$

In typical MD simulation, τ is an input parameter used to adjust the strength of the coupling of the system with the external heat bath. Its value has to be appropriately chosen such that when $\tau \to \infty$ the Berendsen thermostat becomes inactive and the molecular dynamics simulation is sampling a microcanonical ensemble. In this limit, the temperature fluctuations will grow until they reach the typical value of a microcanonical ensemble. However, the temperature will never reach the characteristic values of the canonical ensemble. In contrast, too small values of τ will generate nonphysically low-temperature fluctuations. Besides, when τ is chosen to be equal to the integration timestep Δt, the Berendsen thermostat behaves similarly to the velocity scaling approach discussed above for the isokinetic dynamics method. The suggested value of τ is $\tau \approx 0.1$ ps for MD simulations of condensed-phase systems and the most macromolecular simulations (Berendsen et al. 1984).

It is important to note that the ensemble sampled by the MD simulation using either an isokinetic dynamics method or a Berendsen thermostat approach is not corresponding to a canonical ensemble.

Example 1 Consider a harmonic oscillator with potential energy function

$$U(x) = \frac{k}{2}(x - x_0)^2$$

where $k = 317\,\text{kcal/mol/Å}^2$ and $x_0 = 1.523$ Å, which is a typical force constant and bond length for Csp^3-Csp^3 bond of the MM2 force field (Leach 2001). MD simulations are performed using Berendsen thermostat with integration time step $\Delta t = 1$ fs. The velocity-Verlet algorithm integrate numerically the equations of motion and the input parameter $\tau = 0.01$ ps. The initial velocity of the particle was generated randomly using a Maxwell-Boltzmann distribution at the target temperature $T_0 = 300$ K. While the initial position was randomly chosen.

In Fig. 6.1a we show the scatter plot of the velocity versus displacement of the harmonic oscillator. Besides, Fig. 6.1b shows the running average of the temperature. The results show that the sampling is not a canonical ensemble, but the velocity scaling factor can control the temperature fluctuations to the desired value of the temperature, which is $T_0 = 300$ K, in this case.

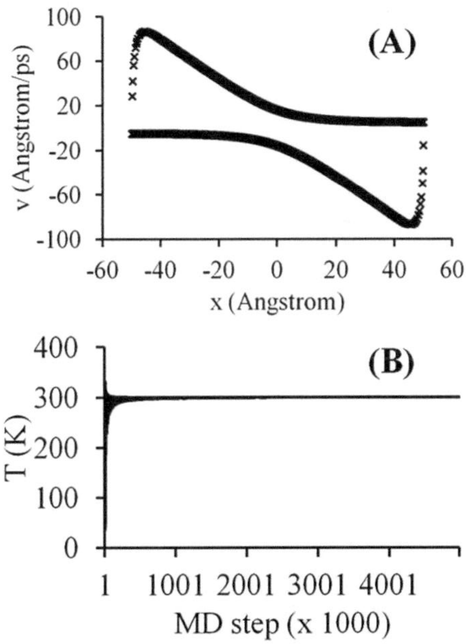

Fig. 6.1 (**a**) The position-velocity scatter plot and (**b**) the running average temperature of an harmonic oscillator using MD simulations with Berendsen thermostat: $\tau = 0.01$ ps, $\Delta t = 1$ fs and $k = 317$ kcal/mol/Å^2 and $x_0 = 1.523$ Å

The Berendsen thermostat is often used to relax a system to the target temperature. When the system's temperature converges to the equilibrium temperature, it is essential from the statistical mechanics' point of view to sampling a correct canonical ensemble.

6.2.2.3 Nosé-Hoover Thermostat

Nosé (1984c) originally introduced the extended dynamical systems method, and then unified by Hoover (1985b).

The extended dynamical systems, as shown below analytically, can sample the canonical ensemble. According to this approach, a heat bath is considered as an integral part of the system by extending the real variables of the system with the so-called thermostat variable s, associated with a new thermostat mass variable, namely $Q > 0$, and its conjugated momentum π. The magnitude of Q controls the temperature fluctuations, representing, thus, the coupling between the reservoir and the real system. The thermostat variable s is a time-scaling parameter, that is, the timescales in the extended system of variables are stretched by the factor s according to

$$dt = \frac{d\tau}{s}$$

Here, we will again consider a system of N particles as in the case of the microcanonical ensemble with the Hamiltonian function $H(\mathbf{r}, \mathbf{p})$. The extended dynamical system consists of one additional degree of freedom with an extended Hamiltonian, namely Nosé Hamiltonian, to control the temperature in an MD simulation given as in (Nosé 1984c):

$$H_N = \sum_{i=1}^{g_N} \left[\frac{(p_i')^2}{2s_i^2 m_i} + \frac{\pi_i^2}{2Q_i} + k_B T \ln(s_i) \right] + U(\mathbf{r}_1, \cdots, \mathbf{r}_N) \tag{6.10}$$

Here, g_N is the number of degrees of freedom of the real system, and T is the target temperature. Note that $\mathbf{p}'$ is the canonical momentum associated with the position variable, $\mathbf{r} = \left(q_1, \cdots, q_{g_N}\right)$. The ($'$) is used to distinguish it from the real momentum given by $\mathbf{p} = \mathbf{p}'/s$. Besides, we have coupled each degree of freedom to a thermostat using in this way a massive number of thermostats approach.

Nosé has shown (Nosé 1984c) that this system generates configurations from a canonical ensemble, under the assumption that the dynamics are ergodic. Besides, Hoover has shown (Hoover 1985b) that the intrinsic time variable must be re-scaled to provide trajectories at fixed spaced points in the real-time variable. It is worth noting that using the configurations separated in time by not fixed intervals may not be an issue for performing ensembles averages. However, it does significantly affect the predictions of the correlation functions. This difficulty is overcome using a real-variable formulation of the equations, the so-called Nosé-Hoover (Nosé 1984c; Hoover 1985b).

Applying Hamilton's equations of motion:

$$\dot{p} = -\frac{\partial H_N}{\partial q}, \qquad \dot{q} = \frac{\partial H_N}{\partial p} \tag{6.11}$$

In Eq. (6.11), q and p represent the generalized coordinate and its conjugate momentum, respectively. Then, we obtain:

$$\begin{aligned}
\frac{dq_i}{d\tau_i} &= \frac{p_i'}{s_i^2 m_i}, \\
\frac{dp_i'}{d\tau_i} &= -\nabla_{q_i} U(q_1, \cdots, q_{g_N}), \\
\frac{ds_i}{d\tau_i} &= \frac{\pi_i}{Q_i}, \\
\frac{d\pi_i}{d\tau_i} &= \frac{(p_i')^2}{s_i^3 m_i} - \frac{k_B T}{s_i}
\end{aligned} \tag{6.12}$$

In Eq. (6.12), the derivatives are with respect to the scaled time τ_i.

These equations represent the dynamics of a system that sample a microcanonical ensemble in the extended system variables, namely $(\mathbf{r}, \mathbf{p}', s, \pi, t')$. However, note that the total energy of the real system is not conserved. That is because of the fluctuations of s, the heat transfers between the system and the heat bath to adjust the temperature of a system.

The equations of motion sample a canonical ensemble in the real system variables, namely $\mathbf{r}$ and $\mathbf{p}'/s$. To show that, we can consider the partition function which, for energy E and Planck's constant h, is defined as,

$$Z = \frac{1}{N! h^{g_N}} \int d\pi \int ds \int d\mathbf{r} \int d\mathbf{p}' \times \\ \times \delta\left(H_{NVE}(\mathbf{r}, \mathbf{p}'/s) + \frac{\pi^2}{2Q} + (g_N + 1) k_B T \ln s - E \right) \tag{6.13}$$

Using the equivalence relation for δ, $\delta\left(f(s)\right) = \delta\left(s - s_0\right)/f'(s)$, where s_0 is the solution of the equation: $f(s) = 0$, and substituting $d\mathbf{p}' = s^{g_N} d\mathbf{p}$, we get

$$Z = \frac{1}{N! h^{g_N}} \int d\pi \int ds \int d\mathbf{r} \int d\mathbf{p} \frac{s^{g_N+1}}{(g_N + 1) k_B T} \times \\ \times \delta\left(s - \exp\left(\frac{-\left[H_{NVE}(\mathbf{r}, \mathbf{p}') + \frac{\pi^2}{2Q} - E \right]}{(g_N + 1) k_B T} \right) \right) \\ = \frac{1}{(g_N + 1) k_B T N! h^{g_N}} \int d\pi \int d\mathbf{r} \int d\mathbf{p}' \times \\ \times \exp\left(\frac{-\left[H_{NVE}(\mathbf{r}, \mathbf{p}') + \frac{\pi^2}{2Q} - E \right]}{k_B T} \right) \tag{6.14}$$

After integration according to π variable, we get:

$$Z = \sqrt{\frac{2\pi Q}{k_B T}} \frac{1}{(g_N + 1)} \frac{1}{N! h^{g_N}} \int d\mathbf{r} \int d\mathbf{p}' \times \\ \times \exp\left(\frac{-\left[H_{NVE}(\mathbf{r}, \mathbf{p}') - E \right]}{k_B T} \right) \\ = \frac{1}{g_N + 1} \sqrt{\frac{2\pi Q}{k_B T}} \exp\left(\frac{E}{k_B T} \right) Z_c \tag{6.15}$$

where

$$Z_c = \frac{1}{N!h^{g_N}} \int d\mathbf{r} \int d\mathbf{p}' \exp\left(\frac{-H_{NVE}(\mathbf{r}, \mathbf{p}')}{k_B T}\right)$$

indicating that constant energy dynamics of the extended Hamiltonian $H_N(\mathbf{r}, s, \mathbf{p}, \pi)$ correspond to constant temperature dynamics of $H(\mathbf{r}, \mathbf{p}')$.

The Nosé equations of motion are continuous, deterministic and time-reversible. However, because a second-order equation describes the time-evolution of the variable s, heat may be absorbed and dissipated by the system periodically yielding nearly periodic temperature fluctuations. The stretched timescale of the Nosé equations causes a sampling of the trajectory at irregular time intervals, which may be impractical for the investigation of the dynamical properties of a system, such as the time correlation functions. Nosé and Hoover showed that the above Nosé equations of motion express in terms of real system variables as follows:

$$p_i = \frac{p_i'}{s_i}, \quad \hat{\pi}_i = \frac{\pi_i}{s_i} \tag{6.16}$$

This was followed by a Sundman time-transformation (Zare and Szebehely 1975) applied to the vector field,

$$\frac{d\tau_i}{dt} = s_i. \tag{6.17}$$

Using this transformation, a new system of non-Hamiltonian dynamical equations for the real variables is obtained as follows:

$$\begin{aligned} \frac{ds_i}{dt} &= s_i^2 \frac{\hat{\pi}_i}{Q_i}, \\ \frac{d\hat{\pi}_i}{dt} &= \frac{1}{s_i}\left(\frac{p_i^2}{m_i} - k_B T\right) - \frac{s_i \hat{\pi}_i^2}{Q_i}, \\ \frac{dq_i}{dt} &= \frac{p_i}{m_i}, \\ \frac{dp_i}{dt} &= -\nabla_{q_i} U(q_1, \cdots, q_{g_N}) - \frac{s_i \hat{\pi}_i}{Q_i} p_i \end{aligned} \tag{6.18}$$

The idea was developed further by Hoover (1985b) who proposed another change of variables form:

$$s_i \hat{\pi}_i \equiv \pi_\eta, \quad \ln s_i \equiv \eta_i$$

for $i = 1, \cdots, g_N$.

According to Hoover (1985b), the Hamiltonian equations of motion can be written as

$$\begin{aligned}\dot{q}_i &= \frac{p_i}{m_i}\,, \\ \dot{p}_i &= F_i - \frac{\pi_{\eta_i}}{Q_i} p_i\,, \\ \dot{\eta}_i &= \frac{\pi_{\eta_i}}{Q_i}\,, \\ \dot{\pi}_{\eta_i} &= \frac{p_i^2}{m_i} - k_B T\end{aligned} \qquad (6.19)$$

for $i = 1, \cdots, g_N$.

Note that the choice of the fictitious mass Q influences the dynamics and the conservation of the extended system energy E_{ext}. In particular, too large values of Q, representing a weak coupling, may cause a poor temperature control (Nose-Hoover thermostat with $Q \to \infty$ is MD which generates a microcanonical ensemble), and the canonical distribution may only be obtained after very long simulation times. Any finite mass is sufficient to guarantee in principle the generation of a canonical ensemble.

On the other hand, too small values, corresponding to tight coupling, may cause the high-frequency temperature oscillations. The variable s may oscillate at a very high frequency, and they may be decoupled with the characteristic frequencies of the real system, yielding a decoupling of the thermostat degrees of freedom from the physical degrees of freedom, characterized by a slow exchange of kinetic energy. Typically, the thermostat mass is as follows:

$$Q = \frac{k_B T}{f^2} \qquad (6.20)$$

where f is the frequency of the heat exchange between the system and the heat bath.

The conserved energy is

$$E_{NH} = \sum_{i=1}^{g_N} \left[\frac{p_i^2}{2m_i} + \frac{\pi_{\eta_i}^2}{2Q_i} + k_B T \eta_i \right] + U\left(q_1, \cdots, q_{g_N}\right) \qquad (6.21)$$

Example 2 For the case study of the Example 1, we show in Fig. 6.2a the total energy E_{NH}, the thermostat energy and the sum of kinetic and potential energy for the harmonic oscillator as a function of the MD time step. Our results indicate perfect conservation of the total energy of the system as depicted in the inset graph. The velocity-Verlet algorithm integrates numerically the equations of motion with a time step $\Delta t = 1$ fs, and the Nosé thermostat mass is determined according to Eq. (6.20) with $f = 100\,\text{ps}^{-1}$. The target temperature was 300 K.

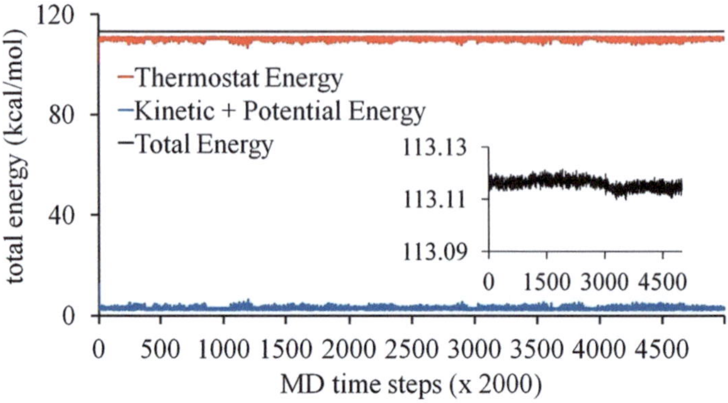

Fig. 6.2 The total energy of the system (E_{NH}) and the running average temperature of an harmonic oscillator using MD simulations with Nosé-Hoover thermostat: $\tau = 0.01$ ps, $\Delta t = 1$ fs and $k = 317$ kcal/mol/Å^2. The block averages are performed every 2000 MD steps

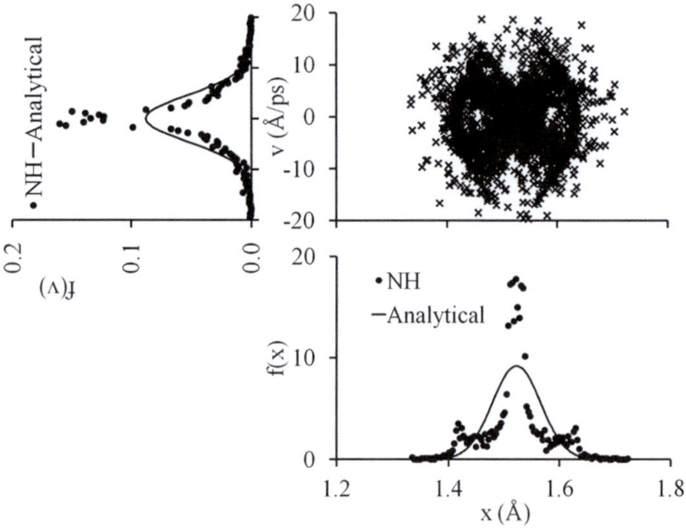

Fig. 6.3 The velocity-position scatter plot. The velocity and position probability density distributions for an harmonic oscillator using MD simulations with Nosé-Hoover thermostat: $\tau = 0.01$ ps, $\Delta t = 1$ fs and $k = 317$ kcal/mol/Å^2

Also, a velocity-position scatter plot is shown in Fig. 6.3 to determine the sampled points. From the comparison of the probability density functions $f(x)$ and $f(v)$ obtained from the simulations (data points) and the analytical curves, a sparse sampling of the canonical ensemble resulted. However, these results show that the sampling is much better than when using the Berendsen thermostat.

It has been shown (Stoffer 1995) that the Sundman time-transformation destroys the canonical symplectic structure, and hence the Nosé-Hoover system, Eq. (6.19) does not have a symplectic structure. However, the dynamics of this system is time-reversible.

6.2.2.4 Nosé-Hoover Chain of Thermostats

The Nosé Hamiltonian generates configurations from the canonical ensemble, assuming that the dynamics is ergodic (Nosé 1984c, 1991; Hoover 1985b). Note that the assumption of the ergodicity can be violated in some cases (Nosé 1991). The Nosé-Hoover equations have further been extended by the so-called Nosé-Hoover chains (Tuckerman et al. 1992), which overcome this ergodicity problem. In the Nosé-Hoover chain of the thermostats approach, a sequence of new thermostats, each one coupled to the previous, is introduced yielding a chain of thermostats. In this approach, the uncontrolled fluctuations in the thermostat degrees of freedom vanish, and hence a better temperature control is achieved. This extension of the original Nosé-Hoover thermostat method generates a canonical distribution assuming that the dynamics it is ergodic. The hypothesis of the ergodicity requires that the dynamics of the system samples the entire phase space, which includes the thermostat variables (Tobias et al. 1993). Nosé-Hoover chain of thermostats dynamics generates the canonical ensemble even in special cases where the standard Nosé-Hoover dynamics failed to do so.

The Nosé-Hoover's Hamiltonian of the system coupled to a chain of the thermostats is as follows:

$$
\begin{aligned}
H_{NC} = \sum_{i=1}^{g_N} \Bigg[\frac{(p_i')^2}{2s_{i,1}^2 m_i} + \sum_{j=1}^{M-1} \frac{\pi_{i,j}^2}{2Q_{i,j} s_{i,j+1}^2} \\
+ \frac{\pi_{i,M}^2}{2Q_{i,M}} + \sum_{j=1}^{M} k_B T \ln s_{i,j} \Bigg] + U\left(q_1, \cdots, q_{g_N}\right)
\end{aligned}
\tag{6.22}
$$

To derive the equations for the chains, we have to start with the standard Nosé-Hoover equations as defined in Eq. (6.19). Now, new $M-1$ more thermostats are introduced, where each one is coupled to the previous, resulting in a Nosé-Hoover chain of thermostats. Tuckerman and Parrinello (1994) derived the equations of motion of a system of N particles for the isothermal dynamics. The following discussion presents the details on the derivation of these equations.

The Hamilton's equations of motion introduced in Chap. 1 can be used to obtain the equations of motions for each degree of freedom i:

$$
\frac{dq}{d\tau_1} = \frac{p'}{s_1^2 m} \tag{6.23}
$$

$$
\begin{aligned}
\frac{dp'}{d\tau_1} &= -\nabla_q U(q) \\
\frac{ds_1}{d\tau_2} &= \frac{\pi_1}{s_2^2 Q_1} \\
\frac{d\pi_1}{d\tau_2} &= \frac{(p')^2}{s_1^3 m} - \frac{k_B T}{s_1} \\
\frac{ds_k}{d\tau_{k+1}} &= \frac{\pi_k}{s_{k+1}^2 Q_k} \\
\frac{d\pi_k}{d\tau_{k+1}} &= \frac{\pi_{k-1}^2}{s_k^3 Q_{k-1}} - \frac{k_B T}{s_k}, \qquad k = 2, \cdots, M-1 \\
\frac{ds_M}{d\tau_M} &= \frac{\pi_M}{Q_M} \\
\frac{d\pi_M}{d\tau_M} &= \frac{\pi_{M-1}^2}{s_M^3 Q_{M-1}} - \frac{k_B T}{s_M}
\end{aligned}
$$

where the relationships between the scaled times and the real time t are given as follows:

$$
\begin{aligned}
d\tau_1 &= s_1 dt \\
d\tau_k &= s_{k-1} s_k dt, \quad (k = 2, \cdots, M-1) \\
d\tau_M &= s_M dt
\end{aligned}
\tag{6.24}
$$

Note that the subscript i in Eq. (6.23) is omitted for simplicity of the appearance. Substituting the real time variables into Eq. (6.23) using the transformations given by Eq.(6.24), we obtain:

$$
\begin{aligned}
\frac{dq}{dt} &= \frac{p'}{s_1 m} \\
\frac{dp'}{dt} &= -s_1 \nabla_q U(q) \\
\frac{ds_1}{dt} &= s_1 \frac{\pi_1}{s_2 Q_1} \\
\frac{d\pi_1}{dt} &= s_2 \frac{(p')^2}{s_1^2 m} - \frac{k_B T}{s_1} \\
\frac{ds_k}{dt} &= s_k \frac{\pi_k}{s_{k+1} Q_k}
\end{aligned}
\tag{6.25}
$$

$$
\begin{aligned}
\frac{d\pi_k}{dt} &= s_{k+1}\frac{\pi_{k-1}^2}{s_k^2 Q_{k-1}} - s_{k+1}k_B T, \quad k = 2, \cdots, M-1 \\
\frac{ds_M}{dt} &= s_M \frac{\pi_M}{Q_M} \\
\frac{d\pi_M}{dt} &= \frac{\pi_{M-1}^2}{s_M^2 Q_{M-1}} - k_B T
\end{aligned}
$$

In the following, we propose some additional transformations of the variables:

$$
\begin{aligned}
p &= \frac{p'}{s_1} \\
\hat{\pi}_k &= \frac{\pi_k}{s_k s_{k+1}}, \quad (k = 1, \cdots, M-1) \\
\hat{\pi}_M &= \frac{\pi_M}{s_M}
\end{aligned}
\tag{6.26}
$$

Substituting the transformations given by Eq. (6.26) into Eq. (6.25), we obtain the equations of motion in the form:

$$
\begin{aligned}
\frac{dq}{dt} &= \frac{p}{m} \\
\frac{dp}{dt} &= -\nabla_q U(q) - \frac{1}{s_1}\frac{ds_1}{dt}p \\
\frac{ds_1}{dt} &= s_1^2 \frac{\hat{\pi}_1}{Q_1} \\
\frac{d\hat{\pi}_1}{dt} &= \frac{1}{s_1}\frac{p^2}{m} - \frac{k_B T}{s_1} - \frac{1}{s_2}\frac{ds_2}{dt}\hat{\pi}_1 - \frac{1}{s_1}\frac{ds_1}{dt}\hat{\pi}_1 \\
\frac{ds_k}{dt} &= s_k^2 \frac{\hat{\pi}_k}{Q_k} \\
\frac{d\hat{\pi}_k}{dt} &= s_{k-1}^2 \frac{\hat{\pi}_{k-1}^2}{s_k Q_{k-1}} - \frac{k_B T}{s_k} - \frac{1}{s_{k+1}}\frac{ds_{k+1}}{dt}\hat{\pi}_k - \frac{1}{s_k}\frac{ds_k}{dt}\hat{\pi}_k \\
& k = 2, \cdots, M-1 \\
\frac{ds_M}{dt} &= s_M^2 \frac{\hat{\pi}_M}{Q_M} \\
\frac{d\hat{\pi}_M}{dt} &= s_{M-1}^2 \frac{\hat{\pi}_{M-1}^2}{s_M Q_{M-1}} - \frac{k_B T}{s_M} - \frac{1}{s_M}\frac{ds_M}{dt}\hat{\pi}_M
\end{aligned}
\tag{6.27}
$$

Using another transformation of the variables as:

$$\eta_k = \ln s_k \qquad (6.28)$$
$$\pi_{\eta_k} = \hat{\pi}_k s_k, \quad (k = 1, \cdots, M)$$

we obtain the equations of motion as follows:

$$\begin{aligned}
\frac{dq}{dt} &= \frac{p}{m} \\
\frac{dp}{dt} &= -\nabla_q U(q) - \frac{\pi_{\eta_1}}{Q_1} p \\
\frac{d\pi_{\eta_1}}{dt} &= \frac{p^2}{m} - k_B T - \frac{\pi_{\eta_2}}{Q_2} \pi_{\eta_1} \\
\frac{d\pi_{\eta_k}}{dt} &= \frac{\pi^2_{\eta_{k-1}}}{Q_{k-1}} - k_B T - \frac{\pi_{\eta_{k+1}}}{Q_{k+1}} \pi_{\eta_k}, \; k = 2, \cdots, M-1, \\
\frac{d\pi_{\eta_M}}{dt} &= \frac{\pi^2_{\eta_{M-1}}}{Q_{M-1}} - k_B T, \\
\frac{d\eta_k}{dt} &= \frac{\pi_{\eta_k}}{Q_k}, \; k = 1, \cdots, M.
\end{aligned} \qquad (6.29)$$

Note that Eq. (6.29) is written for each degree of freedom such that a massive chain of thermostats approach is used.

The conserved energy is given by:

$$E_{NHC} = \sum_{i=1}^{g_N} \left[\frac{p_i^2}{2m_i} + \sum_{j=1}^{M} \left(\frac{\pi^2_{\eta_{i,j}}}{2Q_{i,j}} + k_B T \eta_{i,j} \right) \right] + U\left(q_1, \cdots, q_{g_N}\right) \qquad (6.30)$$

It has been shown (Liu and Tuckerman 2000) that the NHC method has limited use because it is only capable of maintaining appropriate required temperature control in equilibrium. Any perturbation of the system away from the equilibrium state, such as in the presence of external fields or by motion over the high barrier, may result in the method not being able to converge. In such cases, however, a chain of thermostats of a length longer than two can be used, in general, if there is a broad distribution of vibration frequencies (Jang and Voth 1997).

Example 3 Consider again the harmonic oscillator of the Example 1 subject to MD simulation using the chain of thermostats approach. We used the following parameters for the chain: $\tau = 0.01$ ps and the length $M = 3$. For the integration of the equations of motion, we used a velocity-Verlet like an algorithm as discussed in the next chapters with an integration time step of $\Delta t = 1$ fs.

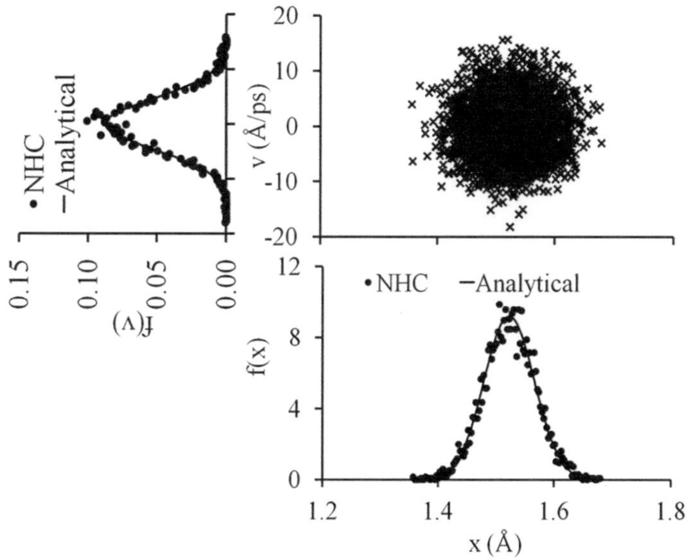

Fig. 6.4 The velocity-position scatter plot. The velocity and position probability density distributions for an harmonic oscillator using MD simulations with Nosé-Hoover chain of thermostats: $\tau = 0.01$ ps, $\Delta t = 1$ fs and $k = 317$ kcal/mol/Å^2. The chain length was $M = 3$

In Fig. 6.4, we have shown as a scatter plot the velocity versus position along with the probability density functions of the speed and position. Our results show a perfect canonical ensemble sampling, and besides, by a direct comparison with analytical curves, it can be seen that the canonical ensemble distribution in velocity and position reproduces.

Furthermore, the numerical integrator algorithm possesses a perfect energy conservation property as indicated by the graphs of the thermostat, the sum of the kinetic and potential, and total energies shown in Fig. 6.5.

6.2.2.5 Nosé-Poincaré Thermostats

An alternative to the Nosé-Hoover method was proposed in Bond et al. (1999). In this method, the real-time differential equations were obtained by using a Poincaré transformation conserving the Hamiltonian structure.

Illustrated in the previous section, too, the traditional real-variable formulation of Nosé-Hoover destroys the symplectic structure associated with the Nosé Hamiltonian. In this section, we will outline a procedure for scaling time while preserving the Hamiltonian structure. The method is formulated through a Poincaré transformation (Zare and Szebehely 1975) of the Hamiltonian $H = H(\mathbf{r}, \mathbf{p})$:

$$H' = f(\mathbf{r}, \mathbf{p})(H - H_0), \qquad f > 0 \tag{6.31}$$

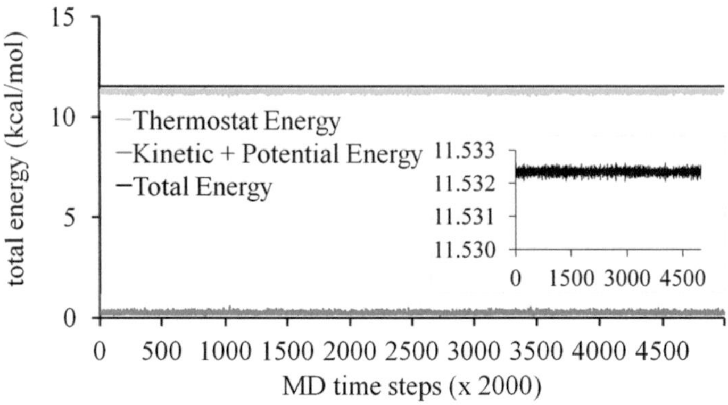

Fig. 6.5 The total energy of the system (E_{NH}) and the running average temperature of an harmonic oscillator using MD simulations with Nosé-Hoover chain of thermostats: $\tau = 0.01$ ps, $\Delta t = 1$ fs and $k = 317\,\text{kcal/mol/Å}^2$. The chain length was $M = 3$. The block averages are performed every 2000 MD steps

where f is a "time scaling" function, and the constant H_0 is the initial value of H. Along the energy slice $H = H_0$, the dynamics of the transformed system are equivalent to the original system, up to a transformation of time. This can be proven by first writing the Hamilton equations of motion

$$\dot{\mathbf{r}}_i = f(\mathbf{r}, \mathbf{p})\frac{\partial H}{\partial \mathbf{p}_i} + (H - H_0)\frac{\partial f}{\partial \mathbf{p}_i}\,, \tag{6.32}$$

$$\dot{\mathbf{p}}_i = -f(\mathbf{r}, \mathbf{p})\frac{\partial H}{\partial \mathbf{r}_i} - (H - H_0)\frac{\partial f}{\partial \mathbf{r}_i}$$

then observe that when $H = H_0$, the equations are the same as the original equations expressed in the real-time variable, t, related to t' by

$$\frac{dt'}{dt} = f(\mathbf{r}, \mathbf{p})$$

Now, we consider the Poincaré transformation, $f(\mathbf{r}, \mathbf{p}) = s$, applied to a slightly modified version of the Nosé extended Hamiltonian in Eq. (6.10):

$$\begin{aligned} H_{NP}(\mathbf{r}, \mathbf{p}', s, \pi) &= s\left(H_N\left(\mathbf{r}, \mathbf{p}', s, \pi\right) - H_0\right) \\ &= s\left(\sum_{i=1}^{N} \frac{(\mathbf{p}_i')^2}{2m_i s^2} + U(\mathbf{r}) + \frac{\pi^2}{2Q} + g_N k_B T \ln s - H_0\right) \end{aligned} \tag{6.33}$$

where g_N is the number of degrees of freedom of the real system, H_N is the Nosé Hamiltonian (see Eq. (6.10)), and H_0 is chosen such that the Nosé-Poincaré Hamiltonian, H_{NP}, is zero when evaluated at the initial conditions.

Nosé-Poincaré Hamiltonian are the same as those given above for the Nosé system (Eq. (6.12)), except that the right-hand side is multiplied by the thermostat variable s:

$$\begin{aligned}
\dot{\mathbf{r}}_i &= \frac{\mathbf{p}'_i}{sm_i}, \\
\dot{\mathbf{p}}'_i &= -s\nabla_i U(\mathbf{r}), \\
\dot{s} &= s\pi/Q, \\
\dot{\pi} &= -\Delta H + \sum_{i=1}^{N} \frac{(\mathbf{p}'_i)^2}{s^2 m_i} - g_N k_B T
\end{aligned} \tag{6.34}$$

Here, the derivatives are with respect to the real time variable t, and $\Delta H = H_N - H_0$:

$$\Delta H = \sum_{i=1}^{N} \frac{(\mathbf{p}'_i)^2}{2m_i s^2} + U(\mathbf{r}) + \frac{\pi^2}{2Q} + g_N k_B T \ln s - H_0 \tag{6.35}$$

The dynamics governed by Eq. (6.34) can be shown to sample from the canonical distribution in the variables $\mathbf{r}$ and $\mathbf{p}'/s$, in a similar manner to that used for Nosé's method, if the modified system is ergodic. In the canonical ensemble, the partition function is obtained as:

$$Z = \frac{1}{N!h^{g_N}} \int ds \int d\pi \int d\mathbf{r} \int d\mathbf{p}' \delta(H_{NP} - 0) \tag{6.36}$$

Using Eqs. (6.33) and (6.10) we get:

$$\begin{aligned}
Z = \frac{1}{N!h^{g_N}} &\int ds \int d\pi \int d\mathbf{r} \int d\mathbf{p}' \\
&\times \delta\left(s\left[\sum_{i=1}^{N} \frac{(\mathbf{p}'_i)^2}{2s^2 m_i} + U(\mathbf{r}) + \frac{\pi^2}{2Q} + g_N k_B T \ln(s) - H_0\right]\right)
\end{aligned} \tag{6.37}$$

We can substitute $\mathbf{p}'/s = \mathbf{p}$, the volume element then becomes $d\mathbf{p}' = s^{g_N} d\mathbf{p}$. No upper limit exists in the momentum space, hence the order of integration of $d\mathbf{p}$ and ds changes as:

$$Z = \frac{1}{N!h^{g_N}} \int d\pi \int d\mathbf{r} \int d\mathbf{p} s^{g_N} \tag{6.38}$$

$$\times\,\delta\left(s\left[\sum_{i=1}^{N}\frac{\mathbf{p}_i^2}{2m_i}+U(\mathbf{r})+\frac{\pi^2}{2Q}+g_Nk_BT\ln(s)-H_0\right]\right)$$

For a function $r(s)$ with a single simple root, $s = s_0$, the equivalence relation for δ becomes: $\delta[r(s)] = \delta[s - s_0]/r'(s_0)$, then,

$$\delta\left(s\left[\sum_{i=1}^{N}\frac{\mathbf{p}_i^2}{2m_i}+U(\mathbf{r})+\frac{\pi^2}{2Q}+g_Nk_BT\ln(s)-H_0\right]\right) \tag{6.39}$$
$$=\frac{1}{g_Nk_BT}\delta\left(s-\exp\left[-\frac{1}{g_Nk_BT}\left(\sum_{i=1}^{N}\frac{\mathbf{p}_i^2}{2m_i}+U(\mathbf{r})+\frac{\pi^2}{2Q}-H_0\right)\right]\right)$$

Substituting Eq. (6.39) into Eq. (6.38) we get:

$$Z=\frac{1}{N!h^{g_N}}\int d\pi\int d\mathbf{r}\int d\mathbf{p}\frac{s^{g_N}}{g_Nk_BT} \tag{6.40}$$
$$\times\,\delta\left(s-\exp\left[-\frac{1}{g_Nk_BT}\left(\sum_{i=1}^{N}\frac{\mathbf{p}_i^2}{2m_i}+U(\mathbf{r})+\frac{\pi^2}{2Q}-H_0\right)\right]\right)$$
$$=\frac{1}{g_Nk_BT}\frac{1}{N!h^{g_N}}\int d\pi\int d\mathbf{r}\int d\mathbf{p}$$
$$\times\exp\left[-\frac{1}{k_BT}\left(\sum_{i=1}^{N}\frac{\mathbf{p}_i^2}{2m_i}+U(\mathbf{r})+\frac{\pi^2}{2Q}-H_0\right)\right]$$

After integrating with respect to π we get:

$$Z=\left(\frac{2\pi Q}{g_N^2k_BT}\right)^{1/2}\exp\left(H_0/k_BT\right)Z_c$$

where Z_c is the partition function of the canonical ensemble:

$$Z_c=\frac{1}{N!h^{g_N}}\int d\mathbf{r}\int d\mathbf{p}\exp\left[-\frac{1}{k_BT}\left(\sum_{i=1}^{N}\frac{\mathbf{p}_i^2}{2m_i}+U(\mathbf{r})\right)\right]$$

The well-known Gaussian integral was used to perform the integration over the thermostat variable π:

$$\int_{-\infty}^{+\infty}\exp\left(-\frac{x^2}{2Qk_BT}\right)dx=\sqrt{2\pi Qk_BT}$$

Again, a sampling with H_{NP} in the scaled time space keeping the energy constant is equivalent to a sampling of the original system at constant temperature T in real-time.

Example 4 Consider the harmonic oscillator of the Example 1 using MD simulation with Nosé-Poincaré thermostat. Equations of motion were integrated using the numerical integrator algorithm in Bond et al. (1999) with an integration time step $\Delta t = 1\,\text{fs}$. The thermostat mass was calculated using the $f = 100\,\text{ps}^{-1}$. The initial velocity of the particle was generated randomly using a Maxwell-Boltzmann distribution at the target temperature of $T_0 = 300\,\text{K}$. While the initial position was random.

In Fig. 6.6, we show the scatter plot of the velocity versus displacement of the harmonic oscillator. It is surprising that the dynamics governed by the Nosé-Poincaré thermostat perform as good as Nosé-Hoover single thermostat in terms of the sampling the canonical distribution, as also can be indicated by direct comparison with analytical probability distributions (see Fig. 6.6) even though they form a symplectic structure. It seems that the time-reversibility of the equations of motion is a substantial property for sampling. However, the Nosé-Poincaré dynamics show an excellent energy conservation property as indicated by the graph in Fig. 6.7, which is because the symplectic dynamical structures have a conserved Hamiltonian function. Our results suggest no drift on the Hamiltonian during the MD simulation run.

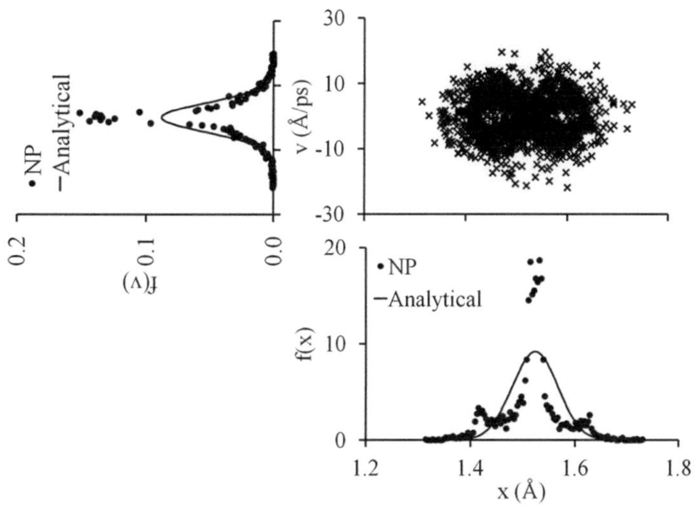

Fig. 6.6 The position-velocity scatter plot and the running average temperature of an harmonic oscillator using MD simulations with Nosé-Poincaré thermostat: $\tau = 0.01\,\text{ps}$, $\Delta t = 1\,\text{fs}$ and $k = 317\,\text{kcal/mol/Å}^2$ and $x_0 = 1.523\,\text{Å}$

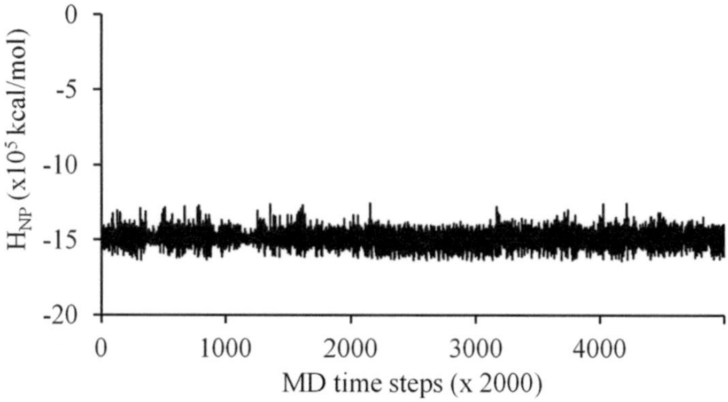

Fig. 6.7 The Hamiltonian of the system (H_{NP}) of the harmonic oscillator using MD simulations with Nosé-Poincaré thermostat: $\tau = 0.01$ ps, $\Delta t = 1$ fs and $k = 317$ kcal/mol/Å^2. The block averages are performed every 2000 MD steps

6.2.2.6 Nosé-Poincaré Chain of Thermostats

In a similar manner to the Nosé equations, thermostatting chains, consisting of M thermostats, can be added to the Nosé-Poincaré equation, with some additional terms as follows,

$$H_{NPC} = s_1 \left[\sum_{i=1}^{N} \frac{(\mathbf{p}_i')^2}{2m_i s_1^2} + U(\mathbf{r}) + \sum_{i=1}^{M-1} \frac{(\pi_i')^2}{2Q_i s_{i+1}^2} \right. \\ \left. + \frac{\pi_M^2}{2Q_M} + g_N k_B T \ln s_1 + \sum_{i=2}^{M} k_B T \ln s_i - H_0 \right] \tag{6.41}$$

where H_0 is Eq. (6.22) evaluated at the initial conditions.

The equations of motion are,

$$\begin{aligned} \dot{\mathbf{r}}_i &= \frac{\mathbf{p}_i'}{m_i s_1} \\ \dot{\mathbf{p}}_i' &= -s_1 \nabla_i U(\mathbf{r}) \\ \dot{s}_1 &= s_1 \frac{\pi_1'}{Q_1 s_2^2} \\ \dot{\pi}_1' &= \sum_{j=1}^{N} \frac{(\mathbf{p}_j')^2}{m_j s_1^2} - g_N k_B T - \Delta H \end{aligned} \tag{6.42}$$

$$\dot{s}_k = s_1 \frac{\pi'_k}{Q_k s_{k+1}^2}$$

$$\dot{\pi}'_k = s_1 \left(\frac{(\pi'_{k-1})^2}{Q_{k-1} s_k^3} - \frac{k_B T}{s_k} \right), \quad k = 2, \cdots, M-1$$

$$\dot{s}_M = \frac{s_1 \pi_M}{Q_M}$$

$$\dot{\pi}_M = s_1 \left(\frac{\pi_{M-1}^2}{Q_{M-1} s_M^3} - \frac{k_B T}{s_M} \right)$$

Here, $\mathbf{p} = \mathbf{p}'/s_1$ represents the momentum of the real particle and, in addition, the derivatives are to the real-time.

A proof shows that these equations generate canonical ensemble on the $(\mathbf{r}, \mathbf{p})$ variables. For that, we write the partition function for this ensemble as

$$Z = \frac{1}{N! h^{g_N}} \int d\pi_M \int d\pi_{M-1} \cdots \int d\pi_1 \int ds_M \cdots \int ds_1 \times \int d\mathbf{p}' \int d\mathbf{r} \delta\left(H_{NPC} - 0\right) \tag{6.43}$$

where $\pi'_k/s_{k+1} \equiv \pi_k$ for $k = 1, \cdots M-1$. Substituting $\mathbf{p}'/s_1 = \mathbf{p}$, the volume element $d\mathbf{p}'$ then becomes $\mathbf{p}' = s_1^{g_N} d\mathbf{p}$. There is no maximum value of the momentum, and hence we can change the order of integration of $d\mathbf{p}$ and ds_1 giving the integral over s_1 as,

$$\int ds_1 \delta(H_{NPC}) = \int ds_1 s_1^{g_N} \times \delta\left(s_1 \left[\sum_{i=1}^{N} \frac{\mathbf{p}_i^2}{2m_i} + U(\mathbf{r}) + \sum_{i=1}^{M-1} \frac{\pi_i^2}{2Q_i} + \frac{\pi_M^2}{2Q_M} + g_N k_B T \ln s_1 + \sum_{i=2}^{M} k_B T \ln s_i - H_0 \right] \right) \tag{6.44}$$

Using the equivalence relation for δ, $\delta[r(s_1)] = \delta[s_1 - x_0]/ \mid r'(x_0) \mid$, where x_0 is the root of $r(s_1) = 0$, to get

$$\int ds_1 \delta(H_{NPC}) = \int ds_1 \frac{s_1^{g_N}}{g_N k_B T} \delta\left(s_1 - \exp\left[-\frac{1}{g_N k_B T} \left(\sum_{i=1}^{N} \frac{\mathbf{p}_i^2}{2m_i} + U(\mathbf{r}) \right. \right.\right. \tag{6.45}$$

$$+ \sum_{i=1}^{M-1} \frac{\pi_i^2}{2Q_i} + \frac{\pi_M^2}{2Q_M} + \sum_{i=2}^{M} k_B T \ln s_i - H_0 \Bigg) \Bigg] \Bigg)$$

$$= \frac{1}{g_N k_B T} \exp \Bigg[-\frac{1}{k_B T} \Bigg(\sum_{i=1}^{N} \frac{\mathbf{p}_i^2}{2m_i} + U(\mathbf{r})$$

$$+ \sum_{i=1}^{M-1} \frac{\pi_i^2}{2Q_i} + \frac{\pi_M^2}{2Q_M} + \sum_{i=2}^{M} k_B T \ln s_i - H_0 \Bigg) \Bigg]$$

The remaining thermostats variables can be integrated out as in the previous section to get the partition function as:

$$Z = C \frac{1}{N! h^{g_N}} \int d\mathbf{p} \int d\mathbf{r} \exp \left(\frac{-\left[\sum_{i=1}^{N} \frac{\mathbf{p}_i^2}{2m_i} + U(\mathbf{r}) \right]}{k_B T} \right) \tag{6.46}$$

where

$$C = \frac{(2\pi)^{\frac{M}{2}} (k_B T)^{\frac{M}{2}-1} \Pi_{i=1}^{M} Q_i^{\frac{1}{2}}}{g_N} \exp(H_0 / k_B T)$$

Here, the following integration was used to simplify the expression for Z:

$$\int_{-\infty}^{+\infty} d\pi_1 \cdots \int_{-\infty}^{+\infty} d\pi_M \exp \left(-\frac{1}{k_B T} \sum_{i=1}^{M} \frac{\pi_i^2}{2Q_i} \right) = \prod_{i=1}^{M} (2\pi Q_i k_B T)^{\frac{1}{2}}$$

Example 5 As an illustration, we considered the harmonic oscillator of the Example 1.

Figure 6.8 shows a scatter plot of the velocity versus displacement of the harmonic oscillator. Again, surprisingly, the dynamics governed by the Nosé-Poincaré chain of thermostats does not reproduce the fluctuations of the velocities and positions at the target temperature, as seen from the comparison with analytical curves. However, the sampled points follow the canonical distribution. Furthermore, the chain of thermostats can control the fluctuations of the temperature around the target value, $T = 300$ K, as shown in Fig. 6.9a.

On the other hand, the Nosé-Poincaré chain of thermostats dynamics have perfect conservation of the Hamiltonian, without drift as can be seen in Fig. 6.9b, which is a property of the symplectic structure of the equations of motion that is preserved by the numerical integrator.

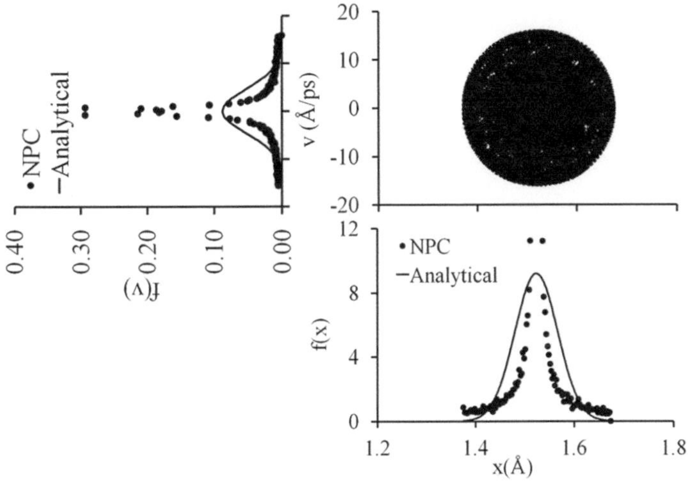

Fig. 6.8 The position-velocity scatter plot and the running average temperature of an harmonic oscillator using MD simulations with Nosé-Poincaré chain of thermostats: $\tau = 0.1$ ps, $\Delta t = 1$ fs and $k = 317$ kcal/mol/Å^2 and $x_0 = 1.523$ Å. Thermostats chain length was $M = 3$

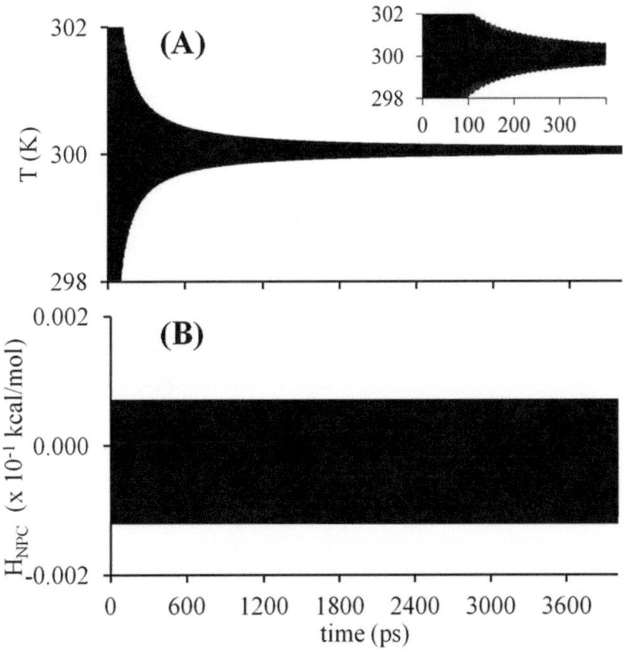

Fig. 6.9 (**a**) The running average temperature of an harmonic oscillator using MD simulations with Nosé-Poincaré chain of thermostats: $\tau = 0.01$ ps, $\Delta t = 1$ fs and $k = 317$ kcal/mol/Å^2. The chain length was $M = 3$. (**b**) The Hamiltonian of the system (H_{NPC}) of the harmonic oscillator using MD simulations with Nosé-Poincaré chain of thermostats: $\tau = 0.1$ ps, $\Delta t = 1$ fs and $k = 317$ kcal/mol/Å^2. Thermostats chain length was $M = 3$. The block averages are performed every 2000 MD steps

6.2.2.7 Gaussian Thermostat

Another alternative approach to control the temperature is to apply Gauss's principle of nonholonomic constraints (Evans et al. 1983; Edberg et al. 1986), which is called the Gaussian thermostat method. In the nonholonomic case, the constraints are defined by the function $g(\mathbf{r}, \mathbf{v}, t) = 0$. The constraint is nonholonomic because g is a function of the velocities. The derivative of g with respect to the time t yields a relation that restricts the accelerations, $\mathbf{a}$, as follows:

$$\begin{aligned}\frac{\partial g(\mathbf{r}_i, \mathbf{v}_i, t)}{\partial t} &= \frac{\partial g}{\partial \mathbf{r}_i}\frac{d\mathbf{r}_i}{dt} + \frac{\partial g}{\partial \mathbf{v}_i}\frac{d\mathbf{v}_i}{dt} + \frac{\partial g}{\partial t} \\ &= \frac{\partial g}{\partial \mathbf{r}_i}\mathbf{v}_i + \frac{\partial g}{\partial \mathbf{v}_i}\mathbf{a}_i + \frac{\partial g}{\partial t}\end{aligned} \tag{6.47}$$

The constant temperature constraint has the following nonlinear form:

$$g(\mathbf{r}, \mathbf{v}, t) = \frac{1}{2}\sum_{i=1}^{N} m_i \mathbf{v}_i^2 - \frac{1}{2} g_N k_B T \tag{6.48}$$

with g_N being the number of degrees of freedom in the system. In Eq. (6.48), T is the target temperature. Gauss's principle yields an equation:

$$\sum_{i=1}^{N} m_i \mathbf{v}_i \cdot \mathbf{a}_i = \sum_{i=1}^{N} \mathbf{F}_i \cdot \mathbf{v}_i = 0 \tag{6.49}$$

To derive the Gaussian equations of motion with the nonholonomic constraint condition, $m_i \mathbf{a}_i$ is substituted by $\mathbf{F}_i - \dot{\xi} m_i \mathbf{v}_i$:

$$\sum_{i=1}^{N} \mathbf{v}_i \cdot \left(\mathbf{F}_i - \dot{\xi} m_i \mathbf{v}_i\right) = \sum_{i=1}^{N} \mathbf{F}_i \cdot \mathbf{v}_i - \dot{\xi} \sum_{i=1}^{N} m_i \mathbf{v}_i^2 = 0$$

Solving it for the time derivative of the friction coefficient, $\dot{\xi}$, gives:

$$\dot{\xi} = \frac{\sum_{i=1}^{N} \mathbf{F}_i \cdot \mathbf{v}_i}{\sum_{i=1}^{N} m_i \mathbf{v}_i^2} \tag{6.50}$$

Note that the equations of motion become for $i = 1, 2, \cdots, N$:

$$\begin{aligned}\dot{\mathbf{q}}_i &= \frac{\mathbf{p}_i}{m_i} \\ \dot{\mathbf{p}}_i &= \mathbf{F}_i - \dot{\xi}\mathbf{p}_i\end{aligned}$$

which is equivalent to adding a dumping term into the equations of motion to keep the kinetic energy constant. Note that $\dot{\xi}$ is a time dependent coefficient, therefore, it is updated at every time step of the molecular dynamics simulations along the velocities and coordinates using Eq. (6.50). The conserved energy associated with Eq. (6.50) is simply given as follows:

$$E_{Gauss} = \frac{1}{2}\sum_i m_i v_i^2 + U(\mathbf{r}_i, \cdots, \mathbf{r}_N) \tag{6.51}$$

6.2.2.8 Langevin Dynamics

The Langevin dynamics is a stochastic method alternative to Newtonian dynamics, used in many biomolecular simulations for various numerical and physical reasons. The Langevin model has been employed to avoid explicit representation of water molecules in the molecular dynamics simulation of macromolecular simulations (Pastor 1994). Other applications of the Langevin model include the treatment of the droplet surface effects (Simonson 1996; Brünger et al. 1982), representation of the hydration shell models in large systems (Beglov and Roux 1994a,b, 1995), and enhancement of the conformation sampling (Loncharich et al. 1992; Derreumaux and Schlick 1995; Klimov and Thirumalai 1997; Doniach and Eastman 1999). Besides, the Langevin model is used to counteract numerical damping while masking mild instabilities of certain long-timestep approaches (Zhang and Schlick 1993, 1994).

The Langevin model (McQuarrie 2000) consists of additional friction and random forces to the systematic forces, which aim to represent a physical simple heat bath for the macromolecule by accounting for molecular collisions. The simplest Langevin equation in a continuous form is given as follows:

$$m\ddot{\mathbf{r}}(t) = -\nabla U(\mathbf{r}(t)) - \gamma m\dot{\mathbf{r}}(t) + \mathbf{R}(t) \tag{6.52}$$

Here, γ denotes the collision parameter (in reciprocal units of time), which is known as the *damping constant*. The random-force vector $\mathbf{R}$ represents a stationary Gaussian process with statistical properties characterized as the following:

$$\langle \mathbf{R}(t) \rangle = 0\,, \tag{6.53}$$

$$\langle \mathbf{R}(t)\mathbf{R}(t')^T \rangle = 2\gamma k_B T m \delta(t - t') \tag{6.54}$$

where k_B is the Boltzmann constant, T is the target temperature, and δ is the usual Dirac symbol.

The magnitude of γ gives the relative strength of the internal forces acting on the system concerning the random (external) forces. Therefore, with increasing γ, we also increase the inertial relative to the diffusive, Brownian-like, regime. The Brownian range is used, by employing stable algorithms, to explore more efficiently configuration phase spaces of the systems with a higher degree of the structure flexibility (Wade et al. 1993; Case 1994; Barth and Schlick 1998c).

The Stoke's law for a hydrodynamic particle of radius a is often used as a physical value of γ for the particle of mass m:

$$\gamma = 6\pi\kappa a/m \tag{6.55}$$

where κ defines the solvent viscosity. For example, the collision frequency of the macromolecule atoms, such as the protein, immersed in a solvent having a viscosity of 1 cP (Centipoise) at room temperature, $\gamma = 50\,\mathrm{ps}^{-1}$ (Pastor et al. 1988), which is in the range of the estimated value for the water ($\gamma = 54.9\,\mathrm{ps}^{-1}$).

An alternative method to choose an appropriate value for γ for a system modeled by the simple Langevin equation is to use the following relation:

$$D_t = k_B T/m\gamma \tag{6.56}$$

where D_t is the experimental translation diffusion constant in the diffusive limit.

6.2.2.9 Remarks

Often, in the simulation of a macromolecular system in an explicit solvent environment, such as water molecules and salt, we encounter distinct sets of degrees of freedom characterized by either very different characteristic frequencies or very different heating rates. In this case, the coupling all the degrees of freedom to a single thermostat may cause different convergence temperatures for the distinct subsets of degrees of freedom of the system. That is because of the different rates of the exchange of kinetic energy between the subset of the degrees of freedom and the thermostat. In particular, in the simulations of the macromolecular systems, this is called "hot solvent" and "cold solute" problem. Because the solvent dynamics is significantly affected by the use of the long-range truncation, in particular, electrostatic interactions, and hence the coupling of all system to a single thermostat may yield an average solute temperature to be lower than the average solvent temperature. In this case, coupling separately the solute and solvent degrees of freedom to two different thermostats can be used to eliminate the problem.

Another problem appears if we are using an MD simulation program that (incorrectly) couples the thermostats directly to the atomic velocities when initially the total linear and angular momenta of the system are not set to zero. In such a situation, the linear and angular momenta of the system are not conserved. Depending on the numerical integrator used, these quantities are not conserved even if initially are set to zero, due to the numerical errors. In this case, the simulations have shown that the thermostat injects kinetic energy from high frequencies to low-frequency degrees of freedom, a phenomenon known as the "flying ice cube effect".

A good practice to perform a simulation in the NVT ensemble is first to run the equilibration with the Berendsen thermostat at a small value of τ (e.g. $\tau = 0.01$ ps).

After the system is equilibrated, τ should be increased to get a good equilibrium run. Then, if a proper canonical ensemble with correct fluctuations is needed at the target temperature, the Nose-Hoover chain of thermostats can be used.

6.2.3 *Isothermal-Isobaric Ensemble*

6.2.3.1 Nosé-Andersen Method

In Andersen (1980), for constant pressure dynamics, the volume V and its conjugate momentum π_V, as additional variables are introduced. The new variables couple to the system dynamics in such a way as to guarantee that the trajectory samples from an isobaric statistical distribution, assuming that the ergodicity hypothesis is satisfied. Similarly, to generate a constant temperature distribution, Nosé (1984c) introduced a new mechanical variable s with a conjugate momentum π_s that couples into the system as described in the previous sections. These two extensions are combined to construct a Hamiltonian system whose trajectories can be shown to sample an isothermal-isobaric ensemble (Nosé 1984b).

This combined Nosé-Andersen Hamiltonian function is given by

$$\begin{aligned} H_{NA} = V^{-2/d} \sum_{i=1}^{N} \frac{(\mathbf{p}_i')^2}{2m_i s^2} + U(V^{1/d}\mathbf{r}') \\ + \frac{(\pi_V')^2}{2s^2 Q_V} + \frac{\pi_s^2}{2Q_s} \\ + (g_N + 1)k_B T \ln s + p_0 V \end{aligned} \tag{6.57}$$

where $\mathbf{p}_i'$ is the conjugate momentum to the scaled position vector $\mathbf{r}_i' = V^{-1/d}\mathbf{r}_i$, p_0 is the external pressure, d is the dimension of the space, and g_N is number of degrees of freedom of the original system. The quantities Q_V and Q_s are the masses of the Andersen "piston" and the Nosé thermostat variables, respectively.

Applying the Hamiltonian equations of motion, the following equations of motion governing this system can be obtained:

$$\begin{aligned} \frac{d\mathbf{r}_i'}{dt'} &= \frac{\mathbf{p}_i'}{s^2 m_i V^{2/d}} \\ \frac{d\mathbf{p}_i'}{dt'} &= -V^{1/d}\nabla_{\mathbf{r}_i} U(\mathbf{r}) \\ \frac{d\pi_V'}{dt'} &= \mathcal{P} - p_0 \end{aligned} \tag{6.58}$$

$$\frac{dV}{dt'} = \frac{\pi'_V}{s^2 Q_V}$$

$$\frac{d\pi_s}{dt'} = V^{-2/d} \sum_{i=1}^{N} \frac{(\mathbf{p}'_i)^2}{s^3 m_i} + \frac{(\pi'_V)^2}{s^3 Q_V} - (g_N + 1)\frac{k_B T}{s}$$

$$\frac{ds}{dt'} = \pi_s / Q_s$$

where $\mathcal{P}$ is the instantaneous pressure given by

$$\mathcal{P} = \frac{1}{dV}\left[\sum_{i=1}^{N} \frac{(\mathbf{p}'_i)^2}{m_i V^{2/d} s^2} + \sum_{i=1}^{N} \frac{\partial U}{\partial \mathbf{r}'_i}\mathbf{r}'_i - (Vd)\frac{\partial U}{\partial V}\right] \tag{6.59}$$

where explicit dependence of the potential U on the volume V is assumed.

The compressibility of the phase space defined in Chap. 1 (Sect. 1.11) is given as:

$$\begin{aligned} \kappa(\mathbf{r}', \mathbf{p}', V, \pi'_V, s, \pi_s, t) = \sum_{i=1}^{g_N} \left(\frac{\partial \dot{x}'_i}{\partial x'_i} + \frac{\partial \dot{p}'_i}{\partial p'_i} \right) \\ + \frac{\partial \dot{\pi}'_V}{\partial \pi'_V} + \frac{\partial \dot{V}}{\partial V} \\ + \frac{\partial \dot{\pi}_s}{\partial \pi_s} + \frac{\partial \dot{s}}{\partial s} = 0 \end{aligned}$$

where $g_N = 3N$ is the number of degrees of freedom.

Therefore, the Jacobian $J(\mathbf{r}', \mathbf{p}', V, \pi'_V, s, \pi_s, t) = 1$, indicating that the volume of phase space is an invariant measure, which is expected since the dynamics governed by Eq. (6.58) are generated by a Hamiltonian function given by Eq. (6.57).

There are two major recognized drawbacks of this approach. First, the time variable in Nosé dynamics appearing in the equations of motion, Eq. (6.58), is not the real time, and hence the trajectory generated by numerically integrating the equations of motion has been transformed back into real-time leading to the configurations that are not spaced at the fixed real-time intervals. Secondly, the Hamiltonian is not separable because the kinetic energy and potential energy in the Hamiltonian function are not only functions of momenta and position variables, respectively. That makes the use of the standard Verlet/leapfrog approaches not applicable (Sanz-Serna and M.P Calvo 1995).

Hoover (1985b) introduced a change of variables and a time re-scaling of the equations of motion, and the new equations of motion obtained in this way generate the same trajectories as the original Nosé Hamiltonian, but in real time.

The following transformations of the variables are first suggested:

$$\mathbf{r}_i = V^{1/d}\mathbf{r}'_i \qquad (6.60)$$
$$dt = dt'/s$$

With these transformations, Eq. (6.58) can be written as:

$$\frac{d\mathbf{r}_i}{dt} = \frac{\mathbf{p}'_i}{m_i s V^{1/d}} + \frac{1}{Vd}\frac{dV}{dt}\mathbf{r}_i \qquad (6.61)$$
$$\frac{d\mathbf{p}'_i}{dt} = -sV^{1/d}\nabla_{\mathbf{r}_i}U(\mathbf{r})$$
$$\frac{d\pi'_V}{dt} = s(\mathcal{P} - p_0)$$
$$\frac{dV}{dt} = \frac{\pi'_V}{sQ_V}$$
$$\frac{d\pi_s}{dt} = \frac{1}{s}\left(\sum_{i=1}^{N}\frac{(\mathbf{p}'_i)^2}{V^{2/d}m_i s^3} + \frac{(\pi'_V)^2}{s^3 Q_V}\right) - (g_N + 1)k_B T$$
$$\frac{ds}{dt} = \frac{s\pi_s}{Q_s}$$

Then, another variable transformation follows, such as

$$\mathbf{p}_i = \frac{\mathbf{p}'_i}{sV^{1/d}}, \qquad (6.62)$$
$$\pi_s = s\hat{\pi}_s,$$
$$\pi'_V = sV\hat{\pi}_V$$

which yields the following equations of motion:

$$\frac{d\mathbf{r}_i}{dt} = \frac{\mathbf{p}_i}{m_i} + \mathbf{r}_i\frac{1}{Vd}\frac{dV}{dt} \qquad (6.63)$$
$$\frac{d\mathbf{p}_i}{dt} = -\nabla_{\mathbf{r}_i}U(\mathbf{r}) - \mathbf{p}_i\frac{1}{Vd}\frac{dV}{dt} - \mathbf{p}_i\frac{1}{s}\frac{ds}{dt}$$
$$\frac{dV}{dt} = V\frac{\hat{\pi}_V}{Q_V}$$
$$\frac{d\hat{\pi}_V}{dt} = \frac{1}{V}(\mathcal{P} - p_0) - \hat{\pi}_V\frac{1}{s}\frac{ds}{dt} - \hat{\pi}_V\frac{1}{V}\frac{dV}{dt}$$

$$\frac{ds}{dt} = \frac{s^2 \hat{\pi}_s}{Q_s}$$

$$\frac{d\hat{\pi}_s}{dt} = \frac{1}{s}\left(\sum_{i=1}^{N} \frac{(\mathbf{p}_i)^2}{m_i} + V^2 \frac{\hat{\pi}_V^2}{Q_V}\right) - (g_N + 1)k_B T/s - \hat{\pi}_s \frac{1}{s}\frac{ds}{dt}$$

To further simplify Eq. (6.63), the following changes on the variables are introduced:

$$\varepsilon = \frac{1}{d} \ln V \,, \quad \pi_\varepsilon = \hat{\pi}_V / d \,,$$

$$\eta = \ln s \,, \quad \pi_\eta = s\hat{\pi}_s$$

Then, the equations of motion given by Eq. (6.63) take the form:

$$\frac{d\mathbf{r}_i}{dt} = \frac{\mathbf{p}_i}{m_i} + \mathbf{r}_i \frac{\pi_\varepsilon}{Q_V} \tag{6.64}$$

$$\frac{d\mathbf{p}_i}{dt} = -\nabla_{\mathbf{r}_i} U(\mathbf{r}) - \mathbf{p}_i \frac{\pi_\eta}{Q_s} - \mathbf{p}_i \frac{\pi_\varepsilon}{Q_V}$$

$$\frac{d\pi_\varepsilon}{dt} = (Vd)(\mathcal{P} - p_0) - \pi_\varepsilon \frac{\pi_\eta}{Q_s}$$

$$\frac{dV}{dt} = Vd \frac{\pi_\varepsilon}{Q_V}$$

$$\frac{d\pi_\eta}{dt} = \sum_{i=1}^{N} \frac{\mathbf{p}_i^2}{m_i} + \frac{\pi_\epsilon^2}{Q_V} - (g_N + 1)k_B T$$

$$\frac{d\eta}{dt} = \frac{\pi_\eta}{Q_s}$$

where the instantaneous pressure is given by:

$$\mathcal{P} = \frac{1}{Vd}\left(\sum_{i=1}^{N} \frac{\mathbf{p}_i^2}{m_i} + \sum_{i=1}^{N} \left[-\nabla_{\mathbf{r}_i} U(\mathbf{r}_1, \cdots, \mathbf{r}_N)\right] \cdot \mathbf{r}_i - (Vd)\frac{\partial U}{\partial V}\right) \tag{6.65}$$

When the long-range interactions, such as $U(r) \propto 1/r^n$, $n \leq 3$, or long-range corrections to short-range potentials are applied, then an explicit dependence of the potential energy on the volume is considered to be present. Cutting off long-range interactions or neglecting long-range corrections in small systems can however give rise to incorrect results (Allen and Tildesley 1989). In our equations, the barostat momentum π_ε has also been coupled to the thermostat momentum π_η.

The Nosé-Hoover dynamics governed by Eq. (6.64) is considered a standard in molecular simulation. However, the change of variables that transformed the

Nosé Hamiltonian to the Nosé-Hoover equations of motion is not a canonical transformation, and hence the total energy is not a Hamiltonian, but it is still a conserved quantity.

It is worth noting, however, that the Hoover's equations (Eq. (6.64)) satisfy the constraint that the volume is greater or equal to zero:

$$V(t) = V(0) \exp\left[\frac{d}{Q_V} \int_0^t dt' \pi_\varepsilon(t')\right]$$

The equations Eq. (6.64) also have a conserved quantity, representing the total energy of the system:

$$H_{NPT} = \sum_{i=1}^{N} \frac{\mathbf{p}_i^2}{2m_i} + \frac{\pi_\varepsilon^2}{2Q_V} + \frac{\pi_\eta^2}{2Q_s} + U(\mathbf{r}, V) + (g_N + 1)k_B T \eta + p_0 V \tag{6.66}$$

For that, calculating the derivative with respect to time t as follows:

$$\frac{dH_{NPT}}{dt} = \sum_{i=1}^{N} \left[\nabla_{\mathbf{p}_i} H_{NPT} \cdot \dot{\mathbf{p}}_i + \nabla_{\mathbf{r}_i} H_{NPT} \cdot \dot{\mathbf{r}}_i\right] \tag{6.67}$$
$$+ \frac{\partial H_{NPT}}{\partial \pi_\eta} \dot{\pi}_\eta + \frac{\partial H_{NPT}}{\partial \eta} \dot{\eta} + \frac{\partial H_{NPT}}{\partial \pi_\varepsilon} \dot{\pi}_\varepsilon + \frac{\partial H_{NPT}}{\partial V} \dot{V}$$

Substituting the equations of motion, Eq. (6.64), we get:

$$\frac{dH_{NPT}}{dt} = 0\,. \tag{6.68}$$

The compressibility of the phase space is defined as:

$$\begin{aligned}
\kappa(\mathbf{r}, \mathbf{p}, V, \pi_\varepsilon, \eta, \pi_\eta, t) &= \sum_{i=1}^{g_N} \left(\frac{\partial \dot{x}_i}{\partial x_i} + \frac{\partial \dot{p}_i}{\partial p_i}\right) \\
&+ \frac{\partial \dot{\pi}_\varepsilon}{\partial \pi_\varepsilon} + \frac{\partial \dot{V}}{\partial V} \\
&+ \frac{\partial \dot{\pi}_\eta}{\partial \pi_\eta} + \frac{\partial \dot{\eta}}{\partial \eta} \\
&= -(g_N + 1)\frac{\pi_\eta}{Q_s} + d\frac{\pi_\varepsilon}{Q_V}
\end{aligned}$$

Therefore, the Jacobian is obtained as (Arnold 1978):

$$J(\mathbf{r}, \mathbf{p}, V, \pi_\varepsilon, \eta, \pi_\eta, t) = \exp\left(\int_{t_0}^{t} \kappa(\mathbf{r}, \mathbf{p}, V, \pi_\varepsilon, \eta, \pi_\eta, t)\, dt\right)$$
$$= V \exp\left(-(g_N + 1)\eta\right)$$

indicating that the volume of phase space is not invariant measure. The Jacobian gives the weights of the phase space volume and it is unity for the (Hamiltonian) systems satisfying the Liouville's theorem (Arnold 1978), as those described by Eq. (6.58). In addition, it represents the transformation from the set of variables $(\mathbf{r}', \mathbf{p}', V, \pi'_V, s, \pi_s)$ where $J = 1$ to the set of variables $(\mathbf{r}, \mathbf{p}, V, \pi_\varepsilon, \eta, \pi_\eta)$, through the variable transformations shown above, with $J(\mathbf{r}, \mathbf{p}, V, \pi_\varepsilon, \eta, \pi_\eta, t) = V \exp\left(-(g_N + 1)\eta\right)$.

The partition function associated with the dynamics can be generated, under the assumption of ergodicity, the Jacobian (J^{-1}), and the conserved quantity as follows (Arnold 1978):

$$Z_{NPT} = \frac{1}{N!h^{g_N}} \int d\pi_\eta \int d\pi_\varepsilon \int d\eta \int dV \int d\mathbf{p} \int_{D(V)} d\mathbf{r} V^{-1} \tag{6.69}$$
$$\times \exp\left[(g_N + 1)\eta\right] \delta(H_{NPT} - E)$$

Using the properties of the δ function as described on the previous sections, we can integrate according to η to get:

$$Z_{NPT} = \frac{\exp\left[E/k_B T\right]}{(g_N + 1)k_B T} \frac{1}{N!h^{g_N}} \int d\pi_\eta \int d\pi_\varepsilon \int dV \int d\mathbf{p} \int_{D(V)} d\mathbf{r} V^{-1} \tag{6.70}$$
$$\times \exp\left[-\frac{1}{k_B T}\left(\sum_{i=1}^{N} \frac{\mathbf{p}_i^2}{2m_i} + U(\mathbf{r}, V)\right.\right.$$
$$\left.\left. + \frac{\pi_\varepsilon^2}{2Q_V} + \frac{\pi_\eta^2}{2Q_s} + p_0 V\right)\right]$$

Here, $D(V)$ denotes the domain defined by the volume. It is shown (Hoover 1985b; Nosé and Klein 1983) that Z_{NPT} is not the isothermal-isobaric partition function, see also discussion in Martyna et al. (1994).

In order to generate the isothermal-isobaric ensemble, the Hoover's equations of motion, Eq. (6.64), are modified as the following (Melchionna et al. 1993; Martyna et al. 1994)

$$\frac{d\mathbf{r}_i}{dt} = \frac{\mathbf{p}_i}{m_i} + (\mathbf{r}_i - \mathbf{r}_{com})\frac{\pi_\varepsilon}{Q_V} \tag{6.71}$$

$$\frac{d\mathbf{p}_i}{dt} = -\nabla_{\mathbf{r}_i} U(\mathbf{r}) - \mathbf{p}_i \frac{\pi_\eta}{Q_s} - \mathbf{p}_i \frac{\pi_\varepsilon}{Q_V}$$

$$\frac{d\pi_\varepsilon}{dt} = (Vd)(\tilde{\mathcal{P}} - p_0) - \pi_\varepsilon \frac{\pi_\eta}{Q_s}$$

$$\frac{dV}{dt} = Vd\frac{\pi_\varepsilon}{Q_V}$$

$$\frac{d\pi_\eta}{dt} = \sum_{i=1}^{N} \frac{\mathbf{p}_i^2}{m_i} + \frac{\pi_\epsilon^2}{Q_V} - (g_N + 1)k_B T$$

$$\frac{d\eta}{dt} = \frac{\pi_\eta}{Q_s}$$

where the center of mass $\mathbf{r}_{com}$ is defined as:

$$\mathbf{r}_{com} = \frac{\sum_{i=1}^{N} m_i \mathbf{r}_i}{\sum_{i=1}^{N} m_i}$$

The instantaneous pressure is calculated as:

$$\tilde{\mathcal{P}} = \frac{1}{Vd}\left(\sum_{i=1}^{N} \frac{\mathbf{p}_i^2}{m_i} + \sum_{i=1}^{N} \left[-\nabla_{\mathbf{r}_i} U(\mathbf{r}_1, \cdots, \mathbf{r}_N)\right] \cdot (\mathbf{r}_i - \mathbf{r}_{com}) - (Vd)\frac{\partial U}{\partial V}\right) \tag{6.72}$$

The equations of motion given by Eq. (6.71) provide the same conserved quantity as the original set (Eq. (6.66)), which can be shown easily by replacing Eq. (6.71) into Eq. (6.67) to find that Eq. (6.68) is satisfied.

Furthermore, the Jacobian is $J = \exp[-(g_N + 1)\eta]$. Thus, replacing this Jacobian into Eq. (6.69), we get:

$$Z_{NPT} = \frac{1}{N!h^{g_N}} \int d\pi_\eta \int d\pi_\varepsilon \int d\eta \int dV \int d\mathbf{p} \int_{D(V)} d\mathbf{r} \times \exp[(g_N + 1)\eta]\, \delta(H_{NPT} - E) \tag{6.73}$$

Similarly, we can integrate according to η, using the properties of the δ function:

$$Z_{NPT} = \frac{\exp[E/k_B T]}{(g_N + 1)k_B T} \frac{1}{N!h^{g_N}} \int d\pi_\eta \int d\pi_\varepsilon \int dV \int d\mathbf{p} \int d\mathbf{r} \times \exp\left[-\frac{1}{k_B T}\left(\sum_{i=1}^{N} \frac{\mathbf{p}_i^2}{2m_i} + U(\mathbf{r})\right.\right. \tag{6.74}$$

$$+ \frac{\pi_\varepsilon^2}{2Q_V} + \frac{\pi_\eta^2}{2Q_s} + p_0 V\Bigg)\Bigg]$$

where the dependence of U on V is omitted.

Then, an integration according to π_ε and π_η yield:

$$Z_{NPT} = \frac{\exp[E/k_BT]}{g_N+1} \frac{2\pi k_BT}{p_0} \sqrt{Q_V Q_s} \frac{1}{N!h^{g_N}} \int d\mathbf{p} \int d\mathbf{r} \tag{6.75}$$
$$\times \exp\left[-\frac{1}{k_BT}\left(\sum_{i=1}^{N} \frac{\mathbf{p}_i^2}{2m_i} + U(\mathbf{r})\right)\right]$$

which shows that an isothermal-isobaric ensemble is generated. In the case of no external forces (Melchionna et al. 1993), $\mathbf{F}_{com} = 0$, it can be seen that $\tilde{\mathcal{P}} = \mathcal{P}$.

Martyna et al. (1994) have proposed alternative equations of motion:

$$\frac{d\mathbf{r}_i}{dt} = \frac{\mathbf{p}_i}{m_i} + \mathbf{r}_i \frac{\pi_\varepsilon}{Q_V} \tag{6.76}$$
$$\frac{d\mathbf{p}_i}{dt} = -\nabla_{\mathbf{r}_i} U(\mathbf{r}) - \mathbf{p}_i \frac{\pi_\eta}{Q_s} - \left(1 + \frac{d}{g_N}\right) \mathbf{p}_i \frac{\pi_\varepsilon}{Q_V}$$
$$\frac{d\pi_\varepsilon}{dt} = Vd(\tilde{\mathcal{P}} - p_0) - \pi_\varepsilon \frac{\pi_\eta}{Q_s} + \frac{d}{g_N} \sum_{i=1}^{N} \frac{\mathbf{p}_i^2}{m_i}$$
$$\frac{dV}{dt} = Vd \frac{\pi_\varepsilon}{Q_V}$$
$$\frac{d\pi_\eta}{dt} = \sum_{i=1}^{N} \frac{\mathbf{p}_i^2}{m_i} + \frac{\pi_\epsilon^2}{Q_V} - (g_N + 1)k_BT$$
$$\frac{d\eta}{dt} = \frac{\pi_\eta}{Q_s}$$

It can be shown that the compressibility of the phase space is determined as:

$$\kappa(\mathbf{r}, \mathbf{p}, V, \pi_\varepsilon, \eta, \pi_\eta, t) = \sum_{i=1}^{g_N} \left(\frac{\partial \dot{x}_i}{\partial x_i} + \frac{\partial \dot{p}_i}{\partial p_i}\right)$$
$$+ \frac{\partial \dot{\pi}_\varepsilon}{\partial \pi_\varepsilon} + \frac{\partial \dot{V}}{\partial V}$$
$$+ \frac{\partial \dot{\pi}_\eta}{\partial \pi_\eta} + \frac{\partial \dot{\eta}}{\partial \eta}$$
$$= -(g_N + 1)\frac{\pi_\eta}{Q_s}$$

Therefore, the Jacobian is obtained as (Arnold 1978):

$$J(\mathbf{r}, \mathbf{p}, V, \pi_\varepsilon, \eta, \pi_\eta, t) = \exp\left(\int_{t_0}^{t} \kappa(\mathbf{r}, \mathbf{p}, V, \pi_\varepsilon, \eta, \pi_\eta, t)\, dt\right)$$
$$= \exp\left(-(g_N + 1)\eta\right)$$

These equations, Eq. (6.76), have the same conserved quantity as Hoover's original set, and the Jacobian is $J = \exp[-(g_N + 1)\eta]$ and as shown above for Eq. (6.71), they generate the isothermal-isobaric partition function.

To asses the differences between the set of equations Eqs. (6.71) and (6.76), we start with analyzing the phase space as in Tobias et al. (1993).

First, we show that the conservation law, $F_{com} = 0$, can affect the volume distribution function generated by the modified equations of motion, Eq. (6.71). In general, satisfying the Liouville equation for the entire distribution is insufficient to guarantee that the individual pieces are properly generated.

The distribution function of the reduced phase space (no positions, V, $\mathbf{p}$ only) generated by the modified equations of motion, Eq. (6.71), is

$$Z_{NPT} = \frac{\exp[E/k_B T]}{(g_N + 1)k_B T} \int d\pi_\eta \int d\pi_\varepsilon \int dV \tag{6.77}$$
$$\times V^{g_N - 1} \exp\left[-\frac{1}{k_B T}\left(\sum_{i=1}^{N} \frac{\mathbf{p}_i^2}{2m_i} + \frac{\pi_\varepsilon^2}{2Q_V} + \frac{\pi_\eta^2}{2Q_s} + p_0 V\right)\right]$$

While for equations Eq. (6.76), the partition functions is:

$$Z_{NPT} = \frac{\exp[E/k_B T]}{(g_N + 1)k_B T} \int d\pi_\eta \int d\pi_\varepsilon \int dV \tag{6.78}$$
$$\times V^{g_N} \exp\left[-\frac{1}{k_B T}\left(\sum_{i=1}^{N} \frac{\mathbf{p}_i^2}{2m_i} + \frac{\pi_\varepsilon^2}{2Q_V} + \frac{\pi_\eta^2}{2Q_s} + p_0 V\right)\right]$$

which is a correct NPT ensemble.

Also for the general case, $F_{com} = 0$ and $\tilde{\mathcal{P}} = \mathcal{P} + \mathbf{r}_{com} \cdot \mathbf{F}_{com}$, the equations of motion, Eq. (6.76), have the same problem (Melchionna et al. 1993).

6.2.3.2 Nosé-Andersen Chain of Thermostats Method

Suppose we have a chain of M thermostats coupled to system of N particles interacting via the potential energy function of the coordinates $U(\mathbf{r}_1, \cdots, \mathbf{r}_N)$. In addition, there is a thermostat chain of length $\mathcal{M}$ coupled to the barostat degrees of freedom. Then, the Nosé-Andersen chain of thermostats Hamiltonian is given by:

$$
\begin{aligned}
H_{NPT}^{chain} &= \sum_{i=1}^{N} \frac{(\mathbf{p}_i')^2}{2s_{1,p}^2 V^{2/d} m_i} + U\left(V^{1/d}\mathbf{r}\right) \\
&+ \sum_{k=1}^{M-1} \frac{\pi_{s_{k,p}}^2}{2Q_{s_{k,p}} s_{k+1,p}^2} + \frac{\pi_{s_{M,p}}^2}{2Q_{s_{M,p}}} + g_N k_B T \ln s_{1,p} + k_B T \sum_{k=2}^{M} \ln s_{k,p} \\
&+ \frac{\pi_V^2}{2s_{1,b}^2 Q_V} + p_0 V \\
&+ \sum_{k=1}^{\mathcal{M}-1} \frac{\pi_{s_{k,b}}^2}{2Q_{s_{k,b}} s_{k+1,b}^2} + \frac{\pi_{s_{\mathcal{M},b}}^2}{2Q_{s_{\mathcal{M},b}}} + k_B T \sum_{k=1}^{\mathcal{M}} \ln s_{k,b}
\end{aligned}
\tag{6.79}
$$

Note that M and $\mathcal{M}$, in practice, do not have to be the same.

Using the Hamiltonian equations for a Hamiltonian system, we get the following equations of motion for the system:

$$
\begin{aligned}
\frac{d\mathbf{r}_i'}{d\tau_{1,p}} &= \frac{\mathbf{p}_i'}{s_{1,p}^2 m_i V^{2/d}} \\
\frac{d\mathbf{p}_i'}{d\tau_{1,p}} &= -V^{1/d} \nabla_{\mathbf{r}_i} U(\mathbf{r}) \\
\frac{d\pi_{s_{1,p}}}{d\tau_{2,p}} &= \sum_{i=1}^{N} \frac{(\mathbf{p}_i')^2}{s_{1,p}^3 V^{2/d} m_i} - g_N \frac{k_B T}{s_{1,p}} \\
\frac{ds_{1,p}}{d\tau_{2,p}} &= \frac{\pi_{s_{1,p}}}{s_{2,p}^2 Q_{s_{1,p}}} \\
\frac{d\pi_{s_{k,p}}}{d\tau_{k+1,p}} &= \frac{\pi_{s_{k-1,p}}^2}{s_{k,p}^3 Q_{k-1,p}} - \frac{k_B T}{s_{k,p}} \\
\frac{ds_{k,p}}{d\tau_{k+1,p}} &= \frac{\pi_{s_{k,p}}}{s_{k+1,p}^2 Q_{s_{k,p}}} \\
k &= 2, \cdots, M-1 \\
\frac{d\pi_{s_{M,p}}}{d\tau_{M,p}} &= \frac{\pi_{s_{M-1,p}}^2}{s_{M,p}^3 Q_{M-1,p}} - \frac{k_B T}{s_{M,p}} \\
\frac{ds_{M,p}}{d\tau_{M,p}} &= \frac{\pi_{s_{M,p}}}{Q_{s_{M,p}}} \\
\frac{d\pi_V}{d\tau_{1,b}} &= \mathcal{P} - p_0
\end{aligned}
\tag{6.80}
$$

$$\frac{dV}{d\tau_{1,b}} = \frac{\pi_V}{s_{1,b}^2 Q_V}$$

$$\frac{d\pi_{s_{1,b}}}{d\tau_{2,b}} = \frac{\pi_V^2}{s_{1,b}^3 Q_V} - \frac{k_B T}{s_{1,b}}$$

$$\frac{ds_{1,b}}{d\tau_{2,b}} = \frac{\pi_{s_{1,b}}}{s_{2,b}^2 Q_{s_{1,b}}}$$

$$\frac{d\pi_{s_{k,b}}}{d\tau_{k+1,b}} = \frac{\pi_{s_{k-1},b}^2}{s_{k,b}^3 Q_{k-1,b}} - \frac{k_B T}{s_{k,b}}$$

$$\frac{ds_{k,b}}{d\tau_{k+1,b}} = \frac{\pi_{s_{k,b}}}{s_{k+1,b}^2 Q_{s_{k,b}}}$$

$$k = 2, \cdots, \mathcal{M} - 1$$

$$\frac{d\pi_{s_{\mathcal{M},b}}}{d\tau_{\mathcal{M},b}} = \frac{\pi_{s_{\mathcal{M}-1},b}^2}{s_{\mathcal{M},b}^3 Q_{\mathcal{M}-1,b}} - \frac{k_B T}{s_{\mathcal{M},b}}$$

$$\frac{ds_{\mathcal{M},b}}{d\tau_{\mathcal{M},b}} = \frac{\pi_{s_{\mathcal{M},b}}}{Q_{s_{\mathcal{M},b}}}$$

where

$$\begin{aligned} d\tau_{1,p} &= s_{1,p}dt \\ d\tau_{k,p} &= s_{k-1,p}s_{k,p}dt\ , \quad k = 2, \cdots, M-1 \\ d\tau_{M,p} &= s_{M,p}dt \\ d\tau_{1,b} &= s_{1,b}dt \\ d\tau_{k,b} &= s_{k-1,b}s_{k,b}dt\ , \quad k = 2, \cdots, \mathcal{M}-1 \\ d\tau_{\mathcal{M},b} &= s_{\mathcal{M},b}dt \end{aligned} \tag{6.81}$$

Substituting the transformation given by Eq. (6.81) along with the position transformation: $\mathbf{r}_i = V^{1/d}\mathbf{r}'_i$ into Eq. (6.82), we get:

$$\begin{aligned} \frac{d\mathbf{r}_i}{dt} &= \frac{\mathbf{p}'_i}{s_{1,p} m_i V^{1/d}} + \frac{1}{Vd}\frac{dV}{dt}\mathbf{r}_i \\ \frac{d\mathbf{p}'_i}{dt} &= -s_{1,p} V^{1/d} \nabla_{\mathbf{r}_i} U(\mathbf{r}) \\ \frac{d\pi_{s_{1,p}}}{dt} &= s_{2,p} \sum_{i=1}^{N} \frac{(\mathbf{p}'_i)^2}{s_{1,p}^2 V^{2/d} m_i} - s_{2,p} g_N k_B T \end{aligned} \tag{6.82}$$

$$\frac{ds_{1,p}}{dt} = s_{1,p}\frac{\pi_{s_{1,p}}}{s_{2,p}Q_{s_{1,p}}}$$

$$\frac{d\pi_{s_{k,p}}}{dt} = s_{k+1,p}\frac{\pi^2_{s_{k-1},p}}{s^2_{k,p}Q_{k-1,p}} - s_{k+1,p}k_BT$$

$$\frac{ds_{k,p}}{dt} = s_{k,p}\frac{\pi_{s_{k,p}}}{s_{k+1,p}Q_{s_{k,p}}}$$

$$k = 2, \cdots, M-1$$

$$\frac{d\pi_{s_{M,p}}}{dt} = \frac{\pi^2_{s_{M-1},p}}{s^2_{M,p}Q_{M-1,p}} - k_BT$$

$$\frac{ds_{M,p}}{dt} = s_{M,p}\frac{\pi_{s_{M,p}}}{Q_{s_{M,p}}}$$

$$\frac{d\pi_V}{dt} = s_{1,b}\left(\mathcal{P} - p_0\right)$$

$$\frac{dV}{dt} = \frac{\pi_V}{s_{1,b}Q_V}$$

$$\frac{d\pi_{s_{1,b}}}{dt} = s_{2,b}\frac{\pi^2_V}{s^2_{1,b}Q_V} - s_{2,b}k_BT$$

$$\frac{ds_{1,b}}{dt} = s_{1,b}\frac{\pi_{s_{1,b}}}{s_{2,b}Q_{s_{1,b}}}$$

$$\frac{d\pi_{s_{k,b}}}{dt} = s_{k+1,b}\frac{\pi^2_{s_{k-1},b}}{s^2_{k,b}Q_{k-1,b}} - s_{k+1,b}k_BT$$

$$\frac{ds_{k,b}}{dt} = s_{k,b}\frac{\pi_{s_{k,b}}}{s_{k+1,b}Q_{s_{k,b}}}$$

$$k = 2, \cdots, \mathcal{M}-1$$

$$\frac{d\pi_{s_{\mathcal{M},b}}}{dt} = \frac{\pi^2_{s_{\mathcal{M}-1},b}}{s^2_{\mathcal{M},b}Q_{\mathcal{M}-1,b}} - k_BT$$

$$\frac{ds_{\mathcal{M},b}}{dt} = s_{\mathcal{M},b}\frac{\pi_{s_{\mathcal{M},b}}}{Q_{s_{\mathcal{M},b}}}$$

Then, another change of the variables can be introduced:

$$\mathbf{p}_i = \frac{\mathbf{p}'_i}{s_{1,p}V^{1/d}}, \tag{6.83}$$

$$
\begin{aligned}
\pi_{s_{k,p}} &= s_{k,p} s_{k+1,p} \hat{\pi}_{s_{k,p}} \,, \quad k = 1, \cdots, M-1 \\
\pi_{s_{M,p}} &= s_{M,p} \hat{\pi}_{s_{M,p}} \,, \\
\pi_V &= s_{1,b} V \hat{\pi}_V \,, \\
\pi_{s_{k,b}} &= s_{k,b} s_{k+1,b} \hat{\pi}_{s_{k,b}} \,, \quad k = 1, \cdots, M-1 \\
\pi_{s_{\mathcal{M},b}} &= s_{\mathcal{M},b} \hat{\pi}_{s_{\mathcal{M},b}}
\end{aligned}
$$

Using these transformations, we obtain the following:

$$
\begin{aligned}
\frac{d\mathbf{r}_i}{dt} &= \frac{\mathbf{p}_i}{m_i} + \frac{1}{Vd}\frac{dV}{dt}\mathbf{r}_i \\
\frac{d\mathbf{p}_i}{dt} &= -\nabla_{\mathbf{r}_i} U(\mathbf{r}) - \frac{1}{dV}\frac{dV}{dt}\mathbf{p}_i - \frac{1}{s_{1,p}}\frac{ds_{1,p}}{dt}\mathbf{p}_i \\
\frac{d\hat{\pi}_{s_{1,p}}}{dt} &= \frac{1}{s_{1,p}} \sum_{i=1}^{N} \frac{(\mathbf{p}_i)^2}{m_i} - \frac{g_N k_B T}{s_{1,p}} - \frac{1}{s_{2,p}}\frac{ds_{2,p}}{dt}\hat{\pi}_{s_{1,p}} - \frac{1}{s_{1,p}}\frac{ds_{1,p}}{dt}\hat{\pi}_{s_{1,p}} \\
\frac{ds_{1,p}}{dt} &= s_{1,p}^2 \frac{\hat{\pi}_{s_{1,p}}}{Q_{s_{1,p}}} \\
\frac{d\hat{\pi}_{s_{k,p}}}{dt} &= s_{k-1,p}^2 \frac{\hat{\pi}_{s_{k-1,p}}^2}{s_{k,p} Q_{k-1,p}} - \frac{k_B T}{s_{k,p}} - \frac{1}{s_{k,p}}\frac{ds_{k,p}}{dt}\hat{\pi}_{s_{k,p}} - \frac{1}{s_{k+1,p}}\frac{ds_{k+1,p}}{dt}\hat{\pi}_{s_{k,p}} \\
\frac{ds_{k,p}}{dt} &= s_{k,p}^2 \frac{\hat{\pi}_{s_{k,p}}}{Q_{s_{k,p}}} \\
& k = 2, \cdots, M-1 \\
\frac{d\hat{\pi}_{s_{M,p}}}{dt} &= \frac{s_{M-1,p}^2}{s_{M,p}} \frac{\hat{\pi}_{s_{M-1,p}}^2}{Q_{M-1,p}} - \frac{k_B T}{s_{M,p}} - \frac{1}{s_{M,p}}\frac{ds_{M,p}}{dt}\hat{\pi}_{s_{M,p}} \\
\frac{ds_{M,p}}{dt} &= s_{M,p}^2 \frac{\hat{\pi}_{s_{M,p}}}{Q_{s_{M,p}}} \\
\frac{d\hat{\pi}_V}{dt} &= \frac{1}{V}(\mathcal{P} - p_0) - \frac{ds_{1,b}}{dt}\frac{\hat{\pi}_V}{s_{1,b}} - \frac{dV}{dt}\frac{\hat{\pi}_V}{V} \\
\frac{dV}{dt} &= V\frac{\hat{\pi}_V}{Q_V} \\
\frac{d\hat{\pi}_{s_{1,b}}}{dt} &= \frac{V^2}{s_{1,b}} \frac{\hat{\pi}_V^2}{Q_V} - \frac{k_B T}{s_{1,b}} - \frac{1}{s_{2,b}}\frac{ds_{2,b}}{dt}\hat{\pi}_{s_{1,b}} - \frac{1}{s_{1,b}}\frac{ds_{1,b}}{dt}\hat{\pi}_{s_{1,b}} \\
\frac{ds_{1,b}}{dt} &= s_{1,b}^2 \frac{\hat{\pi}_{s_{1,b}}}{Q_{s_{1,b}}}
\end{aligned}
\tag{6.84}
$$

$$
\begin{aligned}
\frac{d\hat{\pi}_{s_{k,b}}}{dt} &= s_{k-1,b}^2 \frac{\hat{\pi}^2_{s_{k-1},b}}{s_{k,b}Q_{k-1,b}} - \frac{k_B T}{s_{k,b}} - \frac{1}{s_{k,b}}\frac{ds_{k,b}}{dt}\hat{\pi}_{s_{k,b}} - \frac{1}{s_{k+1,b}}\frac{ds_{k+1,b}}{dt}\hat{\pi}_{s_{k,b}} \\
\frac{ds_{k,b}}{dt} &= s_{k,b}^2 \frac{\hat{\pi}_{s_{k,b}}}{Q_{s_{k,b}}} \\
& \quad k = 2, \cdots, \mathcal{M}-1 \\
\frac{d\hat{\pi}_{s_{\mathcal{M},b}}}{dt} &= \frac{s^2_{\mathcal{M}-1,b}}{s_{\mathcal{M},b}} \frac{\hat{\pi}^2_{s_{\mathcal{M}-1},b}}{Q_{\mathcal{M}-1,b}} - \frac{k_B T}{s_{\mathcal{M},b}} - \frac{1}{s_{\mathcal{M},b}}\frac{ds_{\mathcal{M},b}}{dt}\hat{\pi}_{s_{\mathcal{M},b}} \\
\frac{ds_{\mathcal{M},b}}{dt} &= s^2_{\mathcal{M},b}\frac{\hat{\pi}_{s_{\mathcal{M},b}}}{Q_{s_{\mathcal{M},b}}}
\end{aligned}
$$

To further simplify Eq. (6.84), we introduce the following changes on the variables:

$$
\begin{aligned}
\varepsilon &= \frac{1}{d}\ln V\,, \quad \pi_\varepsilon = \frac{1}{d}\hat{\pi}_V\,, \\
\eta_{k,p} &= \ln s_{k,p}\,, \quad \pi_{\eta_{k,p}} = s_{k,p}\hat{\pi}_{s_{k,p}}\,, \quad k = 1, 2, \cdots, M\,, \\
\eta_{k,b} &= \ln s_{k,b}\,, \quad \pi_{\eta_{k,b}} = s_{k,b}\hat{\pi}_{s_{k,b}}\,, \quad k = 1, 2, \cdots, \mathcal{M}
\end{aligned}
$$

Then, Eq. (6.84) reduces to the following equations of motion:

$$
\begin{aligned}
\frac{d\mathbf{r}_i}{dt} &= \frac{\mathbf{p}_i}{m_i} + \frac{\pi_\varepsilon}{Q_V}\mathbf{r}_i \\
\frac{d\mathbf{p}_i}{dt} &= -\nabla_{\mathbf{r}_i} U(\mathbf{r}) - \frac{\pi_\varepsilon}{Q_V}\mathbf{p}_i - \frac{\pi_{\eta_{1,p}}}{Q_{s_{1,p}}}\mathbf{p}_i \\
\frac{d\pi_{\eta_{1,p}}}{dt} &= \sum_{i=1}^{N}\frac{\mathbf{p}_i^2}{m_i} - g_N k_B T - \frac{\pi_{\eta_{2,p}}}{Q_{s_{2,p}}}\pi_{\eta_{1,p}} \\
\frac{d\eta_{1,p}}{dt} &= \frac{\pi_{\eta_{1,p}}}{Q_{s_{1,p}}} \\
\frac{d\pi_{\eta_{k,p}}}{dt} &= \frac{\pi^2_{\eta_{k-1},p}}{Q_{k-1,p}} - k_B T - \frac{d\pi_{\eta_{k+1,p}}}{Q_{s_{k+1,p}}}\pi_{\eta_{k,p}} \\
\frac{d\eta_{k,p}}{dt} &= \frac{\pi_{\eta_{k,p}}}{Q_{s_{k,p}}} \\
& \quad k = 2, \cdots, M-1 \\
\frac{d\pi_{\eta_{M,p}}}{dt} &= \frac{\pi^2_{\eta_{M-1},p}}{Q_{M-1,p}} - k_B T
\end{aligned}
\tag{6.85}
$$

$$\frac{d\eta_{M,p}}{dt} = \frac{\pi_{\eta_{M,p}}}{Q_{s_{M,p}}}$$

$$\frac{d\pi_\varepsilon}{dt} = Vd\left(\mathcal{P} - p_0\right) - \frac{d\pi_{\eta_{1,b}}}{Q_{s_{1,b}}}\pi_\varepsilon$$

$$\frac{1}{Vd}\frac{dV}{dt} = \frac{\pi_\varepsilon}{Q_V}$$

$$\frac{d\pi_{\eta_{1,b}}}{dt} = \frac{\pi_\varepsilon^2}{Q_V} - k_BT - \frac{\pi_{\eta_{2,b}}}{Q_{s_{2,b}}}\pi_{\eta_{1,b}}$$

$$\frac{d\eta_{1,b}}{dt} = \frac{\pi_{\eta_{1,b}}}{Q_{s_{1,b}}}$$

$$\frac{d\pi_{\eta_{k,b}}}{dt} = \frac{\pi^2_{\eta_{k-1,b}}}{Q_{k-1,b}} - k_BT - \frac{d\pi_{\eta_{k+1,b}}}{Q_{s_{k+1,b}}}\pi_{\eta_{k,b}}$$

$$\frac{d\eta_{k,b}}{dt} = \frac{\pi_{\eta_{k,b}}}{Q_{s_{k,b}}}$$

$$k = 2, \cdots, \mathcal{M} - 1$$

$$\frac{d\pi_{\eta_{\mathcal{M},b}}}{dt} = \frac{\pi^2_{\eta_{\mathcal{M}-1,b}}}{Q_{\mathcal{M}-1,b}} - k_BT$$

$$\frac{d\eta_{\mathcal{M},b}}{dt} = \frac{\pi_{\eta_{\mathcal{M},b}}}{Q_{s_{\mathcal{M},b}}}$$

In order for these equations (Eq. (6.85)) to sample an isothermal-isobaric ensemble distribution, we suggested the following changes (see also Melchionna et al. 1993):

$$\frac{d\mathbf{r}_i}{dt} = \frac{\mathbf{p}_i}{m_i} + \frac{\pi_\varepsilon}{Q_V}\left(\mathbf{r}_i - \mathbf{r}_{com}\right) \tag{6.86}$$

$$\frac{d\mathbf{p}_i}{dt} = -\nabla_{\mathbf{r}_i}U(\mathbf{r}) - \frac{\pi_\varepsilon}{Q_V}\mathbf{p}_i - \frac{\pi_{\eta_{1,p}}}{Q_{s_{1,p}}}\mathbf{p}_i$$

$$\frac{d\pi_{\eta_{1,p}}}{dt} = \sum_{i=1}^{N}\frac{\mathbf{p}_i^2}{m_i} - g_Nk_BT - \frac{\pi_{\eta_{2,p}}}{Q_{s_{2,p}}}\pi_{\eta_{1,p}}$$

$$\frac{d\eta_{1,p}}{dt} = \frac{\pi_{\eta_{1,p}}}{Q_{s_{1,p}}}$$

$$\frac{d\pi_{\eta_{k,p}}}{dt} = \frac{\pi^2_{\eta_{k-1,p}}}{Q_{k-1,p}} - k_BT - \frac{d\pi_{\eta_{k+1,p}}}{Q_{s_{k+1,p}}}\pi_{\eta_{k,p}}$$

$$
\begin{aligned}
\frac{d\eta_{k,p}}{dt} &= \frac{\pi_{\eta_{k,p}}}{Q_{s_{k,p}}} \\
& \quad k = 2, \cdots, M-1 \\
\frac{d\pi_{\eta_{M,p}}}{dt} &= \frac{\pi^2_{\eta_{M-1,p}}}{Q_{M-1,p}} - k_B T \\
\frac{d\eta_{M,p}}{dt} &= \frac{\pi_{\eta_{M,p}}}{Q_{s_{M,p}}} \\
\frac{d\pi_{\varepsilon}}{dt} &= Vd\left(\tilde{\mathcal{P}} - p_0\right) - \frac{d\pi_{\eta_{1,b}}}{Q_{s_{1,b}}}\pi_{\varepsilon} \\
\frac{1}{Vd}\frac{dV}{dt} &= \frac{\pi_{\varepsilon}}{Q_V} \\
\frac{d\pi_{\eta_{1,b}}}{dt} &= \frac{\pi^2_{\varepsilon}}{Q_V} - k_B T - \frac{\pi_{\eta_{2,b}}}{Q_{s_{2,b}}}\pi_{\eta_{1,b}} \\
\frac{d\eta_{1,b}}{dt} &= \frac{\pi_{\eta_{1,b}}}{Q_{s_{1,b}}} \\
\frac{d\pi_{\eta_{k,b}}}{dt} &= \frac{\pi^2_{\eta_{k-1,b}}}{Q_{k-1,b}} - k_B T - \frac{d\pi_{\eta_{k+1,b}}}{Q_{s_{k+1,b}}}\pi_{\eta_{k,b}} \\
\frac{d\eta_{k,b}}{dt} &= \frac{\pi_{\eta_{k,b}}}{Q_{s_{k,b}}} \\
& \quad k = 2, \cdots, \mathcal{M}-1 \\
\frac{d\pi_{\eta_{\mathcal{M},b}}}{dt} &= \frac{\pi^2_{\eta_{\mathcal{M}-1,b}}}{Q_{\mathcal{M}-1,b}} - k_B T \\
\frac{d\eta_{\mathcal{M},b}}{dt} &= \frac{\pi_{\eta_{\mathcal{M},b}}}{Q_{s_{\mathcal{M},b}}}
\end{aligned}
$$

where $\mathbf{r}_{com}$ is the center of mass of the system. In addition, other forms of the equations can be adopted (Martyna et al. 1994). We can propose the following equations, which sample isothermal-isobaric ensemble:

$$
\begin{aligned}
\frac{d\mathbf{r}_i}{dt} &= \frac{\mathbf{p}_i}{m_i} + \frac{\pi_{\varepsilon}}{Q_V}\mathbf{r}_i \qquad (6.87) \\
\frac{d\mathbf{p}_i}{dt} &= -\nabla_{\mathbf{r}_i} U(\mathbf{r}) - \left(1 + \frac{d}{g_N}\right)\frac{\pi_{\varepsilon}}{Q_V}\mathbf{p}_i - \frac{\pi_{\eta_{1,p}}}{Q_{s_{1,p}}}\mathbf{p}_i
\end{aligned}
$$

$$
\begin{aligned}
\frac{d\pi_{\eta_{1,p}}}{dt} &= \sum_{i=1}^{N}\frac{\mathbf{p}_i^2}{m_i} - g_N k_B T - \frac{\pi_{\eta_{2,p}}}{Q_{s_{2,p}}}\pi_{\eta_{1,p}} \\
\frac{d\eta_{1,p}}{dt} &= \frac{\pi_{\eta_{1,p}}}{Q_{s_{1,p}}} \\
\frac{d\pi_{\eta_{k,p}}}{dt} &= \frac{\pi^2_{\eta_{k-1,p}}}{Q_{k-1,p}} - k_B T - \frac{d\pi_{\eta_{k+1,p}}}{Q_{s_{k+1,p}}}\pi_{\eta_{k,p}} \\
\frac{d\eta_{k,p}}{dt} &= \frac{\pi_{\eta_{k,p}}}{Q_{s_{k,p}}} \\
& \quad k = 2, \cdots, M-1 \\
\frac{d\pi_{\eta_{M,p}}}{dt} &= \frac{\pi^2_{\eta_{M-1,p}}}{Q_{M-1,p}} - k_B T \\
\frac{d\eta_{M,p}}{dt} &= \frac{\pi_{\eta_{M,p}}}{Q_{s_{M,p}}} \\
\frac{d\pi_\varepsilon}{dt} &= Vd\left(\mathcal{P} - p_0\right) + \frac{d}{g_N}\sum_{i=1}^{N}\frac{\mathbf{p}_i^2}{m_i} - \frac{d\pi_{\eta_{1,b}}}{Q_{s_{1,b}}}\pi_\varepsilon \\
\frac{1}{Vd}\frac{dV}{dt} &= \frac{\pi_\varepsilon}{Q_V} \\
\frac{d\pi_{\eta_{1,b}}}{dt} &= \frac{\pi_\varepsilon^2}{Q_V} - k_B T - \frac{\pi_{\eta_{2,b}}}{Q_{s_{2,b}}}\pi_{\eta_{1,b}} \\
\frac{d\eta_{1,b}}{dt} &= \frac{\pi_{\eta_{1,b}}}{Q_{s_{1,b}}} \\
\frac{d\pi_{\eta_{k,b}}}{dt} &= \frac{\pi^2_{\eta_{k-1,b}}}{Q_{k-1,b}} - k_B T - \frac{d\pi_{\eta_{k+1,b}}}{Q_{s_{k+1,b}}}\pi_{\eta_{k,b}} \\
\frac{d\eta_{k,b}}{dt} &= \frac{\pi_{\eta_{k,b}}}{Q_{s_{k,b}}} \\
& \quad k = 2, \cdots, \mathcal{M}-1 \\
\frac{d\pi_{\eta_{\mathcal{M},b}}}{dt} &= \frac{\pi^2_{\eta_{\mathcal{M}-1,b}}}{Q_{\mathcal{M}-1,b}} - k_B T \\
\frac{d\eta_{\mathcal{M},b}}{dt} &= \frac{\pi_{\eta_{\mathcal{M},b}}}{Q_{s_{\mathcal{M},b}}}
\end{aligned}
$$

Both Eqs. (6.86) and (6.87) have the same conserved quantity, known as the total energy of the extended system, given as the following:

$$E_{NPT}^{chain} = \sum_{i=1}^{N} \frac{\mathbf{p}_i^2}{2m_i} + U(\mathbf{r}) \tag{6.88}$$
$$+ \sum_{k=1}^{M} \frac{\pi_{\eta_{k,p}}^2}{2Q_{s_{k,p}}} + g_N k_B T \eta_{1,p} + k_B T \sum_{k=2}^{M} \eta_{k,p}$$
$$+ \frac{\pi_\varepsilon^2}{2Q_V} + p_0 V + \sum_{k=1}^{\mathcal{M}} \frac{\pi_{\eta_{k,b}}^2}{2Q_{s_{k,b}}} + k_B T \sum_{k=1}^{\mathcal{M}} \eta_{k,b}$$

Example 6 As an illustration, we considered a system of two coupled particles in three-dimensions with a spring with force constant $k = 317\,\text{kcal/mol/Å}^2$ interacting via the potential:

$$U(\mathbf{r}_1, \mathbf{r}_2) = \frac{k}{2} (\mid \mathbf{r}_1 - \mathbf{r}_2 \mid - x_0)^2$$

where $x_0 = 15.23$ Å. In this simulations we used different chain of thermostats coupled to each degree of freedom of the real system and to barostat. The length for each of the chain of thermostats was $M = 3$. The target pressure was $p_0 = 0$ atm, and target temperature was $T = 300$ K. To each particle was assigned a mass of 12 amu. The barostat mass was taken

$$Q_V = (d + 1) k_B T (\tau_b)^2$$

where $\tau_b = 1$ ps, and the thermostats masses all equal to:

$$Q_s = k_B T (\tau_p)^2$$

with $\tau_p = 0.01$ ps. The integration time step was fixed at $\Delta t = 1$ fs. The numerical integrator algorithm introduced in the next chapters is used to integrate the equations of motion, namely Eq. (6.87). In Fig. 6.10, we show the fluctuations of the instantaneous pressure in atmosphere (Fig. 6.10a), temperature in Kelvin (Fig. 6.10b), total energy components in units of 10× kcal/mol (Fig. 6.10c), and volume $V^{1/d}$ in Å (Fig. 6.10d). We used block averages with a block size of 2000 MD steps. Note that the equations of motion provide an excellent energy conservation (see also Fig. 6.10c) – furthermore, instantaneous pressure and temperature exhibit typical fluctuations around the target values. Besides, gentle fluctuations of the distance between the two particles are observed to adjust the pressure of the system, as shown in Fig. 6.10d.

In Fig. 6.11, we show the probability density functions of the velocities for a single degree of freedom (on average) for a real particle of the system (Fig. 6.11a) and barostat (Fig. 6.11b). Besides, we present the analytical curves. Our results suggest that the molecular dynamics simulations using Eq. (6.87) produce very well fluctuations of an isothermal-isobaric ensemble.

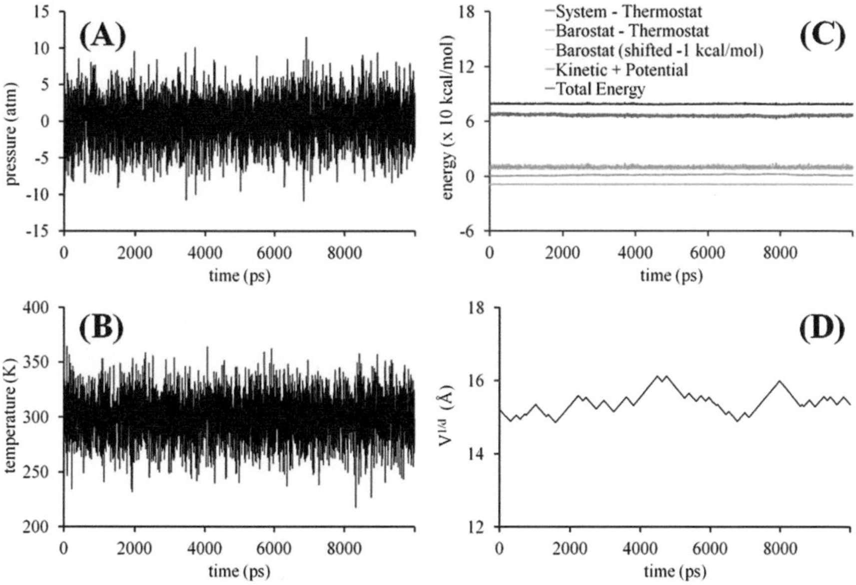

Fig. 6.10 The results of MD simulations of a two coupled particles system with a spring with a force constant $k = 317\,\text{kcal/mol/Å}^2$ and $x_0 = 15.23$ Å using Nosé-Hoover-Andersen chain of thermostats. The chain length was $M = 3$. (**a**) trajectory of the pressure (in atm); (**b**) trajectory of the temperature (in Kelvin); (**c**) trajectory of the total energy of the system (E_{NPT}^{chain}); and (**d**) trajectory of the volume $V^{1/d}$. The block averages are performed every 2000 MD steps. Note that the barostat energy is shifted down by -1.0 kcal/mol for clarity

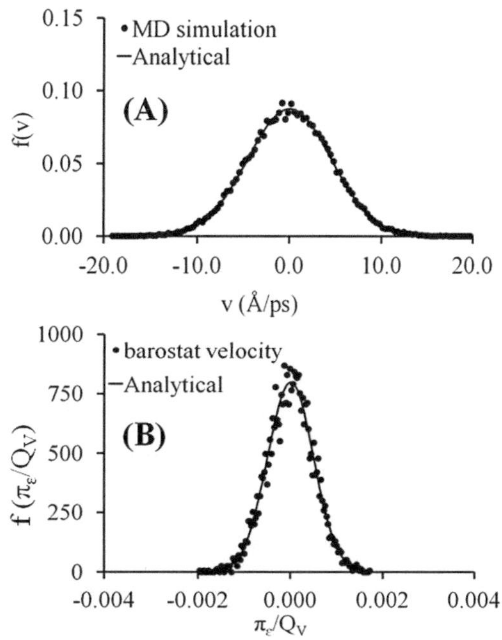

Fig. 6.11 The results of MD simulations of a two coupled particles system with a spring with a force constant $k = 317\,\text{kcal/mol/Å}^2$ and $x_0 = 15.23$ Å using Nosé-Hoover-Andersen chain of thermostats. The chain length was $M = 3$. (**a**) Probability density function of the average velocity particle degree of freedom and (**b**) probability density function of the barostat velocity. The block averages are performed every 2000 MD steps

6.2.3.3 Nosé-Poincaré-Andersen Method

Applying a Poincaré time transformation to the Hamiltonian, representing a combination of the Nosé-Poincaré thermostat with the Andersen method for constant pressure as applied in (Sturgeon and Laird 2000), will give the new Nosé-Poincaré-Andersen Hamiltonian, which can be used to generate the equations of motion for the isothermal-isobaric ensemble molecular dynamics simulation run.

For a system with an Andersen like-piston, Nosé-Poincaré-Andersen Hamiltonian is as follows (Sturgeon and Laird 2000):

$$H'_{NPT} = [H_{NPT} - H_{NPT}(t=0)]\,s \tag{6.89}$$

where H_{NPT} is given by Eq. (6.57), and $H_{NPT}(t=0)$ is the value of H_{NPT} at time $t=0$, chosen such that $\Delta H = H_{NPT} - H_{NPT}(t=0)$ is zero at the start of the simulation.

The equations of motion for this system can be obtained by applying the Hamiltonian equations of motion:

$$\begin{aligned}
\frac{d\mathbf{r}'_i}{dt} &= \frac{\mathbf{p}'_i}{sm_iV^{2/d}} \\
\frac{d\mathbf{p}'_i}{dt} &= -sV^{1/d}\nabla_{\mathbf{r}_i}U\left(V^{1/d}\mathbf{r}'_i\right) \\
\frac{d\pi'_V}{dt} &= s(\mathcal{P} - p_0) \\
\frac{dV}{dt} &= \frac{\pi'_V}{sQ_V} \\
\frac{d\pi_s}{dt} &= \sum_{i=1}^{N}\frac{(\mathbf{p}'_i)^2}{s^2V^{2/d}m_i} + \frac{(\pi'_V)^2}{s^2Q_V} - (g_N+1)k_BT - \Delta H \\
\frac{ds}{dt} &= \frac{\pi_s}{Q_s}
\end{aligned} \tag{6.90}$$

where again $\mathcal{P}$ is the instantaneous pressure given by Eq. (6.65) and the derivatives are with respect to the real time. Here, $\mathbf{p}'_i$, π'_V, and $\mathbf{r}'_i$ are related to the real variables through the following equations:

$$\begin{aligned}
\mathbf{r}_i &= V^{1/d}\mathbf{r}'_i\,, \\
\mathbf{p}_i &= \frac{\mathbf{p}'_i}{sV^{1/d}}\,, \\
\pi_V &= \frac{\pi'_V}{s}
\end{aligned} \tag{6.91}$$

6.2.3.4 Nosé-Poincaré-Andersen Chain of Thermostats

The Nosé-Hoover chain of thermostats equations generate configurations from the correct canonical distribution, given that the dynamics is ergodic, as demonstrated above. The assumption of the ergodicity is provided by the additional degrees of freedom in the chain of thermostats. The same idea can also be applied to the Nosé-Poincaré-Andersen Hamiltonian, resulting in Nosé-Poincaré-Andersen chain of thermostats:

$$\begin{aligned} H_{NPT}^{chain} &= s_1 \left[\sum_{i=1}^{N} \frac{(\mathbf{p}_i')^2}{2s_1^2 V^{2/d} m_i} + U\left(V^{1/d}\mathbf{r}\right) \right. \\ &+ \sum_{k=1}^{M-1} \frac{\pi_{s_k}^2}{2Q_{s_k} s_{k+1}^2} + \frac{\pi_{s_M}^2}{2Q_{s_M}} + \frac{\pi_V^2}{2s_1^2 Q_V} \\ &\left. + (g_N+1)k_B T \ln s_1 + \sum_{k=2}^{M} k_B T \ln s_k + p_0 V - H_0 \right] \end{aligned} \tag{6.92}$$

Here, we are interested in obtaining the correct timescales of the real-variables, $\mathbf{r}$ and $\mathbf{p}/s_1$, hence we employ the following Poincaré transformation:

$$f(\mathbf{p},\mathbf{r}) = s_1$$

M is the length of the chain, and notice that also barostat is coupled to the first thermostat of the chain. Here, H_0 is the value of the Nosé-Andersen chain of thermostats Hamiltonian at $t = 0$. Using Hamilton's equations, we get the equations of motion as follows:

$$\begin{aligned} \frac{d\mathbf{r}_i'}{dt} &= \frac{\mathbf{p}_i'}{s_1 m_i V^{2/d}} \\ \frac{d\mathbf{p}_i'}{dt} &= -s_1 V^{1/d} \nabla_{\mathbf{r}_i} U\left(V^{1/d}\mathbf{r}_i'\right) \\ \frac{d\pi_V}{dt} &= s_1(\mathcal{P} - p_0) \\ \frac{dV}{dt} &= \frac{\pi_V}{s_1 Q_V} \\ \frac{d\pi_{s_1}}{dt} &= \sum_{i=1}^{N} \frac{(\mathbf{p}_i')^2}{s_1^2 V^{2/d} m_i} + \frac{\pi_V^2}{s_1^2 Q_V} - (g_N+1)k_B T - \Delta H \\ \frac{ds_1}{dt} &= \frac{\pi_{s_1} s_1}{Q_{s_1} s_2^2}, \end{aligned} \tag{6.93}$$

$$
\begin{aligned}
\frac{ds_k}{dt} &= \frac{\pi_{s_k} s_1}{Q_{s_k} s_{k+1}^2} \,, \\
\frac{d\pi_{s_k}}{dt} &= \frac{\pi_{s_{k-1}}^2 s_1}{Q_{s_{k-1}} s_k^3} - \frac{k_B T s_1}{s_k} \,, \quad k = 2, \cdots, M-1 \,, \\
\frac{ds_M}{dt} &= \frac{\pi_{s_M} s_1}{Q_{s_M}} \,, \\
\frac{d\pi_{s_M}}{dt} &= \frac{\pi_{s_{M-1}}^2 s_1}{Q_{s_{M-1}} s_M^3} - \frac{k_B T s_1}{s_M}
\end{aligned}
$$

The new thermostats have introduced an implicit coupling to the equations of motion.

6.2.4 Grand Canonical Ensemble

The grand canonical ensemble generated using MD method has been introduced in Cagin and Pettitt (1991a,b), Ji and Pettitt (1994), Weerasinghe and Pettitt (1994), Lo and Palmer (1995), Ji et al. (1992), Palmer and Lo (1994), and Lynch and Pettitt (1997).

The grand canonical ensemble MD simulation aims to use classical equations of motion to study systems in which the temperature, volume, and chemical potential are constant, but the number of particles fluctuates.

That is accomplished by introducing additional variables into the Hamiltonian function that scale particle velocity and couple a fractional particle to the system via a continuous function of a coupling parameter, λ. In this case, a continuous particle number variable ν is defined as

$$\nu = N + \lambda$$

At the moment of the simulation time, N is the number of "whole" particles and λ, which varies between zero and one, represents the extent to which a single particle, the so-called "fractional" particle, is coupled to the rest of the system.

6.2.4.1 Nosé-Hoover Thermostats for Grand Canonical Ensemble

An extension variable, s, is used to scale the particles momenta as described in the previous sections. The Hamiltonian function of the grand canonical ensemble may be written as

$$H_{NH}^{GCE} = \sum_{i=1}^{N} \frac{(\mathbf{p}'_i)^2}{2s^2 m_i} + \frac{(\mathbf{p}'_f)^2}{2s^2 m_f} + \frac{\pi_s^2}{2Q} + \frac{p_\lambda^2}{2W} \tag{6.94}$$
$$+ U(\mathbf{r}_1, \cdots, \mathbf{r}_N) + V(\mathbf{r}_1, \cdots, \mathbf{r}_N : \mathbf{r}_f, \lambda)$$
$$+ (g_N + 1)k_B T \ln(s) + \theta(\lambda)$$

where T is the equilibrium temperature, k_B is the Boltzmann constant, and g_N represents the degrees of freedom for the physical system, similar to the above discussion. The functions $\theta(\lambda)$ and $(g_N + 1)k_B T \ln(s)$ are the potential energy terms for the extension variables. In the Hamiltonian function, Eq. (6.94), the third and fourth terms represent the kinetic energies of the new added variables and allow for the derivation of their equations of motion. The fictitious mass parameters for the extension variables, Q and W, have units of (energy $\times$ time^2). They represent the extent of coupling between the system and a heat bath at the constant temperature. The first and fifth terms in Eq. (6.94) are the Hamiltonian for an N particles system with momenta, $\mathbf{p}'_i$ (with respect to the scaled time t'), and positions, $\mathbf{r}_i$ for $i = 1, 2, \cdots, N$. The second term represents the kinetic energy of the fractional particle whose momentum is given by $\mathbf{p}'_f$ (with respect to the scaled time) and whose position is given by $\mathbf{r}_f$. The sixth term is the fractional particle potential energy.

The fractional particle interactions may be scaled as

$$V(\mathbf{r}_1, \cdots, \mathbf{r}_N : \mathbf{r}_f, \lambda) = f(\lambda) \sum_{i=1}^{N} u(\mathbf{r}_i, \mathbf{r}_f) \tag{6.95}$$

where the function $f(\lambda)$ must obey the boundary conditions

$$f(0) = 0, \quad f(1) = 1$$

The potential energy function of the extension variable characterizing the particle number is given by

$$\theta(\lambda) = -\left[N + h(\lambda)\right]\mu - U_{bias}(\lambda) + h(\lambda)\mu_{id}$$

where μ is the chemical potential, $U_{bias}(\lambda)$ is the bias potential, and μ_{id} is the ideal chemical potential for an $N + 1$ particles system. This form is chosen to provide a sampling from a grand canonical ensemble distribution, as will be show in the following discussion. The functions $h(\lambda)$ and $f(\lambda)$ obey to the same boundary conditions and they are not necessarily identical. The ideal chemical potential may be defined either with or without a rotational motion contribution. Note that the rotational motion will add to the value of the total chemical potential constant term, which, in practice, corresponds to a change in a thermodynamic reference state, and hence it will not affect the simulation trajectories or measured thermodynamic properties of the system.

Equations of motion can be derived from Eq. (6.94) using the Hamiltonian equations:

$$\begin{aligned}
\frac{d\mathbf{r}_i}{dt'} &= \frac{\mathbf{p}'_i}{s^2 m_i}\,, \\
\frac{d\mathbf{p}'_i}{dt'} &= -\nabla_{\mathbf{r}_i} U(\mathbf{r}_1,\cdots,\mathbf{r}_N) - f(\lambda)\nabla_{\mathbf{r}_i} u(\mathbf{r}_i,\mathbf{r}_f)\,, \\
\frac{d\mathbf{r}_f}{dt'} &= \frac{\mathbf{p}'_f}{s^2 m_f}\,, \\
\frac{d\mathbf{p}'_f}{dt'} &= -f(\lambda)\sum_{k=1}^{N} \nabla_{\mathbf{r}_f} u(\mathbf{r}_k,\mathbf{r}_f)\,, \\
\frac{ds}{dt'} &= \frac{\pi_s}{Q}\,, \\
\frac{d\lambda}{dt'} &= \frac{p_\lambda}{W}\,, \\
\frac{d\pi_s}{dt'} &= \sum_{k=1}^{N} \frac{(\mathbf{p}'_k)^2}{s^2 m_k} + \frac{(\mathbf{p}'_f)^2}{s^2 m_f} - (g_N+1)k_B T\,, \\
\frac{dp_\lambda}{dt'} &= -\frac{df(\lambda)}{d\lambda}\sum_{k=1}^{N} u(\mathbf{r}_k,\mathbf{r}_f) - \frac{d\theta(\lambda)}{d\lambda}
\end{aligned} \tag{6.96}$$

These equations are transformed from virtual space to real space by making this variable change (Zare and Szebehely 1975) $dt' = sdt$, and in addition a reformulation in terms of real system variables using the following transformations:

$$\begin{aligned}
\mathbf{p}_i &= \mathbf{p}'_i/s\,,\ i = 1,\cdots,N\,, \\
\mathbf{p}_f &= \mathbf{p}'_f/s\,, \\
\hat{\pi}_s &= \pi_s/s\,, \\
\pi_\lambda &= p_\lambda/s
\end{aligned} \tag{6.97}$$

Then, a new system of non-Hamiltonian equations for the dynamics in the real variables is obtained:

$$\begin{aligned}
\frac{d\mathbf{r}_i}{dt} &= \frac{\mathbf{p}_i}{m_i}\,, \\
\frac{d\mathbf{p}_i}{dt} &= -\nabla_{\mathbf{r}_i} U(\mathbf{r}_1,\cdots,\mathbf{r}_N) - f(\lambda)\nabla_{\mathbf{r}_i} u(\mathbf{r}_i,\mathbf{r}_f) - \frac{s\hat{\pi}_s}{Q}\mathbf{p}_i\,,
\end{aligned} \tag{6.98}$$

$$
\begin{aligned}
\frac{d\mathbf{r}_f}{dt} &= \frac{\mathbf{p}_f}{m_f}\,,\\
\frac{d\mathbf{p}_f}{dt} &= -f(\lambda)\sum_{k=1}^{N}\nabla_{\mathbf{r}_f}u(\mathbf{r}_k,\mathbf{r}_f) - \frac{s\hat{\pi}_s}{Q}\mathbf{p}_f\,,\\
\frac{ds}{dt} &= s^2\frac{\hat{\pi}_s}{Q}\,,\\
\frac{d\lambda}{dt} &= s^2\frac{\pi_\lambda}{W}\,,\\
\frac{d\hat{\pi}_s}{dt} &= \frac{1}{s}\left(\sum_{k=1}^{N}\frac{(\mathbf{p}_k)^2}{m_k} + \frac{(\mathbf{p}_f)^2}{m_f} - g_N k_B T\right) - \frac{s\hat{\pi}_s^2}{Q}\,,\\
\frac{d\pi_\lambda}{dt} &= -\frac{df(\lambda)}{d\lambda}\sum_{k=1}^{N}u(\mathbf{r}_k,\mathbf{r}_f) - \frac{d\theta(\lambda)}{d\lambda} - \frac{s\hat{\pi}_s}{Q}\pi_\lambda
\end{aligned}
$$

Making another change of variables similar to what was proposed in Hoover (1985b):

$$
s\hat{\pi}_s \equiv \pi_\eta\,, \quad \ln s \equiv \eta \tag{6.99}
$$

Then, the second-order differential equations of motion in real space are given by

$$
\begin{aligned}
\dot{\mathbf{r}}_i &= \frac{\mathbf{p}_i}{m_i}\,, \qquad (6.100)\\
\dot{\mathbf{p}}_i &= -\nabla_{\mathbf{r}_i}U(\mathbf{r}_1,\cdots,\mathbf{r}_N) - f(\lambda)\nabla_{\mathbf{r}_i}u(\mathbf{r}_i,\mathbf{r}_f) - \mathbf{p}_i\pi_\eta/Q\,,\\
\dot{\mathbf{r}}_f &= \frac{\mathbf{p}_f}{m_f}\,,\\
\dot{\mathbf{p}}_f &= -f(\lambda)\sum_{k=1}^{N}\nabla_{\mathbf{r}_f}u(\mathbf{r}_k,\mathbf{r}_f) - \mathbf{p}_f\pi_\eta/Q\,,\\
\dot{\eta} &= \frac{\pi_\eta}{Q}\,,\\
\dot{\lambda} &= s^2\frac{\pi_\lambda}{W}\,,\\
\dot{\pi}_\eta &= \sum_{k=1}^{N}\frac{\mathbf{p}_k^2}{m_k} + \frac{\mathbf{p}_f^2}{m_f} - g_N k_B T\,,\\
\dot{\pi}_\lambda &= -\frac{df(\lambda)}{d\lambda}\sum_{k=1}^{N}u(\mathbf{r}_k,\mathbf{r}_f) - \frac{d\theta(\lambda)}{d\lambda} - \pi_\lambda\pi_\eta/Q
\end{aligned}
$$

The grand canonical partition function, $Z_{\mu VT}$, may be written in terms of canonical partition functions,

$$Z_{\mu VT} = \sum_{N=0}^{\infty} Z_{NVT} \exp\left(\frac{1}{k_B T}\mu N\right) \tag{6.101}$$

The MD simulation of the extended system of the variables sample a microcanonical ensemble with partition function given as

$$\begin{aligned} \Xi = & \sum_{N=0}^{\infty} \frac{1}{N! h^{g_N}} \\ & \times \int d\pi_s \int ds \int dp_\lambda \int_0^1 d\lambda \int d\mathbf{p}' \int d\mathbf{r} \int d\mathbf{p}'_f \int d\mathbf{r}_f \\ & \times \delta \left[\sum_{i=1}^{N} \frac{(\mathbf{p}'_i)^2}{2s^2 m_i} + \frac{(\mathbf{p}'_f)^2}{2s^2 m_f} + \frac{\pi_s^2}{2Q} + \frac{p_\lambda^2}{2W} \right. \\ & \left. + U(\mathbf{r}) + V(\mathbf{r} : \mathbf{r}_f, \lambda) + (g_N + 1) k_B T \ln(s) + \theta(\lambda) - E \right] \end{aligned} \tag{6.102}$$

A relationship between this approximate partition function and the exact grand canonical partition function can be obtained by first integrating over the thermostat extension variables s and π_s, following the procedures in Nosé (1984b,c) and Lynch and Pettitt (1997), as described in the previous sections. This involves transforming the momenta into real space using the relationships given by Eq. (6.97), and then applying an equivalence relation for the Dirac delta function as explained in the previous sections. The resulting partition function can then be rewritten as

$$\begin{aligned} \Xi = & \sum_{N=0}^{\infty} \frac{1}{N! h^{g_N}} \frac{\sqrt{2\pi Q k_B T}}{(g_N + 1) k_B T} \exp\left(E / k_B T\right) \\ & \times \int dp_\lambda \int_0^1 d\lambda \int d\mathbf{p} \int d\mathbf{r} \int d\mathbf{p}_f \int d\mathbf{r}_f \\ & \times \exp\left(-H_0^\lambda / k_B T\right) \exp\left(-H_\lambda / k_B T\right) \end{aligned} \tag{6.103}$$

where

$$H_0^\lambda = \sum_{i=1}^{N} \frac{\mathbf{p}_i^2}{2m_i} + \frac{\mathbf{p}_f^2}{2m_f} + \frac{\pi_s^2}{2Q} + U(\mathbf{r}) + V(\mathbf{r} : \mathbf{r}_f, \lambda)$$

and

$$H_\lambda = \frac{p_\lambda^2}{2W} + \theta(\lambda)$$

After, integrating according to p_λ, we get:

$$\Xi = \sum_{N=0}^{\infty} \frac{1}{N!h^{g_N}} \frac{\sqrt{2\pi Q k_B T}\sqrt{2\pi W k_B T}}{(g_N+1)k_B T} \exp\left(E/k_B T\right) \tag{6.104}$$
$$\times \int_0^1 d\lambda \int d\mathbf{p} \int d\mathbf{r} \int d\mathbf{p}_f \int d\mathbf{r}_f$$
$$\times \exp\left(-H_0^\lambda/k_B T\right) \exp\left(-H_\lambda/k_B T\right) \exp\left(-\theta(\lambda)/k_B T\right)$$

The reduced partition function, $\Xi_{N+\lambda}$, is defined by its relationship to the full partition function,

$$\Xi = \sum_{N=0}^{\infty} \int_0^1 \Xi_{N+\lambda} d\lambda$$

The ratio of the reduced partition function for the zero and one configurations is:

$$\frac{\Xi_{N+1}}{\Xi_N} = \frac{\exp\left[-\theta(1)/k_B T\right]}{\exp\left[-\theta(0)/k_B T\right]} \frac{Z_{N+1}}{Z_{N+1}^{id} Z_N^{ex}} \tag{6.105}$$

where Z_{N+1}^{id} and Z_N^{ex} are the kinetic and potential parts of the partition function. The $N+1$ term in the denominator is a direct result of not scaling the kinetic energy of the fractional particle. Substitution of $\theta(1)$ and $\theta(0)$ into Eq. (6.105) gives

$$\frac{\Xi_{N+1}}{\Xi_N} = \left[\exp\left(\left(U_{bias}(1) - U_{bias}(0)\right)/k_B T\right) \exp\left(\mu/k_B T\right) \frac{Z_{N+1}}{Z_N}\right] \tag{6.106}$$

which for $U_{bias}(0) = U_{bias}(1)$ becomes

$$\frac{\Xi_{N+1}}{\Xi_N} = \frac{Z_{(N+1)VT}}{Z_{NVT}} \exp(\mu/k_B T)$$

which is the grand canonical ratio. The bias potential, when calculating the ensemble averages, does not need to be subtracted out, using for example weighted histogram analysis method, as long as this last restriction is satisfied.

6.2.4.2 Nosé-Hoover Chain of Thermostats for Grand Canonical Ensemble

The Hamiltonian function of the grand canonical ensemble using the chain of thermostats approach is written as

$$
\begin{aligned}
H_{NHC}^{GCE} = & \sum_{i=1}^{g_N} \left[\frac{(p_i')^2}{2s_{1,p_i}^2 m_i} \right. \\
& \left. + \sum_{i=1}^{g_N} \sum_{j=1}^{M_p - 1} \frac{\pi_{j,p_i}^2}{2Q_{j,p_i} s_{j+1,p_i}^2} + \frac{\pi_{M_p,p_i}^2}{2Q_{M_p,p_i}} \right] \\
& + \sum_{i=1}^{d} \left[\frac{(p'_{f_i})^2}{2s_{1,f_i}^2 m_{f_i}} + \sum_{j=1}^{M_f - 1} \frac{\pi_{j,f_i}^2}{2Q_{j,f_i} s_{j+1,f_i}^2} + \frac{\pi_{M_f,f_i}^2}{2Q_{M_f,f_i}} \right] \\
& + \left[\frac{p_\lambda^2}{2s_{1,\lambda}^2 W} + \sum_{j=1}^{M_\lambda - 1} \frac{\pi_{j,\lambda}^2}{2Q_{j,\lambda} s_{j+1,\lambda}^2} + \frac{\pi_{M_\lambda,\lambda}^2}{2Q_{M_\lambda,\lambda}} \right] \\
& + U(\mathbf{r}_1, \cdots, \mathbf{r}_N) + V(\mathbf{r}_1, \cdots, \mathbf{r}_N : \mathbf{r}_f, \lambda) + \theta(\lambda) \\
& + \sum_{i=1}^{g_N} \sum_{j=1}^{M_p} k_B T \ln(s_{j,p_i}) \\
& + \sum_{i=1}^{d} \sum_{j=1}^{M_f} k_B T \ln(s_{j,f_i}) \\
& + \sum_{j=1}^{M_\lambda} k_B T \ln(s_{j,\lambda})
\end{aligned}
\tag{6.107}
$$

where g_N is the number of degrees of freedom of the unperturbed system, d is the dimension of the problem and M is the length of the chain of thermostats. Note that we have used a massive chain of thermostats coupled to each degree of freedom of the system, and we have used the same length for the chain of thermostats for both the real and fractional particle. Now, we can use the Hamilton's equations of motion given in Chap. 1. Note that in the following equations, we are going to drop the subscript i for the simplicity of the notation, which runs overall degrees of freedom of the real particles, fractional particle and λ, respectively. Thus, we obtained the following equations for each degree of freedom of the entire system:

$$
\begin{aligned}
\frac{dq}{d\tau_{1,p}} &= \frac{\partial H_{NHC}^{GCE}}{\partial p'} = \frac{p'}{s_{1,p}^2 m} \\
\frac{dp'}{d\tau_{1,p}} &= -\frac{\partial H_{NHC}^{GCE}}{\partial q} = -\nabla_q U(q) - f(\lambda) \nabla_q u(q, q_f) \\
\frac{dq_f}{d\tau_{1,f}} &= \frac{\partial H_{NHC}^{GCE}}{\partial p'_f} = \frac{p'_f}{s_{1,f}^2 m_f}
\end{aligned}
\tag{6.108}
$$

$$\frac{dp'_f}{d\tau_{1,f}} = -\frac{\partial H_{NHC}^{GCE}}{\partial q_f} = -f(\lambda)\nabla_{q_f} u(q, q_f)$$

$$\frac{d\lambda}{d\tau_{1,\lambda}} = \frac{\partial H_{NHC}^{GCE}}{\partial p_\lambda} = \frac{p_\lambda}{s_{1,\lambda}^2 W}$$

$$\frac{dp_\lambda}{d\tau_\lambda} = -\frac{\partial H_{NHC}^{GCE}}{\partial \lambda} = -\frac{df(\lambda)}{d\lambda} u(q, q_f) - \frac{d\theta(\lambda)}{d\lambda}$$

$$\frac{d\pi_{1,p}}{d\tau_{2,p}} = -\frac{\partial H_{NHC}^{GCE}}{\partial s_{1,p}} = \frac{(p')^2}{s_{1,p}^3 m} - \frac{k_B T}{s_{1,p}}$$

$$\frac{ds_{1,p}}{d\tau_{2,p}} = \frac{\partial H_{NHC}^{GCE}}{\partial \pi_{1,p}} = \frac{\pi_{1,p}}{s_{2,p}^2 Q_{s_{1,p}}}$$

$$\frac{d\pi_{k,p}}{d\tau_{k+1,p}} = -\frac{\partial H_{NHC}^{GCE}}{\partial s_{k,p}} = \frac{\pi_{k-1,p}^2}{s_{k,p}^3 Q_{k-1,p}} - \frac{k_B T}{s_{k,p}}$$

$$\frac{ds_{k,p}}{d\tau_{k+1,p}} = \frac{\partial H_{NHC}^{GCE}}{\partial \pi_{s_{k,p}}} = \frac{\pi_{k,p}}{s_{k+1,p}^2 Q_{k,p}}$$

$$k = 2, \cdots, M_p - 1$$

$$\frac{d\pi_{M_p,p}}{d\tau_{M_p,p}} = -\frac{\partial H_{NHC}^{GCE}}{\partial s_{M_p,p}} = \frac{\pi_{M_p-1,p}^2}{s_{M_p,p}^3 Q_{M_p-1,p}} - \frac{k_B T}{s_{M_p,p}}$$

$$\frac{ds_{M_p,p}}{d\tau_{M_p,p}} = \frac{\partial H_{NHC}^{GCE}}{\partial \pi_{M_p,p}} = \frac{\pi_{M_p,p}}{Q_{M_p,p}}$$

$$\frac{d\pi_{1,f}}{d\tau_{2,f}} = -\frac{\partial H_{NHC}^{GCE}}{\partial s_{1,f}} = \frac{(p'_f)^2}{s_{1,f}^3 m_f} - \frac{k_B T}{s_{1,f}}$$

$$\frac{ds_{1,f}}{d\tau_{2,f}} = \frac{\partial H_{NHC}^{GCE}}{\partial \pi_{s_{1,f}}} = \frac{\pi_{1,f}}{s_{2,f}^2 Q_{1,f}}$$

$$\frac{d\pi_{k,f}}{d\tau_{k+1,f}} = -\frac{\partial H_{NHC}^{GCE}}{\partial s_{k,f}} = \frac{\pi_{k-1,f}^2}{s_{k,f}^3 Q_{k-1,f}} - \frac{k_B T}{s_{k,f}}$$

$$\frac{ds_{k,f}}{d\tau_{k+1,f}} = \frac{\partial H_{NHC}^{GCE}}{\partial \pi_{k,f}} = \frac{\pi_{k,f}}{s_{k+1,f}^2 Q_{k,f}}$$

$$k = 2, \cdots, M_f - 1$$

$$\frac{d\pi_{M_f,f}}{d\tau_{M_f,f}} = -\frac{\partial H_{NHC}^{GCE}}{\partial s_{M_f,f}} = \frac{\pi_{M_f-1,f}^2}{s_{M_f,f}^3 Q_{M_f-1,f}} - \frac{k_B T}{s_{M_f,f}}$$

$$\frac{ds_{M_f,f}}{d\tau_{M_f,f}} = \frac{\partial H_{NHC}^{GCE}}{\partial \pi_{M_f,f}} = \frac{\pi_{M_f,f}}{Q_{s_{M_f,f}}}$$

$$\frac{d\pi_{1,\lambda}}{d\tau_{2,f}} = -\frac{\partial H_{NHC}^{GCE}}{\partial s_{1,\lambda}} = \frac{(p_\lambda)^2}{s_{1,\lambda}^3 W} - \frac{k_B T}{s_{1,\lambda}}$$

$$\frac{ds_{1,\lambda}}{d\tau_{2,\lambda}} = \frac{\partial H_{NHC}^{GCE}}{\partial \pi_{1,\lambda}} = \frac{\pi_{1,\lambda}}{s_{2,\lambda}^2 Q_{1,\lambda}}$$

$$\frac{d\pi_{k,\lambda}}{d\tau_{k+1,\lambda}} = -\frac{\partial H_{NHC}^{GCE}}{\partial s_{k,\lambda}} = \frac{\pi_{k-1,\lambda}^2}{s_{k,\lambda}^3 Q_{k-1,\lambda}} - \frac{k_B T}{s_{k,\lambda}}$$

$$\frac{ds_{k,\lambda}}{d\tau_{k+1,\lambda}} = \frac{\partial H_{NHC}^{GCE}}{\partial \pi_{k,\lambda}} = \frac{\pi_{k,\lambda}}{s_{k+1,\lambda}^2 Q_{k,\lambda}}$$

$$k = 2, \cdots, M_\lambda - 1$$

$$\frac{d\pi_{M_\lambda,\lambda}}{d\tau_{M_\lambda,\lambda}} = -\frac{\partial H_{NHC}^{GCE}}{\partial s_{M_\lambda,\lambda}} = \frac{\pi_{M_\lambda-1,\lambda}^2}{s_{M_\lambda,\lambda}^3 Q_{M_\lambda-1,\lambda}} - \frac{k_B T}{s_{M_\lambda,\lambda}}$$

$$\frac{ds_{M_\lambda,\lambda}}{d\tau_{M_\lambda,\lambda}} = \frac{\partial H_{NHC}^{GCE}}{\partial \pi_{M_\lambda,\lambda}} = \frac{\pi_{M_\lambda,\lambda}}{Q_{M_\lambda,\lambda}}$$

where

$$\begin{aligned}
d\tau_{1,p} &= s_{1,p} dt \\
d\tau_{k,p} &= s_{k-1,p} s_{k,p} dt. \quad (k = 2, \cdots, M_p - 1) \\
d\tau_{M_p,p} &= s_{M_p,p} dt \\
d\tau_{1,f} &= s_{1,f} dt \\
d\tau_{k,f} &= s_{k-1,f} s_{k,f} dt, \quad (k = 2, \cdots, M_f - 1) \\
d\tau_{1,\lambda} &= s_{1,\lambda} dt \\
d\tau_{k,\lambda} &= s_{k-1,\lambda} s_{k,\lambda} dt, \quad (k = 2, \cdots, M_\lambda - 1) \\
d\tau_{M_\lambda,\lambda} &= s_{M_\lambda,\lambda} dt
\end{aligned} \tag{6.109}$$

with t being the real time. Substituting these transformations into Eq. (6.108), we obtain:

$$\begin{aligned}
\frac{dq}{dt} &= \frac{p'}{s_{1,p} m} \\
\frac{dp'}{dt} &= -s_{1,p} \nabla_q U(q) - s_{1,p} f(\lambda) \nabla_q u(q, q_f)
\end{aligned} \tag{6.110}$$

$$\begin{aligned}
\frac{dq_f}{dt} &= \frac{p'_f}{s_{1,f} m_f} \\
\frac{dp'_f}{dt} &= -s_{1,f} f(\lambda) \nabla_{q_f} u(q, q_f) \\
\frac{d\lambda}{dt} &= \frac{p_\lambda}{s_{1,\lambda} W} \\
\frac{dp_\lambda}{dt} &= -s_{1,\lambda} \frac{df(\lambda)}{d\lambda} u(q, q_f) - s_{1,\lambda} \frac{d\theta(\lambda)}{d\lambda} \\
\frac{d\pi_{1,p}}{dt} &= s_{2,p} \frac{(p')^2}{s_{1,p}^2 m} - s_{2,p} k_B T \\
\frac{1}{s_{1,p}} \frac{ds_{1,p}}{dt} &= \frac{\pi_{1,p}}{s_{2,p} Q_{1,p}} \\
\frac{d\pi_{k,p}}{dt} &= s_{k+1,p} \frac{\pi_{k-1,p}^2}{s_{k,p}^2 Q_{k-1,p}} - s_{k+1,p} k_B T \\
\frac{ds_{k,p}}{dt} &= s_{k,p} \frac{\pi_{k,p}}{s_{k+1,p} Q_{k,p}} \\
k &= 2, \cdots, M_p - 1 \\
\frac{d\pi_{M_p,p}}{dt} &= s_{M_p,p} \frac{\pi_{M_p-1,p}^2}{s_{M_p,p}^3 Q_{M_p-1,p}} - s_{M_p,p} \frac{k_B T}{s_{M_p,p}} \\
\frac{ds_{M_p,p}}{dt} &= s_{M_p,p} \frac{\pi_{M_p,p}}{Q_{M_p,p}} \\
\frac{d\pi_{1,f}}{dt} &= s_{2,f} \frac{(p'_f)^2}{s_{1,f}^2 m_f} - s_{2,f} k_B T \\
\frac{1}{s_{1,f}} \frac{ds_{1,f}}{dt} &= \frac{\pi_{1,f}}{s_{2,f} Q_{1,f}} \\
\frac{d\pi_{k,f}}{dt} &= s_{k+1,f} \frac{\pi_{k-1,f}^2}{s_{k,f}^2 Q_{k-1,f}} - s_{k+1,f} k_B T \\
\frac{ds_{k,f}}{dt} &= s_{k,f} \frac{\pi_{k,f}}{s_{k+1,f} Q_{k,f}} \\
k &= 2, \cdots, M_f - 1 \\
\frac{d\pi_{M_f,f}}{dt} &= s_{M_f,f} \frac{\pi_{M_f-1,f}^2}{s_{M_f,f}^3 Q_{M_f-1,f}} - s_{M_f,f} \frac{k_B T}{s_{M_f,f}}
\end{aligned}$$

$$
\begin{aligned}
\frac{ds_{M_f,f}}{dt} &= s_{M_f,f}\frac{\pi_{M_f,f}}{Q_{s_{M_f,f}}} \\
\frac{d\pi_{1,\lambda}}{dt} &= s_{2,f}\frac{(p_\lambda)^2}{s_{1,\lambda}^2 W} - s_{2,\lambda}k_B T \\
\frac{1}{s_{1,\lambda}}\frac{ds_{1,\lambda}}{dt} &= \frac{\pi_{1,\lambda}}{s_{2,\lambda}Q_{1,\lambda}} \\
\frac{d\pi_{k,\lambda}}{dt} &= s_{k+1,\lambda}\frac{\pi_{k-1,\lambda}^2}{s_{k,\lambda}^2 Q_{k-1,\lambda}} - s_{k+1,\lambda}k_B T \\
\frac{ds_{k,\lambda}}{dt} &= s_{k,\lambda}\frac{\pi_{k,\lambda}}{s_{k+1,\lambda}Q_{k,\lambda}} \\
& k = 2, \cdots, M_\lambda - 1 \\
\frac{d\pi_{M_\lambda,\lambda}}{dt} &= s_{M_\lambda,\lambda}\frac{\pi_{M_\lambda-1,\lambda}^2}{s_{M_\lambda,\lambda}^3 Q_{M_\lambda-1,\lambda}} - s_{M_\lambda,\lambda}\frac{k_B T}{s_{M_\lambda,\lambda}} \\
\frac{ds_{M_\lambda,\lambda}}{dt} &= s_{M_\lambda,\lambda}\frac{\pi_{M_\lambda,\lambda}}{Q_{M_\lambda,\lambda}}
\end{aligned}
$$

In the following, we propose some other transformations of the variables:

$$
\begin{aligned}
p &= \frac{p'}{s_{1,p}}, \quad p_f = \frac{p'_f}{s_{1,f}}, \quad \pi_\lambda = \frac{p_\lambda}{s_{1,\lambda}} \\
\hat{\pi}_{k,p} &= \frac{\pi_{k,p}}{s_{k,p}s_{k+1,p}}, \quad \hat{\pi}_{k,f} = \frac{\pi_{s_{k,f}}}{s_{k,f}s_{k+1,f}}, \quad \hat{\pi}_{k,\lambda} = \frac{\pi_{s_{k,\lambda}}}{s_{k,\lambda}s_{k+1,\lambda}} \\
& (k = 1, \cdots, M - 1) \\
\hat{\pi}_{M,p} &= \frac{\pi_{M,p}}{s_{M,p}}, \quad \hat{\pi}_{M,f} = \frac{\pi_{M,f}}{s_{M,f}}, \quad \hat{\pi}_{M,\lambda} = \frac{\pi_{M,\lambda}}{s_{M,\lambda}}
\end{aligned}
\tag{6.111}
$$

Substituting the transformations given by Eq. (6.111) into Eq. (6.110), we get:

$$
\begin{aligned}
\frac{dq}{dt} &= \frac{p}{m} \\
\frac{dp}{dt} &= -\nabla_q U(q) - f(\lambda)\nabla_q u(q, q_f) - p\frac{1}{s_{1,p}}\frac{ds_{1,p}}{dt} \\
\frac{dq_f}{dt} &= \frac{p_f}{m_f} \\
\frac{dp_f}{dt} &= -f(\lambda)\nabla_{q_f} u(q, q_f) - p_f\frac{1}{s_{1,f}}\frac{ds_{1,f}}{dt}
\end{aligned}
\tag{6.112}
$$

$$
\begin{aligned}
\frac{d\lambda}{dt} &= \frac{\pi_\lambda}{W} \\
\frac{d\pi_\lambda}{dt} &= -\frac{df(\lambda)}{d\lambda}u(q, q_f) - \frac{d\theta(\lambda)}{d\lambda} - \pi_\lambda \frac{1}{s_{1,\lambda}}\frac{ds_{1,\lambda}}{dt} \\
\frac{d\hat{\pi}_{1,p}}{dt} &= \frac{1}{s_{1,p}}\left[\frac{p^2}{m} - k_B T\right] - \hat{\pi}_{1,p}\frac{1}{s_{1,p}}\frac{ds_{1,p}}{dt} - \hat{\pi}_{1,p}\frac{1}{s_{2,p}}\frac{ds_{2,p}}{dt} \\
\frac{1}{s_{1,p}}\frac{ds_{1,p}}{dt} &= s_{1,p}\frac{\hat{\pi}_{1,p}}{Q_{1,p}} \\
\frac{d\hat{\pi}_{k,p}}{dt} &= s_{k-1,p}^2 \frac{\hat{\pi}_{k-1,p}^2}{s_{k,p}Q_{k-1,p}} - \frac{k_B T}{s_{k,p}} - \frac{1}{s_{k,p}}\frac{ds_{k,p}}{dt} - \frac{1}{s_{k+1,p}}\frac{ds_{k+1,p}}{dt} \\
\frac{1}{s_{k,p}}\frac{ds_{k,p}}{dt} &= s_{k,p}\frac{\hat{\pi}_{k,p}}{Q_{k,p}} \\
& k = 2, \cdots, M_p - 1 \\
\frac{d\hat{\pi}_{M_p,p}}{dt} &= s_{M_p-1,p}^2 \frac{\hat{\pi}_{M_p-1,p}^2}{s_{M_p,p}Q_{M_p-1,p}} - \frac{k_B T}{s_{M_p,p}} - \frac{1}{s_{M,p}}\frac{ds_{M,p}}{dt} \\
\frac{1}{s_{M,p}}\frac{ds_{M_p,p}}{dt} &= s_{M_p,p}\frac{\hat{\pi}_{M_p,p}}{Q_{M_p,p}} \\
\frac{d\hat{\pi}_{1,f}}{dt} &= \frac{1}{s_{1,f}}\left[\frac{p_f^2}{m_f} - k_B T\right] - \hat{\pi}_{1,f}\frac{1}{s_{1,f}}\frac{ds_{1,f}}{dt} - \hat{\pi}_{1,f}\frac{1}{s_{2,f}}\frac{ds_{2,f}}{dt} \\
\frac{1}{s_{1,f}}\frac{ds_{1,f}}{dt} &= s_{1,f}\frac{\hat{\pi}_{1,f}}{Q_{1,f}} \\
\frac{d\hat{\pi}_{k,f}}{dt} &= s_{k-1,f}^2 \frac{\hat{\pi}_{k-1,f}^2}{s_{k,f}Q_{k-1,f}} - \frac{k_B T}{s_{k,f}} - \frac{1}{s_{k,f}}\frac{ds_{k,f}}{dt} - \frac{1}{s_{k+1,f}}\frac{ds_{k+1,f}}{dt} \\
\frac{1}{s_{k,f}}\frac{ds_{k,f}}{dt} &= s_{k,f}\frac{\hat{\pi}_{k,f}}{Q_{k,f}} \\
& k = 2, \cdots, M_f - 1 \\
\frac{d\hat{\pi}_{M_f,f}}{dt} &= s_{M_f-1,f}^2 \frac{\hat{\pi}_{M_f-1,f}^2}{s_{M_f,f}Q_{M_f-1,f}} - \frac{k_B T}{s_{M_f,f}} - \frac{1}{s_{M,f}}\frac{ds_{M,f}}{dt} \\
\frac{1}{s_{M,f}}\frac{ds_{M_f,f}}{dt} &= s_{M_f,f}\frac{\hat{\pi}_{M_f,f}}{Q_{M_f,f}} \\
\frac{d\hat{\pi}_{1,\lambda}}{dt} &= \frac{1}{s_{1,\lambda}}\left[\frac{\pi_\lambda^2}{W} - k_B T\right] - \hat{\pi}_{1,\lambda}\frac{1}{s_{1,\lambda}}\frac{ds_{1,\lambda}}{dt} - \hat{\pi}_{1,\lambda}\frac{1}{s_{2,\lambda}}\frac{ds_{2,\lambda}}{dt}
\end{aligned}
$$

$$
\begin{aligned}
\frac{1}{s_{1,\lambda}}\frac{ds_{1,\lambda}}{dt} &= s_{1,\lambda}\frac{\hat{\pi}_{1,\lambda}}{Q_{1,\lambda}} \\
\frac{d\hat{\pi}_{k,\lambda}}{dt} &= s_{k-1,\lambda}^2\frac{\hat{\pi}_{k-1,\lambda}^2}{s_{k,\lambda}Q_{k-1,\lambda}} - \frac{k_BT}{s_{k,\lambda}} - \frac{1}{s_{k,\lambda}}\frac{ds_{k,\lambda}}{dt} - \frac{1}{s_{k+1,\lambda}}\frac{ds_{k+1,\lambda}}{dt} \\
\frac{1}{s_{k,\lambda}}\frac{ds_{k,\lambda}}{dt} &= s_{k,\lambda}\frac{\hat{\pi}_{k,\lambda}}{Q_{k,\lambda}} \\
& k = 2,\ \cdots,\ M_\lambda - 1 \\
\frac{d\hat{\pi}_{M_\lambda,\lambda}}{dt} &= s_{M_\lambda-1,\lambda}^2\frac{\hat{\pi}_{M_\lambda-1,\lambda}^2}{s_{M_\lambda,\lambda}Q_{M_\lambda-1,\lambda}} - \frac{k_BT}{s_{M_\lambda,\lambda}} - \frac{1}{s_{M,\lambda}}\frac{ds_{M,\lambda}}{dt} \\
\frac{1}{s_{M,\lambda}}\frac{ds_{M_\lambda,\lambda}}{dt} &= s_{M_\lambda,\lambda}\frac{\hat{\pi}_{M_\lambda,\lambda}}{Q_{M_\lambda,\lambda}}
\end{aligned}
$$

We can introduce another change in the variables of the following form:

$$
\pi_{k,x} = s_{k,x}\hat{\pi}_{k,x},\ \ \eta_{k,x} = \ln s_{k,x}
$$

where $x = p,\ f,\ \lambda$ and k runs over the chain length. Then, Eq. (6.112) can be written in the final form as follows:

$$
\begin{aligned}
\frac{dq}{dt} &= \frac{p}{m} \\
\frac{dp}{dt} &= -\nabla_q U(q) - f(\lambda)\nabla_q u(q, q_f) - \frac{\pi_{1,p}}{Q_{1,p}}p \\
\frac{dq_f}{dt} &= \frac{p_f}{m_f} \\
\frac{dp_f}{dt} &= -f(\lambda)\nabla_{q_f} u(q, q_f) - \frac{\pi_{1,f}}{Q_{1,f}}p_f \\
\frac{d\lambda}{dt} &= \frac{\pi_\lambda}{W} \\
\frac{d\pi_\lambda}{dt} &= -\frac{df(\lambda)}{d\lambda}u(q, q_f) - \frac{d\theta(\lambda)}{d\lambda} - \frac{\pi_{1,\lambda}}{Q_{1,\lambda}}\pi_\lambda \\
\frac{d\pi_{1,p}}{dt} &= \frac{p^2}{m} - k_BT - \frac{\pi_{2,p}}{Q_{2,p}}\pi_{1,p} \\
\frac{d\eta_{1,p}}{dt} &= \frac{\pi_{1,p}}{Q_{1,p}} \\
\frac{d\pi_{k,p}}{dt} &= \frac{\pi_{k-1,p}^2}{Q_{k-1,p}} - k_BT - \frac{\pi_{k+1,p}}{Q_{k+1,p}}\pi_{k,p}
\end{aligned}
\tag{6.113}
$$

$$\frac{d\eta_{k,p}}{dt} = \frac{\pi_{k,p}}{Q_{k,p}}$$

$$k = 2, \cdots, M_p - 1$$

$$\frac{d\pi_{M_p,p}}{dt} = \frac{\pi^2_{M_p-1,p}}{s_{M_p,p}Q_{M_p-1,p}} - k_B T$$

$$\frac{d\eta_{M_p,p}}{dt} = \frac{\pi_{M_p,p}}{Q_{M_p,p}}$$

$$\frac{d\pi_{1,f}}{dt} = \frac{p_f^2}{m_f} - k_B T - \frac{\pi_{2,f}}{Q_{2,f}}\pi_{1,f}$$

$$\frac{d\eta_{1,f}}{dt} = \frac{\pi_{1,f}}{Q_{1,f}}$$

$$\frac{d\pi_{k,f}}{dt} = \frac{\pi^2_{k-1,f}}{Q_{k-1,f}} - k_B T - \frac{\pi_{k+1,f}}{Q_{k+1,f}}\pi_{k,f}$$

$$\frac{d\eta_{k,f}}{dt} = \frac{\pi_{k,f}}{Q_{k,f}}$$

$$k = 2, \cdots, M_f - 1$$

$$\frac{d\pi_{M_f,f}}{dt} = \frac{\pi^2_{M_f-1,f}}{s_{M_f,f}Q_{M_f-1,f}} - k_B T$$

$$\frac{d\eta_{M_f,f}}{dt} = \frac{\pi_{M_f,f}}{Q_{M_f,f}}$$

$$\frac{d\pi_{1,\lambda}}{dt} = \frac{\pi_\lambda^2}{W} - k_B T - \frac{\pi_{2,\lambda}}{Q_{2,\lambda}}\pi_{1,\lambda}$$

$$\frac{d\eta_{1,\lambda}}{dt} = \frac{\pi_{1,\lambda}}{Q_{1,\lambda}}$$

$$\frac{d\pi_{k,\lambda}}{dt} = \frac{\pi^2_{k-1,\lambda}}{Q_{k-1,\lambda}} - k_B T - \frac{\pi_{k+1,\lambda}}{Q_{k+1,\lambda}}\pi_{k,\lambda}$$

$$\frac{d\eta_{k,\lambda}}{dt} = \frac{\pi_{k,\lambda}}{Q_{k,\lambda}}$$

$$k = 2, \cdots, M_\lambda - 1$$

$$\frac{d\pi_{M_\lambda,\lambda}}{dt} = \frac{\pi^2_{M_\lambda-1,\lambda}}{s_{M_\lambda,\lambda}Q_{M_\lambda-1,\lambda}} - k_B T$$

$$\frac{d\eta_{M_\lambda,\lambda}}{dt} = \frac{\pi_{M_\lambda,\lambda}}{Q_{M_\lambda,\lambda}}$$

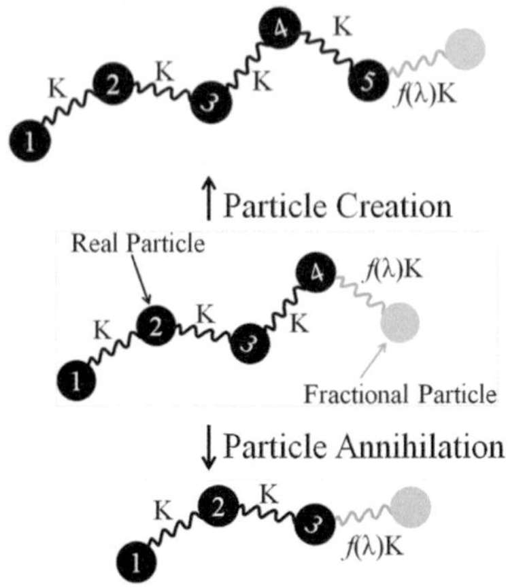

Fig. 6.12 Illustration of the creation and annihilation of a particle in a typical grand canonical ensemble molecular dynamics simulation

Example 7 As an illustration of the grand canonical ensemble molecular dynamics simulation using a chain of thermostats, we considered a chain of particles coupled through springs with force constant $K = 317\,\text{kcal/mol/Å}^2$, as shown in Fig. 6.12.

The potential interaction function is given as:

$$U(\mathbf{r}_1, \cdots, \mathbf{r}_N) = \sum_{i=1}^{N-1} K\left(\mid \mathbf{r}_{i+1} - \mathbf{r}_i \mid -x_0\right)^2 \tag{6.114}$$

where $x_0 = 1.523\,\text{Å}$. The interaction between the real particles and the fractional particles is given by

$$V(\mathbf{r}_1, \cdots, \mathbf{r}_N; \mathbf{r}_f) = f(\lambda)K\left(\mid \mathbf{r}_N - \mathbf{r}_f \mid -x_0\right)^2 \tag{6.115}$$

Thus, the fractional particle is connected to the last real particle of the chain. Here, $f(\lambda)$ is the scaling function of λ given as

$$f(\lambda) = \lambda^3$$

The potential function $\theta(\lambda)$ is given as

$$\theta(\lambda) = -\left(N + h(\lambda)\right)\left(\mu_{ex} + \mu_{id}\right) + h(\lambda)\mu_{id}$$

where

$$h(\lambda) = \lambda^3$$

and μ_{id} is given by

$$\mu_{id} = -k_B T \ln\left(\frac{V}{N+1}\left(\frac{m k_B T}{2\pi\hbar^2}\right)^{3/2}\right)$$

and μ_{ex} is the excess chemical potential, $\mu_{ex} = \mu - \mu_{id}$, where μ is the equilibrium chemical potential. In our simulations, whenever λ equalize one, the fractional particle becomes a real particles, and hence N increases by one, and on the other hand, when $\lambda = 0$, the last real particle of the chain becomes a fractional particle. Each time that a fractional particle becomes a real one, a new fractional particle is generated and a chain of thermostats coupled to the fractional particle is re-initialized. Both when a particle is created or annihilated λ and the chain of thermostat coupled to it are re-initialized. If the fractional particle becomes a real particle, then its velocity is scaled as:

$$\mathbf{v}_{N+1} = \mathbf{v}_f \sqrt{\frac{m_f}{m}}$$

and when the real particle becomes a fractional particle, then

$$\mathbf{v}_f = \mathbf{v}_{N+1} \sqrt{\frac{m}{m_f}}$$

In Fig. 6.13, we show the results of the grand canonical ensemble simulations using a chain of thermostats for each degree of freedom in the system. The thermostat chain length was $M = 3$. To calculate the block averages, we used a block length of 2000 MD steps. The integration time step was $\Delta t = 1$ fs. We used simulations with different excess chemical potentials as input. The equilibrium temperature in all simulations was fixed at $T = 600$ K. The initial number of particles was $N = 8$. In Fig. 6.13a, we show typical fluctuations of the excess chemical potential (in cal/mol) for different runs with fixed μ_{ex} and T. While in Fig. 6.13c–e, we show fluctuations of the temperature for the real particles system, fractional particle, and λ about the equilibrium temperature. Note that the real particle mass was $m = 1$ amu, $m_f = 100$ amu, and $W_\lambda = 500$ amu. The thermostat masses where calculated as

$$Q = k_B T \tau^2$$

with $\tau = 0.001$ ps.

Also, we calculated the average number of the particles as a function of the excess chemical potential, shown in Fig. 6.13b along with the standard deviations calculated using the block averages. Note that for $\mu_{ex} = 12$ cal/mol the curve chain length versus μ_{ex} exhibits a discontinuity, which, perhaps, represents a transition from a folded to the unfolded state. This result is expectable since with increasing the length of the sequence, the chain becomes more flexible.

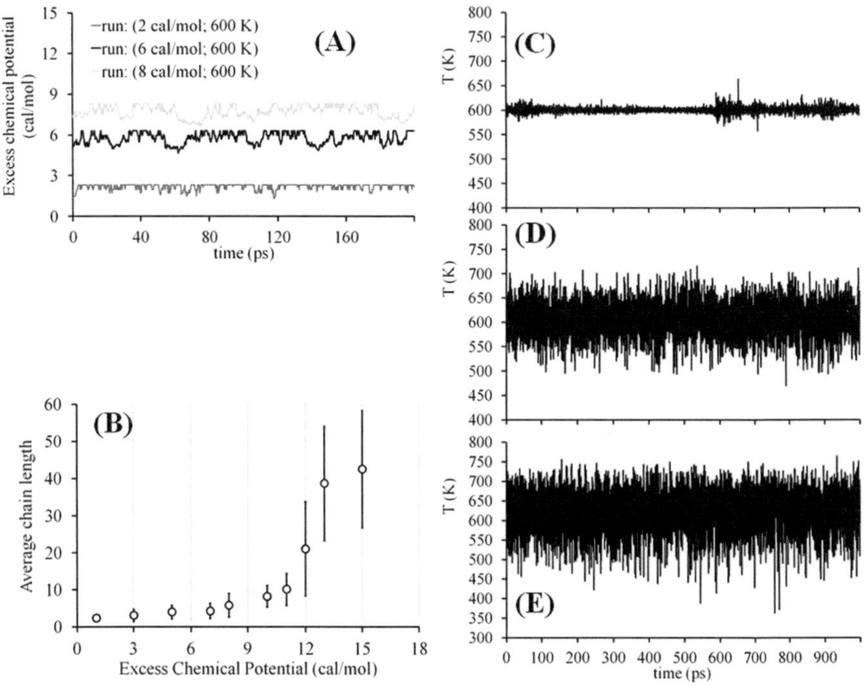

Fig. 6.13 Results of the grand canonical ensemble molecular dynamics simulation: (**a**) Fluctuations of the excess chemical potential (in cal/mol); (**b**) The average length of the chain as a function of excess chemical potential along with standard deviations calculated using block averages with a block length of 2000 MD steps; (**c**), (**d**), and (**e**) Fluctuations of the temperature of real particles system, fractional particle and λ, respectively. Different chains of thermostats are coupled to each degree of freedom of the entire system with a chain length of $M = 3$

6.2.5 *Grand Isothermal-Isobaric Ensemble*

Now, we are going to formulate the equations of motion in a grand isothermal-isobaric ensemble in which μ, p and T are fixed using the Nosé-Hoover formalism.

The following discussion introduces two approaches, namely the Nosé-Andersen thermostat and Nosé-Andersen chain of thermostats approach to control the temperature (T) and pressure (p). Besides, to control the chemical potential (μ) we are again going to employ the above formalism.

6.2.5.1 Nosé-Andersen Thermostats for Grand Isothermal-Isobaric Ensemble

We will suppose that a thermostat with variable (s, π) is coupled to each degree of freedom of the system, namely real particles of the system, fractional particle, λ-parameter, and barostat. Consider we have a N-particles system interacting via the potential energy function $U(\mathbf{r}_1, \mathbf{r}_2, \cdots, \mathbf{r}_N)$. The Hamiltonian function for this system will be written as follows:

$$
\begin{aligned}
H_{NA}^{GIIE} &= \sum_{i=1}^{gN} \left[\frac{(p_i')^2}{2s_{p,i}^2 V^{2/d} m_i} + \frac{\pi_{s_{p,i}}^2}{2Q_{s_{p,i}}} + k_B T \ln s_{p,i} \right] \\
&+ \sum_{i=1}^{d} \left[\frac{(p'_{f,i})^2}{2s_{f,i}^2 V^{2/d} m_{f,i}} + \frac{\pi_{s_{f,i}}^2}{2Q_{s_{f,i}}} + k_B T \ln s_{f,i} \right] \\
&+ \frac{p_\lambda^2}{2s_\lambda^2 W_\lambda} + \frac{\pi_{s_\lambda}^2}{2Q_{s_\lambda}} + k_B T \ln s_\lambda \\
&+ U(q_1', q_2', \cdots, q_{gN}') \\
&+ U'(q_1', q_2', \cdots, q_{gN}' : q'_{f,1} \cdots, q'_{f,d}, \lambda) \\
&+ \theta(\lambda) \\
&+ \frac{p_V^2}{2s_b^2 W_v} + pV + \frac{\pi_{s_b}^2}{2s_b^2 Q_b} + k_B T \ln s_b
\end{aligned}
\tag{6.116}
$$

where the subscript i, running overall degrees of freedom, is omitted for simplicity of appearance. The parameters have the same meaning as previously stated and

$$
q_i' = V^{1/d} q_i, \quad q'_{f,i} = V^{1/d} q_{f,i} \tag{6.117}
$$

In addition, we have introduced a fictitious particle associated with the fluctuations of the volume with mass W_v. Note that we have introduced a thermostat to each degree of freedom of the extended variables system. Using the Hamilton's equations as described in Chap. 1, we obtain:

$$
\begin{aligned}
\frac{dq'}{d\tau_p} &= \frac{p'}{s_p^2 V^{2/d} m} \\
\frac{dp'}{d\tau_p} &= -V^{1/d} \nabla_q U(q') - V^{1/d} f(\lambda) \nabla_q u(q', q_f') \\
\frac{ds_p}{d\tau_p} &= \frac{\pi_{s_p}}{Q_p} \\
\frac{d\pi_{s_p}}{d\tau_p} &= \frac{(p')^2}{s_p^3 V^{2/d} m} - \frac{k_B T}{s_p} \\
\frac{dq'_f}{d\tau_f} &= \frac{p'_f}{s_f^2 V^{2/d} m_f}
\end{aligned}
\tag{6.118}
$$

$$
\begin{aligned}
\frac{dp'_f}{d\tau_f} &= -V^{1/d} f(\lambda) \nabla_{q_f} u(q', q'_f) \\
\frac{ds_f}{d\tau_f} &= \frac{\pi_{s_f}}{Q_f} \\
\frac{d\pi_{s_f}}{d\tau_f} &= \frac{(p'_f)^2}{s_f^3 V^{2/d} m_f} - \frac{k_B T}{s_f} \\
\frac{dq'_\lambda}{d\tau_\lambda} &= \frac{p'_\lambda}{s_\lambda^2 W_\lambda} \\
\frac{dp'_\lambda}{d\tau_\lambda} &= -\frac{df(\lambda)}{d\lambda} u(q', q'_f) - \frac{d\theta(\lambda)}{d\lambda} \\
\frac{ds_\lambda}{d\tau_\lambda} &= \frac{\pi_{s_\lambda}}{Q_\lambda} \\
\frac{d\pi_{s_\lambda}}{d\tau_\lambda} &= \frac{(p'_\lambda)^2}{s_\lambda^3 W_\lambda} - \frac{k_B T}{s_\lambda} \\
\frac{dV}{d\tau_b} &= \frac{p_V}{s_b^2 W_v} \\
\frac{dp_V}{d\tau_b} &= \mathcal{P} - p_0 \\
\frac{ds_b}{d\tau_b} &= \frac{\pi_{s_b}}{Q_b} \\
\frac{d\pi_{s_b}}{d\tau_b} &= \frac{(p_V)^2}{s_b^3 W_v} - \frac{k_B T}{s_b}
\end{aligned}
$$

where $\mathcal{P}$ is the instantaneous pressure, which is given as:

$$
\begin{aligned}
\mathcal{P} = \frac{1}{Vd} \Bigg(& \sum_{i=1}^{gN} \frac{p_i^2}{m_i} + \sum_{i=1}^{d} \frac{p_{f,i}^2}{m_{f,i}} \\
& + \sum_{i=1}^{gN} \left[-\left(\nabla_{q_i} U\right) q_i - \left(\nabla_{q_i} U'\right) q_i \right] - (Vd) \frac{\partial U}{\partial V} - (Vd) \frac{\partial U'}{\partial V} \Bigg)
\end{aligned}
\tag{6.119}
$$

where explicit dependence on the volume is assumed for both U and U'.

The relationships between the real time t and the scaled time for each degree of freedom are given as follows:

$$
d\tau_p = s_p dt, \quad d\tau_f = s_f dt, \quad d\tau_\lambda = s_\lambda dt, \quad d\tau_b = s_b dt \tag{6.120}
$$

where t is the real time. Substituting the transformations given by Eqs. (6.117) and (6.120) into Eq. (6.118), we obtain the equations of motion with respect to the real time t as follows:

$$
\begin{aligned}
\frac{dq}{dt} &= \frac{p'}{s_p V^{1/d} m} + \frac{1}{Vd}\frac{dV}{dt} q \\
\frac{dp'}{dt} &= -s_p V^{1/d} \nabla_q U(q') - s_p V^{1/d} f(\lambda) \nabla_q u(q', q'_f) \\
\frac{ds_p}{dt} &= s_p \frac{\pi_{s_p}}{Q_p} \\
\frac{d\pi_{s_p}}{dt} &= \frac{(p')^2}{s_p^2 V^{2/d} m} - k_B T \\
\frac{dq_f}{dt} &= \frac{p'_f}{s_f V^{1/d} m_f} + \frac{1}{Vd}\frac{dV}{dt} q_f \\
\frac{dp'_f}{dt} &= -s_f V^{1/d} f(\lambda) \nabla_{q_f} u(q', q'_f) \\
\frac{ds_f}{dt} &= s_f \frac{\pi_{s_f}}{Q_f} \\
\frac{d\pi_{s_f}}{dt} &= \frac{(p'_f)^2}{s_f^2 V^{2/d} m_f} - k_B T \\
\frac{dq_\lambda}{dt} &= \frac{p'_\lambda}{s_\lambda W_\lambda} \\
\frac{dp'_\lambda}{dt} &= -s_\lambda \frac{df(\lambda)}{d\lambda} u(q', q'_f) - s_\lambda \frac{d\theta(\lambda)}{d\lambda} \\
\frac{ds_\lambda}{dt} &= s_\lambda \frac{\pi_{s_\lambda}}{Q_\lambda} \\
\frac{d\pi_{s_\lambda}}{dt} &= \frac{(p'_\lambda)^2}{s_\lambda^2 W_\lambda} - k_B T \\
\frac{dV}{dt} &= \frac{p_V}{s_b W_v} \\
\frac{dp_V}{dt} &= s_b(\mathcal{P} - p_0) \\
\frac{ds_b}{dt} &= s_b \frac{\pi_{s_b}}{Q_b}
\end{aligned}
\tag{6.121}
$$

$$\frac{d\pi_{s_b}}{dt} = \frac{(p_V)^2}{s_b^2 W_v} - k_B T$$

Introducing the following changes of the variables

$$p = \frac{p'}{s_p V^{1/d}}, \quad p_f = \frac{p'_f}{s_f V^{1/d}}, \quad p_\lambda = \frac{p'_\lambda}{s_\lambda}, \quad \pi_V = \frac{p_V}{s_b} \tag{6.122}$$
$$\hat{\pi}_p = \frac{\pi_p}{s_p}, \quad \hat{\pi}_f = \frac{\pi_f}{s_f}, \quad \hat{\pi}_\lambda = \frac{\pi_\lambda}{s_\lambda}, \quad \hat{\pi}_b = \frac{\pi_b}{s_b}$$

we obtain these equations of motion with respect to the new variables and the real time:

$$\frac{dq}{dt} = \frac{p}{m} + \frac{1}{Vd}\frac{dV}{dt}q \tag{6.123}$$
$$\frac{dp}{dt} = -\nabla_q U(q) - f(\lambda)\nabla_q u(q, q_f) - \frac{1}{Vd}\frac{dV}{dt}p - \frac{1}{s_p}\frac{ds_p}{dt}p$$
$$\frac{ds_p}{dt} = s_p^2 \frac{\hat{\pi}_{s_p}}{Q_p}$$
$$\frac{d\hat{\pi}_{s_p}}{dt} = \frac{1}{s_p}\frac{p^2}{m} - \frac{k_B T}{s_p} - \frac{1}{s_p}\frac{ds_p}{dt}\hat{\pi}_{s_p}$$
$$\frac{dq_f}{dt} = \frac{p_f}{m_f} + \frac{1}{Vd}\frac{dV}{dt}q_f$$
$$\frac{dp_f}{dt} = -f(\lambda)\nabla_{q_f} u(q, q_f) - \frac{1}{Vd}\frac{dV}{dt}p_f - \frac{1}{s_f}\frac{ds_f}{dt}p_f$$
$$\frac{ds_f}{dt} = s_f^2 \frac{\hat{\pi}_{s_f}}{Q_f}$$
$$\frac{d\hat{\pi}_{s_f}}{dt} = \frac{1}{s_f}\frac{p_f^2}{m_f} - \frac{k_B T}{s_f} - \frac{1}{s_f}\frac{ds_f}{dt}\hat{\pi}_{s_f}$$
$$\frac{dq_\lambda}{dt} = \frac{p_\lambda}{W_\lambda}$$
$$\frac{dp_\lambda}{dt} = -\frac{df(\lambda)}{d\lambda}u(q, q_f) - \frac{d\theta(\lambda)}{d\lambda} - \frac{1}{s_\lambda}\frac{ds_\lambda}{dt}p_\lambda$$
$$\frac{ds_\lambda}{dt} = s_\lambda^2 \frac{\hat{\pi}_{s_\lambda}}{Q_\lambda}$$
$$\frac{d\hat{\pi}_{s_\lambda}}{dt} = \frac{1}{s_\lambda}\frac{p_\lambda^2}{W_\lambda} - \frac{k_B T}{s_\lambda} - \frac{1}{s_\lambda}\frac{ds_\lambda}{dt}\hat{\pi}_{s_\lambda}$$

$$\begin{aligned}
\frac{dV}{dt} &= \frac{\pi_V}{W_v} \\
\frac{d\pi_V}{dt} &= \mathcal{P} - p_0 - \frac{1}{s_b}\frac{ds_b}{dt}\pi_V \\
\frac{ds_b}{dt} &= s_b^2\frac{\hat{\pi}_{s_b}}{Q_b} \\
\frac{d\hat{\pi}_{s_b}}{dt} &= \frac{\pi_V^2}{W_v} - \frac{k_B T}{s_b} - \frac{1}{s_b}\frac{ds_b}{dt}\hat{\pi}_{s_b}
\end{aligned}$$

Another transformation of the variables is suggested:

$$\begin{aligned}
\pi_{\eta_p} &= s_p\hat{\pi}_{s_p}, \quad \eta_p = \ln s_p \\
\pi_{\eta_f} &= s_f\hat{\pi}_{s_f}, \quad \eta_f = \ln s_f \\
\pi_{\eta_\lambda} &= s_\lambda\hat{\pi}_{s_\lambda}, \quad \eta_\lambda = \ln s_\lambda \\
\pi_{\eta_b} &= s_b\hat{\pi}_{s_b}, \quad \eta_b = \ln s_b \\
\varepsilon &= \frac{1}{d}\ln V, \quad \pi_\varepsilon = \frac{\pi_V}{d}
\end{aligned} \tag{6.124}$$

Substituting transformations of type in Eq. (6.124) into Eq. (6.123), we obtain:

$$\begin{aligned}
\frac{dq}{dt} &= \frac{p}{m} + \frac{\pi_\varepsilon}{W_v}q \\
\frac{dp}{dt} &= -\nabla_q U(q) - f(\lambda)\nabla_q u(q, q_f) - \frac{\pi_\varepsilon}{W_v}p - \frac{\pi_{\eta_p}}{Q_p}p \\
\frac{d\eta_p}{dt} &= \frac{\pi_{\eta_p}}{Q_p} \\
\frac{d\pi_{\eta_p}}{dt} &= \frac{p^2}{m} - k_B T \\
\frac{dq_f}{dt} &= \frac{p_f}{m_f} + \frac{\pi_\varepsilon}{W_v}q_f \\
\frac{dp_f}{dt} &= -f(\lambda)\nabla_{q_f} u(q, q_f) - \frac{\pi_\varepsilon}{W_v}p_f - \frac{\pi_{\eta_f}}{Q_f}p_f \\
\frac{d\eta_f}{dt} &= \frac{\pi_{\eta_f}}{Q_f} \\
\frac{d\pi_{\eta_f}}{dt} &= \frac{p_f^2}{m_f} - k_B T
\end{aligned} \tag{6.125}$$

$$
\begin{aligned}
\frac{dq_\lambda}{dt} &= \frac{p_\lambda}{W_\lambda} \\
\frac{dp_\lambda}{dt} &= -\frac{df(\lambda)}{d\lambda}u(q,q_f) - \frac{d\theta(\lambda)}{d\lambda} - \frac{\pi_{\eta_\lambda}}{Q_\lambda}p_\lambda \\
\frac{d\eta_\lambda}{dt} &= \frac{\pi_{\eta_\lambda}}{Q_\lambda} \\
\frac{d\pi_{\eta_\lambda}}{dt} &= \frac{p_\lambda^2}{W_\lambda} - k_B T \\
\frac{d\varepsilon}{dt} &= \frac{\pi_\varepsilon}{W_\upsilon} \\
\frac{d\pi_\varepsilon}{dt} &= Vd\left(\mathcal{P} - p_0\right) - \frac{\pi_{\eta_b}}{Q_b}\pi_\varepsilon \\
\frac{d\eta_b}{dt} &= \frac{\pi_{\eta_b}}{Q_b} \\
\frac{d\pi_{\eta_b}}{dt} &= \frac{\pi_\varepsilon^2}{W_\upsilon} - k_B T
\end{aligned}
$$

It is straightforward to show that the extended system of equations of the variables samples microcanonical ensemble with a partition function given as:

$$
\begin{aligned}
\Sigma_{\mu p T} &= \sum_{N=0}^{\infty} Z_{NpT}\exp\left(\beta\mu N\right) \qquad (6.126)\\
&= \sum_{N=0}^{\infty}\frac{1}{N!h^{g_N}} \\
&\times\left[\prod_{i=1}^{g_N}\int d\pi_{s_{p,i}}\int ds_{p,i}\int dp_i'\int dq_i'\right] \\
&\times\left[\prod_{i=1}^{d}\int d\pi_{s_{f,i}}\int ds_{f,i}\int dp_{f,i}'\int dq_{f,i}'\right] \\
&\times\int d\pi_{s_\lambda}\int ds_\lambda\int dp_\lambda'\int d\lambda \\
&\times\int d\pi_{s_b}\int ds_b\int dp_V\int dV \\
&\times\delta\left\{\sum_{i=1}^{g_N}\left[\frac{(p_i')^2}{2s_{p,i}^2V^{2/d}m_i} + \frac{\pi_{s_{p,i}}^2}{2Q_{s_{p,i}}} + k_B T\ln s_{p,i}\right]\right.
\end{aligned}
$$

$$
\begin{aligned}
&+\sum_{i=1}^{d}\left[\frac{(p'_{f,i})^2}{2s_{f,i}^2 V^{2/d} m_{f,i}}+\frac{\pi_{s_{f,i}}^2}{2Q_{s_{f,i}}}+k_B T \ln s_{f,i}\right] \\
&+\frac{p_\lambda^2}{2s_\lambda^2 W_\lambda}+\frac{\pi_{s_\lambda}^2}{2Q_{s_\lambda}}+k_B T \ln s_\lambda \\
&+U(q'_1, q'_2, \cdots, q'_{g_N}) \\
&+U'(q'_1, q'_2, \cdots, q'_{g_N} : q'_{f,1} \cdots, q'_{f,d}, \lambda)+\theta(\lambda) \\
&\left.+\frac{p_V^2}{2s_b^2 W_v}+pV+\frac{\pi_{s_b}^2}{2s_b^2 Q_b}+k_B T \ln s_b - E\right\}
\end{aligned}
$$

We can integrate the extended variable dynamics and apply the variable transformations as shown above, to get:

$$
\begin{aligned}
\Sigma_{\mu p T} = &\sum_{N=0}^{\infty}\frac{1}{N! h^{g_N}}\frac{1}{(g_N+d+2)k_B T} \\
&\times \prod_{i=1}^{g_N}\left[2\pi Q_{p,i} k_B T\right]^{1/2}\prod_{i=1}^{d}\left[2\pi Q_{f,i} k_B T\right]^{1/2} \\
&\times \left[2\pi Q_\lambda k_B T\right]^{1/2}\left[2\pi Q_b k_B T\right]^{1/2}\exp\left(E/k_B T\right) \\
&\times \left[\prod_{i=1}^{g_N}\int dp_i \int dq_i\right]\left[\prod_{i=1}^{d}\int dp_{f,i}\int dq_{f,i}\right] \\
&\times \int dp_\lambda \int d\lambda \int d\pi_\varepsilon \int dV \\
&\times \exp\left(-H_{0,\lambda}/k_B T\right)\exp\left(-H_\lambda/k_B T\right)
\end{aligned}
\tag{6.127}
$$

where

$$
\begin{aligned}
H_{0,\lambda} = &\sum_{i=1}^{g_N}\frac{p_i^2}{2m_i}+\sum_{i=1}^{d}\frac{p_{f,i}^2}{2m_{f,i}} \\
&+U(q'_1, q'_2, \cdots, q'_{g_N}) \\
&+U'(q'_1, q'_2, \cdots, q'_{g_N} : q'_{f,1} \cdots, q'_{f,d}, \lambda) \\
&+\frac{\pi_\varepsilon^2}{2W_v}+pV
\end{aligned}
\tag{6.128}
$$

and

$$H_\lambda = \frac{p_\lambda^2}{2W_\lambda} + \theta(\lambda) \tag{6.129}$$

This indicates that a grand isothermal-isobaric ensemble is generated.

6.2.5.2 Nosé-Andersen Chain of Thermostats for Grand Isothermal-Isobaric Ensemble

For the Nosé-Andersen chain of thermostats grand isothermal-isobaric ensemble the Hamiltonian function is proposed as follows:

$$\begin{aligned} H_{NAC}^{GIIE} &= \sum_{i=1}^{g_N} \left[\frac{(p_i')^2}{2s_{p_1,i}^2 V^{2/d} m_i} \right. \\ &\quad \left. + \sum_{j=1}^{M-1} \frac{\pi_{p_j,i}^2}{2s_{p_{j+1},i}^2 Q_{p_j,i}} + \frac{\pi_{p_M,i}^2}{2Q_{p_M,i}} + \sum_{j=1}^{M} k_B T \ln s_{p_j,i} \right] \\ &\quad + \sum_{i=1}^{d} \left[\frac{(p'_{f,i})^2}{2s_{f_1,i}^2 V^{2/d} m_{f,i}} \right. \\ &\quad \left. + \sum_{j=1}^{M-1} \frac{\pi_{f_j,i}^2}{2s_{f_{j+1},i}^2 Q_{f_j,i}} + \frac{\pi_{f_M,i}^2}{2Q_{f_M,i}} + \sum_{j=1}^{M} k_B T \ln s_{f_j,i} \right] \\ &\quad + \frac{p_\lambda^2}{2s_{\lambda_1}^2 W_\lambda} \\ &\quad + \sum_{j=1}^{M-1} \frac{\pi_{\lambda_j}^2}{2s_{\lambda_{j+1}}^2 Q_{\lambda_j}} + \frac{\pi_{\lambda_M}^2}{2Q_{\lambda_M}} + \sum_{j=1}^{M} k_B T \ln s_{\lambda_j} \\ &\quad + \frac{p_v^2}{2s_{b_1}^2 W_v} \\ &\quad + \sum_{j=1}^{M-1} \frac{\pi_{b_j}^2}{2s_{b_{j+1}}^2 Q_{b_j}} + \frac{\pi_{b_M}^2}{2Q_{b_M}} + \sum_{j=1}^{M} k_B T \ln s_{b_j} \\ &\quad + U(q_1', q_2', \cdots, q_{g_N}') \\ &\quad + U'(q_1', q_2', \cdots, q_{g_N}' : q_{f,1}' \cdots, q_{f,d}', \lambda) \\ &\quad + \theta(\lambda) + pV \end{aligned} \tag{6.130}$$

where for the simplicity of the appearance, it is assumed the same length of the chain for each degree of the freedom. However, in practice, different length of the chain of the thermostats can be considered. Besides, in the following, we are going to drop off the subscript i and use the Hamilton's equations of motion introduced in Chap. 1, and we can obtain the equations of motion as follows:

$$
\begin{aligned}
\frac{dq'}{d\tau_{p_1}} &= \frac{p'}{s_{p_1}^2 V^{2/d} m} \\
\frac{dp'}{d\tau_{p_1}} &= -V^{1/d}\nabla_q U(q) - V^{1/d} f(\lambda)\nabla_q u(q, q_f) \\
\frac{ds_{p_1}}{d\tau_{p_2}} &= \frac{\pi_{p_1}}{s_{p_2}^2 Q_{p_1}} \\
\frac{d\pi_{p_1}}{d\tau_{p_2}} &= \frac{(p')^2}{s_{p_1}^3 V^{2/d} m} - \frac{k_B T}{s_{p_1}} \\
\frac{ds_{p_k}}{d\tau_{p_{k+1}}} &= \frac{\pi_{p_k}}{s_{p_{k+1}}^2 Q_{p_k}} \\
\frac{d\pi_{p_k}}{d\tau_{p_{k+1}}} &= \frac{\pi_{p_{k-1}}^2}{s_{p_k}^3 Q_{p_{k-1}}} - \frac{k_B T}{s_{p_k}}, \quad k = 2, \cdots, M-1 \\
\frac{ds_{p_M}}{d\tau_{p_M}} &= \frac{\pi_{p_M}}{Q_{p_M}} \\
\frac{d\pi_{p_M}}{d\tau_{p_M}} &= \frac{\pi_{p_{M-1}}^2}{s_{p_M}^3 Q_{p_{M-1}}} - \frac{k_B T}{s_{p_M}} \\
\frac{dq'_f}{d\tau_{f_1}} &= \frac{p'_f}{s_{f_1}^2 V^{2/d} m_f} \\
\frac{dp'_f}{d\tau_{f_1}} &= -V^{1/d} f(\lambda)\nabla_{q_f} u(q, q_f) \\
\frac{ds_{f_1}}{d\tau_{f_2}} &= \frac{\pi_{f_1}}{s_{f_2}^2 Q_{f_1}} \\
\frac{d\pi_{f_1}}{d\tau_{f_2}} &= \frac{(p'_f)^2}{s_{f_1}^3 V^{2/d} m_f} - \frac{k_B T}{s_{f_1}} \\
\frac{ds_{f_k}}{d\tau_{f_{k+1}}} &= \frac{\pi_{f_k}}{s_{f_{k+1}}^2 Q_{f_k}}
\end{aligned}
\tag{6.131}
$$

$$\frac{d\pi_{f_k}}{d\tau_{f_{k+1}}} = \frac{\pi_{f_{k-1}}^2}{s_{f_k}^3 Q_{f_{k-1}}} - \frac{k_B T}{s_{f_k}}, \quad k = 2, \cdots, M-1$$

$$\frac{ds_{f_M}}{d\tau_{f_M}} = \frac{\pi_{f_M}}{Q_{f_M}}$$

$$\frac{d\pi_{f_M}}{d\tau_{f_M}} = \frac{\pi_{f_{M-1}}^2}{s_{f_M}^3 Q_{f_{M-1}}} - \frac{k_B T}{s_{f_M}}$$

$$\frac{d\lambda}{d\tau_{\lambda_1}} = \frac{p_\lambda}{s_{\lambda_1}^2 W_\lambda}$$

$$\frac{dp_\lambda}{d\tau_{\lambda_1}} = -\frac{df(\lambda)}{d\lambda} u(q, q_f) - \frac{d\theta(\lambda)}{d\lambda}$$

$$\frac{ds_{\lambda_1}}{d\tau_{\lambda_2}} = \frac{\pi_{\lambda_1}}{s_{\lambda_2}^2 Q_{\lambda_1}}$$

$$\frac{d\pi_{\lambda_1}}{d\tau_{\lambda_2}} = \frac{p_\lambda^2}{s_{\lambda_1}^3 W_\lambda} - \frac{k_B T}{s_{\lambda_1}}$$

$$\frac{ds_{\lambda_k}}{d\tau_{\lambda_{k+1}}} = \frac{\pi_{\lambda_k}}{s_{\lambda_{k+1}}^2 Q_{\lambda_k}}$$

$$\frac{d\pi_{\lambda_k}}{d\tau_{\lambda_{k+1}}} = \frac{\pi_{\lambda_{k-1}}^2}{s_{\lambda_k}^3 Q_{\lambda_{k-1}}} - \frac{k_B T}{s_{\lambda_k}}, \quad k = 2, \cdots, M-1$$

$$\frac{ds_{\lambda_M}}{d\tau_{\lambda_M}} = \frac{\pi_{\lambda_M}}{Q_{\lambda_M}}$$

$$\frac{d\pi_{\lambda_M}}{d\tau_{\lambda_M}} = \frac{\pi_{\lambda_{M-1}}^2}{s_{\lambda_M}^3 Q_{\lambda_{M-1}}} - \frac{k_B T}{s_{\lambda_M}}$$

$$\frac{dV}{d\tau_{b_1}} = \frac{p_v}{s_{b_1}^2 W_v}$$

$$\frac{dp_v}{d\tau_{b_1}} = \mathcal{P} - p_0$$

$$\frac{ds_{b_1}}{d\tau_{b_2}} = \frac{\pi_{b_1}}{s_{b_2}^2 Q_{b_1}}$$

$$\frac{d\pi_{b_1}}{d\tau_{b_2}} = \frac{p_v^2}{s_{b_1}^3 W_v} - \frac{k_B T}{s_{b_1}}$$

$$\frac{ds_{b_k}}{d\tau_{b_{k+1}}} = \frac{\pi_{b_k}}{s_{b_{k+1}}^2 Q_{b_k}}$$

$$\frac{d\pi_{b_k}}{d\tau_{b_{k+1}}} = \frac{\pi^2_{b_{k-1}}}{s^3_{b_k} Q_{b_{k-1}}} - \frac{k_B T}{s_{b_k}}, \quad k = 2, \cdots, M-1$$

$$\frac{ds_{b_M}}{d\tau_{b_M}} = \frac{\pi_{b_M}}{Q_{b_M}}$$

$$\frac{d\pi_{b_M}}{d\tau_{b_M}} = \frac{\pi^2_{b_{M-1}}}{s^3_{b_M} Q_{b_{M-1}}} - \frac{k_B T}{s_{b_M}}$$

where the relationships between the scaled times and the real time t are given as follows:

$$\begin{aligned}
d\tau_{p_1} &= s_{p_1} dt \\
d\tau_{p_k} &= s_{p_{k-1}} s_{p_k} dt, \quad (k = 2, \cdots, M-1) \\
d\tau_{p_M} &= s_{p_M} dt \\
d\tau_{f_1} &= s_{f_1} dt \\
d\tau_{f_k} &= s_{f_{k-1}} s_{f_k} dt, \quad (k = 2, \cdots, M-1) \\
d\tau_{\lambda_1} &= s_{\lambda_1} dt \\
d\tau_{\lambda_k} &= s_{\lambda_{k-1}} s_{\lambda_k} dt, \quad (k = 2, \cdots, M-1) \\
d\tau_{\lambda_M} &= s_{\lambda_M} dt \\
d\tau_{b_1} &= s_{b_1} dt \\
d\tau_{b_k} &= s_{b_{k-1}} s_{b_k} dt, \quad (k = 2, \cdots, M-1) \\
d\tau_{b_M} &= s_{b_M} dt
\end{aligned} \tag{6.132}$$

Substituting these transformations and the relationships given by Eq. (6.117) into Eq. (6.131), we obtain:

$$\begin{aligned}
\frac{dq}{dt} &= \frac{p'}{s_{p_1} V^{1/d} m} + \frac{1}{Vd}\frac{dV}{dt} q \\
\frac{dp'}{dt} &= -s_{p_1} V^{1/d} \nabla_q U(q) - s_{p_1} V^{1/d} f(\lambda) \nabla_q u(q, q_f) \\
\frac{ds_{p_1}}{dt} &= s_{p_1} \frac{\pi_{p_1}}{s_{p_2} Q_{p_1}} \\
\frac{d\pi_{p_1}}{dt} &= s_{p_2} \frac{(p')^2}{s^2_{p_1} V^{2/d} m} - s_{p_2} k_B T \\
\frac{ds_{p_k}}{dt} &= s_{p_k} \frac{\pi_{p_k}}{s_{p_{k+1}} Q_{p_k}}
\end{aligned} \tag{6.133}$$

$$\frac{d\pi_{p_k}}{dt} = s_{p_{k+1}} \frac{\pi_{p_{k-1}}^2}{s_{p_k}^2 Q_{p_{k-1}}} - s_{p_{k+1}} k_B T, \quad k = 2, \cdots, M-1$$

$$\frac{ds_{p_M}}{dt} = s_{p_M} \frac{\pi_{p_M}}{Q_{p_M}}$$

$$\frac{d\pi_{p_M}}{dt} = \frac{\pi_{p_{M-1}}^2}{s_{p_M}^2 Q_{p_{M-1}}} - k_B T$$

$$\frac{dq_f}{dt} = \frac{p'_f}{s_{f_1} V^{1/d} m_f} + \frac{1}{Vd}\frac{dV}{dt} q_f$$

$$\frac{dp'_f}{dt} = -s_{f_1} V^{1/d} f(\lambda) \nabla_{q_f} u(q, q_f)$$

$$\frac{ds_{f_1}}{dt} = s_{f_1} \frac{\pi_{f_1}}{s_{f_2} Q_{f_1}}$$

$$\frac{d\pi_{f_1}}{dt} = s_{f_2} \frac{(p'_f)^2}{s_{f_1}^2 V^{2/d} m} - s_{f_2} k_B T$$

$$\frac{ds_{f_k}}{dt} = s_{f_k} \frac{\pi_{f_k}}{s_{f_{k+1}} Q_{f_k}}$$

$$\frac{d\pi_{f_k}}{dt} = s_{f_{k+1}} \frac{\pi_{f_{k-1}}^2}{s_{f_k}^2 Q_{f_{k-1}}} - s_{f_{k+1}} k_B T, \quad k = 2, \cdots, M-1$$

$$\frac{ds_{f_M}}{dt} = s_{f_M} \frac{\pi_{f_M}}{Q_{f_M}}$$

$$\frac{d\pi_{f_M}}{dt} = \frac{\pi_{f_{M-1}}^2}{s_{f_M}^2 Q_{f_{M-1}}} - k_B T$$

$$\frac{d\lambda}{dt} = \frac{p_\lambda}{s_{\lambda_1} W_\lambda}$$

$$\frac{dp_\lambda}{dt} = -s_{\lambda_1} \frac{df(\lambda)}{d\lambda} u(q, q_f) - s_{\lambda_1} \frac{d\theta(\lambda)}{d\lambda}$$

$$\frac{ds_{\lambda_1}}{dt} = s_{\lambda_1} \frac{\pi_{\lambda_1}}{s_{\lambda_2} Q_{\lambda_1}}$$

$$\frac{d\pi_{\lambda_1}}{dt} = s_{\lambda_2} \frac{p_\lambda^2}{s_{\lambda_1}^2 W_\lambda} - s_{\lambda_2} k_B T$$

$$\frac{ds_{\lambda_k}}{dt} = s_{\lambda_k} \frac{\pi_{\lambda_k}}{s_{\lambda_{k+1}} Q_{\lambda_k}}$$

$$\frac{d\pi_{\lambda_k}}{dt} = s_{\lambda_{k+1}} \frac{\pi^2_{\lambda_{k-1}}}{s^2_{\lambda_k} Q_{\lambda_{k-1}}} - s_{\lambda_{k+1}} k_B T, \quad k = 2, \cdots, M-1$$

$$\frac{ds_{\lambda_M}}{dt} = s_{\lambda_M} \frac{\pi_{\lambda_M}}{Q_{\lambda_M}}$$

$$\frac{d\pi_{\lambda_M}}{dt} = \frac{\pi^2_{\lambda_{M-1}}}{s^2_{\lambda_M} Q_{\lambda_{M-1}}} - k_B T$$

$$\frac{dV}{dt} = \frac{p_v}{s_{b_1} W_v}$$

$$\frac{dp_v}{dt} = s_{b_1} (\mathcal{P} - p_0)$$

$$\frac{ds_{b_1}}{dt} = s_{b_1} \frac{\pi_{b_1}}{s_{b_2} Q_{b_1}}$$

$$\frac{d\pi_{b_1}}{dt} = s_{b_2} \frac{p_v^2}{s^2_{b_1} W_v} - s_{b_2} k_B T$$

$$\frac{ds_{b_k}}{dt} = s_{b_k} \frac{\pi_{b_k}}{s_{b_{k+1}} Q_{b_k}}$$

$$\frac{d\pi_{b_k}}{dt} = s_{b_{k+1}} \frac{\pi^2_{b_{k-1}}}{s^2_{b_k} Q_{b_{k-1}}} - s_{b_{k+1}} k_B T, \quad k = 2, \cdots, M-1$$

$$\frac{ds_{b_M}}{dt} = s_{b_M} \frac{\pi_{b_M}}{Q_{b_M}}$$

$$\frac{d\pi_{b_M}}{dt} = \frac{\pi^2_{b_{M-1}}}{s^2_{b_M} Q_{b_{M-1}}} - k_B T$$

In the following, we propose some additional transformations of the variables:

$$p = \frac{p'}{s_{p_1} V^{1/d}}, \quad p_f = \frac{p'_f}{s_{f_1} V^{1/d}}, \quad \hat{\pi} = \frac{p_v}{s_{b_1}}, \quad \pi_\lambda = \frac{p_\lambda}{s_{\lambda_1}} \tag{6.134}$$

$$\hat{\pi}_{b_k} = \frac{\pi_{b_k}}{s_{b_k} s_{b_{k+1}}}, \quad \hat{\pi}_{p_k} = \frac{\pi_{p_k}}{s_{p_k} s_{p_{k+1}}}, \quad \hat{\pi}_{f_k} = \frac{\pi_{s_{f_k}}}{s_{f_k} s_{f_{k+1}}}, \quad \hat{\pi}_{\lambda_k} = \frac{\pi_{s_{\lambda_k}}}{s_{\lambda_k} s_{\lambda_{k+1}}}$$

$$(k = 1, \cdots, M-1)$$

$$\hat{\pi}_{b_M} = \frac{\pi_{b_M}}{s_{b_M}}, \quad \hat{\pi}_{p_M} = \frac{\pi_{p_M}}{s_{p_M}}, \quad \hat{\pi}_{f_M} = \frac{\pi_{f_M}}{s_{f_M}}, \quad \hat{\pi}_{\lambda_M} = \frac{\pi_{\lambda_M}}{s_{\lambda_M}}$$

Substituting the transformation given by Eq. (6.134) into Eq. (6.133), we obtain:

$$
\begin{aligned}
\frac{dq}{dt} &= \frac{p}{m} + \frac{1}{Vd}\frac{dV}{dt}q \qquad (6.135)\\
\frac{dp}{dt} &= -\nabla_q U(q) - f(\lambda)\nabla_q u(q, q_f) - \frac{1}{Vd}\frac{dV}{dt}p - \frac{1}{s_{p_1}}\frac{ds_{p_1}}{dt}p\\
\frac{ds_{p_1}}{dt} &= s_{p_1}^2 \frac{\hat{\pi}_{p_1}}{Q_{p_1}}\\
\frac{d\hat{\pi}_{p_1}}{dt} &= \frac{1}{s_{p_1}}\frac{p^2}{m} - \frac{k_B T}{s_{p_1}} - \frac{1}{s_{p_2}}\frac{ds_{p_2}}{dt}\hat{\pi}_{p_1} - \frac{1}{s_{p_1}}\frac{ds_{p_1}}{dt}\hat{\pi}_{p_1}\\
\frac{ds_{p_k}}{dt} &= s_{p_k}^2 \frac{\hat{\pi}_{p_k}}{Q_{p_k}}\\
\frac{d\hat{\pi}_{p_k}}{dt} &= s_{p_{k-1}}^2 \frac{\hat{\pi}_{p_{k-1}}^2}{s_{p_k} Q_{p_{k-1}}} - \frac{k_B T}{s_{p_k}} - \frac{1}{s_{p_{k+1}}}\frac{ds_{p_{k+1}}}{dt}\hat{\pi}_{p_k} - \frac{1}{s_{p_k}}\frac{ds_{p_k}}{dt}\hat{\pi}_{p_k}\\
&\quad k = 2, \cdots, M-1\\
\frac{ds_{p_M}}{dt} &= s_{p_M}^2 \frac{\hat{\pi}_{p_M}}{Q_{p_M}}\\
\frac{d\hat{\pi}_{p_M}}{dt} &= s_{p_{M-1}}^2 \frac{\hat{\pi}_{p_{M-1}}^2}{s_{p_M} Q_{p_{M-1}}} - \frac{k_B T}{s_{p_M}} - \frac{1}{s_{p_M}}\frac{ds_{p_M}}{dt}\hat{\pi}_{p_M}\\
\frac{dq_f}{dt} &= \frac{p_f}{m_f} + \frac{1}{Vd}\frac{dV}{dt}q_f\\
\frac{dp_f}{dt} &= -f(\lambda)\nabla_{q_f} u(q, q_f) - \frac{1}{Vd}\frac{dV}{dt}p_f - \frac{1}{s_{f_1}}\frac{ds_{f_1}}{dt}p_f\\
\frac{ds_{f_1}}{dt} &= s_{f_1}^2 \frac{\hat{\pi}_{f_1}}{Q_{f_1}}\\
\frac{d\hat{\pi}_{f_1}}{dt} &= \frac{1}{s_{f_1}}\frac{p_f^2}{m_f} - \frac{k_B T}{s_{f_1}} - \frac{1}{s_{f_2}}\frac{ds_{f_2}}{dt}\hat{\pi}_{f_1} - \frac{1}{s_{f_1}}\frac{ds_{f_1}}{dt}\hat{\pi}_{f_1}\\
\frac{ds_{f_k}}{dt} &= s_{f_k}^2 \frac{\hat{\pi}_{f_k}}{Q_{f_k}}\\
\frac{d\hat{\pi}_{f_k}}{dt} &= s_{f_{k-1}}^2 \frac{\hat{\pi}_{f_{k-1}}^2}{s_{f_k} Q_{f_{k-1}}} - \frac{k_B T}{s_{f_k}} - \frac{1}{s_{f_{k+1}}}\frac{ds_{f_{k+1}}}{dt}\hat{\pi}_{f_k} - \frac{1}{s_{f_k}}\frac{ds_{f_k}}{dt}\hat{\pi}_{f_k}\\
&\quad k = 2, \cdots, M-1\\
\frac{ds_{f_M}}{dt} &= s_{f_M}^2 \frac{\hat{\pi}_{f_M}}{Q_{f_M}}
\end{aligned}
$$

$$\frac{d\hat{\pi}_{f_M}}{dt} = s_{f_{M-1}}^2 \frac{\hat{\pi}_{f_{M-1}}^2}{s_{f_M} Q_{f_{M-1}}} - \frac{k_B T}{s_{f_M}} - \frac{1}{s_{f_M}} \frac{ds_{f_M}}{dt} \hat{\pi}_{f_M}$$

$$\frac{d\lambda}{dt} = \frac{\pi_\lambda}{W_\lambda}$$

$$\frac{d\pi_\lambda}{dt} = -\frac{df(\lambda)}{d\lambda} u(q, q_f) - \frac{d\theta(\lambda)}{d\lambda} - \frac{1}{s_{\lambda_1}} \frac{ds_{\lambda_1}}{dt} \pi_\lambda$$

$$\frac{ds_{\lambda_1}}{dt} = s_{\lambda_1}^2 \frac{\hat{\pi}_{\lambda_1}}{Q_{\lambda_1}}$$

$$\frac{d\hat{\pi}_{\lambda_1}}{dt} = \frac{1}{s_{\lambda_1}} \frac{p_\lambda^2}{W_\lambda} - \frac{k_B T}{s_{\lambda_1}} - \frac{1}{s_{\lambda_2}} \frac{ds_{\lambda_2}}{dt} \hat{\pi}_{\lambda_1} - \frac{1}{s_{\lambda_1}} \frac{ds_{\lambda_1}}{dt} \hat{\pi}_{\lambda_1}$$

$$\frac{ds_{\lambda_k}}{dt} = s_{\lambda_k}^2 \frac{\hat{\pi}_{\lambda_k}}{Q_{\lambda_k}}$$

$$\frac{d\hat{\pi}_{\lambda_k}}{dt} = s_{\lambda_{k-1}}^2 \frac{\hat{\pi}_{\lambda_{k-1}}^2}{s_{\lambda_k} Q_{\lambda_{k-1}}} - \frac{k_B T}{s_{\lambda_k}} - \frac{1}{s_{\lambda_{k+1}}} \frac{ds_{\lambda_{k+1}}}{dt} \hat{\pi}_{\lambda_k} - \frac{1}{s_{\lambda_k}} \frac{ds_{\lambda_k}}{dt} \hat{\pi}_{\lambda_k}$$

$$k = 2, \cdots, M-1$$

$$\frac{ds_{\lambda_M}}{dt} = s_{\lambda_M}^2 \frac{\hat{\pi}_{\lambda_M}}{Q_{\lambda_M}}$$

$$\frac{d\hat{\pi}_{\lambda_M}}{dt} = s_{\lambda_{M-1}}^2 \frac{\hat{\pi}_{\lambda_{M-1}}^2}{s_{\lambda_M} Q_{\lambda_{M-1}}} - \frac{k_B T}{s_{\lambda_M}} - \frac{1}{s_{\lambda_M}} \frac{ds_{\lambda_M}}{dt} \hat{\pi}_{\lambda_M}$$

$$\frac{dV}{dt} = \frac{\hat{\pi}_v}{W_v}$$

$$\frac{d\hat{\pi}_v}{dt} = (\mathcal{P} - p_0) - \frac{1}{s_{b_1}} \frac{ds_{b_1}}{dt} \pi_v$$

$$\frac{ds_{b_1}}{dt} = s_{b_1}^2 \frac{\hat{\pi}_{b_1}}{Q_{b_1}}$$

$$\frac{d\hat{\pi}_{b_1}}{dt} = \frac{1}{s_{b_1}} \frac{\hat{\pi}_v^2}{W_v} - \frac{k_B T}{s_{b_1}} - \frac{1}{s_{b_2}} \frac{ds_{b_2}}{dt} \hat{\pi}_{b_1} - \frac{1}{s_{b_1}} \frac{ds_{b_1}}{dt} \hat{\pi}_{b_1}$$

$$\frac{ds_{b_k}}{dt} = s_{b_k}^2 \frac{\hat{\pi}_{b_k}}{Q_{b_k}}$$

$$\frac{d\hat{\pi}_{b_k}}{dt} = s_{b_{k-1}}^2 \frac{\hat{\pi}_{b_{k-1}}^2}{s_{b_k} Q_{b_{k-1}}} - \frac{k_B T}{s_{b_k}} - \frac{1}{s_{b_{k+1}}} \frac{ds_{b_{k+1}}}{dt} \hat{\pi}_{b_k} - \frac{1}{s_{b_k}} \frac{ds_{b_k}}{dt} \hat{\pi}_{b_k}$$

$$k = 2, \cdots, M-1$$

$$\frac{ds_{b_M}}{dt} = s_{b_M}^2 \frac{\hat{\pi}_{b_M}}{Q_{b_M}}$$

$$\frac{d\hat{\pi}_{b_M}}{dt} = s_{b_{M-1}}^2 \frac{\hat{\pi}_{b_{M-1}}^2}{s_{b_M} Q_{b_{M-1}}} - \frac{k_B T}{s_{b_M}} - \frac{1}{s_{b_M}} \frac{ds_{b_M}}{dt} \hat{\pi}_{b_M}$$

Another change in the variables is introduced as:

$$\pi_{x_k} = s_{x_k} \hat{\pi}_{x_k}, \ \eta_{x_k} = \ln s_{x_k} \tag{6.136}$$
$$\varepsilon = \frac{1}{d} \ln V, \ \pi_\varepsilon = \frac{\hat{\pi}_v}{d}$$

where $x = p, \ f, \ \lambda, \ b$ and $k = 1, \ 2, \ \cdots, \ M$. Then, substituting these transformations in Eq. (6.135), we can the final form of the equations of motion for the grand isothermal-isobaric ensemble as follows:

$$\frac{dq}{dt} = \frac{p}{m} + \frac{\pi_\varepsilon}{W_v} q \tag{6.137}$$

$$\frac{dp}{dt} = -\nabla_q U(q) - f(\lambda) \nabla_q u(q, q_f) - \frac{\pi_\varepsilon}{W_v} p - \frac{\pi_{p_1}}{Q_{p_1}} p$$

$$\frac{d\eta_{p_1}}{dt} = \frac{\pi_{p_1}}{Q_{p_1}}$$

$$\frac{d\pi_{p_1}}{dt} = \frac{p^2}{m} - k_B T - \frac{\pi_{p_2}}{Q_{p_2}} \pi_{p_1}$$

$$\frac{d\eta_{p_k}}{dt} = \frac{\pi_{p_k}}{Q_{p_k}}$$

$$\frac{d\pi_{p_k}}{dt} = \frac{\pi_{p_{k-1}}^2}{Q_{p_{k-1}}} - k_B T - \frac{\pi_{p_{k+1}}}{Q_{p_{k+1}}} \pi_{p_k}$$
$$k = 2, \cdots, M-1$$

$$\frac{d\eta_{p_M}}{dt} = \frac{\pi_{p_M}}{Q_{p_M}}$$

$$\frac{d\pi_{p_M}}{dt} = \frac{\pi_{p_{M-1}}^2}{Q_{p_{M-1}}} - k_B T$$

$$\frac{dq_f}{dt} = \frac{p_f}{m_f} + \frac{\pi_\varepsilon}{W_v} q_f$$

$$\frac{dp_f}{dt} = -f(\lambda) \nabla_{q_f} u(q, q_f) - \frac{\pi_\varepsilon}{W_v} p_f - \frac{\pi_{f_1}}{Q_{f_1}} p_f$$

$$
\begin{aligned}
\frac{d\eta_{f_1}}{dt} &= \frac{\pi_{f_1}}{Q_{f_1}} \\
\frac{d\pi_{f_1}}{dt} &= \frac{p_f^2}{m_f} - k_B T - \frac{\pi_{f_2}}{Q_{f_2}}\pi_{f_1} \\
\frac{d\eta_{f_k}}{dt} &= \frac{\pi_{f_k}}{Q_{f_k}} \\
\frac{d\pi_{f_k}}{dt} &= \frac{\pi_{f_{k-1}}^2}{Q_{f_{k-1}}} - k_B T - \frac{\pi_{f_{k+1}}}{Q_{f_{k+1}}}\pi_{f_k} \\
& \quad k = 2, \cdots, M-1 \\
\frac{d\eta_{f_M}}{dt} &= \frac{\pi_{f_M}}{Q_{f_M}} \\
\frac{d\pi_{f_M}}{dt} &= \frac{\pi_{f_{M-1}}^2}{Q_{f_{M-1}}} - k_B T \\
\frac{d\lambda}{dt} &= \frac{\pi_\lambda}{W_\lambda} \\
\frac{d\pi_\lambda}{dt} &= -\frac{df(\lambda)}{d\lambda}u(q, q_f) - \frac{d\theta(\lambda)}{d\lambda} - \frac{\pi_{\lambda_1}}{Q_{\lambda_1}}p_\lambda \\
\frac{d\eta_{\lambda_1}}{dt} &= \frac{\pi_{\lambda_1}}{Q_{\lambda_1}} \\
\frac{d\pi_{\lambda_1}}{dt} &= \frac{\pi_\lambda^2}{W_\lambda} - k_B T - \frac{\pi_{\lambda_2}}{Q_{\lambda_2}}\pi_{\lambda_1} \\
\frac{d\eta_{\lambda_k}}{dt} &= \frac{\pi_{\lambda_k}}{Q_{\lambda_k}} \\
\frac{d\pi_{\lambda_k}}{dt} &= \frac{\pi_{\lambda_{k-1}}^2}{Q_{\lambda_{k-1}}} - k_B T - \frac{\pi_{\lambda_{k+1}}}{Q_{\lambda_{k+1}}}\pi_{\lambda_k} \\
& \quad k = 2, \cdots, M-1 \\
\frac{d\eta_{\lambda_M}}{dt} &= \frac{\pi_{\lambda_M}}{Q_{\lambda_M}} \\
\frac{d\pi_{\lambda_M}}{dt} &= \frac{\pi_{\lambda_{M-1}}^2}{Q_{\lambda_{M-1}}} - k_B T \\
\frac{d\varepsilon}{dt} &= \frac{\pi_\varepsilon}{W_v}
\end{aligned}
$$

$$
\begin{aligned}
\frac{d\pi_\varepsilon}{dt} &= Vd\left(\mathcal{P} - p_0\right) - \frac{\pi_{b_1}}{Q_{b_1}}\pi_\varepsilon \\
\frac{d\eta_{b_1}}{dt} &= \frac{\pi_{b_1}}{Q_{b_1}} \\
\frac{d\pi_{b_1}}{dt} &= \frac{\pi_\varepsilon^2}{W_v} - k_B T - \frac{\pi_{b_2}}{Q_{b_2}}\pi_{b_1} \\
\frac{d\eta_{b_k}}{dt} &= \frac{\pi_{b_k}}{Q_{b_k}} \\
\frac{d\pi_{b_k}}{dt} &= \frac{\pi_{b_{k-1}}^2}{Q_{b_{k-1}}} - k_B T - \frac{\pi_{b_{k+1}}}{Q_{b_{k+1}}}\pi_{b_k} \\
& k = 2, \cdots, M-1 \\
\frac{d\eta_{b_M}}{dt} &= \frac{\pi_{b_M}}{Q_{b_M}} \\
\frac{d\pi_{b_M}}{dt} &= \frac{\pi_{b_{M-1}}^2}{Q_{b_{M-1}}} - k_B T
\end{aligned}
$$

To produce the grand isothermal-isobaric ensemble distribution, the following changes are proposed for each degree of freedom i (for $i = 1, 2, \cdots, g_N$):

$$
\begin{aligned}
\frac{dq_i}{dt} &= \frac{p_i}{m_i} + \frac{\pi_\varepsilon}{W_v} q_i \qquad (6.138) \\
\frac{dp_i}{dt} &= -\nabla_{q_i} U(\mathbf{q}) - f(\lambda)\nabla_{q_i} u(\mathbf{q}, q_f) - \left(1 + \frac{d}{g_N}\right)\frac{\pi_\varepsilon}{W_v} p_i - \frac{\pi_{p_1}}{Q_{p_1}} p_i \\
\frac{d\eta_{p_1}}{dt} &= \frac{\pi_{p_1}}{Q_{p_1}} \\
\frac{d\pi_{p_1}}{dt} &= \frac{p_i^2}{m_i} - k_B T - \frac{\pi_{p_2}}{Q_{p_2}}\pi_{p_1} \\
\frac{d\eta_{p_k}}{dt} &= \frac{\pi_{p_k}}{Q_{p_k}} \\
\frac{d\pi_{p_k}}{dt} &= \frac{\pi_{p_{k-1}}^2}{Q_{p_{k-1}}} - k_B T - \frac{\pi_{p_{k+1}}}{Q_{p_{k+1}}}\pi_{p_k} \\
& k = 2, \cdots, M-1 \\
\frac{d\eta_{p_M}}{dt} &= \frac{\pi_{p_M}}{Q_{p_M}}
\end{aligned}
$$

$$
\begin{aligned}
\frac{d\pi_{p_M}}{dt} &= \frac{\pi^2_{p_{M-1}}}{Q_{p_{M-1}}} - k_B T \\
\frac{dq_f}{dt} &= \frac{p_f}{m_f} + \frac{\pi_\varepsilon}{W_v} q_f \\
\frac{dp_f}{dt} &= -f(\lambda)\nabla_{q_f} u(\mathbf{q}, q_f) - \left(1 + \frac{d}{g_N}\right)\frac{\pi_\varepsilon}{W_v} p_f - \frac{\pi_{f_1}}{Q_{f_1}} p_f \\
\frac{d\eta_{f_1}}{dt} &= \frac{\pi_{f_1}}{Q_{f_1}} \\
\frac{d\pi_{f_1}}{dt} &= \frac{p_f^2}{m_f} - k_B T - \frac{\pi_{f_2}}{Q_{f_2}} \pi_{f_1} \\
\frac{d\eta_{f_k}}{dt} &= \frac{\pi_{f_k}}{Q_{f_k}} \\
\frac{d\pi_{f_k}}{dt} &= \frac{\pi^2_{f_{k-1}}}{Q_{f_{k-1}}} - k_B T - \frac{\pi_{f_{k+1}}}{Q_{f_{k+1}}} \pi_{f_k} \\
&\quad k = 2, \cdots, M-1 \\
\frac{d\eta_{f_M}}{dt} &= \frac{\pi_{f_M}}{Q_{f_M}} \\
\frac{d\pi_{f_M}}{dt} &= \frac{\pi^2_{f_{M-1}}}{Q_{f_{M-1}}} - k_B T \\
\frac{d\lambda}{dt} &= \frac{\pi_\lambda}{W_\lambda} \\
\frac{d\pi_\lambda}{dt} &= -\frac{df(\lambda)}{d\lambda} u(\mathbf{q}, q_f) - \frac{d\theta(\lambda)}{d\lambda} - \frac{\pi_{\lambda_1}}{Q_{\lambda_1}} p_\lambda \\
\frac{d\eta_{\lambda_1}}{dt} &= \frac{\pi_{\lambda_1}}{Q_{\lambda_1}} \\
\frac{d\pi_{\lambda_1}}{dt} &= \frac{\pi_\lambda^2}{W_\lambda} - k_B T - \frac{\pi_{\lambda_2}}{Q_{\lambda_2}} \pi_{\lambda_1} \\
\frac{d\eta_{\lambda_k}}{dt} &= \frac{\pi_{\lambda_k}}{Q_{\lambda_k}} \\
\frac{d\pi_{\lambda_k}}{dt} &= \frac{\pi^2_{\lambda_{k-1}}}{Q_{\lambda_{k-1}}} - k_B T - \frac{\pi_{\lambda_{k+1}}}{Q_{\lambda_{k+1}}} \pi_{\lambda_k} \\
&\quad k = 2, \cdots, M-1
\end{aligned}
$$

$$\frac{d\eta_{\lambda_M}}{dt} = \frac{\pi_{\lambda_M}}{Q_{\lambda_M}}$$

$$\frac{d\pi_{\lambda_M}}{dt} = \frac{\pi^2_{\lambda_{M-1}}}{Q_{\lambda_{M-1}}} - k_B T$$

$$\frac{d\varepsilon}{dt} = \frac{\pi_\varepsilon}{W_v}$$

$$\frac{d\pi_\varepsilon}{dt} = Vd\left(\mathcal{P} - p_0\right) + \frac{d}{g_N}\sum_{i=1}^{g_N}\frac{p_i^2}{m_i} - \frac{\pi_{b_1}}{Q_{b_1}}\pi_\varepsilon$$

$$\frac{d\eta_{b_1}}{dt} = \frac{\pi_{b_1}}{Q_{b_1}}$$

$$\frac{d\pi_{b_1}}{dt} = \frac{\pi^2_\varepsilon}{W_v} - k_B T - \frac{\pi_{b_2}}{Q_{b_2}}\pi_{b_1}$$

$$\frac{d\eta_{b_k}}{dt} = \frac{\pi_{b_k}}{Q_{b_k}}$$

$$\frac{d\pi_{b_k}}{dt} = \frac{\pi^2_{b_{k-1}}}{Q_{b_{k-1}}} - k_B T - \frac{\pi_{b_{k+1}}}{Q_{b_{k+1}}}\pi_{b_k}$$

$$k = 2, \cdots, M-1$$

$$\frac{d\eta_{b_M}}{dt} = \frac{\pi_{b_M}}{Q_{b_M}}$$

$$\frac{d\pi_{b_M}}{dt} = \frac{\pi^2_{b_{M-1}}}{Q_{b_{M-1}}} - k_B T$$

6.2.6 Generalized Ensemble

Consider an arbitrary function $F(\mathbf{q}, \mathbf{p})$, which satisfies the conditions of a probability density function in the phase space characterized by the points $(\mathbf{q}, \mathbf{p})$:

$$\int_\Gamma d\mathbf{p}d\mathbf{q}F(\mathbf{q}, \mathbf{p}) = 1 \quad (6.139)$$

$$F(\mathbf{q}, \mathbf{p}) = 1$$

According to Plastino and Anteneodo (1997) and Barth et al. (2003), a canonical probability density function is related to the effective Hamiltonian H_{eff} as follows:

$$F(\mathbf{q}, \mathbf{p}) = \exp\left(-\beta H_{\text{eff}}\right)$$

It is straightforward to obtain that

$$H_{\text{eff}} = -\frac{1}{\beta} \ln F(\mathbf{q}, \mathbf{p}) \tag{6.140}$$

The canonical ensemble sampling with the new Hamiltonian H_{eff} is equivalent to sampling according to the generalized probability density F. Then, the Nosé Hamiltonian, introduced above, can be re-written as

$$H_N^F = -\frac{1}{\beta} \ln F(\mathbf{q}, \mathbf{p}'/s) + \frac{\pi^2}{2Q} + (g_N + 1)k_B T \ln(s) \tag{6.141}$$

Using the Hamiltonian equations of motion and the variable transformations introduced by Hoover as discussed in the previous section, it is found that Nosé-Hoover generalized ensemble equations of motion are given by (Barth et al. 2003):

$$\begin{aligned}
\dot{\mathbf{q}} &= -\frac{k_B T}{F(\mathbf{q}, \mathbf{p})} \nabla_{\mathbf{p}} F(\mathbf{q}, \mathbf{p}) \\
\dot{\mathbf{p}} &= \frac{k_B T}{F(\mathbf{q}, \mathbf{p})} \nabla_{\mathbf{q}} F(\mathbf{q}, \mathbf{p}) - \frac{\pi_\eta}{Q} \mathbf{p} \\
\dot{\eta} &= \frac{\pi_\eta}{Q} \\
\dot{\pi}_\eta &= -\frac{k_B T}{F(\mathbf{q}, \mathbf{p})} \mathbf{p}^T \nabla_{\mathbf{p}} F(\mathbf{q}, \mathbf{p}) - g_N k_B T
\end{aligned} \tag{6.142}$$

In addition, the Nosé Hamiltonian coupled to a chain of thermostats using the effective Hamiltonian is:

$$\begin{aligned}
H_{NC}^F &= -\frac{1}{\beta} \ln F(\mathbf{q}, \mathbf{p}'/s_1) + \sum_{i=1}^{M-1} \frac{\pi_i^2}{2Q_i s_{i+1}^2} \\
&+ \frac{\pi_M^2}{2Q_M} + g_N k_B T \ln(s_1) + \sum_{i=2}^{M} k_B T \ln s_i
\end{aligned} \tag{6.143}$$

Then, the equations of motion using the Nosé-Hoover chain of thermostats variables for the generalized ensemble can be written as

$$\begin{aligned}
\dot{\mathbf{q}} &= -\frac{k_B T}{F(\mathbf{q}, \mathbf{p})} \nabla_{\mathbf{p}} F(\mathbf{q}, \mathbf{p}) \\
\dot{\mathbf{p}} &= \frac{k_B T}{F(\mathbf{q}, \mathbf{p})} \nabla_{\mathbf{q}} F(\mathbf{q}, \mathbf{p}) - \frac{\pi_{\eta_1}}{Q_1} \mathbf{p}
\end{aligned} \tag{6.144}$$

$$
\begin{aligned}
\dot{\eta}_k &= \frac{\pi_{\eta k}}{Q_k}, \quad k = 1, \cdots, M \\
\dot{\pi}_{\eta 1} &= -\frac{k_B T}{F(\mathbf{q}, \mathbf{p})} \mathbf{p}^T \nabla_{\mathbf{p}} F(\mathbf{q}, \mathbf{p}) - g_N k_B T - \pi_{\eta_1} \pi_{\eta_2} / Q_2 \\
\dot{\pi}_{\eta k} &= \frac{\pi_{\eta_{k-1}}^2}{Q_{k-1}} - k_B T - \pi_{\eta_k} \pi_{\eta_{k+1}} / Q_{k+1}, \quad k = 2, \cdots, M-1 \\
\dot{\pi}_{\eta M} &= \frac{\pi_{\eta_{M-1}}^2}{Q_{M-1}} - k_B T
\end{aligned}
$$

In the case of the Nosé-Poincaré transformation, the Hamiltonian for the generalized ensemble can be re-written as

$$
H_{NP}^F = s \left[-\frac{1}{\beta} \ln F(\mathbf{q}, \mathbf{p}'/s) + \frac{\pi^2}{2Q} + (g_N + 1) k_B T \ln(s) \right] \tag{6.145}
$$

We can similarly find the equations of motion for the Nosé-Poincaré generalized ensemble as (Barth et al. 2003)

$$
\begin{aligned}
\dot{\mathbf{q}} &= -\frac{k_B T}{F(\mathbf{q}, \mathbf{p}'/s)} \nabla_{\mathbf{p}'/s} F(\mathbf{q}, \mathbf{p}'/s) \\
\dot{\mathbf{p}}' &= \frac{k_B T}{F(\mathbf{q}, \mathbf{p}'/s)} \nabla_{\mathbf{q}} F(\mathbf{q}, \mathbf{p}'/s) \\
\dot{s} &= s \frac{\pi}{Q} \\
\dot{\pi} &= \frac{k_B T}{s F(\mathbf{q}, \mathbf{p}'/s)} \nabla_{\mathbf{p}'/s} F(\mathbf{q}, \mathbf{p}'/s) - g_N k_B T - \Delta H_N^F
\end{aligned} \tag{6.146}
$$

As mentioned in Barth et al. (2003), the above equations of motion do not prove a separation of the variables that exist in the original form of either the Nosé-Hoover or Nosé-Poincaré equations of motion. Therefore, application of the time reversible schemes yields the nonlinear equations of motion, the solution of each at each time step would require too many calls of the potential energy and forces calculations. That is computationally very expensive. Therefore, two cases of the generalized probability density functions have been proposed (Barth et al. 2003): Separable variable distribution functions and function of the Hamiltonian distribution functions.

6.2.6.1 Separable Distribution Functions

In this case, the generalized probability density function is factorized as (Barth et al. 2003)

$$F(\mathbf{q}, \mathbf{p}) = A(\mathbf{p})B(\mathbf{q})$$

The effective kinetic and potential energies are defined as the following:

$$F(\mathbf{q}, \mathbf{p}) = \exp(-\beta K_{\text{eff}}) \exp(-\beta U_{\text{eff}})$$

we get

$$K_{\text{eff}}(\mathbf{p}) = -\frac{1}{\beta} \ln A(\mathbf{p}) \tag{6.147}$$

$$U_{\text{eff}}(\mathbf{q}) = -\frac{1}{\beta} \ln B(\mathbf{q})$$

Then, the effective Hamiltonian is

$$H_{\text{eff}} = K_{\text{eff}} + U_{\text{eff}}$$

The Nosé-Hoover generalized ensemble equations of motion can be re-written as (Barth et al. 2003):

$$\begin{aligned}
\dot{\mathbf{q}} &= \nabla_{\mathbf{p}} K_{\text{eff}}(\mathbf{p}) \\
\dot{\mathbf{p}} &= -\nabla_{\mathbf{q}} U_{\text{eff}}(\mathbf{q}) - \frac{\pi_\eta}{Q}\mathbf{p} \\
\dot{\eta} &= \frac{\pi_\eta}{Q} \\
\dot{\pi}_\eta &= -\frac{k_B T}{F(\mathbf{q}, \mathbf{p})}\mathbf{p}^T \nabla_{\mathbf{p}} F(\mathbf{q}, \mathbf{p}) - g_N k_B T
\end{aligned} \tag{6.148}$$

The equations of motion for the Nosé-Hoover chain of thermostats using the generalized ensemble can be re-written as

$$\begin{aligned}
\dot{\mathbf{q}} &= \nabla_{\mathbf{p}} K_{\text{eff}}(\mathbf{p}) \\
\dot{\mathbf{p}} &= -\nabla_{\mathbf{q}} U_{\text{eff}}(\mathbf{q}) - \frac{\pi_{\eta_1}}{Q_1}\mathbf{p} \\
\dot{\eta}_k &= \frac{\pi_{\eta_k}}{Q_k}, \quad k = 1, \cdots, M \\
\dot{\pi}_{\eta_1} &= -\frac{k_B T}{F(\mathbf{q}, \mathbf{p})}\mathbf{p}^T \nabla_{\mathbf{p}} F(\mathbf{q}, \mathbf{p}) - g_N k_B T - \pi_{\eta_1}\pi_{\eta_2}/Q_2 \\
\dot{\pi}_{\eta_k} &= \frac{\pi^2_{\eta_{k-1}}}{Q_{k-1}} - k_B T - \pi_{\eta_k}\pi_{\eta_{k+1}}/Q_{k+1}, \quad k = 2, \cdots, M-1
\end{aligned} \tag{6.149}$$

$$\dot{\pi}_{\eta M} = \frac{\pi_{\eta_{M-1}}^2}{Q_{M-1}} - k_B T$$

The Nosé-Poincaré equations of motion for this generalized distribution function follow similarly:

$$\begin{aligned}
\dot{\mathbf{q}} &= \nabla_{\mathbf{p}'/s} K_{\text{eff}}(\mathbf{p}'/s) \\
\dot{\mathbf{p}}' &= -\nabla_{\mathbf{q}} U_{\text{eff}}(\mathbf{q}) \\
\dot{s} &= s\frac{\pi}{Q} \\
\dot{\pi} &= -\nabla_{\mathbf{p}'} K_{\text{eff}}(\mathbf{p}'/s) - g_N k_B T - \Delta H_N^F
\end{aligned} \tag{6.150}$$

Some applications of generalized density for separable distributions are those in which only the coordinate distribution altered through modification of the potential and K_{eff} equals the kinetic energy of the system. For example, the Voter dynamics (Voter 1997) and Tsallis statistics (Andricioaei and Straub 1997).

If the distribution functions are generated by the Hamiltonian functions, the effective Hamiltonian is

$$H_{\text{eff}} = -\frac{1}{\beta} \ln F(H(\mathbf{q}, \mathbf{p})) \equiv f(H(\mathbf{q}, \mathbf{p}))$$

Then, the Nosé-Hoover equations of motion for the generalised ensemble are

$$\begin{aligned}
\dot{\mathbf{q}} &= f'(H(\mathbf{q}, \mathbf{p}))\mathbf{M}^{-1}\mathbf{p} \\
\dot{\mathbf{p}} &= -f'(H(\mathbf{q}, \mathbf{p}))\nabla_{\mathbf{q}} U(\mathbf{q}) - \frac{\pi_\eta}{Q}\mathbf{p} \\
\dot{\eta} &= \frac{\pi_\eta}{Q} \\
\dot{\pi}_\eta &= f'(H(\mathbf{q}, \mathbf{p}))\mathbf{p}^T\mathbf{M}^{-1}\mathbf{p} - g_N k_B T
\end{aligned} \tag{6.151}$$

where

$$f'(H(\mathbf{q}, \mathbf{p})) = \frac{\partial f}{\partial H}$$

Introducing a transformation of time variable as

$$\frac{dt}{d\tau} = \frac{1}{f'(H(\mathbf{q}, \mathbf{p}))}$$

we can re-write the equations as (Barth et al. 2003):

$$
\begin{aligned}
\frac{d\mathbf{q}}{d\tau} &= \mathbf{M}^{-1}\mathbf{p} \\
\frac{d\mathbf{p}}{d\tau} &= -\nabla_{\mathbf{q}} U(\mathbf{q}) - \frac{1}{f'(H(\mathbf{q}, \mathbf{p}))} \frac{\pi_\eta}{Q} \mathbf{p} \\
\frac{d\eta}{d\tau} &= \frac{1}{f'(H(\mathbf{q}, \mathbf{p}))} \frac{\pi_\eta}{Q} \\
\frac{d\pi_\eta}{d\tau} &= \mathbf{p}^T \mathbf{M}^{-1} \mathbf{p} - \frac{1}{f'(H(\mathbf{q}, \mathbf{p}))} g_N k_B T
\end{aligned}
\tag{6.152}
$$

To separate the position and momentum variables, we can again re-write these equations. For that, we can first write the conserved extended energy

$$
E_{\text{ext}} = f(H(\mathbf{q}, \mathbf{p})) + \frac{\pi_\eta^2}{2Q} + g_N k_B T \eta
$$

Assuming that F is a monotonic function, so also f, we can solve it

$$
H(\mathbf{q}, \mathbf{p}) = f^{-1} \left(E_{\text{ext}} - \frac{\pi_\eta^2}{2Q} - g_N k_B T \eta \right)
$$

then a new function ϕ can be introduced as (Barth et al. 2003)

$$
\phi(\eta, \pi_\eta) = \frac{1}{f'}
$$

and the equations are re-written as:

$$
\begin{aligned}
\frac{d\mathbf{q}}{d\tau} &= \mathbf{M}^{-1}\mathbf{p} \\
\frac{d\mathbf{p}}{d\tau} &= -\nabla_{\mathbf{q}} U(\mathbf{q}) - \phi(\eta, \pi_\eta) \frac{\pi_\eta}{Q} \mathbf{p} \\
\frac{d\eta}{d\tau} &= \phi(\eta, \pi_\eta) \frac{\pi_\eta}{Q} \\
\frac{d\pi_\eta}{d\tau} &= \mathbf{p}^T \mathbf{M}^{-1} \mathbf{p} - \phi(\eta, \pi_\eta) g_N k_B T
\end{aligned}
\tag{6.153}
$$

These equations represent the coupling between the thermostat variables, and the positions and momenta are separated, allowing for efficient numerical integration schemes.

The equations of motion for the Nosé-Hoover chain of thermostats using the generalised ensemble can be re-written similarly:

$$
\begin{aligned}
\frac{d\mathbf{q}}{d\tau} &= \mathbf{M}^{-1}\mathbf{p} \\
\frac{d\mathbf{p}}{d\tau} &= -\nabla_{\mathbf{q}} U_{\text{eff}}(\mathbf{q}) - \phi(\eta, \pi_\eta)\frac{\pi_{\eta 1}}{Q_1}\mathbf{p} \\
\frac{d\eta_k}{d\tau} &= \phi(\eta, \pi_\eta)\frac{\pi_{\eta k}}{Q_k}, \quad k = 1, \cdots, M \\
\frac{d\pi_{\eta 1}}{d\eta} &= \mathbf{p}^T\mathbf{M}^{-1}\mathbf{p} - \phi(\eta, \pi_\eta)\left[g_N k_B T - \pi_{\eta_1}\pi_{\eta_2}/Q_2\right] \\
\frac{d\pi_{\eta k}}{d\tau} &= \phi(\eta, \pi_\eta)\left[\frac{\pi^2_{\eta k-1}}{Q_{k-1}} - k_B T - \pi_{\eta_k}\pi_{\eta_{k+1}}/Q_{k+1}\right], \quad k = 2, \cdots, M-1 \\
\frac{\pi_{\eta M}}{d\tau} &= \phi(\eta, \pi_\eta)\left[\frac{\pi^2_{\eta M-1}}{Q_{M-1}} - k_B T\right]
\end{aligned}
\tag{6.154}
$$

The Nosé-Poincaré equations of motion for this generalised distribution function follow similarly:

$$
\begin{aligned}
\frac{d\mathbf{q}}{d\tau} &= \mathbf{M}^{-1}(\mathbf{p}'/s) \\
\frac{d\mathbf{p}'}{d\tau} &= -\nabla_{\mathbf{q}} U(\mathbf{q}) \\
\frac{ds}{d\tau} &= \phi(s, \pi)s\frac{\pi}{Q} \\
\frac{d\pi}{d\tau} &= \phi(s, \pi)\left[\frac{1}{s^2}\mathbf{p}^T\mathbf{M}^{-1}\mathbf{p} - g_N k_B T - \Delta H_N^F\right]
\end{aligned}
\tag{6.155}
$$

Chapter 7
Molecular Mechanics

Many interesting problems that we would like to treat using computational molecular modeling are unfortunately too large to be considered by quantum mechanics (QM). Quantum mechanics methods consider the electronic structure in a molecular system. Even when some of the electrons are omitted, still a large number of particles must be considered, which makes the calculations time-consuming from computations point of view.

Molecular mechanics (MM) methods (also known as *force field* methods) ignore the electronic motion and write the energy function of the system as a function of only nuclear positions. Therefore, MM is used to perform calculations on a system containing a significant number of atoms. In some cases, MM can provide results that are as accurate as high-level QM calculations, but for much faster computation time. However, MM can not provide properties that depend upon the electronic distribution in a molecule.

The first most important assumption in MM is the Born-Oppenheimer approximation, which allows for writing the potential energy as a function of nuclear coordinates only. MM uses simple models of interactions within a system, which include contributions from stretching of bonds, angle bending, and rotations about single bonds. An essential feature of MM is to use a set of parameters, developed for a relatively small number of cases, applied to a broader range of problems. Besides, the parameters generated from data on small molecules may also be used to study much larger molecular systems, such as *macromolecules* (proteins, nucleic acids, polymers, etc.).

7.1 Simple Molecular Mechanics Force Field

There are several MM force fields developed to study molecular systems, such as CHARMM, Amber, Gromacs, MMx, etc. All these force fields can, in general,

H. Kamberaj, *Molecular Dynamics Simulations in Statistical Physics: Theory and Applications*, Scientific Computation,
https://doi.org/10.1007/978-3-030-35702-3_7

be interpreted in terms of a small number of components of the intra- and intermolecular forces within the system.

The functional form for the potential that can be used to model single molecules or ensembles of atoms and/or molecules is given by (Leach 2001)

$$U(\mathbf{r}) = \sum_{\text{bonds}} \frac{k_{i,l}}{2}\left(l_i - l_{i,0}\right)^2 \tag{7.1}$$
$$+ \sum_{\text{angles}} \frac{k_{i,\theta}}{2}\left(\theta_i - \theta_{i,0}\right)^2$$
$$+ \sum_{\text{torsions}} \frac{V_n}{2}\left(1 + \cos(n\omega - \gamma)\right)$$
$$+ \sum_{\text{improper}} \frac{k_{i,\eta}}{2}\left(\eta_i - \eta_{i,0}\right)^2$$
$$+ \sum_{i=1}^{N}\sum_{j=i+1}^{N} 4\varepsilon_{ij}\left[\left(\frac{\sigma_{ij}}{r_{ij}}\right)^{12} - \left(\frac{\sigma_{ij}}{r_{ij}}\right)^{6}\right] \tag{7.2}$$
$$+ \sum_{i=1}^{N}\sum_{j=i+1}^{N} \frac{q_i q_j}{4\pi\epsilon_0\epsilon r_{ij}},$$

where $U(\mathbf{r})$ denotes the potential energy, which is a function of the positions $\mathbf{r}$ of N atoms.

The first term in Eq. (7.1) represents the interaction between pairs of bonded atoms, which is given by a harmonic potential and it represents the increase in energy when the bond length l_i deviated from its equilibrium value $l_{i,0}$. The second term is a summation over all angles formed between three atoms $A - B - C$, where A and C are both bonded to B. The contributions of each bond and angle are characterized by the force constants $k_{i,l}$ and $k_{i,\theta}$ and the reference values $l_{i,0}$ and $\theta_{i,0}$.

The third term in Eq. (7.1) is the torsional potential that models the energy change as a bond rotates, where ω is the torsional angle. The contribution from each torsion angle is characterized by the barrier height V_n, the multiplicity n which gives the number of minimum points in the function as the bond is rotated by 360°, and γ is the phase factor which determines where the torsional angle passes through its minimum value.

The fourth term gives the improper torsion angle energy, used to maintain a certain geometry of the molecule, such as planar geometry. This term depends on the force constant $k_{i,\eta}$ and the torsion angle $\eta_{i,0}$, which represents the reference equilibrium value of the improper torsion angle.

The fifth contribution is the non-bonded term, calculated between all pairs of atoms i and j that are in different molecules or between atoms that are separated

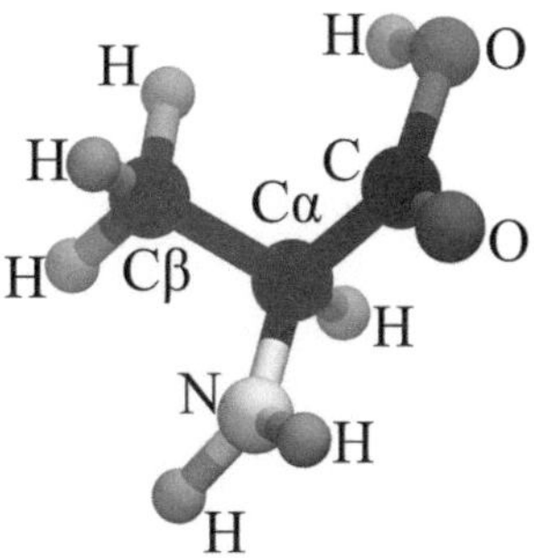

Fig. 7.1 Typical molecular mechanics model for alanine containing twelve stretching bonds, seventeen angle bending, seventeen torsion angles, two improper angles and twenty six non-bonded interactions

by at least three bonds in the same molecule. The non-bonded interactions are usually modeled using a Lennard-Jones (LJ) potential for van der Waals interactions and the Coulombic potential term for electrostatic interactions. The van der Waals interaction is calculated as an LJ potential with appropriate ε_{ij} and σ_{ij} parameters which are the well-depth and collision diameter, respectively. The electrostatic contribution is calculated using the Coulomb's law from partial atomic charges (q_i and q_j) associated with each atom.

To explain every term, we considered the alanine molecule shown in Fig. 7.1. For this molecule, a typical molecular mechanics model would contain twelve stretching bonds (four C-H, two C-C, one C-N, two N-H, two C-O and one O-H bonds), seventeen angle bending (such as C-O-H, O-C-O, O-C-C, C-C-H, C-C-N, C-C-C, C-C-N, N-C-H, C-N-H, and H-N-H angle bending), seventeen torsion angles (such as H-O-C-O, H-O-C-C, O-C-C-H, O-C-C-C, C-C-C-H, O-C-C-N, C-C-N-H, and O-C-C-N torsion angles) and twenty six non-bonded (such as H-O, H-C, H-N, O-C, and O-N non-bonded interactions).

7.2 Features of Molecular Mechanics Force Fields

A molecular mechanics force field is defined not only by the functional form, but also the parameters (such as $k_{i,l}$, $k_{i,\theta}$, V_n, ε_{ij}, σ_{ij}, and q_i) in Eq. 7.1. An essential feature is that two force fields may have the same functional form, but at the same time very different set of parameters. Moreover, different force fields may give approximately the same accuracy in calculations. Furthermore, different force fields cannot be mixed; that is, it is not strictly allowed to divide the energy into individual terms, and take some of the parameters from one force field and mix them with parameters from another force field. However, some of the terms in a molecular mechanics force field are sufficiently independent of the others (e.g., bond stretching and angle bending terms) to make this approximation acceptable in some cases.

The non-bonded interactions determine the thermodynamic equilibrium and processes, such as folding/unfolding, membrane, and micelle formation, ligand-, DNA- and protein-protein binding, solvation (in membrane and water). Therefore, there exist three main problems in the parametrization of a mechanical force field (van Gunsteren et al. 2006): minimal free energy differences and many interactions, entropic effects, and variety of atoms and molecules.

It worth noting that molecular mechanics force fields are *empirical*; that is there is no correct form for a force field. In principle, if one potential energy functional form performs better than another force field, then it is likely that force field will be favored. There have been many efforts to compare the accuracy of different force fields. The potential energy functional form used in a force field is a compromise between accuracy and computational efficiency. Thus, we expect with increasing the computer performance, more complex functional forms will be possible to incorporate into molecular mechanics force fields. Besides this, the new potential functional forms should allow fast calculations of the first and second order derivatives of energy function with respect to atomic coordinates to use methods such as energy minimization and molecular dynamics.

7.3 Molecular Mechanics Force Field Parameters Calibration

After choosing the specific functional form of the molecular mechanics force field, the task of determining appropriate force field set of parameters remain to be performed. The steps involved in this task include the type of data, systems, thermodynamic phase, and properties to be used in the calibration of a specific set of parameters. The choices made from different force fields (for example, CHARMM (Foloppe and MacKerell 2000; MacKerell and Banavali 2000; MacKerell et al. 2004), GROMACS (Hünenberger and van Gunsteren 1997), AMBER (Cornell et al. 1995), MM2 (Allinger 1977), MM3 (Allinger et al. 1990a,b), MM4 (Allinger et al. 1996a,b) and OPLS (Jorgensen and Tirado-Rives 1988)) can be different. Often, the use of small molecules for calibration is more efficient since the parameters set can be transferred to other similar molecular systems. For large macromolecules (such as proteins, DNA, etc.) group of atoms may show different behavior depending on the environmental conditions in folded molecules (e.g., pH, ionic concentration and temperature). Some properties may also depend more strongly only on some force field set of parameters and less strongly (or weakly) on others. In such cases, calibration efforts can be reduced significantly by optimizing only a subset of all set of parameters using a limited number of properties.

For parameterization of a force field there is a vast amount of data to be collected from the existing experimental results or the quantum mechanics calculations. These data include the structure properties, the geometry of the optimized molecular structure, conformation energies, and thermodynamic properties. However, often, there are missing data or sometimes difficult to obtain for some molecular systems. Therefore, the quantum mechanics calculations are often used to provide these data for optimization of the force field parameters. Ab-initio approaches are used to reproduce experimental results, in particular, for small molecular systems. The validation of the force fields calibrated in this way is often a challenge when the comparison with experimental data is performed. In general, this comparison is in the following functional form:

$$U(\mathbf{R}) = \sum_t \sum_i U_t(\Psi_t(\mathbf{R}_i)) \tag{7.3}$$

where t runs over the all types of the interactions (such as those shown in Eq. (7.1)), i runs over all particles representing the molecular structure, U_t is the model of the potential function, and Ψ_t is a scalar function of the Cartesian coordinates of each particle i, $\mathbf{R}_i$, which is a mathematical model of the particular type of interaction as depicted in Eq. (7.1). After the mathematical functional forms, U_t, are determined for each type of interactions, which are often chosen to be simple in their mathematical forms, the next stage includes the optimization of the force field parameters.

There are two traditional optimization approaches, such as the parameterization by trial and error assessment and the least square fitting. Very recently, however, a new approach is introduced using a machine learning method. In the following, we will discuss all these three methods.

7.3.1 Parameterization by Trial and Error Assessment

Usually, the potential interaction types for the parameterization procedure are classified into a hard term, such as stretching bond type and angle bending type, and soft terms, such as torsion angle, van der Waals and electrostatic types. The hard and soft terms are considered separately. In particular, bond stretching parameters for the same kind of atoms are almost identical among all the force fields; therefore, they are often omitted from the calibration, if there are already existing parameters. On the other hand, all the soft terms are calibrated together because they are coupled and hence they influence each other.

The protocol, often, includes the following steps in order:

(i) A range of the parameters for the van der Waals interactions is determined.
(ii) The partial charges are assigned to atoms using electrostatic potential energy surface fitting procedure.
(iii) The torsion potential function is then determined as a function of the torsion angle using a fitting procedure.

The torsion potentials are determined such that the heights of the barriers are reproduced together with relative energies of different conformations, which is done through an iterative procedure until the desired value of accuracy is achieved.

For determination of the torsion potential energy barriers the quantum mechanics calculations are often employed. The calibration protocol includes the following steps:

1. Determine the molecular fragments of a molecular system that represent the rotation bonds of the system.
2. Determine for each fragment the environment from the rest of the molecule close to that fragment.
3. Determine a series of three-dimensional structures obtained by rotating the molecule about the rotation bond at different angles, for example, in constant steps.
4. Assign the parameters for the van der Waals interactions.
5. The partial charges are assigned to each atom at the chosen torsional angle configuration using electrostatic potential energy surface fitting procedure.
6. Perform the quantum mechanics calculations to estimate the energy of the configuration at each torsion angle.
7. Fit the torsion potential energy versus torsion angle curve: $v(\omega) = f(\omega)$, where the parameters for the van der Waals interactions and atomic partial charges are incorporated in calculations.
8. Repeat from (4) to (7) until the desired convergence is achieved.

7.3.2 *Least-Square Fitting*

The least square fitting is another method used for calibration of the force field parameters originally developed by Lifson and Warshel (1968) and Warshel and Lifson (1970). The strategy behind the method is the fitting procedure of the calculated data using the predicted force field parameters to the experimental data and hence determining the set of parameters that give the best fit. The set of experimental data that the force field has to reproduce or calculate (for example, using the quantum mechanics) consist of thermodynamic properties, equilibrium configuration geometries, and vibration frequencies. Assuming that the vector $\mathbf{Y}$ represents the set of properties and the vector $\mathbf{x}$ the set of force field parameters, then the following mean square of errors between the observed and calculated data is subject to a minimization procedure:

$$S^2 = \frac{1}{N_{\text{data}}} \sum_{i=1}^{N_{\text{data}}} \left(Y_i^{\text{obs}} - Y_i^{\text{calc}}(\mathbf{x}) \right)^2 \tag{7.4}$$

The idea behind is that by gently changing the set of parameters to their new values, the error between the observed experimental property vector and the calculated one can be expressed using the Taylor expansion in the following approximation:

$$\mathbf{Y}^{\text{obs}} - \mathbf{Y}^{\text{calc}}(\mathbf{x} + \delta\mathbf{x}) = \Delta\mathbf{Y}(\mathbf{x} + \delta\mathbf{x}) \approx \Delta\mathbf{Y}(\mathbf{x}) + \mathbf{J}\delta\mathbf{x}$$

where $\delta\mathbf{x}$ denotes a small change of the vector of the set of parameters $\mathbf{x}$ and $\mathbf{J}$ is the matrix of the derivatives with elements defined as:

$$J_{ij} = \frac{\partial Y_i}{\partial x_j}$$

In practice, an iterative procedure is used to minimize S^2. The method becomes very efficient when the importance of each experimental data used in the fitting process has to be specified, for example, assigning weights to each contribution term when calculating the mean square of errors:

$$S^2 = \sum_{s=1}^{N_{\text{set}}} \frac{W_s}{N_{\text{data}}^{(s)}} \sum_{i=1}^{N_{\text{data}}^{(s)}} \left(Y_i^{\text{obs},s} - Y_i^{\text{calc},s}(\mathbf{x})\right)^2 \tag{7.5}$$

where $\mathbf{W}$ is the vector of weights for each data set, N_{set} denotes the number of data sets, and $N_{\text{data}}^{(s)}$ is the number of data points in the set s. For example, the thermodynamic property can be considered more important than the vibration spectra and hence the weight of the thermodynamic data set is larger than the weight of the vibration frequencies.

This method is successfully employed to calibrate the parameters of the peptides and proteins force fields using the crystallographic experimental data (Hagler et al. 1974, 1979a,b; Hagler and Lifson 1974). It showed that the hydrogen bonding interactions could be incorporated into the electrostatic and van der Waals interactions.

Note that, often, it is difficult to either fit or interpolate potential energy surfaces using just a few surface points obtained by highly accurate calculations even for small molecules. Therefore, the fitting procedure for calibration of the force field parameters become practically a notoriously tricky protocol. On the other hand, accurate reproduction of various molecular properties (such as conformation changes, free energy of hydration, and so on) is challenging, time-consuming and difficult to automate. Besides, consistency across all the other calibrated molecules can be hard to establish.

7.3.3 Machine-Like Learning Approach

Recently, Grimme (2014) automated parameterization methods have shown a great interest, where the quantum mechanically produced input data are used in a machine learning represented as a black box producing in the output the classical force fields for specific types of the interaction terms. The input quantum mechanical data consist of optimized geometrical structures, atomic partial charges, and covalent bonds. More recently, (Metz et al. 2016) another automated method developed for calculation of the intermolecular potential energy surfaces using the symmetry-adapted perturbation theory calculation as reference data (Misquitta et al. 2005), which was applied to molecular systems up to 42 atoms. In Vleet et al. (2016) and Vandenbrande et al. (2017) a new automated approach is introduced for calibration of the force field parameters using the properties of the atoms in a molecule as input parameters to characterize the chemical space of the molecules.

However, these approaches require some electronic structure based calculations to be used as a reference for optimizations of the parameters of any new molecule to be added. Besides, the screening from a large number of compounds necessary for the calibration of the force field parameters of a new molecule encountered can be computationally very demanding.

Very recently, another novel initiative is introduced for the development of new methods to automate the force field parameterization. These methodologies use the machine learning (ML) to automate the construction of the potentials based on the large datasets of quantum mechanics calculations (Unke and Meuwly 2018; Lubbers et al. 2018). Here, the potential energy surfaces are constructed to reproduce the reference energies from electronic-structure calculations to predict intramolecular interactions without needing to use the usual force field type of possible harmonic approximation. However, these methods are not being able to extrapolate outside the training dataset, but they do very well to interpolate across the training dataset. Therefore, to accurately predict the interactions among many chemical compounds, we have to increase the diversity across the molecules of the training dataset, and hence the variety of the chemical space topology; This is because, in general, these methodologies are governed by experience rather than the principles of physics or chemistry. Thus, to properly incorporate all the laws of physics, one has to increase the amount of the data to be analyzed.

To construct a data-driven model, such as the ML approach, requires specification of the input data with necessary information about the system. Usually, the input data is represented as an array of length N, $\mathbf{X}$. This process is also called *feature description*, and the input data are called *feature descriptors*. Next, the ML approach requires the choice of the predictive algorithm and the determination of specific hyper-parameters to be used for that algorithm, as discussed in the following.

7.3.3.1 Atomic Feature Descriptors

Often, SMILES (Simplified Molecular Input Line Entry System) uses the string representation for molecules (Weininger 1988). Since most of the ML algorithms require a numerical array of data, working with SMILES strings may not be as such convenient. Therefore, a string of data from SMILES has to be decoded and transformed into other descriptors, such as numerical.

Following Unke and Meuwly (2018), the atoms could be encoded by a single integer number, such as H = 1, C = 2, N = 3, and so on, or by the nuclear charge Z, such as H = 1, C = 6, N = 7, and so on. However, this creates a relationship between the input data, namely H < C < N, which could influence on the network performance (Unke and Meuwly 2018). Other encoding methods are also proposed

(Unke and Meuwly 2018) (and the references therein), such as H = [1 0 0 ⋯], C = [0 1 0 ⋯], N = [0 0 1 ⋯], and so on. This fingerprint suffers from the fact that the dimensions of the encoding vector are not the same as the number of elements in the set of atoms from the molecule, and secondly, the atoms belonging to the same periodic table group of elements do not behave in the same manner. Another interesting encoding is by *embedding* (Unke and Meuwly 2018), which is a mapping from a discrete ordered object i to an embedded vector $\mathbf{x}_i \in R^d$ of d-dimensions. Determinations of the embedded dimensions d is discussed in details in the next chapters. But, as discussed in the following, the topological data analysis tools, can also be employed to determine embedded vectors.

Note that the feature descriptors have to do with automation of the chemical perception of the molecule. Therefore, in particular, for force field developments, novel automated chemical perceptions based on the chemical environment are suggested (Mobley et al. 2018).

7.3.3.2 Environment Descriptors

The *sum over bonds* is a numerical descriptor representing a vector of bond types present in a molecule. This vector is constructed with length N_b, where N_b is the number of unique bond types present in the dataset of compounds to be studied. While for each molecule included in the dataset, the descriptor vector contains the integers giving the number of times each bond type is appearing in the molecular structure.

The *E-state* vector is a fixed length descriptor encoding the electrotopological state of a molecule (Hall and Kier 1995). Often, drug design studies use this descriptor.

The so-called *Morgan* encoding or *extended-connectivity* encoding is another topological representation with user-defined length (Rogers and Hahn 2010). This fingerprint represents the local connectivity of group of atoms, in particular, it represents either the presence or absence of the unique groupings of atoms in a molecule.

The geometry of a molecule is often encoded in the so-called *Coulomb matrix*, **C**, which for any two atoms i and j in a molecule is given as (Rupp et al. 2012):

$$C_{ij} = \begin{cases} \dfrac{Z_i^{2.4}}{2}, & i = j \\ \dfrac{Z_i Z_j}{r_{ij}}, & i \neq j \end{cases} \tag{7.6}$$

where Z_i is the atomic number of the atom i and r_{ij} is the distance between the atoms i and j. It takes into account the three-dimensional structure of the molecule and hence for any given molecule it requires the nuclear charge and the Cartesian coordinates of the atomic positions from the equilibrium geometry structure. By definition, the Coulomb matrix is invariant under translation and rotation of the molecule in space; however, it is not invariant under the permutations of the atom order. Therefore, the eigenvalues spectrum of **C** is used as a fingerprint of the molecule, since they are invariant under permutations of both rows and columns.

Another encoding of the environment descriptor is described in Unke and Meuwly (2018) using the density function, where for each atom i the information up to a cutoff radius R_c is given by the neighborhood density function ρ_i as:

$$\rho_i(\mathbf{r}) = \sum_{j\,(r_j \le R_c)} Z_j \delta\left(\mid \mathbf{r} - \mathbf{r}_j \mid\right) \tag{7.7}$$

where the position $\mathbf{r} \in R^3$ is relative to atom i, Z_j is the nuclear charge of the neighboring atom j, and $\mathbf{r}_j$ is the relative position vector of atom j relative to atom i. Here, δ is the Dirac delta function and the sum runs over all neighboring atoms of atom i within the cutoff distance R_c. In contrast to Coulomb matrix, the neighborhood density function is both invariant under rotation and translation, and permutations of the atoms ordering. To obtain a fixed length of the input vector for the ML algorithm, **X**, the density function is approximated in terms of a basis set of fixed dimension as (Unke and Meuwly 2018):

$$\rho_i(\mathbf{r}) = \sum_{k=0}^{K-1} \sum_{l=0}^{L-1} \sum_{m=-l}^{l} c_{klm} \Phi_{klm}(\mathbf{r}) \tag{7.8}$$

where c_{klm} are the expansion coefficients and $\Phi_{klm}(\mathbf{r})$ are the basis set functions:

$$\Phi_{klm}(\mathbf{r}) = P_k(r, R_c) Y_{lm}(\theta, \phi)$$

where $P_k(r, R_c)$ are the radial basis functions and $Y_{lm}(\theta, \phi)$ are the spherical harmonics, where (θ, ϕ) are spherical polar angles. The radial basis functions are as the following:

$$P_k(r, R_c) = S(r, R_c) \exp\left(-\frac{K^2}{R_c^2}\left(r - (k-1)\frac{R_c}{K}\right)^2\right) \tag{7.9}$$

where $S(r, R_c)$ is a cutoff function given as

$$S(r, R_c) = \begin{cases} 1, & r \le R_s \\ 1 - 6\left(\frac{r - R_s}{R_c - R_s}\right)^5 + 15\left(\frac{r - R_s}{R_c - R_s}\right)^4 & \\ -10\left(\frac{r - R_s}{R_c - R_s}\right)^3, & R_s < r < R_c \\ 0, & r \ge R_c \end{cases} \tag{7.10}$$

which provides a smooth first and second derivatives of $P_k(r, R_c)$ at the boundary (i.e., $r = R_c$) (Unke and Meuwly 2018). Here, $R_s = R_c - R_c/K$ is the switching radius, that is the cutoff function $S(r, R_c)$ influences the values of radial basis function $P_k(r, R_c)$ only for $r > R_s$.

For K and L sufficiently large, then the information obtained from the coefficients c_{klm} is close to that encoded in ρ_i. From Eq. (7.8), c_{klm} can be calculated as:

$$c_{klm} = \int \rho_i(\mathbf{r})\Phi_{klm}(\mathbf{r})d^3\mathbf{r} = \sum_{j\ (r_j \leq R_c)} Z_j \Phi_{klm}(\mathbf{r}_j) \tag{7.11}$$

Eq. (7.11) indicates that c_{klm} depend on the orientations of axes of the reference frame and hence they are not rotational invariant. However, these coefficients can be combined to the rotational invariant coefficients as (Unke and Meuwly 2018):

$$a_{kl} = \left(\frac{4\pi}{2l+1} \sum_{m=-l}^{l} (-1)^m c_{klm} c_{kl-m} \right)^{1/2} \tag{7.12}$$

The coefficients a_{kl}, in total $K\,L$ different coefficients, can be added to the atom feature descriptor embedding vector of dimensional N_g to form the descriptor vector $\mathbf{X} \in R^{N_g + K\,L}$, which is the input data vector for the ML algorithm. Typically, $L = K = 7$ and $R_c = 3$ Å were suggested for all datasets considered in Unke and Meuwly (2018).

Very recently, (Faber et al. 2018) the set of interatomic M-body expansion coefficients were proposed to represent the feature descriptors:

$$A_i^{(M)} = \left\{ A_i^{(1)},\ A_i^{(2)},\ A_i^{(3)},\ \cdots,\ A_i^{(M)} \right\} \tag{7.13}$$

This set represents the structural and chemical environment descriptor of the atom i in a molecule containing up to M-body interactions. In general, $A_i^{(m)}$ is expressed as a weighted sum running over all m-body interactions. For example, in the case of 3-body interactions (Faber et al. 2018), the first-order coefficient, $A_i^{(1)}$ describes the chemical composition and it is represented as:

$$A_i^{(1)} = \mathcal{G}(\mathbf{x}_i^{(1)}) \equiv \exp\left(-\frac{(P_i - P_i^{(0)})^2}{2\sigma_P^2} - \frac{(G_i - G_i^{(0)})^2}{2\sigma_G^2} \right) \tag{7.14}$$

where vector $\mathbf{x}_i^{(1)} = (P_i, \sigma_P, G_i, \sigma_G)$ represents the feature descriptor of atom i with the widths σ_P and σ_G characterizing the smearing parameters for the period and the group in the periodic table, respectively. Here, $P_i^{(0)}$ and $G_i^{(0)}$ are two arbitrary variables representing the centers of the Gaussian functions.

The second-order coefficient is:

$$A_i^{(2)} = \mathcal{G}(\mathbf{x}_i^{(1)}) \sum_{j \neq i} \xi_2(d_{ji}) \mathcal{G}(\mathbf{x}_j^{(2)}) \tag{7.15}$$

where vector $\mathbf{x}_j^{(2)} = \left(d_{ji}, \sigma_d, P_j, \sigma_P, G_j, \sigma_G\right)$ represents the environment descriptor of atom i with d_{ji} being the interatomic distance at which a Gaussian with width σ_d and center at $d_{ji}^{(0)}$ is placed. Here, ξ_2 is the 2-body scaling function of the interatomic distance given as a power law.

Similarly, the third-order coefficient is as the following:

$$A_i^{(3)} = \mathcal{G}(\mathbf{x}_i^{(1)}) \sum_{j \neq i} \mathcal{G}(\mathbf{x}_j^{(2)}) \sum_{k \neq i,j} \xi_3(d_{ji}, d_{ki}, \theta_{ijk}) \mathcal{G}(\mathbf{x}_{ijk}^{(3)}) \tag{7.16}$$

where $\mathbf{x}_{ijk}^{(3)} = \left(\theta_{ijk}, \sigma_\theta, P_k, \sigma_P, G_k, \sigma_G\right)$, and $\xi_3(d_{ji}, d_{ki}, \theta_{ijk})$ is the 3-body scaling function, and θ_{ijk} is the bending angle between the position vectors $\mathbf{r}_{ij}$ and $\mathbf{r}_{ik}$. σ_θ is the width of the Gaussian placed at θ_{ijk}.

The scaling functions follow the power laws as (Faber et al. 2018):

$$\xi_2(d_{ji}) = \frac{1}{d_{ji}^{\alpha}} \tag{7.17}$$

$$\xi_3(d_{ji}, d_{ki}, \theta_{ijk}) = \frac{1 - 3\cos(\theta_{ijk})\cos(\theta_{jik})\cos(\theta_{kji})}{(d_{ji} d_{ki} d_{jk})^{\beta}}$$

where $\alpha = 4$ and $\beta = 2$. Here, θ_{jik} is the bending angle between the vectors $\mathbf{r}_{ji}$ and $\mathbf{r}_{jk}$, and θ_{kji} is the bending angle between the vectors $\mathbf{r}_{kj}$ and $\mathbf{r}_{ki}$. The following values of the hyper-parameters were also suggested (Faber et al. 2018):

$$\sigma_P = \sigma_G = 1.6, \;\; \sigma_d = 0.2\,\text{Å}, \;\; \sigma_\theta = \pi$$

This feature descriptors approach is successfully employed to predict intermolecular interactions from the physics-based models, such as atom-distributed multipole electrostatics, charge penetration, repulsion, induction/polarization, and many-body dispersion of the difficult small molecule dimer data sets, water clusters, host-guest complexes, DNA base pairs, the benzene crystal, and amino-acid pairs (Bereau et al. 2018).

7.3.3.3 Machine Learning Approaches

Consider a method to learn a function from a finite dataset $\mathcal{D}$ of input-output pairs, namely $(\mathbf{X}, \mathbf{Y})$, where $\mathbf{X}$ is the feature descriptor input vector for each atom and $\mathbf{Y}$ is the reference output vector for each atom, such as energy, charge, dipole moment, polarization, force, and so on. The dataset is then split into a training dataset $\mathcal{D}_{\text{train}}$ used for learning (or gaining experience) and a validation dataset $\mathcal{D}_{\text{valid}}$ used for testing the knowledge, such that

$$\mathcal{D} = \mathcal{D}_{\text{train}} \cup \mathcal{D}_{\text{valid}}$$

Often, the size of the training dataset is about 80 % of total dataset size.

LASSO

LASSO regression is a method used to minimize the sum of squares of the error between the observed and predicted values as (Hastie et al. 2009):

$$S^2 = \sum_{i=1}^{N_{\text{train}}} \left(Y_i^{\text{obs}} - Y_i^{\text{calc}} \right)^2 + \lambda \sum_{i,j=1}^{N_{\text{train}}} \mid \beta_{ij} \mid \tag{7.18}$$

where N_{train} is the size of the training dataset. Here,

$$\mathbf{Y}^{\text{calc}} = \boldsymbol{\beta}\mathbf{X}$$

where $\boldsymbol{\beta}$ is the matrix of the regression coefficients and $\mathbf{X}$ is the input vector of feature descriptors across the dataset of molecules evaluated. Note that S^2 in Eq. (7.18) is subject to the minimization with respect to λ and $\boldsymbol{\beta}$ with λ being the so-called regularization hyper-parameter.

The second term in Eq. (7.18) represents the regularization term, which is controlled by λ. The regularization is often used to avoid the overfitting of the training dataset because of the similarities in the input descriptors among the molecules, and it is also known as bias-variance trade-off. For large λ values, some of the terms of the matrix $\boldsymbol{\beta}$ will be approximately zero, and then the model can be more straightforward to interpret and more efficiently to evaluate since there are fewer parameters to calibrate.

Kernel-Ridge Method

According to kernel-ridge method, the learning function is as the following (Hastie et al. 2009):

$$Y^{\text{calc}}(x) = \sum_{d=1}^{N_{\text{train}}} K(x, x_d) \sum_{d'=1}^{N_{\text{train}}} (\mathbf{K} + \lambda\mathbf{I})^{-1}_{dd'} \, Y^{\text{obs}}(x_{d'}) \tag{7.19}$$

where **K** is the empirical kernel matrix and **I** is the identity matrix and λ is the regularization hyper-parameter. Kernel-ridge method reduces the value of regression coefficients, $\boldsymbol{\beta}$:

$$\beta_d = \sum_{d'=1}^{N_{\text{train}}} (\mathbf{K} + \lambda \mathbf{I})^{-1}_{dd'} Y^{\text{obs}}(x_{d'}) \tag{7.20}$$

by introducing a second-order regularization term, which is a quadratic penalty on regression coefficients, known as the *kernel trick*, which allows for an efficient evaluation of the model in a high dimensional space.

The method is subject to obtaining the solution for the regression coefficients β_d (for $d = 1, 2, \cdots, N_{\text{train}}$) using the linear regression fit with regularization λ by minimizing the following sum of the square errors between the calculated and observed values:

$$S^2 = \sum_{i=1}^{N_{\text{train}}} \left(Y_i^{\text{obs}} - Y_i^{\text{calc}} \right)^2 \tag{7.21}$$

Gaussian Process Regression

The Gaussian process regression is a Bayesian method, which is very similar to kernel ridge regression, however, not identical. The main assumption of the Gaussian process is that the distribution regarding the noise in the data is known a priory (Rasmussen and Williams 2006). Therefore, it can predict both the mean value of the observations and the uncertainty for each prediction.

In the Gaussian process regression, the learning function is given by Eq. (7.19) for the kernel-ridge method (Zeni et al. 2018), where **K** is a matrix-valued kernel function encoding the correlations of the outputs between any two atomic environments, $K_{dd'} = K(x_d, x_{d'})$, and λ is the regularization hyper-parameter that adjusts the error with the training dataset outputs. Often, it is fixed at 10^{-5} (Zeni et al. 2018). Usually, the prediction efficiency of the method depends on the choice of the kernel function **K**, the hyper-parameters, and the diversity of the training data set. In Zeni et al. (2018) and Bereau et al. (2018), there are proposed different choices for the kernel function in the calculations of the force fields for nanoclusters and alchemical and structural distribution.

Artificial Neural Network

As a contrasting approach to the Gaussian process, the artificial neural networks have emerged recently as the flexible parametric approaches to fit complex pattern data (Anderson 1995). The artificial neural networks are predictive tools consisting of many simple connected nodes that work in parallel. These nodes are also called *neurons*. The connections between neurons are weighted by the so-called *weights*, which are real numbers considered, along with the bias parameters, the primary means of learning in neural networks.

The parameters of an artificial neural network are the number of hidden layers K, number of nodes for each layer k, L_k, and the point-wise non-linear functions neurons f, the so-called transfer functions. For the ith component of the activation vector in the lth layer, post non-linearity and post transformation are denoted by X_i^l and Z_i^l, respectively. For example, for a single-hidden layer neural network:

$$Z_i^1(x) = \sum_{j=1}^{L_1} W_{ij}^1 X_j^1(x) + b_i^1 \tag{7.22}$$

$$X_j^1(x) = f\left(\sum_{k=1}^{n} W_{jk}^0 X_k + b_j^0\right) \tag{7.23}$$

W_{ij}^l and b_j^l are the weights and bias parameter, respectively. For the general case of the K-hidden layers the functional form and the mathematical details are discussed in the following chapter.

Different transfer functions can be suggested (Anderson 1995):

$$f(z) = \frac{1}{1 + \exp(-z)} \tag{7.24}$$

$$f(z) = \frac{1}{1 + z^2} \tag{7.25}$$

$$f(z) = \tanh(z) \tag{7.26}$$

A single-hidden layer neural network with identically independent distributed initial parameters is equivalent to a Gaussian process described above in the case of the infinite network width, that is $L_1 \rightarrow \infty$ (Anderson 1995); This, in turn, allows establishing a Bayesian inference framework for the infinite width neural network. Furthermore, kernel functions can be generated to describe the multi-layer artificial neural networks, which can be used as covariance functions for Gaussian process regression and hence allowing full Bayesian prediction for an artificial neural network (Lee et al. 2018).

In Eq. (7.22), assuming that initial weights and bias parameters are taken as identically independent distributed random variables, that is:

$$W_{ij}^0 \sim \mathcal{G}(0, \sigma_w^2/N_{\text{train}}), \quad b_j^0 \sim \mathcal{G}(0, \sigma_b^2) \tag{7.27}$$

then x_j^1 and $x_{j'}^1$ are independent for $j \neq j'$. In addition, $Z_i^1(x)$ is sum of the identically independent distributed terms, therefore, based on the Center Limit

Theorem in the limit of the infinite network width ($L_1 \to \infty$) it follows that $Z_i^1(x)$ is Gaussian distributed. Moreover, a finite process

$$Z_i^1(\mathbf{X}), \quad \mathbf{X} = (x_1, x_2, \cdots, x_n)$$

has a joint Gaussian distribution, that is, it forms a Gaussian process. Therefore, Z_i^1 is a Gaussian process with mean μ^1 and covariance K^1 (Rasmussen and Williams 2006)

$$Z_i^1 \sim \mathcal{G}(\mu^1, K^1)$$

where

$$\mu^1(x) = E[Z_i^1(x)] = 0$$

and

$$K^1(x, x') = E[Z_i^1(x) Z_i^1(x')] = \sigma_b^2 + \sigma_w^2 C(x, x')$$

with C being the covariance:

$$C(x, x') = E[X_i^1(x) X_i^1(x')] = E\left[f(Z_i^0(x)) f(Z_i^0(x'))\right]$$

For a K-hidden layer neural network, Eq. (7.22) can be generalized for the lth layer as:

$$Z_i^l(x) = \sum_{j=1}^{L_l} W_{ij}^l X_j^l(x) + b_i^l \tag{7.28}$$

$$X_j^l(x) = f\left(\sum_{k=1}^{L_{l-1}} W_{jk}^{l-1} X_k^{l-1} + b_j^{l-1}\right) \tag{7.29}$$

where $Z_i^l(x)$ are identically independent distributed random variables, and $Z_i^l(\mathbf{X})$ forms a Gaussian random process for $L_l \to \infty$: $Z_i^l \sim \mathcal{G}(0, K^l)$, with covariance K^l given as (Anderson 1995)

$$K^l(x, x') = E[Z_i^l(x) Z_i^l(x')] = \sigma_b^2 + \sigma_w^2 E\left[f(Z_i^{l-1}(x)) f(Z_i^{l-1}(x'))\right]$$

which can be re-written in the following recursive form (Lee et al. 2018):

$$K^l(x, x') = \sigma_b^2 + \sigma_w^2 G_f\left(K^{l-1}(x, x'), K^{l-1}(x, x), K^{l-1}(x', x')\right)$$

where G_f is a deterministic function depending on the choice of the function f. This indicates that the artificial neural network can be performed in a series of computations obtaining K^l as in the Gaussian process regression described above. Therefore, there exists an equivalence between the Gaussian process and the artificial neural network in the limit of $L_k \rightarrow \infty$ and that initially the weights and bias parameters are drawn from identically independent distributed random variables by Eq. (7.27).

Therefore, we can use the Gaussian process to do Bayesian training of artificial neural network (Rasmussen and Williams 2006). Following Lee et al. (2018), for that, assume a dataset $\mathcal{D}$ with elements (X_i, Y_i) for $i = 1, 2, \cdots, N_{\text{train}}$ representing the input-reference data-point pairs. The aim is to do a Bayesian prediction of some test point $X^\star$ using the distribution of the outputs $\mathbf{Z}(\mathbf{X})$ as obtained from a trained artificial neural network with probability:

$$\begin{aligned} P(Y^\star|\mathcal{D}, X^\star) &= \int d\mathbf{Z} P(Y^\star|\mathbf{Z}, \mathbf{X}, X^\star) P(\mathbf{Z}|\mathcal{D}) \\ &= \frac{1}{P(\mathbf{Y})} \int d\mathbf{Z} P(Y^\star, \mathbf{Z}|X^\star, \mathbf{X}) P(\mathbf{Y}|\mathbf{Z}) \end{aligned} \tag{7.30}$$

where $Y^\star$ is the predicted output value of the input value $X^\star$. Note that the well-known relation $p(x, y) = p(x|y)p(y)$ is used twice to obtain the final expression. In Eq. (7.30), $P(\mathbf{Y}|\mathbf{Z})$ gives probability of obtaining the reference distribution $\mathbf{Y}$ from the artificial neural network with an output of the distribution $\mathbf{Z}$ from the training dataset, therefore, it represents the error in the output of the artificial neural network, and it can be modeled as noise centered at the output distribution $\mathbf{Z}$ and variance σ_ε^2 with an unbiased estimate as:

$$\sigma_\varepsilon^2 = \frac{1}{N_{\text{train}}} \sum_{i=1}^{N_{\text{train}}} (Z_i - Y_i)^2$$

That is, under the condition of the initial choice of the parameters and assuming that network width is infinite, this implies that process

$$Z_1, Z_2, \cdots, Z_{N_{\text{train}}}, Y^\star$$

is a Gaussian process and hence $P(Y^\star, \mathbf{Z}|X^\star, \mathbf{X}) \sim \mathcal{G}(0, \mathbf{K})$ is a multivariate Gaussian with covariance (Lee et al. 2018)

$$\mathbf{K} = \begin{bmatrix} \mathbf{K}_{\mathcal{D},\mathcal{D}} & \mathbf{K}^T_{X^\star,\mathcal{D}} \\ \mathbf{K}_{X^\star,\mathcal{D}} & \mathbf{K}_{X^\star,X^\star} \end{bmatrix} \tag{7.31}$$

where $X^\star$ is the test point. In Eq. (7.31), $\mathbf{K}_{\mathcal{D},\mathcal{D}}$ is a $N_{\text{train}} \times N_{\text{train}}$ block matrix with elements $K_{ij} = K(X_i, X_j)$ where both X_i and X_j are drawn from $\mathcal{D}$. On the other hand, $\mathbf{K}_{X^\star,\mathcal{D}}$ is a block matrix whose elements are $K_{ij} = K(X^*, X_i)$ with only $X_i \in \mathcal{D}$. The integration in Eq. (7.30) can be performed exactly to get (Lee et al. 2018)

$$P(Y^\star|\mathcal{D}, X^\star) \sim \mathcal{G}(\hat{\mu}, \hat{\mathbf{K}}) \tag{7.32}$$

$$\hat{\mu} = \mathbf{K}_{X^\star,\mathcal{D}}\left(\mathbf{K}_{\mathcal{D},\mathcal{D}} + \sigma_\varepsilon^2\mathbf{I}\right)^{-1}\mathbf{Y}$$

$$\hat{\mathbf{K}} = \mathbf{K}_{X^\star,X^\star} - \mathbf{K}_{X^\star,\mathcal{D}}\left(\mathbf{K}_{\mathcal{D},\mathcal{D}} + \sigma_\varepsilon^2\mathbf{I}\right)^{-1}\mathbf{K}_{X^\star,\mathcal{D}}^T$$

However, the prediction of $P(Y^\star|\mathcal{D}, X^\star)$ and hence the calculation of the mean value and variance of the predicted value $Y^\star$ are under the assumption of the infinite width of the networks to apply the Center Limit Theorem. That is practically difficult to implement. Therefore, in Chap. 8, we are going to introduce an alternative approach for evaluation of $P(Y^\star|\mathcal{D}, X^\star)$ and the mean value and variance of the test data points. This approach is based on the bootstrapping methodology combined with a swarm particle intelligence approach for adding additional higher order terms to the regularization parameters set.

7.3.4 Perspectives of Automated Force Field Parameterization

There are two conventional approaches discussed above and which are often used to calculate the force field parameters, namely the traditional method and the automated procedure.

Figure 7.2 shows the methodology schematically for parameterization of the force fields using a traditional approach. For every molecule, the input vector uses either experimental or quantum mechanically generated data (which can computationally be very expensive). Often, in particular for large molecular systems, a fragmentation of the molecule needs to be done, and then the feature descriptions

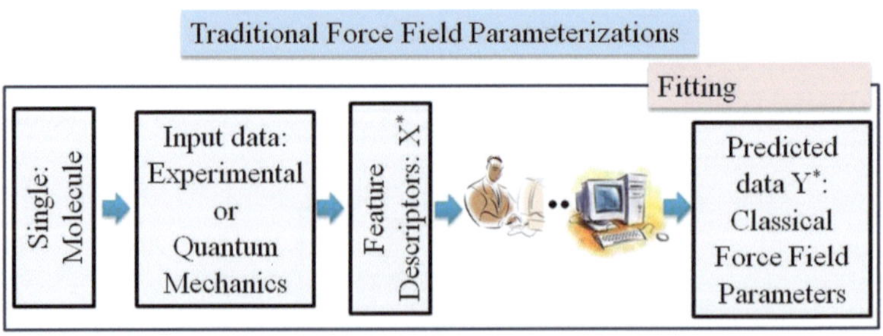

Fig. 7.2 A diagram of the general protocol for parameterization of the classical force fields using the traditional approach

of the molecule are characterized, which are related to the type of interaction terms used in the force fields. Then, the next step requires both human and computer efforts to perform a fitting to obtain the force field parameters, which is the most cumbersome task in terms of computing power and time. Therefore, in practice, in computer simulations of the (bio)molecular systems establishing the force field parameters become the most critical and difficult task, and the traditional approach may not be the most efficient methodology, though it may be the most accurate one. However, the high demands in terms of the human and computer efforts, often, require to sacrifice the accuracy for efficiency. Therefore, the automated approaches are becoming very popular as a new methodology in predicting the force field parameters.

Figure 7.3 shows the methodology schematically for parameterization of the force fields using an automated machine-like learning approach. In this approach, first a training data set of molecules, $\mathcal{D}$ is created consisting of N_{train} pairs (X_i, Y_i) for $i = 1, 2, \cdots, N_{\text{train}}$, where the vector $\mathbf{X}$ denote the feature descriptors and $\mathbf{Y}$ the reference values. The feature descriptors are used as input for training a *black box*, representing a *machine learning* (e.g., an artificial neural network) by minimizing the error between the output data from the machine, $\mathbf{Z}$ and the reference values $\mathbf{Y}$. The final aim of the training a machine-like learning engine is to obtain an estimate of the probability $P(Y^\star|\mathcal{D}, X^\star)$ to predict the output $Y^\star$ of a trained machine, such as neural network, for any input test data-point $X^\star$. This calculation is now an automated process since the black box is trained to predict the output

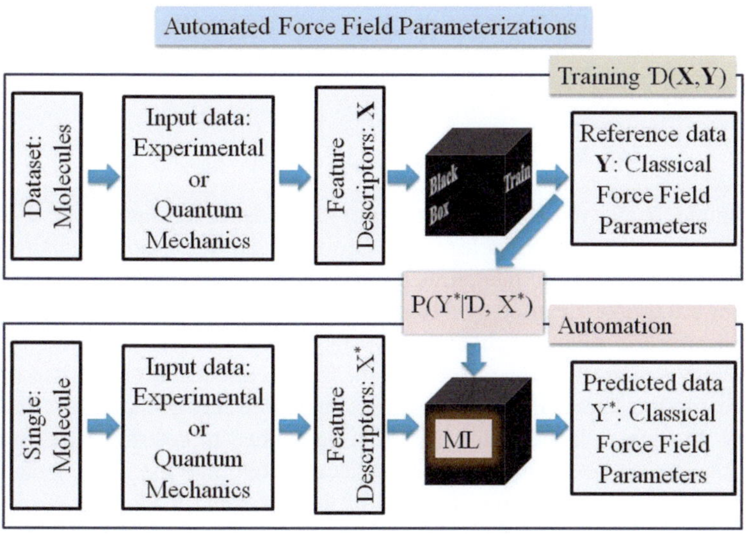

Fig. 7.3 A diagram of the general protocol for parameterization of the classical force fields using the automated machine learning approach

value described by the probability $P(Y^\star|\mathcal{D}, X^\star)$, which makes the parameterization of the force fields a very efficient automation process.

However, based on the above discussion, the accuracy in estimation of that probability, namely $P(Y^\star|\mathcal{D}, X^\star)$, is a data-driven process, and the evaluation of $Y^\star$ depends on the trained dataset. In particular, it depends on the diversity of the feature descriptors for the dataset of molecules and the size of the dataset. Both the feature descriptors diversity of the compounds and the size of the dataset are interconnected; however, a large size dataset is practically difficult to be established due to the lack of the experimental data, and quantum mechanics data may be expensive to obtain. On the other hand, the diversity of the feature descriptors of the compound database is essential to increase the range of the test data that can be predicted since the machine learning methodology works very well in interpolating the new data points, but suffers on extrapolating new data outside the range covered by the training dataset. Therefore, one of the critical future developments of the automated machine learning methodologies is the choice of the training dataset and the feature descriptors of the chemical compounds. Therefore, in the following, the topological data analysis tools is employed to analyze the feature descriptors of the molecules.

7.3.4.1 Topological Data Analysis

The topological data analysis (TDA) is a field dealing with the topology of the data to understand and analyze large and complex datasets (Carlsson 2009; Edelsbrunner and Harer 2010). In the following discussion, we assume that we are analyzing a dataset represented by a vector of feature descriptors of length N and each data point has a dimension D:

$$\mathbf{X} = \{\mathbf{x}_1, \mathbf{x}_2, \cdots, \mathbf{x}_N\}, \quad \mathbf{x}_i = \{x_{i1}, x_{i2}, \cdots, x_{iD}\} \tag{7.33}$$

For example, N may represent the number of molecules in the dataset and d number of specific features for each molecule. Moreover, in our discussion, we assume that the data of the dataset are hidden in a "black box", for example, a *database*, and also, they are about to be used by a machine learning, which is another "black box". In such a situation, knowing about the topology of the data (e.g., the sparsity of the data points) is of great interest. Note that the TDA is applicable even when the user has access to the data, that is, the structure of the molecules of the dataset is known a priory. In such a situation, the TDA can be applied to determine the topology of the key feature descriptors for each molecule. Note that the TDA is employed to reveal the intrinsic persistent features of the DNA and RNA (Xia et al. 2015; Mamuye et al. 2016). Therefore, the construction of the topological spaces upon the input data of a machine learning approach can be applied for each dimension separately, namely to the time series of the form $\mathbf{X}_d = \{x_{1d}, x_{2d}, \cdots, x_{Nd}\}$, or for each molecular structure, namely $\mathbf{X}_k = \{x_{k1}, x_{k2}, \cdots, x_{kD}\}$. But, it can also apply to both dimensions at the same time, for instance, by constructing the input data in the form of the following time series obtained by aligning feature descriptors of the molecular structures in one dimension:

$$\mathbf{X} = \{x_{11}, \cdots, x_{1D}, x_{21}, \cdots, x_{2D}, \cdots, x_{N1}, \cdots, x_{ND}\} \tag{7.34}$$

In that case, the input vector of the feature descriptors is a time series of length $N_{\text{train}} = ND$.

Then, to determine the topological space for this dataset, we first define a distance $\sigma > 0$. The Vietoris-Rips simplicial complex $R(\mathbf{X}, \sigma)$ or simply Rips complex for each $k = 1, 2, \cdots$ as a k-simplex of vertices $\mathbf{X}_i^k = \{x_{i_1}, x_{i_2}, \cdots, x_{i_k}\}$ such that they satisfy the condition that the mutual distances between any pair of the vertices is less than σ:

$$d(x_{i_k}, x_{i_l}) \leq \sigma, \quad \forall x_{i_k}, x_{i_l} \in \mathbf{X}_i^k \tag{7.35}$$

With other words, a k-simplex is part of a $R(\mathbf{X}, \sigma)$ for every set of k data points that are distinct from each-other at a resolution σ and hence the Rips complexes form a filtration of the data from the dataset at a resolution σ. That is, for any two values of the resolution σ and σ' such that $\sigma < \sigma'$, then

$$R(\mathbf{X}, \sigma) \subseteq R(\mathbf{X}, \sigma')$$

where $\subseteq$ denotes the subset.

All the vertices of a k-simplex can be connected in a two-dimensional space by undirected edges forming a graph, which can have different two-dimensional shapes. Figure 7.4 illustrates how to build simplicial complexes using a set of point cloud data by increasing the resolution value σ.

The k-simplex dataset points form a loop that is called *hole*. By increasing the resolution σ, the shapes grow, and some of the holes die, and some new holes are

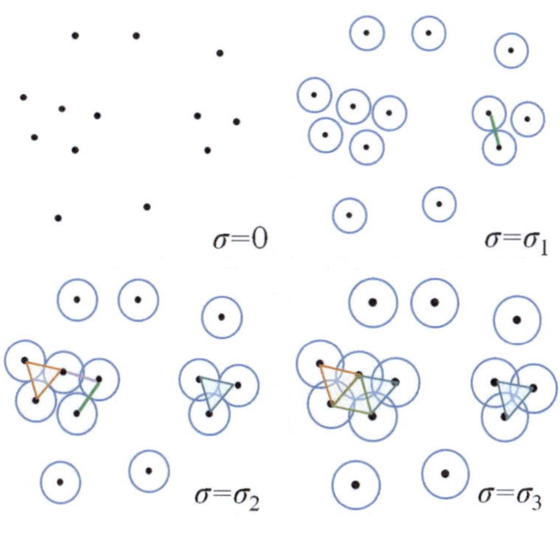

Fig. 7.4 An illustration of building the simplicial complexes by increasing the resolution value σ

born. This process is the so-called σ loop expansion. The interval between birth and death of a hole is called *persistence interval* indicating whether a hole is structurally relevant or just a noise into the data.

Persistent homology (PH) is an essential tool of TDA, which aims to construct a topological space gradually upon the input dataset, which is done by growing shapes based on the input data. Persistent homology measures in this way the persistence interval of the topological space. The features will be identified as persistent if after the last iteration they are still present.

This procedure is analog to systematic coarse-graining and is of crucial importance for any attempt at capturing natural feature descriptors in terms of a few relevant degrees of freedom, and thus they form the essential philosophical basis of a dataset for the machine learning approaches.

I may argue that the fundamentals of the PH notions on the relevance or irrelevance of perturbations in the data analysis are crucial, and the persistence homology can be considered as necessary as the renormalization group theory in statistical physics when applied to equilibrium phenomena in understanding the relevant or irrelevant interactions. In this analogy, the resolution scaling σ on the topological data analysis can be considered similar to the characteristic correlation length scale that determines the judgment of the strong interactions and correlations renormalization group theory (Täuber 2012).

7.3.4.2 Hardware Versus Software Machine Learning Approach

The advanced digital technologies use binary bits that take two values, either zero or one, and are stored using modern technologies at room temperature. On the other hand, quantum circuits use q-bits that are a superposition of the states zero and one and require novel technologies to work at cryogenic temperatures. Recently, (Camsari et al. 2017, 2019) classical *p-bits*, standing for *probabilistic-bits*, is defined as a building unit between the bits and q-bits. Note that the concept of the probabilistic computers is also described by Feynman that inspired developments of quantum computers (Feynman 1982). While the basic building block of the modern digital electronics is the transistor, used to represent deterministic bits (namely zero and one), it is argued (Camsari et al. 2019) that the existing technologies of digital electronics can be used to create the basic hardware building blocks of probabilistic computers, called *p-circuits*. Furthermore, possible applications that probabilistic computer can be used is discussed.

Note that we already discussed above several interesting problems, such as automated machine learning approaches of the force field parameters, which may involve use of large probabilistic networks. In particular, the p-bit concept relates to that of the binary stochastic neuron (BSN) used in the machine learning (Camsari et al. 2019). In BSN, the response Z_i to an input X_i is given as:

$$Z_i = \text{sign}\,(\tanh X_i - r) \tag{7.36}$$

where sign function is defined as

$$\operatorname{sign}(x) = \begin{cases} -1, \ x < 0 \\ +1, \ x \geq 0 \end{cases}$$

and r is a random number between -1 and $+1$. Then, the binary state "0" is obtained for $m = -1$ and state "1" for $m = +1$. The probabilistic network is then obtained when combined as:

$$X_i = \sum_j W_{ji} Z_j + b_i \tag{7.37}$$

where W_{ij} is the weight and b_i is the bias parameter. Equations (7.36) and (7.37) are the basic building blocks of the software algorithms in a machine learning approach similarly to Eq. (7.22).

Efforts have been made to develop hardware basic building blocks to accelerate the operations described by Eqs. (7.36) and (7.37) in hardware level (Hu et al. 2016; Camsari et al. 2019). In particular, Camsari et al. (2017, 2019) propose that the three-terminal p-bits provide a hardware accelerator for Eq. (7.36). These hardware basic building blocks can serve for building a probabilistic computer, which like a quantum computer can be used to accelerate the quantum computations, the probabilistic computer can be used to accelerate machine learning approaches using stochastic neural networks.

Figure 7.5 shows the p-bits (neurons) interconnected to each other via a network of resistances (synapse) constructing the building block architecture used in a hardware stochastic neural network (Camsari et al. 2017). The equations describing the synapse and the p-bit are given as (Camsari et al. 2017, 2019) (and the references therein):

$$\underbrace{\frac{V_j^{\text{out}}}{V_{\text{DD}}/2}}_{Z_j} = \operatorname{sign}\left(\tanh \underbrace{\left(\frac{V_j^{\text{in}}}{V_0} \right)}_{X_j} - r \right) \tag{7.38}$$

$$\underbrace{\frac{V_i^{\text{in}}}{V_0}}_{X_i} = \sum_j \underbrace{\frac{R_0}{R_{ji}}}_{W_{ji}} \underbrace{\frac{V_j^{\text{out}}}{V_{\text{DD}}/2}}_{Z_j} \tag{7.39}$$

$$\underbrace{V_i^{\text{in}}}_{X_i} = \sum_j \underbrace{\frac{R_{\text{ref}}}{R_{ji}}}_{W_{ji}} \underbrace{\bar{V}_j^{\text{out}}}_{Z_j} \tag{7.40}$$

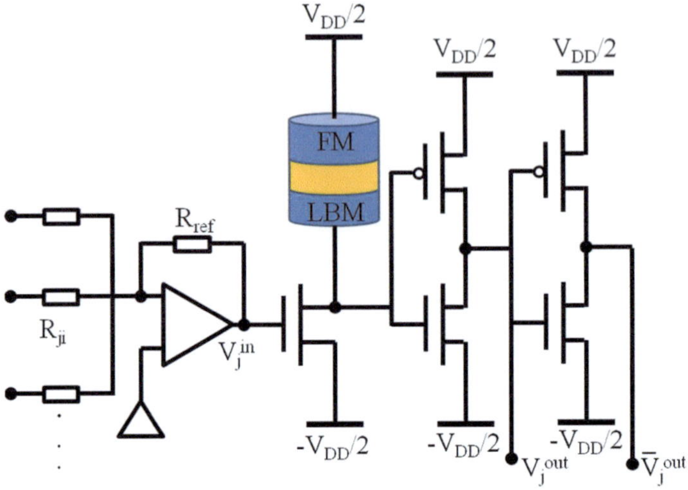

Fig. 7.5 The digital circuit of the p-bit and synapse represented by a stochastic neural network in hardware (Camsari et al. 2017) (and the references therein). The fixed layer ferromagnetic (FM) presents a stable magnet with large energy barrier; the free layer is a circular low-barrier magnet (LBM) with a height approximately $0\,k_B T$, whose magnetization varies due to thermal noise

where R_0 is the unit resistor that is used to change the inverse temperature β as

$$\beta = \frac{V_{DD} R_{ref}}{2 V_0 R_0}$$

and V_0 is the transistor voltage parameter, $V_0 = 40$ mV defining the stochastic window width of the p-bit ($\tanh(V_j^{in}/V_0)$). For the synaptic connections is used either V^{out} or $\bar{V}^{out}$ depending on the sign of the interconnection W_{ji}. Note that similar control is achieved using a network of capacitors, replacing the resistors (Camsari et al. 2019).

Full on-chip implementation of the p-bit could have many applications (Camsari et al. 2019), including a low-power and an efficient hardware accelerator for use in machine learning applications. In particular, the network of p-bits could be useful to accelerate learning algorithms, where the network weights and bias parameters are trained offline by a learning algorithm in software, and then the hardware can be used to perform inference tasks efficiently (Camsari et al. 2019) (and the references therein). A natural application of these stochastic circuits is in automation of the force field parameterizations that inspire the use of machine learning algorithms as discussed above. Moreover, soon, a complex multi-chip synapse/p-bit combination (Camsari et al. 2019) could also be more useful, for example, in molecular dynamics simulation.

7.4 Classical Force Fields Interaction Types

7.4.1 *Bond Stretching*

There are many functional forms to model the potential energy curve for a typical bond. One of this include the Morse potential, which has the form:

$$v(l) = D_e \left[1 - \exp\left(-(l - l_0)\right)\right]^2 \tag{7.41}$$

D_e is the depth of the potential energy minimum and $a = \omega\sqrt{\mu/2D_e}$, where μ is the reduced mass and ω is the frequency of the bond vibration, which is related to the stiffness constant of bond stretching k, by $\omega = \sqrt{k/\mu}$. $l_0$is the reference value of the bond (or equilibrium bond length.) Fig. 7.6 shows the Morse potential functional form. Note that the Morse potential is not often used in molecular simulations since it requires more computational efforts and it also needs three parameters to parametrise. In additional, at temperatures considered in molecular dynamics simulations it is not common that bonds deviate significantly from their equilibrium values. More simple expressions are often used, such as this using Hooke's law formula, where the energy is a functional of the square of the displacement from an equilibrium bond length l_0:

$$v(l) = \frac{k}{2}\left(l - l_0\right)^2 \tag{7.42}$$

The functional form is shown in Fig. 7.6.

The reference bond length or sometimes called equilibrium bond length represents the value of bond length at minimum energy structure when all the other terms in the force field also contribute. Typical values of the force constants for bond stretching are very high in magnitude, indicating that the forces between bonded atoms are very strong and significantly high energy is required to deviate the bonded atoms far from their equilibrium distance.

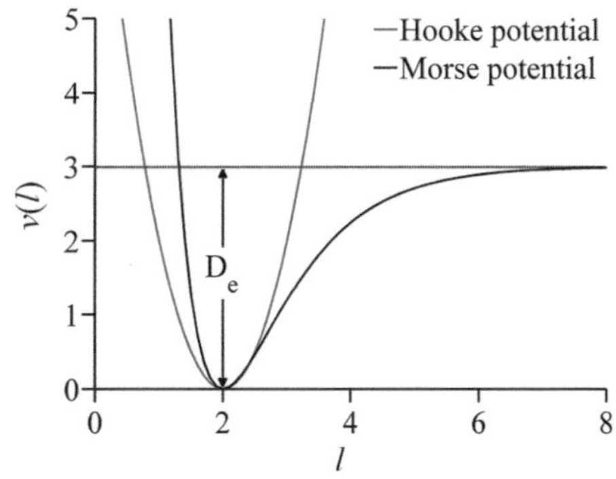

Fig. 7.6 The Morse potential and Hooke's law potential of bond stretching

The Hooke's law functional form is another practical shape of the potential energy, which is used to model the bottom of the potential well at a distance that corresponds to bonding in ground-state molecules. However, it is less accurate away from the equilibrium, as indicated in Fig. 7.6. Higher terms can also be included in the expression of the Hooke's law functional form, which results in a better approximation to the Morse functional form:

$$v(l) = \frac{k}{2}(l - l_0)^2 \left[1 - k'(l - l_0) - k''(l - l_0)^2 - k'''(l - l_0)^3 + \cdots\right]$$

7.4.2 *Angle Bending*

The Hooke's law, or often called harmonic potential, is the formula used to describe the deviations of angles from their equilibrium values. This is given as

$$v(\theta) = \frac{k}{2}(\theta - \theta_0)^2 \tag{7.43}$$

Two parameters are used to characterise this term, the force constant (k) and the equilibrium angle θ_0. Compare to bond stretching, rather less energy is required to deviate the angle away from the equilibrium value. Thus, the values of the force constants of angle bending are smaller than the force constants of bond stretching. Similar to bond stretching, higher order terms can also be included into the expression for the energy function of the angle bending, such as:

$$v(\theta) = \frac{k}{2}(\theta - \theta_0)^2 \left[1 - k'(\theta - \theta_0) - k''(\theta - \theta_0)^2 - k'''(\theta - \theta_0)^3 + \cdots\right]$$

7.4.3 *Torsional Angle*

The torsional angle term is considered as *soft* degree of freedom, in contrast to bond stretching and angle bending terms which are considered as *hard* degrees of freedom, because much less energy is required to cause a significant deviation from the equilibrium value.

Most of the molecular mechanics force fields use a torsional potential functional form that is expressed as a cosine series expansion, such as

$$v(\omega) = \sum_{n=0}^{N} \frac{V_n}{2}\left[1 + \cos(n\omega - \gamma)\right] \tag{7.44}$$

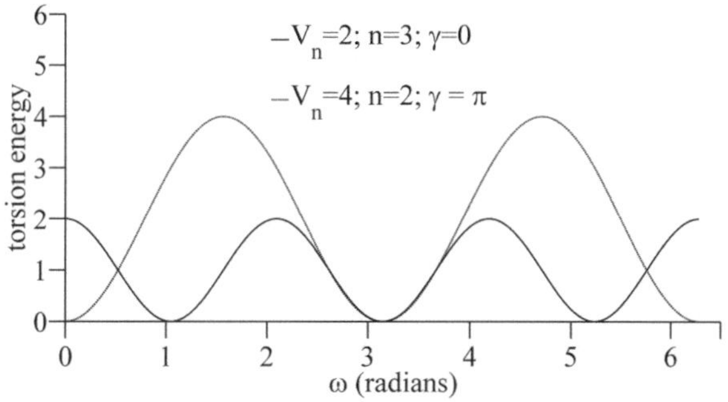

Fig. 7.7 The torsion energy plotted for different values of V_n, n and γ

where ω is the torsional angle formed by a quartet of atoms $A - B - C - D$ in the system. There also exists an equivalent expression of the following form:

$$v(\omega) = \sum_{n=0}^{N} C_n \cos(\omega)^n \tag{7.45}$$

The parameter V_n in Eq. (7.44) is also called barrier height, which gives a contribution to barrier height due to the rotation. In Eq. (7.44), n is the multiplicity, which characterizes the number of minimum points in the function as the bond rotates through 360°, γ is the phase factor that represents the position where the torsion angle passes through the minimum value. For example, $n = 3$ and $\gamma = 0$ would give a rotational profile with minima at torsion angles of ±60° and 180° and a maximum at ±120° and 0°; for $n = 2$ and $\gamma = \pi$ the rotational profile will result with minima at 0° and 180°. That is indicated in Fig. 7.7. As can be seen, a larger value of V_n would give a more significant barrier of the torsion energy. Molecular mechanics force fields often contain just one term from the cosine series expansion, but there are also cases when it is found necessary to include more than one term, for instance, in AMBER force field. Another feature of the force fields is that energy profile for rotation about a bond depends upon the atom types of the two atoms of the central bond and not upon the type of terminal atoms. For example, all torsional angles in which the central bond is between two sp^3- hybridized carbon atoms (H-C-C-H, C-C-C-C, H-C-C-C) are assigned the same torsional parameters, with some exclusion, such as O-C-C-O torsion angle.

7.4.4 *The van der Waals Potential*

The electrostatic interactions cannot include all non-bonded interactions in a system. For example, in rare cases, where all multipole moments are zero, there are no dipole-dipole and dipole-induced dipole interactions. However, there exist interactions between atoms, manifested on the existence of different phases, such as

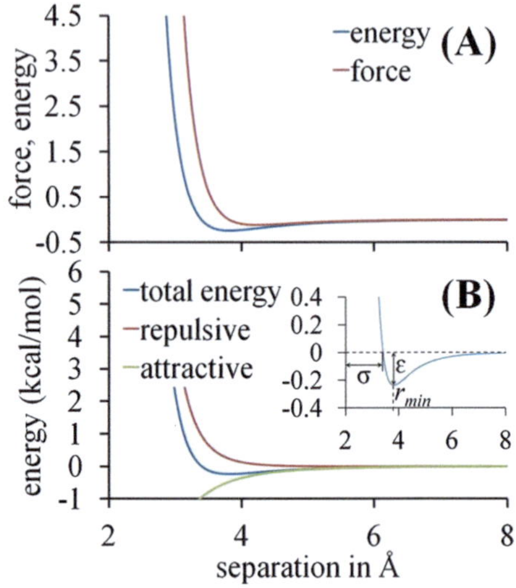

Fig. 7.8 (**a**) The van der Waals interaction between two isolated argon atoms. (**b**) The attractive and repulsive parts of the total van der Waals interaction energy. Energy is given in kcal/mol, and the force (in Newton) is scaled by 10^{13} to show it in the same plot with energy. For argon $\sigma = 3.40$ Å and $\varepsilon = 0.99$ kJ/mol

liquid and solid phases. Van der Waals first introduced that deviation from the ideal gas phase.

Experimental investigations of the interaction between two isolated rare gas atoms, for example, argon atoms, using the molecular beam experiment, reveal the following interaction energy as a function of the separation between atoms as shown in Fig. 7.8. Some standard features include that interaction energy goes to zero at the infinite separation between the atoms. As the distance reduces, the energy decreases, passing through a minimum value of approximately 3.82 Å. The energy increases very rapidly as the separation decreases further.

Figure 7.8 shows the force between the two atoms, calculated as the minus derivative of the interaction energy with respect to the separation distance. Other experimental methods used to provide evidence of the nature of van der Waals interactions include gas imperfections, spectroscopic techniques, and measurements of the transport properties.

The van der Waals interactions are considered to arise from a glance between attractive and repulsive forces. Attractive forces are long-range, so they act at long distances between atoms; whereas repulsive forces are short-range, so they operate at short distances between atoms.

The attractive term is due to *dispersive* forces, and can be explained using quantum mechanics (London 1930). The dispersive force is due to instantaneous dipoles which arise during the fluctuations in the electronic clouds. The instantaneous dipole in a molecule can, in turn, induce a dipole in the neighboring atoms, giving rise to an attractive inductive effect.

Figure 7.8 shows that a small decrease in the distance between two isolated pairs of atoms causes a significant increase in the interaction energy. The basis of this increase is quantum mechanical, and it is understood in terms of the Pauli principle, which does not allow for having any two electrons of the system with the same set of quantum numbers. The interaction is because of having two electrons with the same spin, and hence it is short-range interaction with repulsive forces, referred to as *exchange forces* or *overlap forces*. The effect of exchange is to reduce the electrostatic repulsion between two paired electrons, so not allowing them to occupy the same internuclear region of space. This reduction of the electron density in the internuclear region results in repulsion between the two nuclei. At very short distances between two nuclei, the interaction energy varies as $1/r$ due to the nuclear repulsion, and at more considerable distances, the energy decays exponentially as $\exp(-2r/a_0)$, where a_0 is the Bohr radius.

The van der Waals potential for a non bonded distance r_{ij} has the common form of a Lennard-Jones in a macromolecular force field given as:

$$V_{\mathrm{LJ}}(r_{ij}) = 4\varepsilon_{ij}\left(\frac{\sigma_{ij}^{12}}{r_{ij}^{12}} - \frac{\sigma_{ij}^{6}}{r_{ij}^{6}}\right) \tag{7.46}$$

where ε_{ij} is the well depth and σ_{ij} is the collision diameter, which is the distance between the two atoms for which the energy is zero. These two parameters are depicted in Fig. 7.8b.

Often, the Lennard-Jones functional form of interaction can also be expressed in terms of the distance between the two atoms for which the energy is minimum, r_{min}. This is calculated by equalising to zero the derivative of the energy (Eq. (7.46)) with respect to the separation r_{ij} and solving for r_{ij}. This yields an expression for $r_{min} = 2^{1/6}\sigma$, then Eq. (7.46) can be re-written as:

$$V_{\mathrm{LJ}}(r_{ij}) = \varepsilon_{ij}\left(\frac{r_{ij,min}^{12}}{r_{ij}^{12}} - 2\frac{r_{ij,min}^{6}}{r_{ij}^{6}}\right) \tag{7.47}$$

or

$$V_{\mathrm{LJ}}(r_{ij}) = \frac{A_{ij}}{r_{ij}^{12}} - \frac{B_{ij}}{r_{ij}^{6}} \tag{7.48}$$

where the repulsive ($A_{ij} = \varepsilon_{ij} r_{ij,min}^{12}$) and attractive ($B_{ij} = 2\varepsilon_{ij} r_{ij,min}^{6}$) coefficients depend on the type of the interacting atoms i and j. The following rules are often used:

$$\varepsilon_{ij} = \sqrt{\varepsilon_i \cdot \varepsilon_j} \tag{7.49}$$

$$r_{ij,min} = \frac{r_{i,min} + r_{j,min}}{2}$$

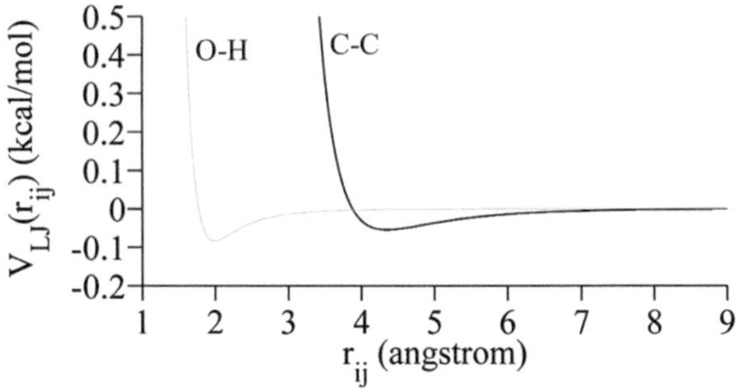

Fig. 7.9 The Lennard-Jones energy (in kcal/mol) for C-C and O-H non-bounded terms

For example, in Fig. 7.9, we have plotted the Lennard-Jones interaction energy for C-C and O-H non-bonded terms for which

$$r_{C,min} = 2.1750\ \text{Å};\ \varepsilon_C = -0.0550$$

$$r_{O,min} = 1.7682\ \text{Å};\ \varepsilon_C = -0.1521$$

$$r_{H,min} = 0.2245\ \text{Å};\ \varepsilon_C = -0.0460$$

The 6-12 potential is an excellent choice for rare gases, but may not be for other systems, such as hydrocarbons. However, it is widely used for large biomolecular systems due to computational simplicity, since the term r^{-12} can be very efficient calculated by taking the square of the term r^{-6}. Moreover, the term r^{-6} can also be derived from the square of the distance without the need to calculate the square root of it, which can be computationally very expensive.

It is interesting to note that other powers can be used as well for the repulsive part of the potential. For example, the power of 9 and 10 result in less steep curves and are still used in some force fields. The original form of the Lennard-Jones potential is given by

$$v(r) = k\varepsilon\left[\left(\frac{\sigma}{r}\right)^n - \left(\frac{\sigma}{r}\right)^m\right], \quad k = \frac{n}{n-m}\left(\frac{n}{m}\right)^{\frac{m}{n-m}} \tag{7.50}$$

The particular form given by expression (7.46) can be obtained for $n = 12$ and $m = 6$.

7.4.5 Electrostatic Potential

The last term in Eq. (7.1) represents the electrostatic interaction energy between the non-bonded atoms in the system, which is given using the Coulombic law between any two charges q_i and q_j:

$$V_{Elec}(r_{ij}) = \frac{1}{4\pi\varepsilon_0}\frac{q_i q_j}{\varepsilon r_{ij}} \tag{7.51}$$

where q_i and q_j represent the partial charges assigned to every atom in the molecule, ε_0 is the permittivity of the free space:

$$\varepsilon_0 = 8.8542 \times 10^{-12}\,\frac{C^2}{N \cdot m^2}$$

Here, r_{ij} denotes the distance between the two charges and ε is the dielectric constant. In Eq. (7.51), the factor

$$k_e \equiv \frac{1}{4\pi\varepsilon_0}$$

is also called Coulomb's constant, which is introduced to adjust the conversion of the units from the standard SI units into force field units, namely kcal/mol. Often, in the common used force fields, the units of the charges are the electrostatic charge units (esu), which is a unit per electron charge

$$1\,\text{esu} = 1\,\frac{\text{C}}{1.6 \times 10^{-19}\,\text{C}}$$

It can be found that

$$k_e = 332\,\frac{\text{kcal}}{\text{mol}} \cdot \frac{\text{Å}}{\text{esu}^2}$$

Note that because of the $1/r_{ij}$ dependence concerning the distance between the charges, r_{ij}, the electrostatic energy decays slower to zero with increasing the distance between charges in comparison with Lennard-Jones energy. In Fig. 7.10, we show the electrostatic energy (in kcal/mol) for the non-bonded interactions O $\cdots$ O and O $\cdots$ H of the water molecule. Besides, the electrostatic interactions are particularly important in the non-bonded interactions of biomolecular systems,

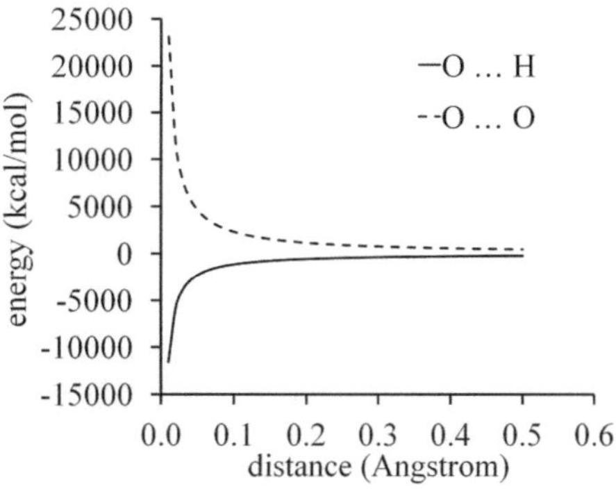

Fig. 7.10 The Coulomb energy (in kcal/mol) for O $\cdots$ O and O $\cdots$ non-bounded terms

such as the proteins, DNA, RNA, and their complexes, since they establish the stability of conformations in the solvent environments. However, the computation of these interactions requires high efforts during the force and energy calculations. Typically, the complexity of the electrostatic interactions scales as N^2, where N is the number of charged atoms. To reduce the computation efforts in a calculation of the electrostatic interactions more sophisticated methods are used in typical molecular dynamics simulations. Usually, a cutoff radius, R_c, is defined prior to simulation, typically 10–12 Å, which counts for direct calculations of the short-range electrostatic interactions ($r_{ij} \leq R_c$) using the Coulombic form. Then, the long-range electrostatic interactions ($r_{ij} > R_c$) are computed using fast-electrostatic interactions algorithms, such as the Ewald and the Particle Mesh Ewald approaches, which are characterized by a computation complexity $\mathcal{O}(N)$ (Ewald 1921; Luty et al. 1994, 1995; Hardy et al. 2011).

In Eq. (7.51), the dielectric constant plays the role of a reduction of the electrostatic interaction if the charged atoms are placed in a medium different from that of a vacuum. In other words, the electrostatic interactions occurring in a polarizable medium are weaker than those in the vacuum. That is due to the screening of the charges from the permanent dipoles of the medium. Often, a distance-dependent dielectric constant function is used to characterize the dielectric properties of the medium, such as:

$$\varepsilon = Dr$$

where D is a constant.

There also exist other function of the distance dependence dielectric constants, such as the sigmoidal function (Mehler 1996):

$$\varepsilon(r) = D_0 \exp(\kappa r) \tag{7.52}$$

where κ is screening parameter of the ionic strength of the solvent in dimensions of the inverse of the length and D_0 a free parameter. This expression is used to take into account different ionic concentrations.

Other functions have also been proposed, such as the following distance-dependent dielectric function (Mehler and Guarnieri 1999):

$$\varepsilon(r) = \frac{\varepsilon_w + D_0}{1 + k \exp(-\kappa(\varepsilon_w + D_0)r)} - D_0 \tag{7.53}$$

where ε_w is the dielectric constant of water, D_0 and κ are free parameters, and k is a fixed constant. These dielectric function forms are mainly used to take into accounts the decay of the electrostatic interactions when the charges separate at distances of several Angstroms, which are considered moderate distances, and to obtain a range of electrostatic interactions in solvent close their values in the vacuum if the charges are too close to each other.

Chapter 8
Slow Collective Variables of Macromolecular Systems

This chapter aims to discuss different methods used to determine the frequency spectrum of the motions in a macromolecular system, namely the *normal modes*, *principal components analysis*, and the *time-lagged auto-encoder* machine learning approach. In general, the normal modes method is successfully used to estimate the force constants, inter-atomic distances, and bond dissociation energies of molecular structures from the molecular vibrational spectra of small molecules. This technique, along with the method of the principal components analysis, is often used to obtain the global motions in the macromolecular systems along the most emphasized principal components. Whereas, the time-lagged auto-encoder method is a new method used for determining the slow collective coordinates using the machine learning approaches, such as artificial neural networks. In this chapter, we will discuss an improved modified version of the artificial neural network, called *Bootstrapping Swarm Artificial Neural Network*. Finally, in this chapter, we are going to introduce an approach of how to derive the equations of motion in the reduced essential subspace of the slow collective variables using the harmonic bath coupling of these variables with the environmental fast degrees of freedom of the system.

8.1 Normal Modes

The basis for determining the force constants, inter-atomic distances, and bond dissociation energies of molecular structures are molecular vibrational spectra of small molecules (Lifson and Warshel 1968). The vibrational bond motions represent the small deviations from the equilibrium states. In particular, all possible vibrational motions of a molecule can be characterized as a superposition of different basic types of vibration (called *normal modes*). For a molecule of N atoms, there are $3N$ degrees of freedom to describe the motion of atoms in a three-

H. Kamberaj, *Molecular Dynamics Simulations in Statistical Physics: Theory and Applications*, Scientific Computation,
https://doi.org/10.1007/978-3-030-35702-3_8

dimensional space related to atomic coordinates. Omitting the three translational and three rotational degrees of freedom of the molecule as a whole, then there are $3N - 6$ normal modes necessary to describe the internal motion of molecules.

Experimental methods used to detect the vibrational energy levels of a molecule include the spectroscopic techniques, such as the vibrational absorption of infrared radiation (IR) and Raman scattering. IR spectroscopy is a powerful technique that captures information on the transitions between vibrational quantum states since these transitions lead to absorption and emission of infrared radiation. IR wavelength is in the range 1–100 μm. IR transitions can occur if there exists a change in the dipole moment of the molecule during the transition. On the other hand, Raman spectroscopy, which also captures the transitions between vibrational levels, occurs when the polarizability of the molecular system changes during the transition.

These two techniques are considered complementary methods, since some transitions that have a change in the dipole moment absorb light, and others have a change in polarizability and scatter light. For small symmetric molecules, the observed transitions are complementary, whereas, for large and asymmetric molecules, both, Raman and IR spectra are mostly the same.

The units of normal modes are *hertz* (Hz), that is, the inverse of seconds. However, often they are also reported in *wavenumbers*, that is, the number of waves per centimeter. Higher the frequency, more energetically expensive is the deformation along that particular mode. For example, bond-stretching modes have, in general, higher frequencies than angle-bending modes, which in turn have higher frequencies than torsion angle modes.

8.1.1 General Theory of Normal Modes

Based on the description in Goldstein (2002), if we consider a conservative system in which the potential energy is a function of coordinates only, the *equilibrium* state of the system is characterized by

$$Q_i = -\left(\frac{\partial V}{\partial q_i}\right)_0 \tag{8.1}$$

where $q_1, q_2, \cdots, q_f$ are the generalized coordinates of the system, which do not involve the time explicitly. Equation (8.1) indicates that potential energy has an extremum at the equilibrium configuration of the system, $q_{01}, q_{02}, \cdots, q_{0f}$. If the system is initially at the equilibrium position, with zero initial velocities ($\dot{q}_i$), then the system will continue in equilibrium indefinitely. An equilibrium position is classified as *stable* if a small disturbance of the system from equilibrium produces only a small bounded motion about the equilibrium position. The equilibrium is *unstable* if a small disturbance will eventually result in an unbounded motion. For example, a pendulum at rest is in stable equilibrium, but the egg that is standing on an end is an unstable equilibrium (Goldstein 2002). Following the discussion, when the extremum of V is a minimum, the equilibrium is a stable one. That is, if the system is at equilibrium and it has deviated from this position, then this disturbance will produce an increase in energy by an amount of dE above the equilibrium value V. Since the system is conservative, based on the conservation law of energy, the velocities must then decrease and with time become zero, which is a bound motion. On the other hand, if V decreases due to a deviation from equilibrium, the velocities (and so the kinetic energy) increase indefinitely, resulting in an unbounded motion.

Now, we consider the motion of the system close to the configuration of stable equilibrium. Because the deviations from the equilibrium are too small, we can expand the potential energy function in the Taylor series about the equilibrium and keeping only the lowest-order energy terms:

$$V(q_1, \cdots, q_f) = V(q_{01}, \cdots, q_{0f}) + \left(\frac{\partial V}{\partial q_i}\right)_0 \eta_i + \frac{1}{2}\left(\frac{\partial^2 V}{\partial q_i \partial q_j}\right)_0 \eta_i \eta_j + \cdots \tag{8.2}$$

where $\eta_i = q_i - q_{0i}$ is the deviation of the generalized coordinate from the equilibrium position and the usual summation convention is used. Terms linear in η_i vanish automatically due to equilibrium conditions. The first term in the series is the potential energy of the equilibrium position, and by shifting the arbitrary zero of potential to be the equilibrium minimum potential energy value, then

$$V(q_1, \cdots, q_f) = \frac{1}{2}\left(\frac{\partial^2 V}{\partial q_i \partial q_j}\right)_0 \eta_i \eta_j = \frac{1}{2} V_{ij} \eta_i \eta_j \tag{8.3}$$

where the second derivatives of V have been denoted by V_{ij} depend only on the equilibrium values of the q_i. Based on the definition, it can be seen that V_{ij} are symmetric, $V_{ij} = V_{ji}$.

The kinetic energy is written as the following summation:

$$T = \frac{1}{2}\sum_{i,j} m_{ij}\dot{q}_i\dot{q}_j = \frac{1}{2}\sum_{i,j} m_{ij}\dot{\eta}_i\dot{\eta}_j \tag{8.4}$$

where the mass coefficients m_{ij} are in general functions of the coordinates q_k, but they may be expanded in a Taylor series about the equilibrium configuration:

$$m_{ij}(q_1, \cdots, q_f) = m_{ij}(q_{01}, \cdots, q_{0f}) + \sum_k \left(\frac{\partial m_{ij}}{\partial q_k}\right)_0 \eta_k + \cdots$$

Denoting the values of m_{ij} at equilibrium as T_{ij}, we get an alternative expression as:

$$T = \frac{1}{2}\sum_{i,j} T_{ij}\dot{\eta}_i\dot{\eta}_j \tag{8.5}$$

where T_{ij} is symmetric, $T_{ij} = T_{ji}$.

The Lagrangian of the system is:

$$L = \frac{1}{2}\sum_{i,j} T_{ij}\dot{\eta}_i\dot{\eta}_j - \frac{1}{2}\sum_{i,j} V_{ij}\eta_i\eta_j$$

Applying the Lagrangian's equations of motion, we get the following equations of motion:

$$\sum_{j=1}^{f}\left(T_{ij}\ddot{\eta}_j + V_{ij}\eta_j\right) = 0 \tag{8.6}$$

for $i = 1, \cdots, f$.

The equations of motion (Eq. (8.6)) are linear differential equations with constant coefficients. The real part of the general form of the solution is given by

$$\eta_j = \sum_{k=1}^{f} A_{jk}\alpha_k \cos(\omega_k t + \delta_k) \tag{8.7}$$

where $A_{jk}\alpha_k$ is the real amplitude of the oscillation for each coordinate η_j. Then, the second derivative of Eq. (8.7) with respect to time is

$$\ddot{\eta}_j = -\sum_{k=1}^{f} A_{jk}\alpha_k\omega_k^2 \cos(\omega_k t + \delta_k)$$

Substituting this expression and the expression given by Eq. (8.7) into Eq. (8.6), we get

$$\sum_{j=1}^{f}\left(-T_{ij}\sum_{k=1}^{f} A_{jk}\alpha_k\omega_k^2\cos(\omega_k t + \delta_k) + V_{ij}\sum_{k=1}^{f} A_{jk}\alpha_k\cos(\omega_k t + \delta_k)\right) = 0 \tag{8.8}$$

or

$$\sum_{k=1}^{f}\left[\sum_{j=1}^{f} -T_{ij}A_{jk}\omega_k^2 + \sum_{j=1}^{f} V_{ij}A_{jk}\right]\alpha_k\cos(\omega_k t + \delta_k) = 0 \tag{8.9}$$

This equation holds for all time t if

$$\left(\sum_{j=1}^{f} T_{ij}A_{jk}\right)\omega_k^2 = \sum_{j=1}^{f} V_{ij}A_{jk} \tag{8.10}$$

for $k = 1, 2, \cdots, f$. By introducing the matrices, $\mathbf{T}$ with elements T_{ij}, $\mathbf{V}$ with elements V_{ij}, $\mathbf{A}$ with elements A_{jk} and the diagonal matrix $\boldsymbol{\Lambda}$ with diagonal elements $\Lambda_{kk} = \omega_k^2$, we can write Eq. (8.10) in a matrix form as

$$\mathbf{TA}\boldsymbol{\Lambda} = \mathbf{VA} \tag{8.11}$$

Following Levitt et al. (1985), before we determine $\boldsymbol{\Lambda}$ and $\mathbf{A}$ by solving Eq. (8.11), a normalisation condition is established. For that Q_k is defined as

$$Q_k = \alpha_k \cos(\omega_k t + \delta_k)$$

Then Eq. (8.7) is written as

$$\eta_j = \sum_{k=1}^{f} A_{jk} Q_k \equiv \mathbf{AQ} \tag{8.12}$$

where $\mathbf{Q}$ is a vector with elements Q_k. This expression gives a relation between the generalised coordinates $\boldsymbol{\eta}$ and the coordinates $\mathbf{Q}$. Then, the potential energy and the kinetic energy can be written as

$$V = \frac{1}{2}\mathbf{Q}^T\mathbf{A}^T\mathbf{VAQ} \tag{8.13}$$
$$T = \frac{1}{2}\dot{\mathbf{Q}}^T\mathbf{A}^T\mathbf{TA}\dot{\mathbf{Q}}$$

By choosing the normalisation condition as (Levitt et al. 1985)

$$\mathbf{A}^T\mathbf{TA} = \mathbf{1} \tag{8.14}$$

then, the kinetic energy can be written as

$$T = \sum_{k=1}^{f} \frac{1}{2}\dot{Q}_k^2 \equiv \frac{1}{2}\dot{\mathbf{Q}}^T\dot{\mathbf{Q}}$$

If we multiply both sides of Eq. (8.11) by $\mathbf{A}^T$ and using the normalisation condition, Eq. (8.14), we get

$$\boldsymbol{\Lambda} = \mathbf{A}^T\mathbf{VA} = \mathbf{A}^T\mathbf{TA}\boldsymbol{\Lambda} \tag{8.15}$$

Thus, the potential energy can be written as

$$V = \frac{1}{2}\mathbf{Q}^T\boldsymbol{\Lambda}\mathbf{Q} = \frac{1}{2}\sum_{k=1}^{f}\Lambda_{kk}Q_k^2 = \frac{1}{2}\sum_{k=1}^{f}\omega_k^2 Q_k^2$$

It can be seen that both potential and kinetic energy are simple sums of squares of coordinates Q_k and its time derivative $\dot{q}_k$, hence the coordinates Q_k are normal coordinates and mass scaled one, which is indicated by the form of the kinetic energy.

8.1.2 Dynamical Behavior of System

We can search for solution of Eq. (8.11) using the normalization condition, Eq. (8.14), using standard methods. Then, the dynamics of the system are characterized by equation

$$\eta_j = \sum_{k=1}^{f} A_{jk}\alpha_k \cos(\omega_k t + \delta_k)$$

or

$$q_j = q_{0j} + \sum_{k=1}^{f} A_{jk}\alpha_k \cos(\omega_k t + \delta_k) \tag{8.16}$$

where δ_k is the phase of the k-th normal mode of motion, and ω_k is an angular frequency which is determined as: $\omega_k = \sqrt{\Lambda_{kk}}$. In general, ω_k, α_k and δ_k will depend on the initial positions and velocities at $t = 0$.

For the Eq. (8.16) to be applied, we first need to determine $\mathbf{q}_0$, which represents the coordinates at the thermodynamic equilibrium state. The thermodynamic equilibrium state is the configuration for which the potential energy of the system has a minimum value, which can be determined using numerical methods for minimizing general functions (Andricioaei and Straub 1996b, 1998).

As it has been suggested elsewhere (Levitt et al. 1985), the derivative of the potential energy with respect to generalized coordinates, $\partial V/\partial q_j$, should be calculated analytically, then the elements V_{ij} of the matrix $\mathbf{V}$ are calculated numerically as

$$V_{ij} = \frac{1}{\epsilon}\left[\left(\frac{\partial V}{\partial q_i}\right)_{q_j=q_{0j}+\epsilon} - \left(\frac{\partial V}{\partial q_i}\right)_{q_j=q_{0j}}\right] \tag{8.17}$$

For calculation of the matrix $\mathbf{T}$, one can start with expression of the kinetic energy in terms of the time derivative of the Cartesian coordinates:

$$T = \frac{1}{2}\sum_{i=1}^{f} m_i \dot{\mathbf{r}}_i$$

Then, small changes of the Cartesian coordinates $\delta \mathbf{r}_i$ can be expressed in terms of the small changes on the generalized coordinates δq_i as

$$\delta \mathbf{r}_i = \sum_{k=1}^{f} \frac{\partial \mathbf{r}_i}{\partial q_k} \delta q_k$$

Assuming that these small changes occur in time interval δt, then

$$\dot{\mathbf{r}}_i = \sum_{k=1}^{f} \frac{\partial \mathbf{r}_i}{\partial q_k} \frac{\delta q_k}{\delta t} = \sum_{k=1}^{f} \frac{\partial \mathbf{r}_i}{\partial q_k} \dot{q}_k$$

Thus, the kinetic energy can be written as

$$\begin{aligned} T &= \frac{1}{2} \sum_{i=1}^{f} m_i \sum_{k=1}^{f} \frac{\partial \mathbf{r}_i}{\partial q_k} \dot{q}_k \sum_{j=1}^{f} \frac{\partial \mathbf{r}_i}{\partial q_j} \dot{q}_j \\ &= \frac{1}{2} \sum_{k=1}^{f} \sum_{j=1}^{f} \left(\sum_{i=1}^{f} m_i \frac{\partial \mathbf{r}_i}{\partial q_k} \cdot \frac{\partial \mathbf{r}_i}{\partial q_j} \right) \dot{q}_k \dot{q}_j \\ &= \frac{1}{2} \sum_{k=1}^{f} \sum_{j=1}^{f} T_{jk} \dot{q}_k \dot{q}_j \end{aligned} \tag{8.18}$$

where T_{jk} are given as:

$$T_{jk} = \sum_{i=1}^{f} m_i \frac{\partial \mathbf{r}_i}{\partial q_k} \cdot \frac{\partial \mathbf{r}_i}{\partial q_j}$$

Since the kinetic energy should include the overall translational and rotational kinetic energy, when calculating $\partial \mathbf{r}_i / \partial q_k$, the coordinate $\mathbf{r}_i$ must be calculated before the overall translational and rotational degrees of freedom are removed from the system.

The algorithm for calculation of normal modes includes the following steps:

(1) Define the Cartesian coordinates as a function of the generalized coordinates q_i.
(2) Minimize the potential energy with respect to the coordinates q_i to determine q_{0i}.
(3) Calculate the matrices $\mathbf{V}$ and $\mathbf{T}$ using Eqs. (8.17) and (8.18).
(4) Solve equations of motion given by Eq. (8.11) with respect to $\mathbf{A}$ and $\boldsymbol{\Lambda}$.
(5) Express the motion of the generalised coordinates $\mathbf{q}$ in terms of the normal coordinates $\mathbf{Q}$ using Eqs. (8.12) and (8.16).

8.1.3 Time Averaged Properties

The trajectory of every atomic motion depends on the initial conditions, positions and velocities, which determine the amplitude α_k and phase δ_k of each normal mode. Time averaged properties along the trajectory depend only on the amplitudes α_k and, thus, are of more general importance. For example, let Δp_i be some property of the system along the i-th degree of freedom that is a linear function of the change in the generalized coordinate Δq_k:

$$\Delta p_i = \sum_{j=1}^{f} P_{ij} \Delta q_j$$

Then, the correlation coefficients of that property between two different degrees of freedom of the system i and j is

$$\begin{aligned} \langle \Delta p_i(\tau) \Delta p_j(\tau) \rangle &= \sum_{k=1}^{f} \sum_{m=1}^{f} P_{ik} P_{jm} \langle \Delta q_k(\tau) \Delta q_m(\tau) \rangle \\ &= \frac{1}{2} \sum_{n=1}^{f} \sum_{k=1}^{f} \sum_{m=1}^{f} P_{ik} P_{jm} A_{kn} A_{mn} \alpha_n^2 \\ &= \frac{1}{2} \sum_{n=1}^{f} P'_{in} P'_{jn} \alpha_n^2 \end{aligned} \tag{8.19}$$

where

$$\begin{aligned} P'_{in} &= \sum_{k=1}^{f} P_{ik} A_{kn} \\ P'_{jn} &= \sum_{m=1}^{f} P_{jm} A_{mn} \end{aligned} \tag{8.20}$$

8.1.4 Thermal Amplitudes

The amplitude α_k for every normal mode k depends on the temperature T (Levitt et al. 1985). The average potential energy of every normal mode is

$$\langle V \rangle_k = \frac{1}{2} \omega_k^2 \langle Q_k^2(\tau) \rangle$$

$$
\begin{aligned}
&= \frac{1}{2}\omega_k^2\alpha_k^2\langle\cos^2(\omega_k\tau + \delta_k)\rangle \\
&= \frac{1}{4}\omega_k^2\alpha_k^2 \\
&= \frac{1}{2}k_B T
\end{aligned}
$$

for classical dynamics. Thus,

$$\langle Q_k^2(\tau)\rangle = k_B T/\omega_k^2$$

where $\langle Q_k^2(\tau)\rangle$ is the classical mean square fluctuation of Q_k. Moreover,

$$\alpha_k = \sqrt{\frac{2k_B T}{\omega_k^2}}$$

Hence, since all the time averaged properties depend on α, they also depend on the temperature T.

8.2 Principal Components Analysis

The principal component analysis (PCA) (Karhunen 1947) has often been used to reduce the number of degrees of freedom of the biomolecular systems from the MD simulations (Brooks et al. 1995), also known as quasi-harmonic analysis (Karplus and Jushick 1981; Ichiye and Karplus 1991). In addition, the method has been employed to describe molecular dynamics trajectories in terms of a small number of variables, namely essential degrees of freedom, responsible for all relevant structural transitions in these molecules (Go 1990; Kitao et al. 1991; Garcia 1992; Amadei et al. 1993; Aalten et al. 1993), obtained as the directions with significant non-zero eigenvalues calculated from the covariance matrix of the atomic position fluctuations of the MD trajectory (Amadei et al. 1993). As such, the PCA method is proposed as a method for reducing the phase space of proteins for long-time molecular dynamics (Grubmüller 1995). In PCA, we can determine a small number of essential modes and then project equations of motion on the resulting low-dimensional phase space to obtain a new smaller set of the differential equations on the reduced phase space (Lange et al. 2006; Stepanova 2007; Kamberaj 2011), which have the form of well-known generalized Langevin equation (Albers et al. 1971).

The detailed algorithm for PCA is as the following, according to Janezic and Brooks (1995) and Brooks et al. (1995) and, in particular, to Kamberaj (2017). First, we have to remove the translational and rotational motions of all system, then the structure of a system of N atoms at each time step of the MD trajectory is represented by single vector

$$\mathbf{q}(t) = (q_1(t),\ q_2(t),\ \cdots,\ q_{3N}(t))^T$$

where

$$q_i(t) \equiv q_i(t) - \langle q_i(t) \rangle, \quad (i = 1,\ 2,\ \cdots,\ 3N)$$

are representing the fluctuations of atomic positions. These representative points are distributed in the phase space during a MD trajectory production. Based on the quasi-harmonic approach (Karplus and Jushick 1981), the distribution of the fluctuations of the coordinates can be described by a multivariate Gaussian distribution with covariance matrix $\boldsymbol{\sigma}$ defined as

$$\sigma_{ij} = \langle q_i(t) q_j(t) \rangle$$

where $\langle \cdots \rangle$ denotes an ensemble average. The covariance matrix can also be expressed in terms of the mass weighted coordinates,

$$\tilde{q}_i(t) = \sqrt{m_i} q_i(t)$$

to obtain the mass weighted covariance matrix as

$$\tilde{\sigma}_{ij} = \sqrt{m_i m_j} \langle q_i(t) q_j(t) \rangle$$

Both, $\tilde{\boldsymbol{\sigma}}$ and $\boldsymbol{\sigma}$ are $3N \times 3N$ symmetric matrices, which are then diagonalised to obtain a new set of coordinates,

$$\mathbf{X} = (x_1,\ X_2,\ \cdots,\ X_{3N})^T$$

which are given as

$$X_j = \sum_{i=1}^{3N} E_{ij} q_i$$

with $\mathbf{E}$ being the orthogonal matrix whose i-th column is the i-th eigenvector of $\tilde{\boldsymbol{\sigma}}$, which are also called *principal components* (PC). The eigenvalues of $\tilde{\boldsymbol{\sigma}}$, λ_i represent the mean-square fluctuations along each principal axis:

$$\lambda_i = \langle X_i^2(t) \rangle = \mathbf{e}_i^T \tilde{\boldsymbol{\sigma}} \mathbf{e}_i$$

where $\mathbf{e}_i$ is the i-th eigenvector of $\tilde{\boldsymbol{\sigma}}$.

In the coarse-grained model, M coordinates are selected (Grubmüller 1995) with the largest eigenvalues, namely

$$\mathbf{c}(t) = (X_1(t), \cdots, X_M(t))^T$$

These coordinates are called *collective coordinates*. This choice is based on the observation that the remaining X_i $(i = M+1, \cdots, 3N)$ with small eigenvalues describe localized, high frequency, nearly harmonic vibration modes with small amplitudes, which are expected not to reflect the conformational transitions. In contrast, the dynamics of M conformational coordinates is slow, essentially non-harmonic and it is known to dominate the collective motion in proteins (Go 1990; Kitao et al. 1991; Garcia 1992; Amadei et al. 1993; Aalten et al. 1993; Grubmüller 1995). Therefore, it is assumed that with this choice we capture the relevant degrees of freedom of the protein dynamics. The number M of the conformational degrees of freedom, which are explicitly considered, will determine the level of coarse-graining.

8.2.1 Diffusive Motion in a Protein

The diffusive motion in proteins is described here by the Green-Kubo expression, which relates the correlation function integrals to the transport coefficients (Balucani and Zoppi 1994):

$$D_m(t) = \frac{1}{d} \int_0^{\tau} d\tau \langle V_m(t_0) V_m(t_0 + \tau) \rangle_{t_0}$$

where $\mathbf{V}$ denotes the velocity vector of the collective coordinates

$$\mathbf{V} = (\dot{c}_1, \cdots, \dot{c}_M)^T \tag{8.21}$$

d is the dimension and $\langle \cdots \rangle_{t_0}$ denotes an ensemble average over many starting times t_0 and t is the MD simulation time. In Eq. (8.21), the term

$$C_m^{VV}(t) = \langle V_m(t_0) V_m(t_0 + \tau) \rangle_{t_0}$$

represents the velocity auto-correlation function.

A double diffusion model (Amadei et al. 1999a), is used to characterize diffusive motion of collective coordinates, that is, short time diffusion D_0 within a configuration space region which can be approximated by a single harmonic well, followed by a diffusion between harmonic well regions that can be approximated by a long time diffusion constant D_∞. In this model a theoretical curve can be used to model the short time decay of the negative tail of the auto-correlation function

$$C_{VV}^{fit}(t) = C_0 \exp\left(-\frac{t-\tau_0}{\tau_c}\right) \tag{8.22}$$

where τ_0 corresponds to fast relaxation time, i.e. the time when first $C_{VV}(t)$ is negative, τ_c is the slow relaxation time and C_0 is a constant. Equation (8.22) is valid only for $t \geq \tau_0$. Integrating expression (8.22), a general expression for the double diffusion model is found (Amadei et al. 1999a):

$$D^{fit}(t) = (D_0 - D_\infty)\exp\left(-\frac{t-\tau_c}{\tau_c}\right) + D_\infty \tag{8.23}$$

which similarly is valid only for $t \geq \tau_0$. In Eq. (8.23), the first term describes the short time diffusion and D_∞ the long-time diffusion constant, and $D_0 > D_\infty$. First τ_0 can be defined from $C_{VV}(t)$ as the time when it first becomes negative, then using a nonlinear least squares fitting procedure of $D(t)$, τ_c and D_∞ can be calculated (Kamberaj 2011). The short time diffusion coefficient is calculated as $D_0 = D(\tau_c)$.

8.2.2 *Stability of PCA*

To measure the fraction of the atomic fluctuations captured by a given subset of the principal components, e.g. $\{\mathbf{e}_i\}$, the quantity $\Gamma(M)$ is used as proposed in Lange and Grubmüller (2006)

$$\Gamma(M) = \langle \frac{|| P(\mathbf{q}) ||^2}{|| \mathbf{q} ||^2} \rangle \tag{8.24}$$

where $|| \cdot ||$ denotes the norm of the vector. $|| P(\mathbf{q}) ||$ is given as

$$|| P(\mathbf{q}) || = \left(\sum_{i=1}^{M} X_i^2\right)^{1/2}$$

and

$$|| \mathbf{q} || = \left(\sum_{i=1}^{3N} X_i^2\right)^{1/2}$$

The full-length trajectory is divided into time windows with length, let say T_w, then the subspaces from these time windows are computed to obtain measures of the similarity, $\Gamma_i(M)$. The similarity can then be defined as an average over all number of the time windows, N_w:

$$\langle \Gamma(M) \rangle = \frac{1}{N_w} \sum_{i=1}^{N_w} \Gamma_i(M) \tag{8.25}$$

The standard deviation of the similarity is estimated as (Lange and Grubmüller 2006)

$$\sigma_\Gamma(M) = \left(\frac{1}{N_w - 1} \sum_{i=1}^{N_w} (\Gamma_i(M) - \langle \Gamma_i(M) \rangle)^2 \right)^{1/2} \tag{8.26}$$

Note that on writing Eqs. (8.25) and (8.26) we have assumed that all subspaces are equivalent, that is, they have the same statistical weights. But this should depend on the length of the time window of each fragment. To examine the convergence of similarity on each time window, we considered it, both as a function of trajectory length t and subspace dimension M, $\Gamma(t, M)$, where

$$t = T_w,\ 2T_w,\ \cdots,\ T$$

with T being the length of full trajectory and $M = 1,\ 2,\ 3,\ \cdots$.

The convergence establishes for sufficiently large PCA subspaces. Besides, this convergence is obtained only if a relatively long MD simulation is obtained, which is not always available. However, in the context of our discussion, the quality of the chosen PCA subspace has to be established based on the short run of MD simulations available (Lange and Grubmüller 2006). The inner product matrix **P** discussed elsewhere (Amadei et al. 1993) is a good alternative for comparing the basis vectors of two subspaces. However, the disadvantage of it is that all directions are weighted equally, which is not the case when measuring Γ because in the last one the ensemble average ensures that less essential directions of the subspace also have less contribution to Γ.

Lange and Grubmüller (2006) suggested to compute the mutual similarity $\tilde{\Gamma}$ for two adjacent fragments of equal length: a PCA was carried out for the first fragment, and the similarity is calculated, with the ensemble average $\langle \cdots \rangle$ replaced by an average overall configuration of the second fragment. This measure, $\tilde{\Gamma}$, as well as the similarity, Γ, itself, depends on the chosen subspace dimension M and will never reach unity. Instead, this measure allows one to judge how accurate a PCA subspace of a particular dimension might describe the right ensemble.

Usually, PCA is carried out on subsets of the protein atoms, such as C_α or backbone atoms.

8.3 Equations of Motion of Collective Coordinates

In this section, we will derive a smaller set of the differential equations from the projections of the main dynamics equations of motion of a system. In general, these equations can be used to describe the dynamics of the so-called collective coordinates **X** similar to Newton's equations describing the motion of **y** coordinates,

$$\mathbf{m}\ddot{\mathbf{y}} = -\frac{\partial U(\mathbf{y})}{\partial \mathbf{y}} \tag{8.27}$$

Here, **m** represents a diagonal matrix of the particle masses along each degree of freedom and $U(\mathbf{y})$ is the potential energy function of **y**. The projection of the force can be written as:

$$\mathbf{F}_{\mathbf{y}} = \left(-\frac{\partial U}{\partial y_1}, -\frac{\partial U}{\partial y_2}, \dots, -\frac{\partial U}{\partial y_{3N}}\right)^T$$

This force is acting along each of the degrees of freedom of the system, determined by the eigenvectors of the correlation matrix **C** as

$$\mathbf{F}_{\tilde{\mathbf{X}}} = -\mathbf{E}^T \frac{\partial U}{\partial \mathbf{y}} = \mathbf{E}^T \mathbf{m}\ddot{\mathbf{y}}, \tag{8.28}$$

where the relation given by Eq. (8.27) is used. Besides, we can write that

$$\mathbf{F}_{\tilde{\mathbf{X}}} = \tilde{\mathbf{M}}^{\text{eff}} \ddot{\tilde{\mathbf{X}}}$$

where $\tilde{\mathbf{M}}^{\text{eff}}$ is an effective mass matrix. It can easily be shown using the relation $\mathbf{X} = \mathbf{E}^T \mathbf{y}$ that we can express this effective mass matrix as

$$\tilde{\mathbf{M}}^{\text{eff}} = \mathbf{E}^T \mathbf{m} \mathbf{E}. \tag{8.29}$$

In this form, $\tilde{\mathbf{M}}^{\text{eff}}$ is not diagonal matrix, but it is completely determined in terms of the diagonal matrix representing the particles' masses and principal components of the correlation matrix. Hence, there exist a reference frame, where the effective mass matrix is a diagonal matrix. Note that if all the real degrees of freedom of system have the same mass, which is often the case,[1] then the effective mass matrix, $\tilde{\mathbf{M}}^{\text{eff}}$, is diagonal by definition. Moreover, to have a diagonal effective mass matrix, $\tilde{\mathbf{M}}^{\text{eff}}$, it is also suitable from the computations point of view, as it will be shown in the following. Denoting **V** a $3N \times 3N$ orthogonal matrix of eigenvectors, which represents the space in which $\tilde{\mathbf{M}}^{\text{eff}}$ is diagonal, we can obtain

[1] This could be the case when, for example, only the C_α atoms are considered in the PCA.

$$\mathbf{M}^{\text{eff}} = \mathbf{V}^T \tilde{\mathbf{M}}^{\text{eff}} \mathbf{V},$$

Now, $\mathbf{M}^{\text{eff}}$ is a diagonal matrix, where each diagonal element corresponds to the effective mass of the coordinate $\tilde{X}_i$, $(i = 1, 3N)$. Here, of course, we consider only M of the principal components, corresponding to the largest eigenvalues of the correlation matrix $\mathbf{C}$, and hence only the first M diagonal elements of $\mathbf{M}^{\text{eff}}$ are taken into account. Besides, we project the collective coordinates vector $\mathbf{X}$ on the principal frame of the matrix $\tilde{\mathbf{M}}^{\text{eff}}$, which is denoted as $\mathbf{X} \equiv \mathbf{V}^T\mathbf{X}$. Here, we consider the first M columns of $\mathbf{V}^T$ only.

In general, in the PCA analysis, we consider a smaller number of the degrees freedom $(M \ll 3N)$, called here collective coordinates. These collective coordinates can be used to characterize the collective motion of the macromolecules, such as a protein, and they determine the essential subspace of the macromolecules – the degrees of freedom coupled to the other system's degrees of freedom. For example, if a macromolecule immerses in a solvent environment, then the solvent degrees of freedom and the slow degrees of freedom of the macromolecule from the so-called *environmental degrees of freedom*.

In the following, we will present a harmonic bath model (Kamberaj 2011), which, in general, can be employed to describe the weak coupling to the environmental degrees of freedom (Benguria and Kac 1981; Ford and Kac 1987; Ford et al. 1988; Hänggi and Ingold 2005). We will apply the classical framework, and we will show that based on this formalism we can derive the so-known generalized Langevin equation (Benguria and Kac 1981; Ford and Kac 1987; Ford et al. 1988).

8.3.1 Bath of Harmonic Oscillators

According to this approach, all the other degrees of freedom named here environmental degrees of freedom, coupled with the essential degrees of freedom, represent slow degrees of freedom of the system and the solvent degrees of freedom. A set of harmonic oscillators here replaces them. Let us consider $X_1, \cdots, X_M$ degrees of freedom of interest, coupled linearly to the bath, then the Hamiltonian of this system is written in the form

$$H = \sum_{m=1}^{M} \frac{P_m^2}{2M_m^{\text{eff}}} + U(X_1, \cdots, X_M) + \sum_i \left[\frac{p_i^2}{2m_i} + \frac{K_i x_i^2}{2} + \sum_{m=1}^{M} \left(\Gamma_{i,m} x_i X_m + \frac{\mu_{i,m} X_m^2}{2} \right) \right], \tag{8.30}$$

In Eq. (8.30), the first two terms represent the kinetic and the potential energy function of the collective coordinates. While the other terms, namely the third, fourth and fifth, represent the kinetic and the potential energy of the bath (the index i

runs over all the harmonic oscillators), and the linear coupling between the collective coordinates and the bath of the harmonic oscillator, respectively. The last term, in Eq. (8.30), represents the potential energy of the collective coordinates in the bath environment. In Eq. (8.30), P_m denotes the conjugate momentum of X_m, p_i is the conjugate momentum of the harmonic oscillator coordinate x_i, m_i is the harmonic oscillator mass, and K_i is the force constant of the harmonic oscillator, related to the harmonic frequency, ω_i as $\omega_i^2 = K_i/m_i$. $\Gamma_{i,m}$ is the coupling constant between the bath and the coordinate X_m and $\mu_{i,m}$ is the force constant of the harmonic potential energy of the collective coordinate in the bath, for which we will give an explicit expression in the following discussion.

The equations of motion will be obtained by solving the Hamilton's equations of motion represented in Chap. 1

$$\dot{Q} = \frac{\partial H}{\partial P}, \quad \dot{P} = -\frac{\partial H}{\partial Q} \tag{8.31}$$

Here, P is the conjugate momentum of the canonical coordinate Q. Using the harmonic bath Hamiltonian, Eq. (8.30), the equations of motion can be written as:

$$\begin{aligned} \dot{X}_m &= P_m/M_m^{\text{eff}}, \\ \dot{P}_m &= -\frac{\partial U}{\partial X_m} - \sum_i \left(\Gamma_{i,m} x_i + \mu_{i,m} X_m\right), \\ \dot{x}_i &= p_i/m_i, \\ \dot{p}_i &= -K_i x_i - \sum_{m=1}^{M} \Gamma_{i,m} X_m. \end{aligned} \tag{8.32}$$

As the second derivatives, respectively $\ddot{X}_m$ and $\ddot{x}_i$, the equations of motion (Eq. (8.32)) can be written as

$$\begin{aligned} M_m^{\text{eff}} \ddot{X}_m &= -\frac{\partial U}{\partial X_m} - \sum_i \left(\Gamma_{i,m} x_i + \mu_{i,m} X_m\right), \\ m_i \ddot{x}_i &= -K_i x_i - \sum_{m=1}^{M} \Gamma_{i,m} X_m. \end{aligned} \tag{8.33}$$

In general, the system of the equations given by Eq. (8.33), can be solved in this way: first, solving the second equation with respect to x_i, and then replacing it into the first differential equation that will give a second order differential equation with respect to X_m.

The second order differential equation with respect to x_i (see Eq. (8.33)) represents a driven undamped classical harmonic oscillator equation. In Kamberaj

(2011), it is suggested to solve it by taking the Laplace transform of both sides giving:

$$(s^2+\omega_i^2)\tilde{x}_i(s) = sx_i(0) + \dot{x}(0) - \sum_{m=1}^{M} \frac{\Gamma_{i,m}}{m_i}\tilde{X}_m(s) \tag{8.34}$$

Here, $\tilde{x}_i(s)$ and $\tilde{X}_m(s)$ are the Laplace transforms of $x_i(t)$ and $X_m(t)$, respectively. Dividing both sides by $s^2+\omega_i^2$ and taking the inverse Laplace transform, the solution, $x_i(t)$, of the Eq. (8.34) with respect to time t can be obtained as (Kamberaj 2011):

$$x_i(t) = x_i(0)\int_0^\infty \frac{se^{st}}{s^2+\omega_i^2}ds + \dot{x}_i(0)\int_0^\infty \frac{e^{st}}{s^2+\omega_i^2}ds \tag{8.35}$$
$$- \sum_{m=1}^{M} \frac{\Gamma_{i,m}}{m_i}\int_0^\infty \frac{\tilde{X}_m}{s^2+\omega_i^2}ds.$$

Using the convolution theorem of the Laplace transform as:

$$\int_0^\infty e^{-st}dt\int_0^t f(\tau)g(t-\tau)d\tau = \tilde{f}(s)\tilde{g}(s),$$

we can obtain an expression for $x_i(t)$ as (Kamberaj 2011)

$$x_i(t) = x_i(0)\cos\omega_i t + \frac{\dot{x}_i(0)}{\omega_i}\sin\omega_i t \tag{8.36}$$
$$- \sum_{m=1}^{M} \frac{\Gamma_{i,m}}{m_i\omega_i}\int_0^t d\tau X_m(\tau)\sin\omega_i(t-\tau).$$

The friction kernel can also be obtained as a function of the velocity $\dot{X}_m$ by writing

$$\int_0^t d\tau X_m(\tau)\sin\omega_i(t-\tau) = \frac{1}{\omega_i}\left[X_m(t) - X_m(0)\cos\omega_i t\right. \tag{8.37}$$
$$\left. - \int_0^t d\tau \dot{X}_m(\tau)\cos\omega_i(t-\tau)\right].$$

and then $x_i(t)$, given by the Eq. (8.36), into the first expression of Eq. (8.33) and taking $\mu_{i,m} \equiv \lambda_{i,m}^2/(m_i\omega_i^2)$ (Kamberaj 2011). These manipulations yield a second order differential equation for X_m in this form

$$M_m^{\text{eff}} \ddot{X}_m = -\frac{\partial U}{\partial X_m} - \int_0^t d\tau \dot{X}_m(\tau)\xi_m(t-\tau) + R_m(t), \tag{8.38}$$

where

$$\xi_m(t) = \sum_i \frac{\Gamma_{i,m}^2}{m_i \omega_i^2} \cos \omega_i t, \tag{8.39}$$

is the so-called *dynamic friction kernel* and

$$R_m(t) = -\sum_i \Gamma_{i,m} \left[\left(x_i(0) + \frac{\Gamma_{i,m}}{m_i \omega_i^2} X_m(0) \right) \cos \omega_i t \right. \tag{8.40}$$
$$\left. + \frac{p_i(0)}{m_i \omega_i} \sin \omega_i t, \right]$$

is the so-called *random force* (Kamberaj 2011). Equation (8.38) is the so-called generalized Langevin equation in the literature (Benguria and Kac 1981; Ford and Kac 1987; Ford et al. 1988). Note that this equation for each coordinate X_m, see also Eq. (8.38), represents the dynamics of an one-dimensional particle moving under the force $-\frac{\partial U}{\partial X_m}$, driven by the random force $R_m(t)$ with a non-local in time dumping term $-\int_0^t d\tau \dot{X}_m(\tau)\xi_m(t-\tau)$, which depends on entire history of X_m.

The random term force $R_m(t)$ under the context of the formalism discussed here depends completely on the dynamics of the bath. However, for a large bath with many degrees of freedom, it might not be interesting following the trajectory in the phase space of all degrees of freedom, then we can define $R_m(t)$ as a random force. The stochastic behavior of the $R_m(t)$ can also be interpreted from the fact that it depends on the initial values of $x_i(0)$ and $X_m(0)$, Eq. (8.40). Hence, we can model $R_m(t)$ as a random process satisfying certain conditions. From the expression of $R_m(t)$, Eq. (8.40), we see that $R_m(t)$ does not depend on the $X_m(t)$, except at $t = 0$, $X_m(0)$.

It can easily be found that

$$\langle \dot{X}_m(0) R_m(t) \rangle = 0,$$

where $\langle \cdots \rangle$ denotes the ensemble average. This correlation arises from the fact that $R_m(t)$ does not depend on $\dot{X}_m(0)$. It can also be seen that

$$\langle R_m(0) R_m(t) \rangle = k_B T \xi_m(t),$$

where k_B is the Boltzmann's constant and T is the temperature of the system, which is the well known *fluctuation dissipation theorem*.

The friction kernel term $\int_0^t d\tau \dot{X}_m(\tau)\xi(t-\tau)$, which is also called memory kernel function since it depends on the time evolution of X_m. That, in physics terms, indicates that the bath requires a finite time to respond to any changes on the motion of the collective coordinate (X_m), which in turn indicates how the bath acts back to the system. Thus, the force of the bath upon the X_m depends on what the coordinate X_m did on the past. However, we expect that fluctuations of $\dot{X}_m$ with time decay fast to zero after a certain interval of time τ_0. Hence, what the coordinate did earlier that τ_0 will no longer affect the force acting on it, and the integral can be written as $\int_{\tau_0}^t d\tau \dot{X}_m(\tau)\xi_m(t-\tau)$. That indicates that also the kernel function $\xi_m(t)$ decays to zero with time. Next section shows that ξ_m exhibits a power law decay to zero.

We can consider two separate cases for the time behavior of the kernel function or dynamical friction coefficient $\xi_m(t)$. The first we consider the case when the bath responds infinitely quickly to any changes on the motion of the coordinate X_m, which can be seen as the case when $M_m^{\text{eff}} \ll m_i$. Then the bath retain no memory to what the coordinate X_m did on the past, and we can take the kernel function to be δ-function in time

$$\xi_m(t) = 2\xi_{0,m}\delta(t),$$

where $\xi_{0,m}$ is the static friction coefficient and is defined as

$$\xi_{0,m} = \int_0^\infty \xi_m(t)dt. \tag{8.41}$$

The friction kernel term becomes

$$\int_0^t d\tau \dot{X}_m(\tau)\xi_m(t-\tau) = 2\xi_{0,m} \int_0^t d\tau \dot{X}_m(\tau)\delta(t-\tau) = 2\xi_{0,m}\dot{X}_m(t),$$

and Eq. (8.38) can be written as

$$M_m^{\text{eff}}\ddot{X}_m = -\frac{\partial U}{\partial X_m} - 2\xi_{0,m}\dot{X}_m + R_m(t), \tag{8.42}$$

which is known as the Langevin equation and it is often used to describe Brownian motion (Coffey et al. 1996).

The second case would be the case of a bath which responds slowly to the fluctuations of the coordinate X_m, i.e. $m_i \ll M_m^{\text{eff}}$. In this case we can write $\xi_m(t) = \xi_m(t=0)$, which is constant all the time. Then the friction kernel integral becomes

$$\int_0^t d\tau \dot{X}_m(\tau)\xi_m(t-\tau) = \xi_m(0)\left(X_m(t) - X_m(0)\right),$$

and the Eq. (8.38) can be written as

$$M_m^{\text{eff}} \ddot{X}_m = -\frac{\partial}{\partial X_m}\left(U + \frac{1}{2}\xi_m(0)\left(X_m(t) - X_m(0)\right)^2\right) + R_m(t). \tag{8.43}$$

As one can see, the friction term is added to the potential energy function as a harmonic term, which has the effect of trapping the system in a local configuration space. As one can see, this case does not show any practical interest. In terms of the dynamical behavior of the collective coordinate X_m, it indicates that the motion between different potential wells has a very long relaxation time.

In the following, we are presenting a method for determining the dynamic friction as in Kamberaj (2011).

8.3.2 Dynamic Friction Coefficient

To calculate the dynamic friction coefficient $\xi_m(t)$, we use the Eq. (8.38) and the auto-correlation functions of the collective coordinates velocity, namely $\mathbf{V}$. First, we multiply both sides of the Eq. (8.38) by $V_m(0)$ and take the ensemble average. This yields:

$$M_m^{\text{eff}} \langle V_m(0)\dot{V}_m(t)\rangle = -\langle V_m(0)\frac{\partial U}{\partial X_m}\rangle - \int_0^t d\tau \langle V_m(0)V_m(\tau)\rangle \xi_m(t-\tau) \tag{8.44}$$
$$+ \langle V_m(0)R_m(t)\rangle$$

which can further be written as

$$M_m^{\text{eff}} \dot{C}_{VV}^m(t) = C_{FV}^m(t) - \int_0^t d\tau C_{VV}^m(\tau)\xi_m(t-\tau), \tag{8.45}$$

since $\langle V_m(0)R_m(t)\rangle = 0$, as shown above. In Eq. (8.45), C_{VV}^m denotes the velocity auto-correlation function and C_{FV}^m is the cross-correlation function between the force $(-dU/dX_m)$ and the velocity (V_m). The potential of the mean force, $U(X_m)$ is determined from the equilibrium distribution function of the system by integrating out all degrees of freedom, except the coordinate X_m, i.e., (Frenkel and Smit 2001)

$$e^{-U(X_m)/k_B T} = p(X_m) \tag{8.46}$$
$$= \frac{1}{Z_0}\int\int d\mathbf{P}d\mathbf{X}\, e^{-H(\mathbf{X},\mathbf{P})/k_B T}\delta(X_m - \tilde{X}(\mathbf{XP})),$$

Here, $p(X_m)$ is the equilibrium distribution function of the coordinate X_m, Z_0 is the partition function, and $\delta(X)$ is the Dirac-delta function, which guarantees that the integrand in Eq. (8.46) is non-zero only when the coordinate X has the desired value, i.e., $\tilde{X}(\mathbf{X}, \mathbf{P}) = X_m$.

The equilibrium MD simulations enable calculation of the equilibrium distribution function $p(X_m)$. It is often taken to be proportional to the logarithm of the binned histogram of the coordinate X_m sampled along the MD trajectory, and hence the potential of mean force is given by Kamberaj (2011)

$$U(X_m) = -k_B T \ln\left(p(X_m)\right). \tag{8.47}$$

Note that the equilibrium MD simulations sample only a restricted region of the coordinate X_m domain of interest, usually within the vicinity of the potential of mean force minimum, and the direct application of Eq. (8.46) is, in practice, a crude approximation. To properly sample the regions of the phase space, which are energetically more difficult to reach, we may need to guide the system towards those regions by employing more accurate methods, as discussed in Park and Schulten (2004). Besides, Eq. (8.47) gives, in fact, the Gibbs free energy, G, in constant N, P, T MD simulations, which is related to the potential energy according to $G = U - TS$ (if we ignore the fluctuations of the term pV), where S is the entropy and T the temperature. Our approximation, $\partial U/\partial X = \partial G/\partial X$ is valid only in the cases when the entropy changes very slowly with X, and hence the term TS is to a good approximation a constant.

Calculating the correlation functions from MD simulations and substituting them into the integral Eq. (8.45), followed by numerical integration, gives $\xi_m(t)$. Alternatively, one can evaluate the Laplace transformation of both sides of Eq. (8.45), giving

$$\tilde{\xi}_m(s) = -\frac{M_m^{\text{eff}}\left(s\tilde{C}_{VV}(s) - C_{VV}(0)\right) - \tilde{C}_{FV}(s)}{\tilde{C}_{VV}(s)} \tag{8.48}$$

Here, $\tilde{\xi}_m(s)$, $\tilde{C}_{VV}(s)$ and $\tilde{C}_{FV}(s)$ denote the Laplace transformations of $\xi_m(t)$, $C_{VV}(t)$ and $C_{FV}(t)$, respectively. Then, $\xi_m(t)$ can be solved via Laplace inversion, which exists if $\tilde{C}_{VV}(s) \neq 0$.

To calculate the static friction coefficient $\xi_{0,m}$, we can consider the integral given by Eq. (8.41) as a function of time and taking its limit for $t \to \infty$

$$\xi_{0,m} = \lim_{t\to\infty} \int_0^t d\tau \xi_m(\tau). \tag{8.49}$$

Alternatively, one can use the Einstein diffusion equation (Einstein 1926; Chandrasekhar 1949; Islam 2004), which relates the diffusion coefficient, D_m, of the collective coordinate X_m with $\xi_{0,m}$ as

$$D_m = k_B T/\xi_{0,m}. \tag{8.50}$$

Moreover, one can use the Green-Kubo equation relating the integrals of the correlation function to the transport coefficients (Balucani and Zoppi 1994):

$$D_m = \frac{1}{d} \lim_{t\to\infty} \int_0^t d\tau \langle V_m(t_0) V_m(t_0+\tau)\rangle, \tag{8.51}$$

where d is the number of diffusion dimensions. Equation (8.51) can be derived from the Einstein formula (Einstein 1926), which relates the diffusion coefficient with the average square displacement of X_m as

$$D_m = \frac{1}{2d} \lim_{t\to\infty} \frac{1}{t} \langle \mid X_m(t_0+t) - X_m(t_0) \mid^2 \rangle, \tag{8.52}$$

In both, Eqs. (8.51) and (8.52), $\langle \cdots \rangle$ denotes an ensemble average over many starting times t_0 and t is the time length of the molecular dynamics trajectory. Recently, diffusion calculation results have been presented for Lennard-Jones liquids (Yulmetyev et al. 2003; Mokshin et al. 2005). It is interesting to note that Eqs. (8.51) and (8.52) are valid for the limit of long simulation run ($t \to \infty$). Therefore, D_m cannot be calculated locally in time (short trajectory), but only globally, i.e. the molecular dynamics simulations of the system have to run for a very long time to capture all possible correlations between slow and fast modes on the system.

Equation (8.51) or Eq. (8.52) describes the relations between the diffusion and velocity auto-correlation function or with the mean square displacement in the case of the simple diffusion models, where $C_{VV}(t)$ is a rapidly decreasing function converging to zero within an approximate time interval of τ_c. Furthermore, the integral in Eq. (8.51) in a typical simple model of diffusion is a rapidly increasing function converging to a positive value reached at a time interval of τ_c. Kamberaj (2011) noticed that the velocity auto-correlation functions of the collective coordinates show a double long-time decay behavior. First, there was a fast decrease to a negative value within a short time interval of τ_0, then a slow decay to zero of a negative tail for a longer time interval of τ_c. Note that the dense liquids also show this slowly decaying negative tail of the velocity auto-correlation function (Wainwright et al. 1971; Gallo et al. 2000), and the same behavior is also observed in describing the kinetics of the essential subspace of protein motion (Amadei et al. 1999b).

According to the harmonic bath model, described in the previous section, the short-time decay to zero of the negative tail of C_{VV} can be interpreted as a slowing force acting on the collective coordinates, $X_1, \cdots, X_M$, by the harmonic bath, which is the time needed by the bath to equilibrate. That, of course, will result in slow motion of the collective coordinates and thus decrease of their diffusions. In contrary, the fast decrease to a negative value of the C_{VV} can be seen as a free motion of the collective coordinate from the bath, and the short interval τ_0 is the time that bath needs to respond to X_m.

Here, we are suggesting the same theoretical framework as in Amadei et al. (1999b), which corresponds to a double diffusion model, i.e., a short time diffusion D_0 within a configuration space region which can be approximated by a single harmonic well, followed by a diffusion between harmonic well regions that can be approximated by a long time diffusion constant D_∞, as discussed above (Kamberaj 2011).

It is important to mention that the full dynamics of the collective coordinates, similarly to the internal degrees of freedom of the system, can not be described by this model to full equilibrium because, due to the time limits, the coordinates have not sampled the whole available phase space. Therefore, one can not be sure that D_∞ describes the dynamical behavior of $D(t)$ at the limit of a very long time.

8.4 Analyzing Slow Collective Variables Using Machine Learning Approach

In this section, we will introduce an improved version of the auto-encoder machine learning approach to the algorithm of determining the collective variables from molecular dynamics simulation data.

Machine Learning (ML) approach provides a potential method to predict the properties of a system using decision-making algorithms, based on some predefined features characterizing these properties of the system. There exist different ML methods used to predict missing data and discover new patterns during the data mining process (McCulloch and Pitts 1943). Neural networks method considers a large training dataset, and then it tries to construct a system, which is made up of rules for recognizing the patterns within the training data set by a learning process.

In general, for an ANN with K hidden layers (see also Fig. 8.1), the output Y_i is defined as

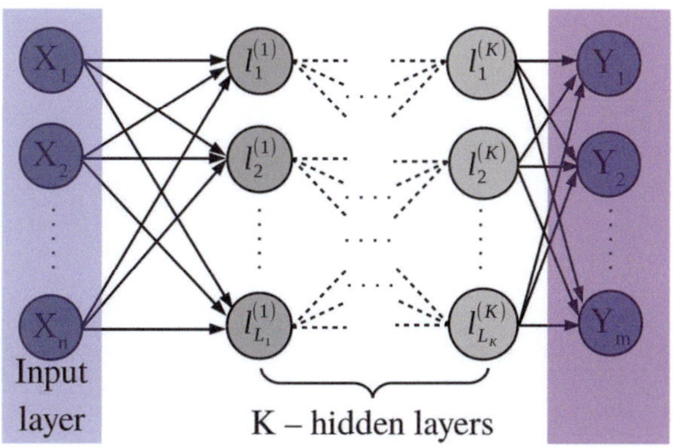

Fig. 8.1 Illustration diagram of an artificial neural network (ANN). It is characterized by an input vector of dimension n, K hidden layers of $l_{L_1}^{(1)}, l_{L_2}^{(2)}, \cdots, l_{L_K}^{(K)}$ neurons each, and an output vector of dimension m

$$Y_i = f\left(\underbrace{\sum_{l_K=1}^{L_K} f\left(\underbrace{\sum_{l_{K-1}=1}^{L_{K-1}} f\left(\underbrace{\cdots f\left(\underbrace{\sum_{l_2=1}^{L_2} f\left(\underbrace{\sum_{l_1=1}^{L_1} f\left(\underbrace{\sum_{j=1}^{n} X_j W_{jl_1} + b_{l_1}}_{\text{input layer}} \right) W_{l_1 l_2} + b_{l_2}}_{\text{1st hidden layer}} \right) \cdots}_{\text{2nd hidden layer}} \right) \cdots}_{\cdots} \right) W_{l_{K-1} l_K} + b_{l_K}}_{(K-1)\text{th hidden layer}} \right) W_{l_K i} + b_i}_{K\text{th hidden layer}} \right) \tag{8.53}$$

Here, **W** and **b** are considered as free parameters, which need to be optimized for a given training data used as inputs and given outputs, which are known. To optimize these parameters the so-called loss function is minimized using Gradient Descent method (Qian 1999):

$$S(\mathbf{W}, \mathbf{b}) = \sum_{i=1}^{m} \left(Y_i^0 - Y_i \right)^2 \tag{8.54}$$

where $\mathbf{Y}^0$ represent the true output vector. For that, the gradients of $S(\mathbf{W}, \mathbf{b})$ with respect to **W** and **b** are calculated (Qian 1999):

$$\Delta\mathbf{W} = -\left(\frac{\partial S(\mathbf{W}, \mathbf{b})}{\partial \mathbf{W}} \right)_{\mathbf{b}} \tag{8.55}$$

$$\Delta\mathbf{b} = -\left(\frac{\partial S(\mathbf{W}, \mathbf{b})}{\partial \mathbf{b}} \right)_{\mathbf{W}}$$

In order to avoid overfitting, which is one of pitfalls of the machine learning approaches (Srivastava et al. 2014), the following regularization terms have been introduced in literature:

$$\Delta'\mathbf{W} = \gamma_w\,(\Delta\mathbf{W} + \gamma_1\mathbf{W}) \tag{8.56}$$
$$\Delta'\mathbf{b} = \gamma_w\,(\Delta\mathbf{b} + \gamma_1\mathbf{b})$$

where γ_w is called learning rate for the gradient and γ_1 is called the regulation strength.

Because the Gradient Descent method often converges to a local minimum, it provides a local optimization to the problem. To avoid this pitfall, we are going to introduce a new approach, called here as Bootstrapping Swarm Artificial Neural Network (BSANN).

8.4.1 Bootstrapping Swarm Artificial Neural Network

The standard ANN method deals with random numbers, which are used to initialize the parameters **W** and **b**; therefore, the optimal solution of the problem will be different for different runs. In particular, we can say that there exists an uncertainty in the calculation of the optimal solution (i.e., in determining **W** and **b**.) To calculate these uncertainties in the estimation of the optimal parameters, **W** and **b**, we introduce a new approach, namely bootstrapping artificial neural network based on the method proposed by Gerhard Paass (1993). In this approach, M copies of the same neural network are run independently using different input vectors. Then, at regular intervals, we swap optimal parameters (i.e., **W** and **b**) between the two neighboring neural networks. Figure 8.2 shows the layout of this configuration.

Furthermore, to achieve a good sampling of the phase space extended by the vectors **W** and **b**, we introduce two other regularization terms similar to the swarm-particle sampling approach. First, we define two vectors for each neural network, namely $\mathbf{W}_n^{\mathrm{Lbest}}$ and $\mathbf{b}_n^{\mathrm{Lbest}}$, which represent the best local optimal parameters for each neural network n. In addition, we also define $\mathbf{W}^{\mathrm{Gbest}}$ and $\mathbf{b}^{\mathrm{Gbest}}$, which represent the global best optimal parameters among all neural networks.

Then, the expressions in Eq. (8.56) are modified by introducing these two regularization terms as the following:

$$\begin{aligned}
\Delta''\mathbf{W}_n &= \gamma_w\,(\Delta\mathbf{W}_n + \gamma_1\mathbf{W}_n \\
&\quad - \gamma_2 U(0,1)\left(\mathbf{W}_n - \mathbf{W}_n^{\mathrm{Lbest}}\right) \\
&\quad - \gamma_3 U(0,1)\left(\mathbf{W}_n - \mathbf{W}^{\mathrm{Gbest}}\right)) \\
\Delta''\mathbf{b}_n &= \gamma_w\,(\Delta\mathbf{b}_n + \gamma_1\mathbf{b}_n \\
&\quad - \gamma_2 U(0,1)\left(\mathbf{b}_n - \mathbf{b}_n^{\mathrm{Lbest}}\right) \\
&\quad - \gamma_3 U(0,1)\left(\mathbf{b}_n - \mathbf{b}^{\mathrm{Gbest}}\right))
\end{aligned} \tag{8.57}$$

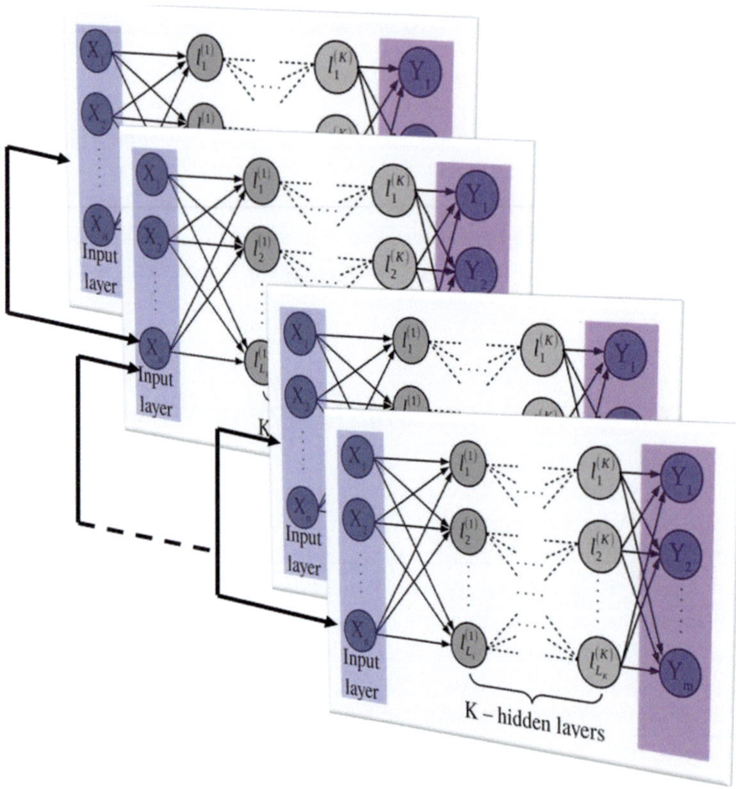

Fig. 8.2 Layout of the Bootstrapping Swarm Artificial Neural Network (BSANN). It is characterized by M different input vectors each of dimension n, K hidden layers of $l_{L_1}^{(1)}, l_{L_2}^{(2)}, \cdots, l_{L_K}^{(K)}$ neurons each, and M different output vectors each of dimension m. Every two neighboring neural networks communicate regularly with each other by swapping the optimized parameters

for each neural network configuration n, $n = 1, 2, \cdots, M$. Here, $U(0, 1)$ is a random number between zero and one, and γ_2 and γ_3 represent the strength of biases toward the local best optimal parameters and global best optimal parameters, respectively. The first term indicates the individual knowledge of each neural network and the second bias term the social knowledge among the neural networks. This method is called here, Bootstrapping Swarm Artificial Neural Network (BSANN). Then, the weights, $\mathbf{W}_n$, and biases, $\mathbf{b}_n$, for each neural network n are updated at each iteration step according to:

$$\mathbf{W}_n^{\text{new}} = \mathbf{W}_n^{\text{old}} + \Delta''\mathbf{W}_n \tag{8.58}$$

$$\mathbf{b}_n^{\text{new}} = \mathbf{b}_n^{\text{old}} + \Delta''\mathbf{b}_n$$

8.4.2 Related Work

Recently, neural network method has seen a broad range of applications in molecular modeling. In Lubbers et al. (2018), a hierarchical interacting particle neural network approach is introduced using quantum models to predict molecular properties. In this approach different hierarchical regularization terms have been introduced to improve the convergence of the optimized parameters. While in Gastegger et al. (2018), the machine learning like-potentials are used to predict molecular properties, such as enthalpies or potential energies. A discussion about the degree to which the general features included in characterizing the chemical space of molecules to improve the predictions of these models is in Goh et al. (2018) and Collins et al. (2018). Tuckerman and co-workers (Schneider et al. 2017) used a stochastic neural network technique to fit high-dimensional free energy surfaces characterized by reduced subspace of collective coordinates. While very recently (Kamath et al. 2018), a comparison study has been performed between neural network approach and Gaussian process regression to fit the potential energy surfaces. One of the recognized problems in using machine learning approaches in predicting free energy surfaces is the inaccurate representation of general features of the surface topology by the training data. To improve on this, a combination of metadynamics molecular dynamics with neural network chemical models have also been proposed (Herr et al. 2018). It is worth noting that in the prediction of free energy surfaces, an accurate representation of the reduced subspace can be important. For that, Wehmeyer and Noé (2018) have used the time-lagged auto-encoder to determine essential degrees of freedom of dynamical data.

Machine learning approaches have also been in the field of drug-design, for instance, in predicting drug-target interactions (Chen et al. 2018), and it is a promising approach. In particular, the method is used in combination with molecular dynamics to predict the ligand-binding mechanism to purine nucleoside phosphorylase (Decherchi et al. 2015) by capturing the mechanism of drug-target binding modes accurately.

8.4.3 Time-Lagged Auto-encoder Approach

Consider the vector $\mathbf{Q}_T$, which represents T configurations of a trajectory of the system:

$$\mathbf{Q}_T = \{\mathbf{q}(0) \to \mathbf{q}(1) \to \cdots \to \mathbf{q}(T-1)\}$$

where $\mathbf{q}(t)$ (for $t = 0, 1, \cdots, T-1$) represents the coordinates of a configuration of the system of $3N$ degrees of freedom (with N being the number of atoms):

$$\mathbf{q}(t) = (q_1(t), q_2(t), \cdots, q_{3N}(t))$$

This forms a Markovian chain of the states of a stationary stochastic random process visited by the system during molecular dynamics simulation. The problem is to find a reduced M-dimensional space ($M < 3N$), which compresses the data. This is suggested here by determining an encoding function as

$$E : R^{3N} \rightarrow R^{M} \tag{8.59}$$

and a decoding function as the following:

$$D : R^{M} \rightarrow R^{3N} \tag{8.60}$$

Eq. (8.59) provides a non-linear mapping using the Bootstrapping Swarm Artificial Neural Network of the Cartesian coordinates $\mathbf{q}(t)$ as:

$$\mathbf{X}(t) = E(\mathbf{q}(t))$$

where $\mathbf{X}(t)$ is an M-dimensional vector in the essential subspace of slow collective variables:

$$\mathbf{X}(t) = (X_1(t),\ X_2(t),\ \cdots,\ X_M(t))$$

Then, similarly, using the non-linear mapping D we obtain an approximate time-lagged signal, $\tilde{\mathbf{q}}(t+\tau)$:

$$\tilde{\mathbf{q}}(t+\tau) = D(\mathbf{X}(t))$$

This aims on average to minimize the error using the variation principle:

$$S = \min_{E,\,D} \sum_{t=0}^{T-1-\tau} ||\mathbf{q}(t+\tau) - D\left(E\left(\mathbf{q}(t)\right)\right)||^2 \tag{8.61}$$

Here, τ is the time-lag of the input signal $\mathbf{q}$, and the approach is called time-lagged auto-encoder. For $\tau = 0$, the approach represents the standard auto-encoder method. Note that both the input and output signal of the encoder-decoder non-linear neural network is the trajectory $\mathbf{q}$ in the Cartesian space and the output signal of the encoder, which is the input signal for the decoder, represent the slow collective variables $\mathbf{X}$.

The following steps have been suggested (Wehmeyer and Noé 2018) to create the input signals. First, two new signals are reconstructed using the Cartesian space vectors:

$$\mathbf{x}(t) = \mathbf{q}(t) - \frac{1}{T-\tau} \sum_{k=0}^{T-1-\tau} \mathbf{q}(k) \tag{8.62}$$

$$\mathbf{y}(t) = \mathbf{q}(t+\tau) - \frac{1}{T-\tau} \sum_{k=0}^{T-1-\tau} \mathbf{q}(k+\tau)$$

The convariance matrices are constructed as the following:

$$\mathbf{C}_1 = \frac{1}{T-\tau}\sum_{t=0}^{T-1-\tau} \mathbf{x}(t)\mathbf{x}'(t) \tag{8.63}$$

$$\mathbf{C}_2 = \frac{1}{T-\tau}\sum_{t=0}^{T-1-\tau} \mathbf{y}(t)\mathbf{y}'(t)$$

where with (′) is denoted the transpose of a vector. Then, both signals **x** and **y** are whitened as the following (Wehmeyer and Noé 2018):

$$\hat{\mathbf{x}}(t) = \mathbf{C}_1^{-\frac{1}{2}}\mathbf{x}(t) \tag{8.64}$$

$$\hat{\mathbf{y}}(t) = \mathbf{C}_2^{-\frac{1}{2}}\mathbf{y}(t)$$

These two signals are the input and the output, respectively, of the encoder-decoder approach, which aims to define the non-linear functions E and D (that represent the BSANN algorithm), such that, the following reconstructed error is minimum:

$$\hat{S} = \min_{E,D} \sum_{t=0}^{T-1-\tau} ||\hat{\mathbf{y}}(t) - D\left(E\left(\hat{\mathbf{x}}(t)\right)\right)||^2 \tag{8.65}$$

Chapter 9
Information Theory and Statistical Mechanics

In this chapter, we will discuss some of the elements of the information theory measures. In particular, we will introduce the so-called Shannon and relative entropy of a discrete random process and Markov process. Then, we will discuss the relationship between the entropy using the thermodynamic view and information theory view. Also, we will introduce the transfer entropy as a measure of the information flow and discuss its relationship with mechanics and thermodynamics.

9.1 Random Walks in Macromolecular Systems

To determine the structure of a macromolecular system, such as a protein, DNA, RNA, and their complexes, often the atomic coordinates are used. In this context, the following vector presents the structure:

$$(\mathbf{r}_1, \mathbf{r}_2, \cdots, \mathbf{r}_N)$$

where $\mathbf{r}_i = (x_i, y_i, z_i)$ is the position vector of the atom i of a N atoms macromolecular system. Often, these coordinates are obtained from the X-ray crystallography experimental data and are known as a deterministic description of the macromolecular structure. The X-ray data represents a static three-dimensional structure view of the macromolecule, which is often considered as the starting point of the functional dynamics of this macromolecule in a biological environment. In general, with the functional dynamics, we understand the macromolecular ensemble, which is also called the statistical description of the structure of a macromolecular system.

The coarse-grained (CG) models have become popular with the particle-based computational models for macromolecular systems (Voth 2008). These models

H. Kamberaj, *Molecular Dynamics Simulations in Statistical Physics: Theory and Applications*, Scientific Computation,
https://doi.org/10.1007/978-3-030-35702-3_9

showed a great promise in overcoming the problems arising from the time and size scale limitations for biomolecular systems (Ueda et al. 1978; McCammon et al. 1980; Bahar and Jernigan 1997; Irbäck et al. 2000; Smith and Hall 2001a,b; Oldziej et al. 2004; Tozzini 2005; Tozzini and McCammon 2005; Tozzini et al. 2006). The primary challenge with coarse-grained protein models is in how to obtain an effective potential energy function that captures the physics of the actual potential for the space and time resolution of interest, and hence that they can be as predictive as the atomistic models.

In a CG model, the particles referred to as *sites*, correspond to groups of one or more atoms. Thus, a CG model is a transformation of an atomistic structure into a CG representation of the same structure. This transformation, in general, may be tailored by including more detailed critical features of a specific system by omitting other details considered not that essential. This transformation consists of a specification of the number, types, and connectivity of the sites used to describe the CG model. In general, each site associates with a particular atomic group, which determines the nature of the site, and then the sites connect via the CG bonds related to the chemical bonds between associated atomic groups.

For a transformation from the atomistic model to a CG model, the transformation matrix $\mathbf{T}$ can be determined, which maps the atomic configuration $\mathbf{r}$ into the configuration $\mathbf{X}$ of the CG model (see Fig. 9.1). For instance, for the site s, the Cartesian coordinates $\mathbf{X}$ are determined as a linear combination of the Cartesian coordinates $\mathbf{r}_i$:

$$\mathbf{X} = \mathbf{T}(\mathbf{r}) = \sum_i c_{si}\mathbf{r}_i \tag{9.1}$$

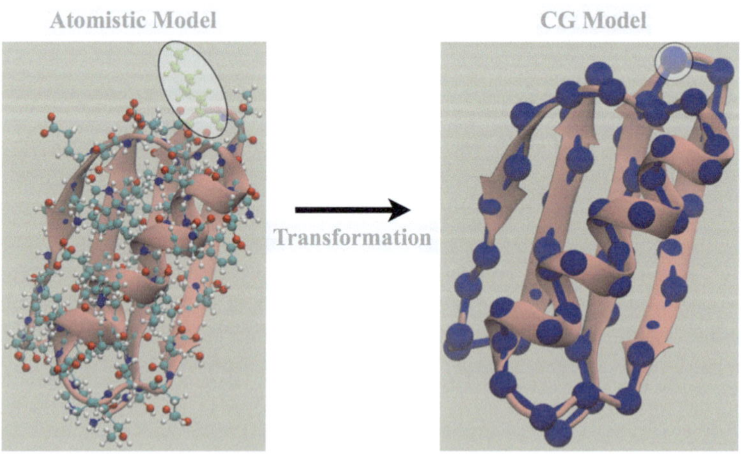

Fig. 9.1 Illustration of the atomistic structure $\mathbf{r}$ to a CG structure, $\mathbf{X}$. The ribbons indicate the underlying protein fold

where c_{si} are positive coefficients. If $c_{si} = m_{si}/\sum_j m_{sj}$, then $\mathbf{X}$ corresponds to the center of mass of the associated atomic group.

Equation (9.1) gives the transformation used for a variety of CG models, with explicit or implicit solvent (Shi et al. 2006). Some models may explicitly treat the internal structure (Murtola et al. 2009; Zhang and Muthukumar 2009; Dama et al. 2013) or anisotropy of sites (Gay and Berne 1981).

The transformation is responsible for defining the CG model, but also for determining the accuracy and efficiency of the model. An ideal CG model will preserve the features that are necessary to describe both the phenomena of interest and accurately include the effects of the fluctuation motions of the system. Besides, it should omit unnecessary details to provide significant gains in computational efficiency and filter out the high frequency and low amplitude fluctuations, only weakly coupled to the slower collective motions. Moreover, it should allow for computationally efficient treatment of the physical forces that govern the phenomena of interest.

A general approach in designing high-resolution CG transformations is based on atomistic structures (Tschöp et al. 1998; Canutescu et al. 2003; Rotkiewicz and Skolnick 2008) or potential energy (Gopal et al. 2010; Maciejczyk et al. 2010) instead of free energy. In the case of large biomolecular systems and their complexes the lower CG models are reconstructed using the correlated fluctuations (Gohlke and Thorpe 2006; Voth 2008; Sinitskiy et al. 2012; Potestio et al. 2009; Stepanova 2007; Kamberaj 2011) or density maps determined using the experimental data (Shih et al. 2006; Arkhipov et al. 2008).

We start, following Kamberaj (2018), assuming that a Markovian chain of states creates, the probability of obtaining a trajectory in the configuration space is as:

$$\begin{aligned} P\left(\mathbf{X}_T\right) &= P\left(\mathbf{x}_0\right)\prod_{t=0}^{T-1}\pi\left(\mathbf{x}_t \rightarrow \mathbf{x}_{t+1}\right) \\ &= P\left(\mathbf{X}_{T-1}\right)\pi\left(\mathbf{x}_{T-1} \rightarrow \mathbf{x}_T\right) \end{aligned} \tag{9.2}$$

Here, the vector $\mathbf{X}_T$ represents T configurations of a trajectory of the system:

$$\mathbf{X}_T = \{\mathbf{x}_0 \rightarrow \mathbf{x}_1 \rightarrow \cdots \rightarrow \mathbf{x}_{T-1}\}$$

The initial configuration $\mathbf{x}_0$ is obtained from a canonically distributed with an initial energy of the system $E(\mathbf{x}_0)$:

$$P\left(\mathbf{x}_0\right) = \exp\left(-\beta E(\mathbf{x}_0)\right)$$

with β being the inverse temperature.

In Eq. (9.2), $\pi(\mathbf{x}_t \rightarrow \mathbf{x}_{t+1})$ is the propagation probability at each time step, which depends on the details of deterministic or stochastic dynamics. In general, the Markovian transition probability $\pi(\mathbf{x}_t \rightarrow \mathbf{x}_{t+1})$ can have any distribution that conserves the Boltzmann distribution. Here, $\pi(\mathbf{x}_t \rightarrow \mathbf{x}_{t+1})$ represents the action characterized by dynamical system, which produces a Boltzmann distribution in the phase space of variables. In the general case of the Newtonian dynamics, we can write:

$$\pi(\mathbf{x}_t \rightarrow \mathbf{x}_{t+1}) = \delta(\mathbf{x}_{t+1} - \Phi_{\Delta t}(\mathbf{x}_t))$$

where δ is the delta function and $\Phi_{\Delta t}(\mathbf{x}_t)$ is the discrete flow map of one time step Δt propagation operator. In this case, a trajectory can be generated using an initial state sampled from some canonical distribution and then propagating in time using usual Hamiltonian dynamics. Note that for Hamiltonian dynamics is easy to find a time-reversible discrete flow map.

First, we will briefly introduce the method of embedding parameters used to reconstruct the phase space of a dynamical system from the time series representing the trajectories of the components of the system, such as a macromolecule, as described in Kamberaj and van der Vaart (2009b). Let $\mathbf{X}$ be the time vector for a process characterizing a coarse-grained component of the dynamical system. Time is considered discrete with $t_k = k\Delta t$, where k is an integer. Then $\mathbf{X} = \{x(t_k)\}_{k=0,\cdots,T-1}$ is a time discrete process, with T being the total number of time snapshots. For simplicity, we are going to use the notation: $x(t_k) \equiv x_k$.

To characterize the dynamics of the discrete process, we use the time delayed embedding method (Packard et al. 1980; Takens 1981; Grassberger and Procaccia 1983; Sauer et al. 1991). In this method, a state vector in a m-dimensional space of a discrete process $x(t)$ is obtained as:

$$\mathbf{x}_k^{\mu} = \left(x_{k-(m-1)\tau}, x_{k-(m-1)\tau+\tau}, \cdots, x_k\right)$$

where τ is the time lag, which, in general, is a multiple of Δt, and $k = (m-1)\tau, \cdots, T-1$. Here, the superscript μ represents the pair of embedded parameters (m, τ). The hope here is that the vector $\mathbf{x}_k^{\mu}$ will reconstruct the phase space of the original component of the dynamical system in Cartesian coordinates $\mathbf{r}$.

9.2 Optimization of Embedding Parameters

With the proper values of the m and τ embedding parameters, a smooth discrete time process is defined which reconstructs the underlying dynamics. The correct choice of these two parameters is crucial for the proper characterization of the structure of the time series (Kennel et al. 1992; Abarbanel and Kennel 1993; Cellucci et al. 2003). The mathematical concepts for defining the state vector dimension m have been reviewed in details in Noakes (1991) and Sauer et al. (1991). There are several methods proposed for estimating the optimal embedding parameters m and τ simultaneously (Kennel et al. 1992; Abarbanel and Kennel 1993). These methods try to minimize the number of false nearest neighbors. A comparison of these methods is in Cellucci et al. (2003), as well as an approach that combines the *global false nearest neighbor* (GFNN) method for the calculation of m, with a separate process for determining the time shift (Cellucci et al. 2003).

In GFNN method, the time at which the auto-correlation function has its first zero is used as the time lag τ.

Using the time lag τ defined above for the false nearest neighbors method, for a state vector

$$\mathbf{x}_k^{\mu} = \left(x_{k-(m-1)\tau}, x_{k-(m-1)\tau+\tau}, \cdots, x_k\right)$$

the nearest neighbor is

$$\mathbf{x}_k^{\mu,\mathrm{NN}} = \left(x_{k-(m-1)\tau}^{\mathrm{NN}}, x_{k-(m-1)\tau+\tau}^{\mathrm{NN}}, \cdots, x_k^{\mathrm{NN}}\right)$$

where $k = k_0, \cdots, T-1$ with $k_0 = (m-1)\tau$. The Euclidean distance between these two points in m-dimensional space is given by

$$R_k^m = \mid \mathbf{x}_k^{\mu} - \mathbf{x}_k^{\mu,\mathrm{NN}} \mid = \left(\sum_{i=0}^{m-1}\left(x_{k-i\tau} - x_{k-i\tau}^{\mathrm{NN}}\right)^2\right)^{1/2}. \tag{9.3}$$

The distance between these two points in the $(m+1)$-dimensional space is

$$R_k^{m+1} = \left((R_k^m)^2 + \left(x_{k-m\tau} - x_{k-m\tau}^{\mathrm{NN}}\right)^2\right)^{1/2}. \tag{9.4}$$

This distance is normalized against the distance in m-dimensional space (Abarbanel 1996):

$$\gamma_k^m = \left(\frac{(R_k^{m+1})^2 - (R_k^m)^2}{(R_k^m)^2}\right)^{1/2} = \frac{\mid x_{k-m\tau} - x_{k-m\tau}^{\mathrm{NN}} \mid}{R_k^m}. \tag{9.5}$$

γ_k^m is compared to a threshold value R_{tol}, which is determined a priori (Abarbanel 1996; Cellucci et al. 2003) and recommended to be 15 (Abarbanel 1996; Cellucci et al. 2003). If γ_k^m exceeds R_{tol} then $\mathbf{x}_k^{\mu,\mathrm{NN}}$ is a false nearest neighbor of $\mathbf{x}_k^{\mu}$ in the m-dimensional space and f_{FNN}, the frequency of the false nearest neighbors, is increased by one. The value of m is increased until f_{FNN} approaches zero.

9.3 Shannon and Relative Entropy

Consider a random process $\mathbf{X}^{\mu}$ as described above, which represents the dynamics of a component of the macromolecular system, such as the trajectory of the center of mass of one of the residues of the protein. Let us denote with $p(\mathbf{x}_k^{\mu})$ the probability mass function of the state k. Besides, this represents a Markovian process. By definition (Thomas and Joy 2006), the Shannon entropy of the random process $\mathbf{X}^\mu$ is as:

$$H_p(\mathbf{X}^\mu) = -\sum_{k=k_0}^{T-1} p(\mathbf{x}_k^\mu) \ln p(\mathbf{x}_k^\mu) \tag{9.6}$$

where 'ln' is the natural logarithm, and hence H is measured in natural units of information (the so-called *nats*). From the information theory point of view, Shannon entropy expresses the uncertainty on describing the random variable $\mathbf{X}^\mu$, or the information required on average to completely describe the random variable $\mathbf{X}^\mu$. In Eq. (9.6), the summation runs over all possible state vectors of $\mathbf{X}^\mu$.

If we assume that we do not know the real probability distribution of the process, but we use another probability distribution, let us say, $\tilde{p}(\mathbf{x}_k^\mu)$ (for $k = k_0, \cdots, T-1$) to describe the process, then the so-called *relative entropy* is given as (Thomas and Joy 2006):

$$D(p||\tilde{p}) = \sum_{k=k_0}^{T-1} p(\mathbf{x}_k^\mu) \ln \frac{p(\mathbf{x}_k^\mu)}{\tilde{p}(\mathbf{x}_k^\mu)} \tag{9.7}$$

which represents the average value of the logarithm likelihood function $\frac{p(\mathbf{x}_k^\mu)}{\tilde{p}(\mathbf{x}_k^\mu)}$ according to the true probability distribution $\mathbf{p}$. The relative entropy is also known as Kullback-Leibler distance (Kullback 1959).

It is easy to show that

$$H_{\tilde{p}}(\mathbf{X}^\mu) = H_p(\mathbf{X}^\mu) + D(p||\tilde{p}) \tag{9.8}$$

which indicates that the uncertainty of the random process $\mathbf{X}^\mu$ increases if we use the guess probability mass function $\tilde{p}(\mathbf{x}_k^\mu)$ to describe $\mathbf{X}^\mu$ instead of the true probability distribution $p(\mathbf{x}_k^\mu)$. Besides, it can be shown that

$$D(p||\tilde{p}) \geq 0$$

with equality if and only if $p(\mathbf{x}_k^\mu) = \tilde{p}(\mathbf{x}_k^\mu)$ for all states k.

9.4 Relationship with the Second Law of Thermodynamics

As mentioned above, the Shannon entropy measures the uncertainty on describing the random variable X. From the statistical mechanics, the Boltzmann's entropy is as:

$$S = k_B \ln \Omega \tag{9.9}$$

where k_B is the Boltzmann's constant and Ω is the number of microstates. Based on the variation principle of the entropy (discussed in the Chap. 2), the entropy of an isolated system is a non-decreasing quantity. Therefore, the maximum entropy corresponds to the macroscopic states with the maximum number of the microscopic states. In other words, the disordered systems (those with the largest number of microstates) have higher values of the entropy. If we make a correspondence that "uncertainty" indicates "disorder", then this shows that there is a connection between the Shannon entropy and Boltzmann's entropy in statistical mechanics. Indeed, if we assume that all the microstates are equally probable, then

$$p = \frac{1}{\Omega}$$

determines the probability of observing the system in one of the microstates. Replacing this expression in Eq. (9.6), definition of the Shannon entropy, we get

$$H(X) = -\sum_{x \in \mathcal{X}} \frac{1}{\Omega} \ln \frac{1}{\Omega} = \ln \Omega$$

This indicates that, up to a constant (i.e., k_B), these two definitions are equivalent.

Another view of this equivalence between the second law of thermodynamics and the entropy function introduced by Shannon is as the following. Based on the variation principle, the entropy for an isolated system is a non-decreasing quantity (see also Chap. 2).

Now, let us consider two probability distributions, namely $P(\mathbf{X}_T)$ and $P'(\mathbf{X}_T)$, on the state space of the same Markovian chain at some time T, which can be obtained using two different initial configuration probabilities, let us say $P(\mathbf{x}_0)$ and $P'(\mathbf{x}_0)$, or it may correspond to the case of a system having two possible energy levels $E_0(\mathbf{X})$ and $E_1(\mathbf{X})$. Then, the relative entropy between the joint probability mass functions at time T and $T+1$ is given as:

$$\begin{aligned} & D\left(p(\mathbf{x}_T, \mathbf{x}_{T+1}) || p'(\mathbf{x}_T, \mathbf{x}_{T+1})\right) \\ &= D\left(p(\mathbf{x}_T) || p'(\mathbf{x}_T)\right) + D\left(\pi(\mathbf{x}_T \rightarrow \mathbf{x}_{T+1}) || \pi(\mathbf{x}_T \rightarrow \mathbf{x}_{T+1})\right) \\ &= D\left(p(\mathbf{x}_{T+1}) || p'(\mathbf{x}_{T+1})\right) + D\left(\pi(\mathbf{x}_{T+1} \rightarrow \mathbf{x}_T) || \pi'(\mathbf{x}_{T+1} \rightarrow \mathbf{x}_T)\right) \end{aligned} \tag{9.10}$$

Since

$$\begin{aligned} D\left(\pi(\mathbf{x}_T \rightarrow \mathbf{x}_{T+1}) || \pi(\mathbf{x}_T \rightarrow \mathbf{x}_{T+1})\right) &= 0 \\ D\left(\pi(\mathbf{x}_{T+1} \rightarrow \mathbf{x}_T) || \pi'(\mathbf{x}_{T+1} \rightarrow \mathbf{x}_T)\right) &\geq 0 \end{aligned} \tag{9.11}$$

then

$$D\left(p(\mathbf{x}_{T+1}) || p'(\mathbf{x}_{T+1})\right) \leq D\left(p(\mathbf{x}_T) || p'(\mathbf{x}_T)\right) \tag{9.12}$$

Eq. (9.12) indicates that the two distributions get closer and closer as the time passes. If we calculate the entropies, we get these relationships:

$$D\left(p(\mathbf{x}_{T+1})||p'(\mathbf{x}_{T+1})\right) = H\left(p'(\mathbf{x}_{T+1})\right) - H\left(p(\mathbf{x}_{T+1})\right) \tag{9.13}$$
$$D\left(p(\mathbf{x}_{T})||p'(\mathbf{x}_{T})\right) = H\left(p'(\mathbf{x}_{T})\right) - H\left(p(\mathbf{x}_{T})\right)$$

If we assume that the probability distribution $P'(\mathbf{X}_T)$ is stationary, that is $p'(\mathbf{x}_{T+1}) = p'(\mathbf{x}_T) = p$, then

$$H\left(p'(\mathbf{x}_{T+1})\right) = H\left(p'(\mathbf{x}_{T})\right) \equiv H\left(p\right)$$

Therefore, we obtain:

$$D\left(p(\mathbf{x}_{T+1})||p'(\mathbf{x}_{T+1})\right) = H\left(p\right) - H\left(p(\mathbf{x}_{T+1})\right) \tag{9.14}$$
$$D\left(p(\mathbf{x}_{T})||p'(\mathbf{x}_{T})\right) = H\left(p\right) - H\left(p(\mathbf{x}_{T})\right)$$

Combining Eqs. (9.12) and (9.14), we get

$$H\left(p(\mathbf{x}_{T+1})\right) \geq H\left(p(\mathbf{x}_{T})\right) \tag{9.15}$$

which indicates that Shannon entropy is a non-decreasing function with time. This is the second observation that shows the relationship between the Shannon entropy and the second law of thermodynamics. In other words, since the uniform stationary microstates are characterized by a maximum entropy, then they are most probable microstates, in agreement with statistical mechanics view. Moreover, if we start the Markovian chain from a uniform stationary distribution, then the principle of the entropy indicates that the system will tend to the same state, in other words, the entropy remains constant.

Now, consider a system having the microstates $\{\mathbf{x}_k^\mu\}_{k=k_0}^{T-1}$, which have two energy levels $E_0(\mathbf{X}^\mu)$ and $E_1(\mathbf{X}^\mu)$. Denoting with $Z(\beta)$ the partition function at the inverse temperature $\beta = 1/k_B T$, then

$$Z_0(\beta) = \sum_k \exp\left(-\beta E_0(\mathbf{x}_k^\mu)\right) \tag{9.16}$$
$$Z_1(\beta) = \sum_k \exp\left(-\beta E_1(\mathbf{x}_k^\mu)\right)$$

The Boltzmann-Gibbs probability distribution functions are given as

$$P_0(\beta, \mathbf{X}^\mu) = \frac{1}{Z_0(\beta)} \exp\left(-\beta E_0(\mathbf{X}^\mu)\right) \tag{9.17}$$
$$P_1(\beta, \mathbf{X}^\mu) = \frac{1}{Z_1(\beta)} \exp\left(-\beta E_1(\mathbf{X}^\mu)\right)$$

The relative entropy between P_0 and P_1 is given as

$$0 \leq D(P_0||P_1) = \sum_k P_0(\beta, \mathbf{x}_k^{\mu}) \ln \frac{\frac{1}{Z_0(\beta)} \exp\left(-\beta E_0\left(\mathbf{x}_k^{\mu}\right)\right)}{\frac{1}{Z_1(\beta)} \exp\left(-\beta E_1\left(\mathbf{x}_k^{\mu}\right)\right)} \tag{9.18}$$

$$= \ln Z_1 - \ln Z_0 + \beta \langle E_1(\mathbf{X}^{\mu}) - E_0(\mathbf{X}^{\mu}) \rangle_0$$

where $\langle \cdots \rangle_0$ denotes an ensemble average with respect to the probability P_0. Equation (9.18) can be arranged in the form

$$\langle E_1(\mathbf{X}^{\mu}) - E_0(\mathbf{X}^{\mu}) \rangle_0 - (F_1 - F_0) = k_B T D(P_0||P_1) \geq 0 \tag{9.19}$$

where F_i (for $i = 0,\ 1$) is the free energy related to P_i: $F_i = k_B T \ln P_i$. This could correspond physically to the case when the system for $t < 0$ is at equilibrium at the state with energy E_0, then some external work has been done on the system or additional energy injected into the system at $t = 0$, $W(\mathbf{x}_k^{\mu}) = E_1(\mathbf{x}_k^{\mu}) - E_0(\mathbf{x}_k^{\mu})$, and the system experiences a transition from the state E_0 to E_1, irreversibly. Then, the expectation of this external work taken with P_0, which represents the equilibrium distribution, is

$$W(\mathbf{X}^{\mu}) = \langle E_1(\mathbf{X}^{\mu}) - E_0(\mathbf{X}^{\mu}) \rangle_0$$

Sign of Eq. (9.19) indicates that the process is irreversible, and that the total entropy of the system and the environment increases.

9.5 Transfer Entropy

The Granger causality (Granger 1969) concept is usually used to characterize the dependence of one variable Y measured over time on another variable X measured synchronously. This concept is initially being used to define the direction of interaction by estimating the contribution of X in predicting Y. However, there exist many other variations of this concept, such as the linear approaches in the time and frequency domain.

The information theory measure of transfer entropy quantifies the statistical coherence between two processes that evolve in time. Schreiber (2000) introduced the transfer entropy as the deviation from the independence of the state transition (from the previous state to the next state) of an information destination X from the (previous) state of an information source Y.

To characterize the dynamics of the two random Markovian processes, we will use two vectors. For the first process, the vector is $\mathbf{X}^{\mu_x} = \{\mathbf{x}_k^{\mu_x}\}_{k=k_{0x}}^{T-1}$ where

$$\mathbf{x}_k^{\mu_x} = \left\{x_{k-(m_x-1)\tau_x}, x_{k-(m_x-1)\tau_x+\tau_x}, \cdots, x_{k-\tau_x}, x_k\right\}$$

Here, $k_{0x} = (m_x - 1)\tau_x$, and the dynamics of the second process is characterized by the vector $\mathbf{Y}^{\mu_y} = \{\mathbf{y}_k^{\mu_y}\}_{k=k_{0y}}^{T-1}$:

$$\mathbf{y}_k^{\mu_y} = \left\{y_{k-(m_y-1)\tau_y}, y_{k-(m_y-1)\tau_y+\tau_y}, \cdots, y_{k-\tau_y}, y_k\right\}$$

where $k_{0y} = (m_y - 1)\tau_y$. In the following discussion, $\mathbf{x}_k^{\mu_x+1}$ represents the vector

$$\mathbf{x}_k^{\mu_x+1} = \left\{x_{k-(m_x-1)\tau_x}, x_{k-(m_x-1)\tau_x+\tau_x}, \cdots, x_{k-\tau_x}, x_k, x_{k+\delta}\right\}$$

Similarly, $\mathbf{y}_k^{\mu_y+1}$ represents the vector

$$\mathbf{y}_k^{\mu_y+1} = \left\{y_{k-(m_y-1)\tau_y}, y_{k-(m_y-1)\tau_y+\tau_y}, \cdots, y_{k-\tau_y}, y_k, y_{k+\delta}\right\}$$

Note that m and τ are characteristics of each random process. The choice of m and τ is crucial in order to reconstruct the dynamical structure of the random processes as discussed above. For clarity, we denote $\mu + 1 \equiv (m + 1, \tau)$ and $k_0 = \max\left((m_x - 1)\tau_x, (m_y - 1)\tau_y\right)$.

By definition (Schreiber 2000), the transfer entropy, $T_{Y\to X}$, is given as:

$$T_{Y\to X} = \sum_k p(x_{k+\delta}, \mathbf{x}_k^{\mu_x}, \mathbf{y}_k^{\mu_y}) \log_2 \frac{p(x_{k+\delta} \mid \mathbf{x}_k^{\mu_x}, \mathbf{y}_k^{\mu_y})}{p(x_{k+\delta} \mid \mathbf{x}_k^{\mu_x})} \tag{9.20}$$

where k is a time index. Similarly, the transfer entropy, $T_{X\to Y}$, is given as:

$$T_{X\to Y} = \sum_k p(\mathbf{x}_k^{\mu_x}, \mathbf{y}_k^{\mu_y}, y_{k+\delta}) \log_2 \frac{p(\mathbf{x}_k^{\mu_x}, y_{k+\delta} \mid \mathbf{y}_k^{\mu_y})}{p(y_{k+\delta} \mid \mathbf{y}_k^{\mu_y})} \tag{9.21}$$

This formulation of the transfer entropy represents a dynamical measure, as a generalization of the entropy rate to more than one element to form a mutual information rate (Schreiber 2000).

In Eq. (9.20), $p(x_{k+\delta}, \mathbf{x}^{\mu_x}, \mathbf{y}_k^{\mu_y})$ is the joint probability distribution of observing future value $x_{k+\delta}$ and the histories of $\mathbf{x}_k^{\mu_x}$ and $\mathbf{y}_k^{\mu_y}$, $p(x_{k+\delta} \mid \mathbf{x}_k^{\mu_x}, \mathbf{y}_k^{\mu_y})$ is the conditional probability of observing the future value of X given the past values of both X and Y, and $p(x_{k+\delta} \mid \mathbf{x}_k^{\mu_x})$ is the conditional probability of observing the future of X knowing its past. From Eq. (9.20), we can write

$$\begin{aligned} T_{Y\to X} &= \sum_k p(x_{k+\delta}, \mathbf{x}_k^{\mu_x}, \mathbf{y}_k^{\mu_y}) \Big[\log_2 p(x_{k+\delta} \mid \mathbf{x}_k^{\mu_x}, \mathbf{y}_k^{\mu_y}) \\ &\quad - \log_2 p(x_{k+\delta} \mid \mathbf{x}_k^{\mu_x})\Big] \\ &= \sum_k p(x_{k+\delta}, \mathbf{x}_k^{\mu_x}, \mathbf{y}_k^{\mu_y}) \log_2 \frac{p(x_{k+\delta}, \mathbf{x}_k^{\mu_x}, \mathbf{y}_k^{\mu_y})}{p(\mathbf{x}_k^{\mu_x}, \mathbf{y}_k^{\mu_y})} \end{aligned} \tag{9.22}$$

$$
\begin{aligned}
&-\sum_{k} p(x_{k+\delta}, \mathbf{x}_k^{\mu_x}, \mathbf{y}_k^{\mu_y}) \log_2 \frac{p(x_{k+\delta}, \mathbf{x}_k^{\mu_x})}{p(\mathbf{x}_k^{\mu_x})} \\
&= \sum_{k} p(x_{k+\delta}, \mathbf{x}_k^{\mu_x}, \mathbf{y}_k^{\mu_y}) \log_2 p(x_{k+\delta}, \mathbf{x}_k^{\mu_x}, \mathbf{y}_k^{\mu_y}) \\
&- \sum_{k} p(x_{k+\delta}, \mathbf{x}_k^{\mu_x}, \mathbf{y}_k^{\mu_y}) \log_2 p(\mathbf{x}_k^{\mu_x}, \mathbf{y}_k^{\mu_y}) \\
&- \sum_{k} p(x_{k+\delta}, \mathbf{x}_k^{\mu_x}, \mathbf{y}_k^{\mu_y}) \log_2 p(x_{k+\delta}, \mathbf{x}_k^{\mu_x}) \\
&+ \sum_{k} p(x_{k+\delta}, \mathbf{x}_k^{\mu_x}, \mathbf{y}_k^{\mu_y}) \log_2 p(\mathbf{x}_k^{\mu_x}) \\
&= H(\mathbf{X}^{\mu_x+1}) - H(\mathbf{X}^{\mu_x}) - H(\mathbf{X}^{\mu_x+1}, \mathbf{Y}^{\mu_y}) + H(\mathbf{X}^{\mu_x}, \mathbf{Y}^{\mu_y})
\end{aligned}
$$

Thus, it can be seen that

$$
\begin{aligned}
T_{Y\rightarrow X} &= H(\mathbf{X}^{\mu_x+1}) - H(\mathbf{X}^{\mu_x}) - H(\mathbf{X}^{\mu_x+1}, \mathbf{Y}^{\mu_y}) + H(\mathbf{X}^{\mu_x}, \mathbf{Y}^{\mu_y}) \\
&\equiv I\left(X_{+\delta}; \mathbf{Y}^{\mu_y} \mid \mathbf{X}^{\mu_x}\right)
\end{aligned}
\tag{9.23}
$$

where $I\,(X_{+\delta}; \mathbf{Y}^{\mu_y} \mid \mathbf{X}^{\mu_x})$ is the conditional mutual information as previously stated in literature (Gourévitch and Eggermont 2007). This indicates that the transfer entropy, $T_{Y\rightarrow X}$ can be interpreted as the average amount of information contained in the source about next state $X_{+\delta}$ of the variable X that was not already contained in the past of X.

Similarly, $T_{X\rightarrow Y}$ is given as

$$
\begin{aligned}
T_{X\rightarrow Y} &= H(\mathbf{Y}^{\mu_y+1}) - H(\mathbf{Y}^{\mu_y}) - H(\mathbf{X}^{\mu_x}, \mathbf{Y}^{\mu_y+1}) + H(\mathbf{X}^{\mu_x}, \mathbf{Y}^{\mu_y}) \\
&\equiv I\left(Y_{+\delta}; \mathbf{X}^{\mu_x} \mid \mathbf{Y}^{\mu_y}\right)
\end{aligned}
\tag{9.24}
$$

To determine a local transfer entropy measure, we first note that Eq. (9.20) is summed over all possible state transition tuples $\left(x_{k+\delta}, \mathbf{x}_k^{\mu_x}, \mathbf{y}_k^{\mu_y}\right)$, weighted by the probability of observing each such tuple, $p(x_{k+\delta}, \mathbf{x}_k^{\mu_x}, \mathbf{y}_k^{\mu_y})$. Therefore, we can write:

$$
\begin{aligned}
T_{Y\rightarrow X} &= \sum_{k} p(x_{k+\delta}, \mathbf{x}_k^{\mu_x}, \mathbf{y}_k^{\mu_y}) \log_2 \frac{p(x_{k+\delta} \mid \mathbf{x}_k^{\mu_x}, \mathbf{y}_k^{\mu_y})}{p(x_{k+\delta} \mid \mathbf{x}_k^{\mu_x})} \\
&= \langle \log_2 \frac{p(x_{k+\delta} \mid \mathbf{x}_k^{\mu_x}, \mathbf{y}_k^{\mu_y})}{p(x_{k+\delta} \mid \mathbf{x}_k^{\mu_x})} \rangle
\end{aligned}
$$

where $\langle \cdots \rangle$ indicates the average with probability $p(x_{k+\delta}, \mathbf{x}_k^{\mu_x}, \mathbf{y}_k^{\mu_y})$ and

$$t_{Y \to X}(k+\delta) = \log_2 \frac{p(x_{k+\delta} \mid \mathbf{x}_k^{\mu_x}, \mathbf{y}_k^{\mu_y})}{p(x_{k+\delta} \mid \mathbf{x}_k^{\mu_x})} \tag{9.25}$$

is the local transfer entropy. Equation (9.25) can further be simplified as:

$$\begin{aligned} t_{Y \to X}(k+\delta) &= \log_2 p(x_{k+\delta} \mid \mathbf{x}_k^{\mu_x}, \mathbf{y}_k^{\mu_y}) - \log_2 p(x_{k+\delta} \mid \mathbf{x}_k^{\mu_x}) \\ &= \log_2 \frac{p(x_{k+\delta}, \mathbf{x}_k^{\mu_x}, \mathbf{y}_k^{\mu_y})}{p(\mathbf{x}_k^{\mu_x}, \mathbf{y}_k^{\mu_y})} - \log_2 \frac{p(x_{k+\delta}, \mathbf{x}_k^{\mu_x})}{p(\mathbf{x}_k^{\mu_x})} \\ &= \log_2 p(\mathbf{x}_k^{\mu_x+1}, \mathbf{y}_k^{\mu_y}) - \log_2 p(\mathbf{x}_k^{\mu_x}, \mathbf{y}_k^{\mu_y}) \\ &- \log_2 p(\mathbf{x}_k^{\mu_x+1}) + \log_2 p(\mathbf{x}_k^{\mu_x}) \end{aligned} \tag{9.26}$$

Thus, the local transfer entropy, $t_{Y \to X}(k)$, is given by

$$\begin{aligned} t_{Y \to X}(k+\delta) = &\left[\log_2 p(\mathbf{x}_k^{\mu_x+1}, \mathbf{y}_k^{\mu_y}) - \log_2 p(\mathbf{x}_k^{\mu_x}, \mathbf{y}_k^{\mu_y})\right] \\ &- \left[\log_2 p(\mathbf{x}_k^{\mu_x+1}) - \log_2 p(\mathbf{x}_k^{\mu_x})\right] \end{aligned} \tag{9.27}$$

Introducing the definition of the local condition entropy as the following:

$$\begin{aligned} h(x_{k+\delta} \mid \mathbf{x}_k^{\mu_x}, \mathbf{y}_k^{\mu_y}) &= -\left(\log_2 p(\mathbf{x}_k^{\mu_x+1}, \mathbf{y}_k^{\mu_y}) - \log_2 p(\mathbf{x}_k^{\mu_x}, \mathbf{y}_k^{\mu_y})\right) \\ h(x_{k+\delta} \mid \mathbf{x}_k^{\mu_x}) &= -\left(\log_2 p(\mathbf{x}_k^{\mu_x+1}) - \log_2 p(\mathbf{x}_k^{\mu_x})\right) \end{aligned} \tag{9.28}$$

we obtain

$$t_{Y \to X}(k+\delta) = h(x_{k+\delta} \mid \mathbf{x}_k^{\mu_x}) - h(x_{k+\delta} \mid \mathbf{x}_k^{\mu_x}, \mathbf{y}_k^{\mu_y}) \tag{9.29}$$

Similarly, for the local transfer entropy, $t_{X \to Y}(k+\delta)$, can be obtained

$$\begin{aligned} t_{X \to Y}(k+\delta) = &\left[\log_2 p(\mathbf{x}_k^{\mu_x}, \mathbf{y}_k^{\mu_y+1}) - \log_2 p(\mathbf{x}_k^{\mu_x}, \mathbf{y}_k^{\mu_y})\right] \\ &- \left[\log_2 p(\mathbf{y}_k^{\mu_y+1}) - \log_2 p(\mathbf{y}_k^{\mu_y})\right] \end{aligned} \tag{9.30}$$

Again, introducing the local condition entropy as:

$$\begin{aligned} h(y_{k+\delta} \mid \mathbf{x}_k^{\mu_x}, \mathbf{y}_k^{\mu_y}) &= -\left(\log_2 p(\mathbf{x}_k^{\mu_x}, \mathbf{y}_k^{\mu_y+1}) - \log_2 p(\mathbf{x}_k^{\mu_x}, \mathbf{y}_k^{\mu_y})\right) \\ h(y_{k+\delta} \mid \mathbf{y}_k^{\mu_x}) &= -\left(\log_2 p(\mathbf{y}_k^{\mu_y+1}) - \log_2 p(\mathbf{y}_k^{\mu_y})\right) \end{aligned} \tag{9.31}$$

the local transfer entropy becomes:

$$t_{X\to Y}(k+\delta) = h(y_{k+\delta} \mid \mathbf{y}_k^{\mu_x}) - h(y_{k+\delta} \mid \mathbf{x}_k^{\mu_x}, \mathbf{y}_k^{\mu_y}) \tag{9.32}$$

9.6 Relationship Between Transfer Entropy and Thermodynamics

To derive a relationship between the transfer entropy and the laws of the thermodynamics, we are going to use the formalism introduced in Prokopenko et al. (2013) and Prokopenko and Lizier (2014). For that, first we will consider a physical system in a non-equilibrium thermodynamic state, but close to equilibrium. The state vector $\mathbf{x}_t \in R^f$ represents the macroscopic state of the system at any time t, where f denotes the total number of macroscopic parameters characterizing the state, such as temperature, pressure, the number density of each component of the physical system. The set of all these vectors forms the phase space, and every macrostate can be formed by many microstate vectors, each having the same probability being at this macrostate.

The dynamics of the macrostates is defined by the transition probability $p(x_{t+\delta} \mid \mathbf{x}_t)$, which gives the propagation probability with δ time step. In general, it will depend on the details of deterministic or stochastic dynamics. Based on Prokopenko et al. (2013) and Prokopenko and Lizier (2014), it is postulated that

$$p(x_{t+\delta} \mid \mathbf{x}_t) = \frac{\Omega_r^{(1)}}{Z_1} \tag{9.33}$$

where Z_1 is a normalization factor representing the partition function that depends on the macrostate $\mathbf{x}_t$, and $\Omega_r^{(1)}$ is the ratio of the total number of microstates of all the macrostates of the state $\mathbf{x}_{t+\delta}$ and $\mathbf{x}_t$:

$$\Omega_r^{(1)} = \frac{\Omega_{t+\delta}}{\Omega_t}$$

Thus, using the Boltzmann definition of the entropy: $S = k_B \ln \Omega$, we obtain

$$\Delta S = S(\mathbf{x}_{t+\delta}) - S(\mathbf{x}_t) = k_B \ln \Omega_r^{(1)}$$

where ΔS represents the change on the physical system entropy during the transition from the state at t to the state at $t+1$. Therefore, Eq. (9.33) can be written in the following form:

$$p(x_{t+\delta} \mid \mathbf{x}_t) = \frac{1}{Z_1} \exp\left(-\beta\,(T\,\Delta S)\right) \tag{9.34}$$

where T is the thermodynamic temperature.

Now, let us consider that the physical system is interacting with surrounding due to its coupling with this surrounding, however, the nature of this coupling is not known. Let $\mathbf{y}_t$ be the state characterizing the macrostate of the surrounding at t, $p(x_{t+\delta} \mid \mathbf{x}_t, \mathbf{y}_t)$ is representing the transition probability in the presence of the environment, which is obtained by sampling both $\mathbf{X}$ and $\mathbf{Y}$. Then, the second postulate is given as follows (Prokopenko et al. 2013; Prokopenko and Lizier 2014):

$$p(x_{t+\delta} \mid \mathbf{x}_t, \mathbf{y}_t) = \frac{\Omega_r^{(2)}}{Z_2} \tag{9.35}$$

where Z_2 is a normalization factor, and $\Omega_r^{(2)}$ is the ratio of the total number of the microstates of the state $\mathbf{x}_{t+\delta}$ and $\mathbf{x}_t$ in the context of $\mathbf{y}_t$, which is related to the change of the internal entropy of the system in the presence of the surrounding:

$$\Delta\sigma_y = \sigma(\mathbf{x}_{t+1})_y - \sigma(\mathbf{x}_t)_y = k_B \ln \Omega_r^{(2)}$$

Therefore, Eq. (9.35) can be written as:

$$p(x_{t+\delta} \mid \mathbf{x}_t, \mathbf{y}_t) = \frac{1}{Z_2} \exp\left(-\beta\left(T\Delta\sigma_y\right)\right) \tag{9.36}$$

Substituting Eqs. (9.34) and (9.36) into Eq. (9.25), we obtain:

$$t_{Y\to X}(k+\delta) = \log_2\left(\frac{Z_1}{Z_2}\right) + \frac{1}{k_B \ln 2}\left(\Delta\sigma_y - \Delta S\right) \tag{9.37}$$

Using the variation principle of the entropy (Prokopenko et al. 2013; Prokopenko and Lizier 2014), the change on the entropy ΔS equals the sum of the entropy change due to the coupling with the surrounding environment ΔS_{ext} and the internal entropy production inside the physical system $\Delta\sigma_y$:

$$\Delta S = \Delta S_{\text{ext}} + \Delta\sigma_y \tag{9.38}$$

Using Eqs. (9.37) and (9.38), we obtain:

$$t_{Y\to X}(k+\delta) = \log_2\left(\frac{Z_1}{Z_2}\right) - \frac{1}{k_B \ln 2}\Delta S_{\text{ext}} \tag{9.39}$$

For a closed system, the external entropy production, ΔS_{ext}, is given as (Prokopenko et al. 2013; Prokopenko and Lizier 2014):

$$\Delta S_{\text{ext}} = \int \frac{\delta q}{T} \tag{9.40}$$

where q is the heat given to the system by the surrounding and T is the temperature of the system. If the transition from the state k to the state $k+\delta$ is reversible, then

$\Delta S = \Delta S_{\text{ext}}$, and hence $\Delta\sigma_y = 0$. If the process is irreversible, then $\Delta S > \Delta S_{\text{ext}}$, and hence $\Delta\sigma_y > 0$. Thus, we can write that

$$\Delta S \geq \int \frac{\delta q}{T}$$

where the equality stands for reversible transition only. For small fluctuations close to the equilibrium, $Z_1 \approx Z_2$, then Eq. (9.39) can be written as:

$$t_{Y\to X}(k+\delta) = -\frac{\Delta S_{\text{ext}}}{k_B \ln 2} \tag{9.41}$$

The sign minus indicates the direction of the heat flow, and hence entropy production, attributed to the surrounding Y and the local transfer entropy is opposite.

For processes close to the equilibrium, both reversible or irreversible, at constant temperature, we obtain

$$t_{Y\to X}(k+\delta) = -\frac{\Delta q_{\text{ext}}}{k_B T \ln 2} \tag{9.42}$$

where Δq_{ext} is the heat flow to the system from the surrounding in the context of the source Y, given as (Prokopenko and Lizier 2014)

$$\Delta q_{\text{ext}} = \int \delta q_{\text{ext}}$$

Following Prokopenko and Lizier (2014), Eq. (9.42) indicates that if the heat flows to the system from the surrounding, i.e. $\Delta q_{\text{ext}} > 0$, then $t_{Y\to X}(k+\delta) < 0$ and a measure on Y misinforms about the macroscopic state transition $k+\delta$, and hence the production entropy due to the interaction with surrounding increases. On the other hand, if the heat flows to the surrounding from the system, i.e. $\Delta q_{\text{ext}} < 0$, then $t_{Y\to X}(k+\delta) > 0$, and we can say that a measure of the source Y improves our predictions about the macroscopic transition state of X. It is trivial to show that for thermally isolated transitions, that is $\Delta q_{\text{ext}} = 0$, we obtain $t_{Y\to X}(k+\delta) = 0$, indicating that X and Y are independent, in other words, X and Y do not interact with each other.

If the process is not isothermic, then, in general,

$$\int \frac{\delta q_{\text{ext}}}{T} \neq \frac{\Delta q_{\text{ext}}}{T}$$

Furthermore, for non-equilibrium transition states (i.e., $Z_1 \neq Z_2$), Eq. (9.39) can be written as

$$t_{Y\to X}(k+\delta) = \log_2\left(\frac{Z_1}{Z_2}\right) - \frac{1}{k_B \ln 2}\int \frac{\delta q_{\text{ext}}}{T} \tag{9.43}$$

It can easily be found that for $Z_1 \leq Z_2$ and assuming that the system dissipates heat to the surrounding, we have $\log_2\left(\frac{Z_1}{Z_2}\right) \leq 0$, and hence we obtain

$$t_{Y\to X}(k+\delta) \leq -\frac{1}{k_B \ln 2}\int \frac{\delta q_{\text{ext}}}{T} \tag{9.44}$$

In this case the local transfer entropy is positive. We can further write that

$$\int \frac{\delta q_{\text{ext}}}{T} \leq -(k_B \ln 2)\, t_{Y\to X}(k+\delta) \tag{9.45}$$

which indicates that $\int \frac{\delta q_{\text{ext}}}{T}$ is negative and has as an upper bound the value $-(k_B \ln 2)\, t_{Y\to X}(k+\delta)$.

While for $Z_1 \geq Z_2$ and if the system absorbs heat from the surrounding, we have $\log_2\left(\frac{Z_1}{Z_2}\right) \leq 0$, which yields

$$t_{Y\to X}(k+\delta) \geq -\frac{1}{k_B \ln 2}\int \frac{\delta q_{\text{ext}}}{T} \tag{9.46}$$

and the local transfer entropy is negative. Or, we can write that

$$\int \frac{\delta q_{\text{ext}}}{T} \geq -(k_B \ln 2)\, t_{Y\to X}(k+\delta) \tag{9.47}$$

Eq. (9.47) indicates that $\int \frac{\delta q_{\text{ext}}}{T}$ is positive and has a lower bound equal to $-(k_B \ln 2)\, t_{Y\to X}(k+\delta)$.

These two cases indicate that in absolute value we can write that

$$\left| \int \frac{\delta q_{\text{ext}}}{T} \right| \geq (k_B \ln 2)\, |\, t_{Y\to X}(k+\delta)\, | \tag{9.48}$$

Furthermore, for isothermal processes, we can write:

$$|\, \Delta q_{\text{ext}}\, | \geq (k_B T \ln 2)\, |\, t_{Y\to X}(k+\delta)\, | \tag{9.49}$$

which indicates a linearity on the temperature T. Now, consider that physical system X interacts with two surroundings in the context of the source Y. The first one, is at colder temperature T_c and the second surrounding environment is hold at a higher temperature T_h. During the interaction with the first environment the system dissipates heat

$$\Delta q_{\text{ext}} \leq -(k_B T_c \ln 2)\, t_{Y\to X}(k+\delta) \tag{9.50}$$

which is negative. On the other hand, during the interaction with the second surrounding environment, the system absorbs heat

$$\Delta q_{\text{ext}} \geq -(k_B T_h \ln 2)\, t_{Y\rightarrow X}(k+\delta) \tag{9.51}$$

which is positive. Then, the net heat flow is given as:

$$\mid \Delta q_{\text{net,ext}} \geq\mid -(k_B(T_h - T_c)\ln 2)\, t_{Y\rightarrow X}(k+\delta) \mid \tag{9.52}$$

or

$$\mid \Delta q_{\text{net,ext}} \mid\geq (k_B(T_h - T_c)\ln 2) \mid t_{Y\rightarrow X}(k+\delta) \mid \tag{9.53}$$

These results indicate an essential relationship between the local transfer entropy and thermodynamics, that is $t_{Y\rightarrow X}(k+\delta)$ is proportional to $-\Delta S_{\text{ext}}$, which is the external entropy production during the irreversible transition in the context of the source Y.

9.7 A Statistical Mechanics Point of View of Transfer Entropy

In this section, we are going to introduce a rigorous theory for the information transfer between dynamical system components using statistical mechanics developed in Liang and Kleeman (2005) and Liang (2013).

Consider the dynamical system characterized by the vector $\mathbf{X}$ in a phase space Ω:

$$\mathbf{X} = \{x_1, x_2, \cdots, x_N\}$$

which is given as follows:

$$\frac{d\mathbf{X}}{dt} = \mathbf{F}(\mathbf{X}) \tag{9.54}$$

The joint differential entropy between x_1 and x_2 is given as

$$h(t) = -\int\int_{\Omega} \rho(\mathbf{X}, t) \ln \rho(\mathbf{X})\, dx_1\, dx_2 \tag{9.55}$$

where $\rho(\mathbf{X}, t)$ is the phase space distribution function of the ensemble, which satisfies the Liouville equation (Goldstein 2002):

$$\frac{d\rho}{dt} = \frac{\partial\rho}{\partial t} + \mathbf{F}\cdot\nabla\rho = 0$$

and $\rho(\mathbf{X}, t)$ is a conserved quantity. It can be found that rate change of the joint differential entropy is

$$\frac{dh(t)}{dt} = \langle \nabla \cdot \mathbf{F} \rangle \tag{9.56}$$

which indicates that the change on the differential entropy depends on the compressibility of the phase space. We can now determine the marginal differential entropy of the variable x_1 as:

$$h_1(t) = -\int \rho_1(x_1, t) \ln \rho_1(x_1, t)\, dx_1 \tag{9.57}$$

where $\rho_1(x_1, t)$ is the marginal distribution function of x_1 defined as

$$\rho_1(x_1, t) = \int \rho(x_1, x_2, t)\, dx_2$$

Then, the change on the marginal differential entropy can be determined as

$$\frac{dh_1(t)}{dt} = -\int\int \rho(x_1, x_2, t) \left(\frac{F_1}{\rho_1}\right)\left(\frac{\partial \rho_1}{\partial x_1}\right) dx_1\, dx_2 \tag{9.58}$$

which can further be written as

$$\frac{dh_1(t)}{dt} = \frac{dH_1^*}{dt} + \frac{d}{dt}(H_1 - H_1^*) \tag{9.59}$$

where

$$\frac{dH_1^*}{dt} = \int\int \rho(x_1, x_2, t) \frac{\partial F_1}{\partial x_1}\, dx_1\, dx_2$$

Then, the transfer entropy from x_2 component to x_1 is defined as:

$$\begin{aligned} T_{2\to 1} &= \frac{d(H_1 - H_1^*)}{dt} \\ &= -\int\int \rho(x_1, x_2, t) \left(\frac{F_1}{\rho_1}\frac{\partial \rho_1}{\partial x_1} + \frac{\partial F_1}{\partial x_1}\right) dx_1\, dx_2 \\ &= -\int\int \rho_{2|1}(x_2|x_1, t) \frac{\partial\left(F_1 \rho_1(x_1, t)\right)}{\partial x_1}\, dx_1\, dx_2 \end{aligned} \tag{9.60}$$

where $\rho_{2|1}(x_2|x_1, t)$ is the conditional distribution function defined as

$$\rho_{2|1}(x_2|x_1, t) = \frac{\rho(x_1, x_2, t)}{\rho_1(x_1, t)}$$

Consider the transformations (discrete maps):

$$\Phi : \begin{cases} x_1(t+\Delta t) = x_1(t) + \Delta t F_1(\mathbf{X}) \\ x_2(t+\Delta t) = x_2(t) + \Delta t F_2(\mathbf{X}) \\ \ldots\ldots \\ x_N(t+\Delta t) = x_N(t) + \Delta t F_N(\mathbf{X}) \end{cases}$$

Using the Frobenius-Perron operator:

$$\mathcal{P}\rho(x_1, x_2) = \rho[\Phi^{-1}(x_1, x_2)] \mid J^{-1} \mid$$

where

$$J_{ij}^{-1} = \left[\frac{\partial^2(\Phi^{-1}(x_1, x_2))}{\partial x_i \partial x_j} \right]$$

for $i,\ j = 1,\ 2$, and $\mid J^{-1} \mid$ is the determinant of the inverse Jacobian matrix, J^{-1}. The entropy increase after applying an invertible mapping Φ is:

$$\Delta H = -\int\int \mathcal{P}\rho \ln \mathcal{P}\rho dx_1 dx_2 + \int\int \rho \ln \rho dx_1 dx_2 = \langle \ln \mid J \mid \rangle$$

Consider now the entropy transfer, e.g., X_2 to X_1 first; the entropy of X_1 increases as (Liang and Kleeman 2005; Liang 2013)

$$\begin{aligned} \Delta H_1 = &-\int_{\Omega_1} \left(\int_{\Omega_2} \mathcal{P}\rho dx_2 \right) \ln \left(\int_{\Omega_2} \mathcal{P}\rho dx_2 \right) dx_1 \\ &+ \int_{\Omega_1} \rho_1 \ln \rho_1 dx_1 \\ &= (\Delta H_1 - \Delta H_1^*) + \Delta H_1^* = T_{2\to 1} + \Delta H_1^* \end{aligned} \tag{9.61}$$

It can be noticed that when Φ_1 is invertible, then, $\Delta H_1^* = \langle \ln \mid J_1 \mid \rangle$, and when Φ_1 is independent of X_2, then $\Delta H_1^* = \Delta H_1$, i.e. the asymmetry property.

Because of the asymmetry, information transfer is distinctly different from the transfer of other quantities, such as energy. In other words, the information is not lost in one component for another component to receive it. That is, according to Liang (2013), it could be that $T_{j\to i} = 0$, for example if Φ_i (or equivalently F_i for the continuous case) is independent of x_j, but in mean time, $T_{i\to j}$ needs not to be zero unless Φ_j (or equivalently F_j for the continuous case) is independent of x_i.

As an illustration, one can consider the Baker transformation, which is an area-preserving chaotic map $\Phi : \Omega \to \Omega$ (Liang 2013):

$$\Phi(x_1, x_2) = \begin{cases} \left(2x_1, \frac{x_2}{2}\right), & 0 \le x_1 \le \frac{1}{2},\ 0 \le x_2 \le 1 \\ \left(2x_1 - 1, \frac{1}{2}x_2 + \frac{1}{2}\right), & \frac{1}{2} < x_1 \le 1,\ 0 \le x_2 \le 1 \end{cases}$$

where Ω is a unit square $\Omega = [0, 1] \times [0, 1]$. The Jacobian J is

$$J = \det\left(\frac{\partial\left(\Phi_1(x_1), \Phi_2(x_2)\right)}{\partial\left(x_1, x_2\right)}\right) = 1$$

Following Liang (2013), the entropy is a conserved quantity because of the area-conserving property of the map, which means that

$$\Delta H = \langle \ln J \rangle = 0$$

indicating that this is an invertible map with inverse map given by (Liang 2013):

$$\Phi^{-1}(x_1, x_2) = \begin{cases} \left(\frac{x_1}{2}, 2x_2\right), & 0 \le x_1 \le 1,\ 0 \le x_2 < \frac{1}{2} \\ \left(\frac{x_1 + 1}{2}, 2x_2 - 1\right), & 0 \le x_1 \le 1,\ \frac{1}{2} \le x_2 \le 1 \end{cases}$$

The projection operator $\mathcal{P}$ can be found following Liang (2013) as:

$$\begin{aligned} \mathcal{P}\rho(x_1, x_2) &= J^{-1}\rho\left(\Phi^{-1}(x_1, x_2)\right) \\ &= \begin{cases} \rho\left(\frac{x_1}{2}, 2x_2\right), & 0 \le x_1 \le 1,\ 0 \le x_2 < \frac{1}{2} \\ \rho\left(\frac{x_1 + 1}{2}, 2x_2 - 1\right), & 0 \le x_1 \le 1,\ \frac{1}{2} \le x_2 \le 1 \end{cases} \end{aligned} \tag{9.62}$$

Next, the integration with respect to x_2 of $\mathcal{P}\rho(x_1, x_2)$ in Eq. (9.62) allows obtaining the marginal density of x_1, ρ_i at time $\tau + 1$ (Liang 2013):

$$\begin{aligned} (\mathcal{P}\rho)_1(x_1) &= \int_0^{1/2} \rho\left(\frac{x_1}{2}, 2x_2\right) dx_2 + \int_{1/2}^{1} \rho\left(\frac{x_1 + 1}{2}, 2x_2 - 1\right) dx_2 \\ &= \frac{1}{2}\int_0^1 \left(\rho\left(\frac{x_1}{2}, x_2\right) + \rho\left(\frac{x_1 + 1}{2}, x_2\right)\right) dx_2 \\ &= \frac{1}{2}\left(\rho_1\left(\frac{x_1}{2}\right) + \rho_1\left(\frac{x_1 + 1}{2}\right)\right) \end{aligned} \tag{9.63}$$

where a change of integration variable is employed, respectively, $x_2 \to 2x_2$ in the first integral and $x_2 \to 2x_2 - 1$ in the second integral.

If the coordinate x_2 is fixed, then one dimensional map is obtained (Liang 2013): $\Phi_1 : \Omega_1 \to \Omega_1$, where $\Omega_1 = [0, 1]$ and

$$\Phi_1(x_1) = 2x_1 \pmod 1$$

Therefore,

$$\begin{aligned}(\mathcal{P}^*\rho)_1(x_1) &= \frac{\partial}{\partial x_1} \int\limits_{\Phi_1^{-1}([0,x_1])} \rho(x)dx \qquad (9.64)\\ &= \frac{\partial}{\partial x_1} \int\limits_0^{x_1/2} \rho(x)dx + \frac{\partial}{\partial x_1} \int\limits_{1/2}^{(x_1+1)/2} \rho(x)dx \\ &= \frac{1}{2}\left(\rho_1\left(\frac{x_1}{2}\right) + \rho_1\left(\frac{x_1+1}{2}\right)\right)\end{aligned}$$

where * denotes projection along the dimension x_1, while keeping dimension x_2 constant, and $\Phi_1^{-1}([0, x_1])$ is the counterimage of $[0, x_1]$ (Liang 2013):

$$\Phi_1^{-1}([0, x_1]) = \left[0, \frac{x_1}{2}\right] \cup \left[\frac{1}{2}, \frac{x_1+1}{2}\right]$$

Thus, the transfer entropy from 2 to 1 is given by:

$$T_{2\to 1} = \Delta H_1 - \Delta H_1^*$$

where

$$\begin{aligned}\Delta H_1 &= -\int (\mathcal{P}\rho)_1(x_1) \ln ((\mathcal{P}\rho)_1(x_1))\, dx_1 + \int \rho_1(x_1) \ln \rho_1(x_1) dx_1 \qquad (9.65)\\ \Delta H_1^* &= -\int (\mathcal{P}^*\rho)_1(x_1) \ln \left((\mathcal{P}^*\rho)_1(x_1)\right) dx_1 + \int \rho_1(x_1) \ln \rho_1(x_1) dx_1\end{aligned}$$

Since, $(\mathcal{P}\rho)_1(x_1) = (\mathcal{P}^*\rho)_1(x_1)$ as indicated by Eqs. (9.63) and (9.64), then $\Delta H_1 = \Delta H_1^*$, and thus

$$T_{2\to 1} = 0$$

Now, the information flow in the opposite direction ($T_{1\rightarrow 2}$) can be calculated as described in Liang (2013). For that, the marginal density is first evaluated:

$$(\mathcal{P}\rho)_2(x_2) = \int_0^1 \mathcal{P}\rho(x_1, x_2)dx_1 \tag{9.66}$$

$$= \begin{cases} \int_0^1 \rho\left(\frac{x_1}{2}, 2x_2\right) dx_1, & 0 \leq x_2 < \frac{1}{2} \\ \int_0^1 \rho\left(\frac{x_1+1}{2}, 2x_2 - 1\right) dx_1, & \frac{1}{2} \leq x_2 \leq 1 \end{cases}$$

Then, the marginal entropy increase of x_2 is obtained by

$$\begin{aligned} \Delta H_2 = &-\int_0^1 (\mathcal{P}\rho)_2(x_2) \ln\left((\mathcal{P}\rho)_2(x_2)\right) dx_2 \\ &+ \int_0^1 \rho_2(x_2) \ln\left(\rho_2(x_2)\right) dx_2 \\ = &-\int_0^1 \int_0^1 \mathcal{P}\rho(x_1, x_2) \ln\left(\int_0^1 \mathcal{P}\rho(s, x_2)ds\right) dx_1 dx_2 \\ &+ \int_0^1 \int_0^1 \rho(x_1, x_2) \ln\left(\int_0^1 \rho(s, x_2)ds\right) dx_1 dx_2 \\ = &-\ln 2 \\ &+ \int_0^1 \int_0^{1/2} \rho(x_1, x_2) \ln\left(\int_0^1 \rho(s, x_2)ds\right) dx_1 dx_2 \\ &+ \int_0^1 \int_{1/2}^1 \rho(x_1, x_2) \ln\left(\int_0^1 \rho(s, x_2)ds\right) dx_1 dx_2 \\ &- \int_0^1 \int_0^{1/2} \rho(x_1, x_2) \ln\left(\int_0^{1/2} \rho(s, x_2)ds\right) dx_1 dx_2 \\ &- \int_0^1 \int_{1/2}^1 \rho(x_1, x_2) \ln\left(\int_{1/2}^1 \rho(s, x_2)ds\right) dx_1 dx_2 \end{aligned} \tag{9.67}$$

where the term $-\ln 2$ is because the interval of x_2 values in Eq. (9.66) is not a closed interval on $x_2 = 1/2$. Equation (9.67) can be arranged in a compact form (Liang 2013):

$$\Delta H_2 = -\ln 2 + I_1 + I_2 \tag{9.68}$$

where

$$I_1 = \int_0^1 \int_0^{1/2} \rho(x_1, x_2) \ln \left(\int_0^1 \rho(s, x_2) ds \right) dx_1 dx_2$$
$$- \int_0^1 \int_0^{1/2} \rho(x_1, x_2) \ln \left(\int_0^{1/2} \rho(s, x_2) ds \right) dx_1 dx_2$$
$$I_2 = \int_0^1 \int_{1/2}^1 \rho(x_1, x_2) \ln \left(\int_0^1 \rho(s, x_2) ds \right) dx_1 dx_2$$
$$- \int_0^1 \int_{1/2}^1 \rho(x_1, x_2) \ln \left(\int_{1/2}^1 \rho(s, x_2) ds \right) dx_1 dx_2$$

In addition, fixing x_1, and evaluating the Jacobian J_2 (Liang 2013):

$$J_2 = \det \left(\frac{\partial \Phi(x_1, x_2)}{\partial x_2} \right)_{x_1} = \frac{1}{2}$$

The marginal entropy increase of x_2 keeping x_1 fixed is obtained as

$$\Delta H_2^* = \langle \ln J_2 \rangle = -\ln 2 \tag{9.69}$$

Thus, the information flow from 1 to 2 is given by:

$$T_{1\to 2} = \Delta H_2 - \Delta H_2^* = I_1 + I_2 > 0$$

indicating that $T_{2\to 1} = 0$, but in mean time $T_{1\to 2} > 0$, which is the asymmetry property of the information flow.

Furthermore, ΔH_1^* is due to the expansion or contraction of the phase space in dimension X_1, while keeping dimension X_2 unchanged. Even though an N-dimension system is invertible, its components may not be so. As elucidated above, while X_1 gains information from X_2, X_2 might have nothing to do with X_1. For a non-Hamiltonian system (Tuckerman et al. 2001), however, the compressibility does not need to vanish ($\mid J \mid \neq 1$).

9.8 Mutual Information

Consider again two time series as the following:

$$\{x_k^{\mu_x}\}_{k=k_0,\cdots,T-1}; \quad \{y_k^{\mu_y}\}_{k=k_0,\cdots,T-1}$$

which are representing two random processes. Here, X and Y are considered *fully* independent of each other if the following relationship is satisfied between their fluctuation probabilities:

$$p(\mathbf{x}_k^{\mu_x}, \mathbf{y}_k^{\mu_y}) = p(\mathbf{x}_k^{\mu_x})p(\mathbf{y}_k^{\mu_y}), \tag{9.70}$$

otherwise they are dependent. A continuous measure of the *distance* between the probability distributions $p(\mathbf{x}_k^{\mu_x}, \mathbf{y}_k^{\mu_y})$ and $p(\mathbf{x}_k^{\mu_x})p(\mathbf{y}_k^{\mu_y})$, can be introduced using the so-called Kullback-Leibler distance (Kullback and Leibler 1951; Kullback 1959, 1987):

$$\mathrm{I}(X;\ Y) \equiv K_D = \sum_{k=k_0}^{T-1} p(\mathbf{x}_k^{\mu_x}, \mathbf{y}_k^{\mu_y}) \log_2 \frac{p(\mathbf{x}_k^{\mu_x}, \mathbf{y}_k^{\mu_y})}{p(\mathbf{x}_k^{\mu_x})p(\mathbf{y}_k^{\mu_y})}, \tag{9.71}$$

which equals the *mutual information* between two time series.

Equation (9.71) can also be expressed in terms of the Shannon information entropy (S) (Shannon and Weaver 1949) as the following:

$$\begin{aligned} \mathrm{I}(X;\ Y) &= \sum_k p(\mathbf{x}_k^{\mu_x}, \mathbf{y}_k^{\mu_y}) \log_2 p(\mathbf{x}_k^{\mu_x}, \mathbf{y}_k^{\mu_y}) \\ &\quad - \sum_k p(\mathbf{x}_k^{\mu_x}, \mathbf{y}_k^{\mu_y}) \log_2 p(\mathbf{x}_k^{\mu_x})p(\mathbf{y}_k^{\mu_y}) \\ &= -H(\mathbf{X}^{\mu_x}, \mathbf{Y}^{\mu_y}) - \sum_k p(\mathbf{x}_k^{\mu_x}, \mathbf{y}_k^{\mu_y}) \log_2 p(\mathbf{x}_k^{\mu_x}) \\ &\quad - \sum_k p(\mathbf{x}_k^{\mu_x}, \mathbf{y}_k^{\mu_y}) \log_2 p(\mathbf{y}_k^{\mu_y}) \\ &= H(\mathbf{X}^{\mu_x}) + H(\mathbf{Y}^{\mu_y}) - H(\mathbf{X}^{\mu_x}, \mathbf{Y}^{\mu_y}), \end{aligned} \tag{9.72}$$

where the sum is over all states.

9.9 Symbolic Analysis

To calculate the Shannon entropy we need to label all the possible states visited by the system. In general, we could use the binning method, in which we define the range of fluctuation amplitudes for each degree of freedom and then, divide the range into bins and enumerate the entries (n_k) for each bin (k), which will give us the probability of visiting each state (i.e. bin). The disadvantage of this approach is that one has to define the bin width, which introduces a new free parameter and thus increasing the complexity of the method. Other used methods include, for example, kernel density estimators (Moon et al. 1995), or k-nearest neighbor distances approach (Kraskov et al. 2004). In particular, the last method is considered to be an improvement of the mutual information estimators (Kraskov et al. 2004).

In this section, we are going to follow a different approach, namely *symbolic* method, based on coarse-graining the time series into symbols (Lehrman et al. 1997; Rechester and White 1991a,b; Bandt and Pompe 2002; Staniek and Lehnertz 2008; Kamberaj and van der Vaart 2009a). In particular, we use the symbolization technique proposed in Kamberaj and van der Vaart (2009a), which we found to be computationally very robust and at the same time maximizing the information content about the real-time series. According to this method, a symbolic sequence

$$\left(\hat{X}_0, \hat{X}_1, \cdots, \hat{X}_{N-1}\right)$$

is created associated with the time series

$$(X_0, X_1, \cdots, X_{N-1})$$

through a process, called here *coarse-graining* (Kamberaj and van der Vaart 2009a) (and the references therein). In the coarse-graining, all information concerning the dynamics of series is suitably encoded using a partitioning of phase space. The time-series $(X_0, X_1, \cdots, X_{N-1})$ is converted into a symbolic sequence using the following rule

$$\hat{X}_j = \hat{S}_k, \quad \text{if } X_k^c < X_j < X_{k+1}^c \tag{9.73}$$

where $\left(X_0^c, X_1^c, \cdots, X_D^c\right)$ is a given set of $D + 1$ critical points, and

$$\left(\hat{S}_0, \hat{S}_1, \cdots, \hat{S}_{D-1}\right)$$

is a set of D symbols, here the numbers 0, 1, 2, and so on. Here, D is chosen such that the Kraft inequality is satisfied:

$$\sum_k D^{-m_k} \leq 1$$

where the sum runs over all state vectors, and m_k is the length of each state vector, which for a single time series is considered to be equal for every state vector.

The new state vector generated this way,

$$\hat{\mathbf{X}}_k^{\mu} \equiv \left(\hat{S}_1^{(k)}, \hat{S}_2^{(k)}, \cdots, \hat{S}_m^{(k)}\right)^T,$$

represents symbolic state vector, which is a subset of numbers from 0 to $D - 1$. Concatenation of the symbols of a sequence of length m yields the word W_k:

$$W_k = \left(\hat{X}_{k-(m-1)\tau} \cdots \hat{X}_{k-\tau} \hat{X}_k\right)$$

A particular sequence of symbols $\{\hat{X}_0, \hat{X}_1, \cdots, \hat{X}_{N-1}\}$ is uniquely characterized by the words W_k for $k = k_0, \cdots, N-1$. The probability of finding a particular value of W_k is calculated from the simulation data, and used to compute the Shannon entropy

$$H(\hat{\mathbf{X}}^{\mu}) = -\sum_k p(W_k)\log_2 p(W_k) \tag{9.74}$$

Since the time series $\{X_0, X_1, \cdots, X_{N-1}\}$ is mapped onto the symbolic sequence $\{\hat{X}_0, \hat{X}_1, \cdots, \hat{X}_{N-1}\}$ uniquely (i.e., the symbolic representation is injective) the entropies $H(\hat{\mathbf{X}}^{\mu})$ and $H(\mathbf{X}^{\mu})$ coincides (Bonanno and Mega 2004).

We obtain the critical points $\{X_d^c\}_{d=0}^D$ for a particular series by maximizing the entropy for all possible partitions. Increasing the number of critical points will initially increase the information entropy, but after a sufficient number of critical points, the information entropy plateaus. At this point the optimum number of critical points has been reached, a further increase will not increase the accuracy of the calculation, but it does slow down the computation. In our implementation, we optimize critical points by maximizing the Shannon entropy through a Monte Carlo approach (Kamberaj and van der Vaart 2009a). Similarly, the joint Shannon information entropy of two discrete symbolic processes $\{\hat{X}_0, \hat{X}_1, \cdots, \hat{X}_{N-1}\}$ and $\{\hat{Y}_0, \hat{Y}_1, \cdots, \hat{Y}_{N-1}\}$ is calculated as

$$H(\hat{\mathbf{X}}^{\mu_x}, \hat{\mathbf{Y}}^{\mu_y}) = -\sum_k p(\bar{W}_k)\log_2 p(\bar{W}_k)$$

where $\bar{W}_k$ is the concatenation of two words, $W_k^{(x)}$ and $W_k^{(y)}$, representing the words of processes X and Y, respectively.

In general, the length of discrete processes and the number of states are limited by the sampling. To correct for the finite sampling of the time discrete processes, we use (Grassberger 1988):

$$H(\hat{\mathbf{X}}^{\mu}) = \frac{1}{\ln 2}\left(\ln N - \frac{1}{N}\sum_k n_k \psi(n_k)\right), \tag{9.75}$$

where sum is over all states, n_k is the frequency of observing state k, and $\psi(x)$ is the derivative of Gamma function Γ with respect to x. Note that the division by ln 2 is necessary to convert into bits the units of Shannon entropy in Eq. (9.75).

The symbolic transfer entropy can be written as:

$$\hat{T}_{\hat{Y}\to\hat{X}} = H(\hat{\mathbf{X}}^{\mu_x+1}) - H(\hat{\mathbf{X}}^{\mu_x}) - H(\hat{\mathbf{X}}^{\mu_x+1}, \hat{\mathbf{Y}}^{\mu_y}) + H(\hat{\mathbf{X}}^{\mu_x}, \hat{\mathbf{Y}}^{\mu_y}) \tag{9.76}$$

From Eq. (9.30), we define the symbolic local transfer entropy as:

$$\begin{aligned}\hat{t}_{\hat{Y}\to\hat{X}}(k+\delta) = &\left[\log_2 p(\hat{\mathbf{x}}_k^{\mu_x+1}, \hat{\mathbf{y}}_k^{\mu_y}) - \log_2 p(\hat{\mathbf{x}}_k^{\mu_x}, \hat{\mathbf{y}}_k^{\mu_y})\right] \\ &- \left[\log_2 p(\hat{\mathbf{x}}_k^{\mu_x+1}) - \log_2 p(\hat{\mathbf{x}}_k^{\mu_x})\right]\end{aligned} \tag{9.77}$$

The symbolic mutual information, $\hat{I}_{XY}$, is calculated as

$$\hat{I}(X;\, Y) = H(\hat{\mathbf{X}}^{\mu_x}) + H(\hat{\mathbf{Y}}^{\mu_y}) - H(\hat{\mathbf{X}}^{\mu_x}, \hat{\mathbf{Y}}^{\mu_y}). \tag{9.78}$$

We will use a generalized time coarse-grained correlation coefficient similar to that proposed elsewhere (Joe 1989):

$$R_{XY}^{SMI} = \left[1 - \exp\left(-2\,(\ln 2)\,\hat{I}(X;\, Y)/g\right)\right]^{1/2}, \tag{9.79}$$

where g is the space dimension. For a Gaussian joint probability distribution $p(\hat{\mathbf{x}}_k^{\mu_x}, \hat{\mathbf{y}}_k^{\mu_y})$, R_{XY}^{SMI} coincides with the Pearson correlation coefficient, $R_{XY}^{Pearson}$ (Cellucci et al. 2005).

Chapter 10
Practical Aspects of Molecular Dynamics Simulations

In this chapter, we will introduce some practical aspects of molecular dynamics simulations, such as designing the constraints (e.g., SHAKE), periodic boundary conditions, spherical cutoffs, treatment of the long-range interactions (in particular, electrostatic interactions), and identifying the equilibrium states of the simulations.

For more about this topic, the reader can consider the material in Leach (2001), Allen and Tildesley (1989) and Frenkel and Smit (2001).

10.1 Designing Constraints for Molecular Dynamics Simulations

Realization of a molecular dynamics simulation has to account for both, the physical nature of the system under study and the available computational power. System size, time step, and total time duration must be selected so that the calculation can finish within a reasonable period. On the other hand, the simulation run has to be long enough to be able to capture phenomena at the time scales of the studied natural processes. Besides, statistically, the timescales span by the simulations should match the kinetics of the natural process. Most scientific works about the dynamics of proteins and DNA use data from simulations spanning nanoseconds to microseconds. In general, to obtain the simulations at these timescales, we will need several CPU-days to CPU-years also depending on the system size. Besides, parallel algorithms allow sharing the computation workload among CPUs, and special algorithms for running in the computer's GPU (Stone et al. 2011; Phillips et al. 2014).

During a classical MD simulation, the most computationally intensive task is the evaluation of the potential and forces as a function of the particles internal coordinates, where the most expensive are the evaluation of the non-bonded or non-

H. Kamberaj, *Molecular Dynamics Simulations in Statistical Physics: Theory and Applications*, Scientific Computation,
https://doi.org/10.1007/978-3-030-35702-3_10

covalent terms. Common molecular dynamics simulations scale by $\mathcal{O}(N^2)$, where N is the number of particles if all pair-wise electrostatic and van der Waals interactions account explicitly. This computational cost reduces by employing electrostatics methods, such as Particle Mesh Ewald ($\mathcal{O}(N \log(N))$) or good spherical cutoff techniques ($\mathcal{O}(N)$).

The integration time-step is another factor that determines total CPU time required for a simulation, which gives in typical MD simulation the time length between evaluations of the potential and forces. The timestep has to be small enough to avoid numerical errors, as a rule, lower than the fastest vibrational frequency in the system. In typical classical MD simulation, the time step is of the order of 1 femtosecond. This value may extend by using algorithms, such as SHAKE (Ryckaert et al. 1977), which fix the vibrations of the fastest atoms (e.g., hydrogens) into place. Multiple time scale methods allow for extended times between updates of slower long-range forces (Tuckerman et al. 1992).

For performing the simulations of the macromolecules in a solvent environment, often a choice should be made between explicit solvent and implicit solvent. Usually, the explicit solvent models of the force fields, such as the TIP3P and SPC/E waters, are computationally expensive, while implicit solvents use a mean-field approach, and hence they are usually less costly. Typically, explicit solvent models require the inclusion of about ten times more particles in the simulation. Besides, the explicit models are such that they allow to reproduce specific properties of the solute macromolecules and their kinetics.

In the molecular dynamics simulations of the macromolecules, the simulation box size must be large enough to avoid boundary condition artifacts. Usually, the boundary conditions are treated by either having fixed values at the edges or by applying periodic boundary conditions in which one side of the simulation loops back to the opposite side. In practice, this mimics a bulk phase.

10.2 Initial Configuration

The initial configuration for the system is chosen before we perform MD simulations. That is very crucial for the success of the MD simulation since it often depends on the arrangement of the particles. Usually, the initial configuration for a system at equilibrium is close to the desired state. Also, it is recommended that the energy of the initial configuration does not contain terms with high-energy interactions. Often, to remove such terms, energy minimization can be performed before the start of MD simulations to minimize the system to the closest local minimum.

If there exists an experimental structure of the system, then this could be used. If experimental structures are not available, then, to simulate homogeneous liquid systems, for example, containing a large number of molecules, lattice-like structures can be chosen as a starting configuration. Lattice structures can be selected from one of the common crystallographic lattices. The most common one is the face-centered cubic lattice, which contains $4n^3$ points, where $n = 2, 3, 4, \cdots$, resulting in an initial configuration of 32, 108, 256, $\cdots$, atoms or molecules. The size of the lattice is also chosen appropriately; often it is selected such that the density is lower than of the system under study. If the system under investigation is composed of molecules, then it is also necessary to assign an orientation for each molecule. For linear molecules, a face-centered cubic lattice with molecules regularly oriented along the four diagonals of the unit cell (for example, the solid structure of CO_2). For non-linear molecules, the orientation can be chosen at random or by making small random changes from the direction in a regular lattice. For high-density structures, this may result in non-physical overlaps and instability of simulations, especially in the case of large molecules. Therefore, in such cases, it is suggested to chose an initial configuration close to the expected equilibrium distribution. For example, all molecules aligned approximately in the same direction initially, as in the case of the liquid crystals. It is important noting that merely placing the molecules at random can usually give rise to high-energy overlaps and numerical instabilities.

For MD simulations of inhomogeneous systems, for example, the systems containing a solute molecule or complex macromolecules immersed in a solvent, the starting configuration of the solute may be obtained from an experimental technique (usually X-ray crystallography or NMR) or generated theoretically using modeling methods. The coordinates of the solvent molecules are added to give an appropriate solvent density in normal conditions. (Note that for some solvent molecules the coordinates of the X-Ray structures can be used as a starting point.) In practice, these solute coordinates, obtained from previously equilibrated pure solvent configurations, immerse in the solvent, removing the solvent molecules in close contact with the solute. Typically, for the inhomogeneous systems, some minimization steps are required priory of MD simulations to equilibrate the system to the nearest local minima.

10.3 Periodic Boundary Conditions

In computer simulations, the maximum size of the system is limited by the available storage on the computer or by the speed of execution of the program. As an example, if we consider that one mole of a liquid will contain approximately 10^{23} particles we are far from being able to study this system. Mainly if we are interested in studying the properties of the bulk fluid, a small number of particles we can simulate, if not all of them, would be within the influence of the walls of the boundary. *Periodic boundary conditions* enable a simulation to be performed using a relatively small number of particles, in such a way that every particle experiences force as if it was in bulk fluid phase. In the two-dimensional each box is surrounded by eight neighbors as is shown in Fig. 10.1; in three dimensions each box would have 26 nearest neighbors.

Fig. 10.1 Illustration of the periodic boundary conditions in two dimensions. The graph is produced using VMD program (Humphrey et al. 1996)

The coordinates of the particles in the image boxes can be computed merely by adding or subtracting integral multiples of the box sides. If a particle leaves the box during the simulation, then it is replaced by an imaged particle that enters from the opposite side, as illustrated in Fig. 10.1. The number of particles within the central box thus remain constant. The cubical cell and its close relation, the parallelepiped, are the simplest periodic systems to program and to visualize. However, a cell of a different shape might be more appropriate for a given simulation. It is often sensible to choose a periodic cell that reflects the specified property of the system. For example, allowing the cell to change the shape during a molecular dynamics or Monte Carlo simulation in the smectic phase might affect the packing. Periodic boundaries are widely used in computer simulations, but they do have some drawbacks (Allen and Tildesley 1989). A limitation of the periodic cell is that it is not possible to achieve fluctuations that have a wavelength larger than the length of the cell. The range of the interactions present in the system is also important; if the scale over which the interactions act is larger than the cell size then there should be no problems.

10.4 Potential Cutoffs and the Minimum Image Convention

The most time-consuming part of the computer simulations is the calculation of the energies and forces. For example, for a pairwise model of the potential such as the Lennard-Jones, the time for calculation of the forces is of order N^2 for a N particles system. In principle, the interactions are calculated for each pair of particles in the system. In practice, for many of the interaction models, this is not necessary since the potentials fall off very rapidly with distance as is shown in Fig. 6.6 (Chap. 6).

In such cases, a cut-off applies to the interaction potential with the minimum image convention (Allen and Tildesley 1989). In the minimum image convention, each particle interacts at most with just one image of every other particle in the system, which is repeated infinitely via periodic boundary conditions. When the cut-off has employed the interactions between all pairs of particles that are further apart than the cut-off value are modified using some function $f(r)$, taking into account the closest image.

For example, the Lennard-Jones and electrostatic interactions between a pair of atoms i and j can be re-written as

$$V_{\mathrm{LJ}}(r_{ij}) + V_{\mathrm{Elec}}(r_{ij}) = f(r_{ij})\left[4\varepsilon_{ij}\gamma_{ij}^{\mathrm{LJ}}\left(\frac{\sigma_{ij}^{12}}{r_{ij}^{12}} - \frac{\sigma_{ij}^{6}}{r_{ij}^{6}}\right) + \gamma_{ij}^{\mathrm{Elec}} k_e \frac{q_i q_j}{\varepsilon r_{ij}}\right] \tag{10.1}$$

where $f(r_{ij})$ is either the switch or shift function. In Eq. (10.1), $\gamma_{ij}^{\mathrm{LJ}}$ and $\gamma_{ij}^{\mathrm{Elec}}$ are scaling factors, which are constants defined a priory of simulations. They are often chosen such that $\gamma_{ij}^{\mathrm{LJ}} = \gamma_{ij}^{\mathrm{Elec}} = 0$ for atoms separated by less than three bonds, and they are either one or less than one, depending on the force field, for atoms separated by three bonds, the so-called 1-4 interactions, and moreover, they are both one for atoms separated by more than three bonds, the so-called non-bonded atoms.

If simple truncation of the potential energy function are used, then the switch function is given as:

$$f(r_{ij}) = \begin{cases} 1, & r_{ij} < R_c \\ 0, & r_{ij} \geq R_c \end{cases} \tag{10.2}$$

where R_c is the cutoff distance. It can be seen that a function given by Eq. (10.2) introduces a discontinuity in the force, and hence it does not guarantee the conservation of the energy. Therefore, in practice, different molecular dynamics simulation engines have introduced the so-called switch and/or shift function in either potential energy or force calculations. In this approach, the distance-dependent function $f(r_{ij})$ is determined in terms of the cutoff distance R_c and the so-called switch distance R_s, such as

$$f(r_{ij}) = \begin{cases} 1, & r_{ij} < R_c \\ S(r_{ij}), & R_c \leq r_{ij} \leq R_s \\ 0, & r_{ij} > R_s \end{cases} \tag{10.3}$$

where, now, $S(r_{ij})$ is a smooth function decaying to zero between R_c and R_s. In general, $S(r_{ij})$ can be designed different for the Lennard-Jones and the electrostatic interactions, therefore, Eq. (10.1) can also be written as:

$$V_{\text{LJ}}(r_{ij}) + V_{\text{Elec}}(r_{ij}) = \left[4\varepsilon_{ij} f^{\text{LJ}}(r_{ij})\gamma_{ij}^{\text{LJ}} \left(\frac{\sigma_{ij}^{12}}{r_{ij}^{12}} - \frac{\sigma_{ij}^{6}}{r_{ij}^{6}} \right) \right. \\ \left. + f^{\text{Elec}}(r_{ij})\gamma_{ij}^{\text{Elec}} k_e \frac{q_i q_j}{\varepsilon r_{ij}} \right] \tag{10.4}$$

We can design the so-called *potential switch* functions, $S(r_{ij})$, to be a polynomial function of the distance such that potential energy of non-bonded interactions slowly decays to zero in the interval $[R_c, R_s]$, where $V(r)$, for $r \leq R_c$, is unchanged and $V(R_s) = 0$. In this case, the polynomial function is chosen with such degree that both energy function and its gradient are continuous functions. Such function is given in the following as designed in CHARMM program (Brooks et al. 2009) (based on the CHARMM's manual):

$$f(r_{ij}) = \begin{cases} 1, & r_{ij} < R_c \\ \dfrac{(R_s^2 - r_{ij}^2)^2(R_s^2 + 2r_{ij}^2 - 3R_c^2)}{(R_s^2 - R_c^2)^3}, & R_c \leq r_{ij} \leq R_s \\ 0, & r_{ij} > R_s \end{cases} \tag{10.5}$$

From Eq. (10.5), it can easily be seen that $S(r_{ij})$ decays smoothly to zero as r_{ij} increases from R_c to R_s. In addition, the derivative of $f(r_{ij})$ with respect to the separation r_{ij} is given as:

$$\frac{\partial f(r_{ij})}{\partial r_{ij}} = \begin{cases} 0, & r_{ij} < R_c \\ \dfrac{-12r_{ij}(R_s^2 - r_{ij}^2)(r_{ij}^2 - R_c^2)}{(R_s^2 - R_c^2)^3}, & R_c \leq r_{ij} \leq R_s \\ 0, & r_{ij} > R_s \end{cases} \tag{10.6}$$

which indicates that at both R_c and R_s the derivative is equal to zero. Hence, both the potential function and its derivative are continuous functions.

Other function modifications that can be used are the so-called *potential shift* functions, which are used to avoid the sudden changes on the potential energy function or its gradient. For example, typical potential shift function has the form:

$$f(r_{ij}) = \begin{cases} \left(1 - \left(\frac{r_{ij}}{R_s}\right)^2\right)^2, & r_{ij} \le R_s \\ 0, & r_{ij} > R_s \end{cases} \tag{10.7}$$

and its gradient with respect to r_{ij} is given as the following:

$$\frac{\partial f(r_{ij})}{\partial r_{ij}} = \begin{cases} -4\frac{r_{ij}}{R_s^2}\left(1 - \left(\frac{r_{ij}}{R_s}\right)^2\right), & r_{ij} \le R_s \\ 0, & r_{ij} > R_s \end{cases} \tag{10.8}$$

It can be seen that both potential function and its gradient decay smoothly to zero at $r_{ij} = R_s$, and hence they are continuous functions of the distance.

Another form of the potential shift function is given as the following:

$$f(r_{ij}) = \begin{cases} \left(1 - \frac{r_{ij}}{R_s}\right)^2, & r_{ij} \le R_s \\ 0, & r_{ij} > R_s \end{cases} \tag{10.9}$$

The gradient with respect to r_{ij} is given as the following:

$$\frac{\partial f(r_{ij})}{\partial r_{ij}} = \begin{cases} -\frac{2}{R_s}\left(1 - \frac{r_{ij}}{R_s}\right), & r_{ij} \le R_s \\ 0, & r_{ij} > R_s \end{cases} \tag{10.10}$$

These two potential shift functions are better suited to electrostatic potential interactions when used without the Ewald summation to correct for long-range interactions. For Lennard-Jones potential interactions, it is recommended to use the potential switch functions of the form given by Eq. (10.5). Note that the potential shift function of the form as in Eq. (10.9) used for the electrostatic interactions as follows:

$$V'_{\text{Elec}}(r_{ij}) = k_e \frac{q_i q_j}{\varepsilon r_{ij}} f(r_{ij})$$

yields the following gradient of the potential

$$\begin{aligned} \frac{\partial V'_{\text{Elec}}(r_{ij})}{\partial r_{ij}} &= -k_e \frac{q_i q_j}{\varepsilon r_{ij}^2}\left(1 - \frac{r_{ij}}{R_s}\right)^2 - k_e \frac{q_i q_j}{\varepsilon r_{ij}} \frac{2}{R_s}\left(1 - \frac{r_{ij}}{R_s}\right) \\ &= -k_e \frac{q_i q_j}{\varepsilon r_{ij}^2} + k_e \frac{q_i q_j}{\varepsilon R_s^2} \\ &= \frac{\partial V_{\text{Elec}}(r_{ij})}{\partial r_{ij}} + k_e \frac{q_i q_j}{\varepsilon R_s^2} \end{aligned} \tag{10.11}$$

where $\dfrac{\partial V_{\text{Elec}}(r_{ij})}{\partial r_{ij}}$ represents the gradient of unmodified potential energy function, and the second term is a constant term, which represents the force shift. That is why sometimes this method is also called *force shift*.

To provide a smooth separation between the fast and slow forces, in particular in multiple-time step integration schemes, the so-called *force switching* functions are also used. The force switching function is applied as the following (Schlick 2010):

$$F_{\text{fast}}^{\text{LJ, Elec}}(r_{ij}) = f(r_{ij})F^{\text{LJ, Elec}}(r_{ij}) \tag{10.12}$$
$$F_{\text{slow}}^{\text{LJ, Elec}}(r_{ij}) = \left(1 - f(r_{ij})\right) F^{\text{LJ, Elec}}(r_{ij})$$

where the switching function is defined as follows:

$$f(r_{ij}) = \begin{cases} 1, & r_{ij} < R_c \\ 1 + \left(S(r_{ij})\right)^2 \left(2S(r_{ij}) - 3\right), & R_c \le r_{ij} \le R_s \\ 0, & r_{ij} > R_s \end{cases} \tag{10.13}$$

where

$$S(r) = \frac{r^2 - R_c^2}{R_s^2 - R_c^2}$$

When periodic boundary conditions are used, the cut-off must not be so large that a particle sees its image or the same particle twice: the cut-off is limited to no more than half of the length of the cell in the case of the cubical cell; for rectangular cells, it must be smaller than half of the length of the shortest side.

10.5 Neighbor Lists

The use of the cut-off may not significantly reduce the time required to compute the number of interactions. That is because to decide if every pair of particles is close enough to calculate their interaction energy, we need to figure the distance between the particles in each pair in the system. Verlet (Allen and Tildesley 1989) proposed a method that would significantly decrease the time of calculation of the energy and forces, by keeping a stored list of the nearest neighbors to molecule i which are all particles within the cut-off distance, together with all particles that are slightly further away than the cut-off distance (see Fig. 10.2). For the simulations of the multi-sites molecules, the upper limit on the cut-off may also be affected by the size of the molecules. That is most efficiently done using a large pointer array P of neighbor lists as described in Allen and Tildesley (1989) and Leach (2001) and illustrated in Fig. 10.2.

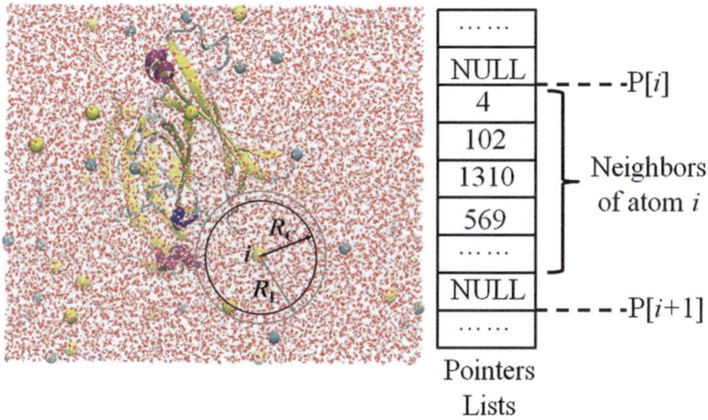

Fig. 10.2 Illustration of the pointer and neighbor arrays used to implement the neighbor list

The pointer array P indicates where in the neighboring list array the first and the last neighbor for that atom is located. The neighbors of the atom i are stored in elements $P[i]$ through $P[i] \rightarrow$ first and $P[i] \rightarrow$ last of the pointer array $P[i]$ until $P[i] \rightarrow \text{next} = NULL$, which indicate that the neighbor list of atom i finishes and the neighbor list of the next atom starts. Once the neighbor lists create, the program does not check through all atoms but only those within the list for atom i and the computation time is of order N (number of atoms in the system). The nearest neighboring lists to atoms are updated once a specified criterion satisfies, such as one of the atoms displaces more than $R_{\mathrm{L}} - R_{\mathrm{C}}$. The computation time required to update it is of order N^2. Usually, the difference $R_{\mathrm{L}} - R_{\mathrm{C}}$ is in the range $1-2$ Å. It is worth noting that large values of $R_{\mathrm{L}} - R_{\mathrm{C}}$ yield a less often update of the neighbor lists; however, it increases the length of the neighbor lists for the atoms. On the other hand, small values of $R_{\mathrm{L}} - R_{\mathrm{C}}$ produces short neighbor lists for the atoms, however, more often requires to update the neighbor lists. Therefore, practically, it is a trade off the choice of R_{L} and R_{C}.

10.6 Cell Lists

The cell lists is another approach used to reduce the computational efforts for calculation of the energy and forces in the MD simulations (Allen and Tildesley 1989). In this approach, the simulation box is partitioned into cells, as shown in Fig. 10.3, with a side length of each cell taken about the cut-off distance or slightly bigger. Then, each atom of a cell interacts only with atoms of the same cell or the other neighbor cells. Since the time for assigning the atoms to cells has a computation time complexity of $\mathcal{O}(N)$, where N is the number of atoms, then the

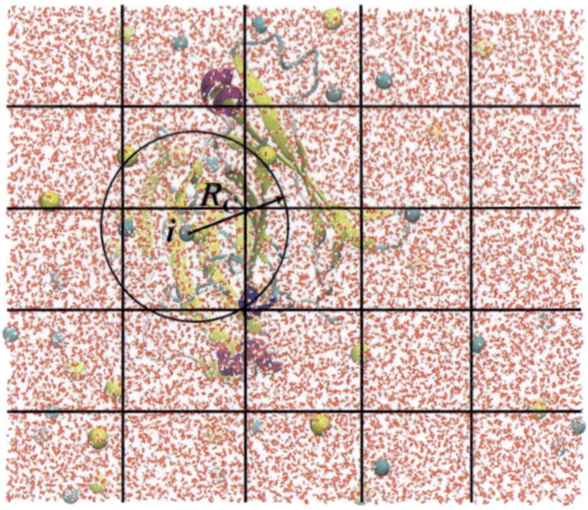

Fig. 10.3 Illustration of the simulation box partition used to implement the cell list

complexity of the energy and force calculations is of order $\mathcal{O}(N)$. Each time that an atom crosses the cells, the list of the atoms in the corresponding cells updates.

It is worth noting that the combination of the neighbor lists and cell lists approaches is also possible.

10.7 Long-Range Forces

Electrostatic interactions are critical, especially, in biomolecular systems. Compared to van der Waals and covalent interactions, their range is relatively long, because of electrostatic interactions between molecules or parts of molecules at a distance r from each other decrease only slowly with increasing value of r. In particular, interaction energy, W, between two charged molecules is proportional to r^{-1}, while the corresponding force, $\mathbf{F} = -\nabla W$, is proportional to r^{-2}. The interactions between a neutral molecule with a dipole moment and a charged molecule is proportional to r^{-2}, and the corresponding force is proportional to r^{-3}. Furthermore, the interaction energy between two neutral molecules with dipole moments is proportional to r^{-3}, while the corresponding force is proportional to r^{-4}.

By considering in this multipole expansion, the quadrupole moments or octupole moments, it can be shown that even the interaction between two neutral molecules without dipole moments, but with quadrupole moments is proportional to r^{-5}, so it has a longer range than the van der Waals interactions, which is proportional to r^{-6}. If we consider the electrostatic interaction energy of either a single charge, dipole, or quadrupole with all the charges, dipoles, and quadrupoles surrounding it, we have to integrate the electrostatic interaction $U_{\text{elec}}(r)4\pi r^2$ from r to infinity, where $4\pi r^2 dr$ is the volume of the spherical shell between r and $r + dr$ surrounding the central charge, dipole, or quadrupole:

$$\int_0^\infty U_{\text{elec}}(r)4\pi r^2 dr \tag{10.14}$$

which converges if the following condition is satisfied:

$$U_{\text{elec}}(r) \propto r^{-n}, \quad n > 3$$

Note that the total electrostatic energy of ionic systems depends on the spatial boundary conditions that restrict the range of the integral Eq. (10.14) in practical calculations. That indicates that decrease of the pair interaction energies and forces with interatomic separation r will affect how the long-range interactions are treated in the force and energy calculations during simulations.

10.7.1 Ewald Summation Method

The Ewald summation method was introduced by Ewald (1921). Based on this method, the sum of the long-range interactions between particles and all their infinite periodic images are calculated efficiently. The total electrostatic interaction energy of a system of N particles in a cubic box of size L and infinite replicas in periodic boundary condition is given by (Toukmaji and Board 1996)

$$U = \frac{1}{2} \sum_{n=0}^{\infty}{}' \sum_{i=1}^{N} \sum_{j=1}^{N} \frac{q_i q_j}{4\pi \epsilon_0 \mid \mathbf{r}_i - \mathbf{r}_j + \mathbf{n}L \mid} \tag{10.15}$$

where q_i is the charge of the i-th particle and $\mathbf{n}$ is the cell-coordinate vector

$$\mathbf{n} = n_1 L\mathbf{i} + n_2 L\mathbf{j} + n_3 L\mathbf{k}$$

where $\mathbf{i}, \mathbf{j}, \mathbf{k}$ are Cartesian coordinate unit vectors. The origin cell is located at $\mathbf{n} = (0, 0, 0)$ with image cells located at $L\mathbf{n}$ intervals in all three dimensions as $\mathbf{n}$ goes to infinity, as shown in Fig. 10.4.

The sign ($'$) in the first sum indicates that the terms with $i = j$ are not included for $\mathbf{n} = 0$.

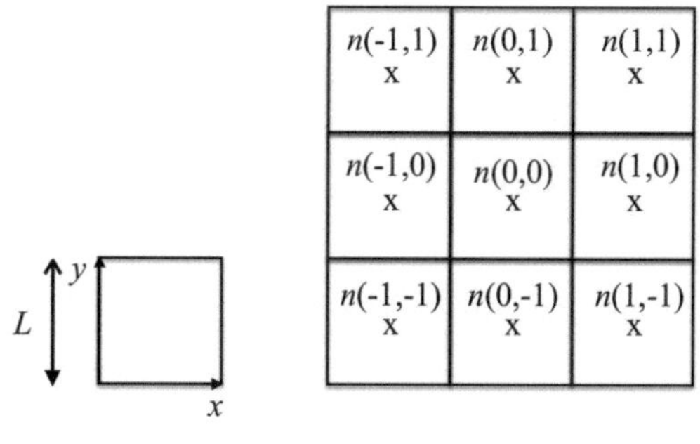

Fig. 10.4 In a 2D system (**a**) the unit cell coordinates and (**b**) a 3 × 3 periodic lattice built from unit cells as in Toukmaji and Board (1996)

According to the Ewald method, the potential energy of Eq. (10.15) can be expressed as a sum three terms

$$U_{\text{Ewald}} = U_r + U_m + U_0 \tag{10.16}$$

where U_r is the real space sum, U_m is the reciprocal sum and U_0 is the self-term, a constant term. These three terms are given, respectively as:

$$U_r = \frac{1}{2}\sum_{i,j}^{N'}\sum_{n=0}^{\infty}\frac{q_i q_j}{4\pi\epsilon_0}\frac{\text{erfc}(\alpha r_{ij,n})}{r_{ij,n}} \tag{10.17}$$

$$U_m = \frac{1}{2\pi V}\sum_{i,j}^{N}\frac{q_i q_j}{4\pi\epsilon_0}\sum_{m\neq 0}\frac{\exp\left(-(\pi m/\alpha)^2\right)\cos\left(2\pi\mathbf{m}\cdot(\mathbf{r}_i-\mathbf{r}_j)\right)}{m^2} \tag{10.18}$$

$$U_0 = \frac{-\alpha}{\sqrt{\pi}}\sum_{i=1}^{N}\frac{q_i^2}{4\pi\epsilon_0} \tag{10.19}$$

where

$$r_{ij,n} = \mid \mathbf{r}_i - \mathbf{r}_j + \mathbf{n}L \mid$$

V denotes the volume of the simulation box, $\mathbf{m} = (l, j, k)$ is the reciprocal-space vector, and erfc is the complementary error function, given as

$$\text{erfc}(x) = \frac{2}{\sqrt{\pi}}\int_x^{\infty}\exp(-t^2)dt$$

It should be noted that this summation involving the error function converges very fast and for distances larger than a cutoff value its value can be omitted. Specifically, α is chosen such that the only terms in the series of U_r are those for which $\mid \mathbf{n} \mid = 0$.

The reciprocal sum U_m also converges much faster than the first term U_r. But the number of terms increases with the width α. Hence, there is a need for a balance between the real-space and reciprocal-space sums. Here, the first term U_r converges more rapidly for large α, whereas the second U_m converges more rapidly for small α. Following the discussion in Toukmaji and Board (1996), α is large for small systems such that the real-space sum extends up to the nearest neighbors of the original cell. In Toukmaji and Board (1996) (and the reference therein), a value of $\alpha = \sqrt{\pi}/L$ is suggested, which provides an equal rate of convergence for both direct and reciprocal-space terms. However, for large systems (typically, $N > 10^4$ (Toukmaji and Board 1996)), using minimum-image convention with a cutoff radius smaller than $L/2$, larger α are used. Based on our experience, for simulations of macromolecular systems in explicit solvent using CHARMM program (Brooks et al. 2009) with a cutoff radius in the range $10 - 12$ Å, typical value of α is $\alpha = 5/L$.

The self-term U_0 is a correction term that cancels out the interaction of each of the introduced artificial counter-charges with itself.

Another term is also added depending on the medium that surrounds the sphere of simulation box. If the surrounding medium is a vacuum, i.e. with a relative permittivity 1, then the following term is added

$$U_{corr} = \frac{2\pi}{3V} \mid \sum_{i=1}^{N} \frac{q_i}{4\pi\epsilon_0}\mathbf{r}_i \mid^2 \tag{10.20}$$

This term is zero, if the surrounding medium is a conductor, i.e. with an infinite relative permittivity.

The Ewald sum is an accurate way for how to treat the long-range forces in an MD simulation. It has been used in simulations of highly charged systems and to other systems as well, such as protein, DNA and lipid bilayers where the electrostatic effects are significant.

The Ewald summation is computationally very expensive to implement. If α is allowed to vary, the algorithm can be made to scale as $N^{3/2}$. However, the problem can be that for these α, the range of electrostatic interactions are incompatible with the scope of the van der Waals interactions. There have been other methods introduced to speed up computational demands for calculation of the reciprocal space terms. For example, using the polynomial approximations, which, however, do not solve unfavorable N^2 scaling. Another proposed method is to use Fast Fourier Transform (FFT) for calculation of the reciprocal space summation, which scales as $N \ln N$. This method combined with using a large value of α, such that interatomic interactions are negligible for r_{ij} greater than some cutoff distance (typically, 9 Å), can reduce the real-space summation to order of N and the order of the entire algorithm is $N \ln N$.

A known approach for implementation of FFT method is particle-mesh method (Hockney and Eastwood 1988; Darden et al. 1993) or alternative variants, such as particle-particle-particle-mesh method (Luty et al. 1994, 1995). Unified version of these algorithms has also been discussed as an alternative for increasing the accuracy (Deserno and Holm 1998a,b).

The Ewald method has been widely used to simulate highly polar or charged systems, and it is often used to study different types of solid-state materials. Of particular interest, are the applications of the method for analyzing large molecular systems, such as proteins, DNA and their complexes (York et al. 1994; Darden et al. 1999). The nature of DNA systems make the use of the particle-mesh Ewald method particularly crucial for dealing with long-range electrostatic interactions during the simulations, producing more stable trajectories (Cheatham et al. 1995).

10.7.2 A Physical Perspective of Ewald Method

If the system is neutral, i.e. $\sum_{i=1}^{N} q_i = 0$, then Eq. (10.15) can be written as the following (Toukmaji and Board 1996):

$$\sum_{n} \frac{1}{|\mathbf{n}|} F(\mathbf{n}) + \sum_{m} \frac{1}{|\mathbf{m}|} (1 - F(\mathbf{m})) \tag{10.21}$$

Since $F(\mathbf{n})$ decays fast as $\mathbf{n} \to \infty$, the first series in the real space converges rapidly. Moreover, in the second series, the term $(1 - F(\mathbf{m}))/ \mid \mathbf{m} \mid$ in reciprocal space is a smooth continuous function, therefore, its Fourier transform decays rapidly.

Physically, each point partial charge is enveloped by a Gaussian charge density distribution of equal magnitude but opposite sign with a charge density as follows (Allen and Tildesley 1989):

$$\rho_i(\mathbf{r}) = \frac{q_i \alpha^3}{\sqrt{\pi^3}} \exp\left(-\alpha^2 r^2\right)$$

Here, α is a positive parameter representing the width of the distribution, and $\mathbf{r}$ is the position vector with respect to the center of the distribution. The Gaussian charge distribution surrounding each charge acts as a screening of the charge-charge interactions between the neighboring point-charges, and hence reducing the net interaction to a short-range interaction. Therefore, we expect that the summations over the charges and their images in real space converge rapidly.

In reciprocal space, in contrast to the real space, another Gaussian charge distribution is added around each charge of the same sign as the point charge. Then, the sum is performed in the reciprocal space using the Fourier transforms.

10.7.3 Choice of Ewald Summation Parameters

There is three parameter that determines the convergence of each term in Eq. (10.16). The first parameter is n_{max}, which is an integer that determines the range of the real-space sum by controlling the maximum number of image cells. The second parameter is m_{max}, which also is an integer that determines the summation range in reciprocal space and the number of vectors in this space. The third parameter is α, defining the Ewald convergence parameter that gives the rate of relative convergence between the real and reciprocal spaces. From Eq. (10.17), it can be seen that as $\alpha \to \infty$, the function $\operatorname{erfc}(\alpha x) \to 0$, that is large values of α yield a narrow Gaussian distribution, and hence the real-space sum converges faster. That is because a smaller number of n vectors are generated, n_{max} is small sufficiently that the sum converges rapidly. In contrast, small values of α make the reciprocal-space sum to converge faster, because

$$\alpha \to 0, \ \exp(-x/\alpha) \to 0$$

and hence only a small number of m_{max} is sufficient for the reciprocal-space sum to converge rapidly. Typically, $m_{max} = 5$ is sufficient enough to make the reciprocal-space sum to converge at a reasonable speed.

The input parameters that will determine the choice of the above parameters are as follows (Toukmaji and Board 1996):

1. Number of point-charges, N: larger systems will require a large value of α and R_c to limit the number of pairs charge-charge such that the real-space sum converges faster.
2. Desired accuracy: Larger values of R_c (or, n_{max} and m_{max}) increases the accuracy, but increases the computation cost of the short-range interactions.
3. CPU time: Increasing the value of α yields less work done from the computational point of view on the calculation of the real sum, which is the time-consuming part.
4. Cutoff radius, R_c: Smaller values of the cutoff, R_c, which means larger values of α, needs to be taken such that the real-space sum converges faster with a reasonable number of n space vectors.

It can be seen that the choice of these parameters is a trade-off between the accuracy and computational efforts. The following are suggested for α for a system size of $N \leq 10{,}000$ (Toukmaji and Board 1996)

$$\alpha = \frac{1}{2}\left(N^{1/3} + \frac{\beta L}{2R_c}\right)$$

where β is an adjustable parameter that depends on the system taken to be $\beta = 4$ for Sodium Chloride. Other values are also suggested, such as (Toukmaji and Board 1996):

$$\alpha \approx \frac{3.5}{R_c}$$

or

$$\alpha \approx \sqrt{-\ln \delta}$$

where δ is the expected accuracy ensuring that the maximum term neglected is $\mathcal{O}(\delta)$.

10.7.4 Improved Ewald Summation

To improve the convergence of the Ewald equations, Eq. (10.17), the following changes have suggested, which replace the double loop in the reciprocal-space with a single loop sum:

$$U_m = \frac{1}{2\pi V} \sum_{m \neq 0} \frac{\exp\left(-(\pi m/\alpha)^2\right)}{m^2} \sum_{i,j}^{N} \frac{q_i q_j}{4\pi \epsilon_0} Re \left\{\exp\left(2\pi i \mathbf{m} \cdot (\mathbf{r}_i - \mathbf{r}_j)\right)\right\} \tag{10.22}$$

$$= \frac{k_e}{2\pi V} \sum_{m \neq 0} \frac{\exp\left(-(\pi m/\alpha)^2\right)}{m^2} \sum_{i,j}^{N} q_i q_j \cos\left(2\pi \mathbf{m} \cdot (\mathbf{r}_i - \mathbf{r}_j)\right)$$

where

$$k_e = \frac{1}{4\pi \epsilon_0}$$

Using the trigonometric relation

$$\cos(\alpha - \beta) = \cos\alpha \cos\beta + \sin\alpha \sin\beta$$

we obtain

$$U_m = \frac{k_e}{2\pi V} \sum_{m \neq 0} \frac{\exp\left(-(\pi m/\alpha)^2\right)}{m^2} \tag{10.23}$$

$$\times \sum_{i,j}^{N} q_i q_j \left[\cos\left(2\pi \mathbf{m} \cdot \mathbf{r}_i\right) \cos\left(2\pi \mathbf{m} \cdot \mathbf{r}_j\right)\right.$$

$$\left. + \sin\left(2\pi \mathbf{m} \cdot \mathbf{r}_i\right) \sin\left(2\pi \mathbf{m} \cdot \mathbf{r}_j\right)\right]$$

$$= \frac{k_e}{2\pi V} \sum_{m \neq 0} \frac{\exp\left(-(\pi m/\alpha)^2\right)}{m^2} \left[\left(\sum_{i=1}^{n} q_i \cos(2\pi \mathbf{m} \cdot \mathbf{r}_i)\right)^2\right.$$

$$\left. + \left(\sum_{i=1}^{n} q_i \sin(2\pi \mathbf{m} \cdot \mathbf{r}_i)\right)^2\right]$$

Denoting $S(\mathbf{m})$ the structure factor:

$$S(\mathbf{m}) = \sum_{i=1}^{N} q_i \exp(2\pi i \mathbf{m} \mathbf{r}_i)$$

and that the complex conjugate is

$$S^*(\mathbf{m}) = \sum_{i=1}^{N} q_i \exp(-2\pi i \mathbf{m} \mathbf{r}_i)$$

we obtain

$$S(\mathbf{m})S^*(\mathbf{m}) \equiv \mid S(\mathbf{m}) \mid^2 = \left[\sum_{i=1}^{N} q_i \cos(2\pi \mathbf{m}\mathbf{r}_i)\right]^2 + \left[\sum_{i=1}^{N} q_i \sin(2\pi \mathbf{m}\mathbf{r}_i)\right]^2$$

Then, Eq. (10.23) can be written as

$$U_m = \frac{k_e}{2\pi V} \sum_{m \neq 0} \frac{\exp\left(-(\pi m/\alpha)^2\right)}{m^2} \mid S(\mathbf{m}) \mid^2 \tag{10.24}$$

It can be seen that the double sum over i and j in Eq. (10.22), which has a time complexity $\mathcal{O}(N^2)$, is transformed in a single sum over i, which has a time complexity of $\mathcal{O}(N)$.

10.7.5 Particle-Particle Particle-Mesh Ewald

In the following, we will introduce the Particle-Particle Particle-Mesh (PPPM) method developed (Hockney and Eastwood 1981; Luty et al. 1994; Rajagopal and Needs 1994) as described in Toukmaji and Board (1996). Based on this method, the long-range interactions are the sum of two terms: the so-called short-range interaction, which includes particle-particle interaction up to a cutoff distance, and the second term, named as the reference, which includes the so-called long-range interaction that is a smooth function and it is approximated on a mesh grid.

According to Luty et al. (1994), as a charge distribution for the Ewald sum is used a sphere with a uniform decreasing density, $\sigma(\mathbf{r})$, given as:

$$\sigma(\mathbf{r}) = \begin{cases} \dfrac{48}{\pi a^4}\left(\dfrac{a}{2} - r\right), & r < \dfrac{a}{2} \\ 0, & r \geq \dfrac{a}{2} \end{cases} \tag{10.25}$$

where a is an adjustable parameter. The short-range interaction energy between two particles is calculated using the cutoff $R_c = 0.7a$ as (Toukmaji and Board 1996)

$$U_{\mathrm{SR}}\left(\xi_{ij}\right) = k_e \left(\frac{2}{a\xi_{ij}} - \frac{1}{70a} \sum_{n=-1}^{7} C_n \xi_{ij}^n\right), \quad 0 \leq \xi_{ij} < 2 \tag{10.26}$$

where $\xi_{ij} = 2r_{ij}/a$ and

$$C_{-1\cdots7} = \begin{cases} (0,\ 208,\ 0,\ -112,\ 0,\ -14,\ -8,\ 3), & 0 \leq \xi_{ij} \leq 1 \\ (12,\ 128,\ 224,\ -448,\ 280,\ -56,\ -14,\ 8,\ -1), & 1 < \xi_{ij} \leq 2 \end{cases}$$

While the long-range potential interaction energy is calculated in reciprocal-space as

$$\hat{U}_{\mathrm{LR}}(\mathbf{k}) = \hat{\rho}(\mathbf{k})\hat{G}(\mathbf{k}) \tag{10.27}$$

where $\cdot\hat{\cdot}\cdot$ denotes the Fourier transform of the function and $\hat{G}(\mathbf{k})$ is the so-called *influence function* given as

$$\hat{G}(\mathbf{k}) = \frac{\hat{\sigma}(\mathbf{k})}{\varepsilon_0 k^2}$$

The long-range term is computed following these steps (Toukmaji and Board 1996):

1. Create a three-dimension grid and assign each point-charge to this grid. Then, based on this assignment we can calculate the charge distribution at each grid point $\rho(\mathbf{r})$, which depends on both charge distribution $\sigma(\mathbf{r})$ and assignment function in the grid.
2. Calculate $\hat{\rho}(\mathbf{k})$ using the Fourier transform, then using Eq. (10.27), we calculate $\hat{U}_{\mathrm{LR}}(\mathbf{k})$. The real-space of the long-range term, $U_{\mathrm{LR}}(\mathbf{r})$ is calculated using the inverse Fourier transform of $\hat{U}_{\mathrm{LR}}(\mathbf{k})$.
3. Calculate the grid-defined electrostatic forces using numerical differentiation, for example, by employing 4-point central differentiated method.
4. Interpolate the electrostatic potential function from the grid point positions to particle positions using the same function in the step (1).

It has been shown that the PPPM algorithm has a time complexity of the computation $\mathcal{O}(N \log N)$.

It is worth noting that the influence function $\hat{G}(\mathbf{k})$ is system and configuration specific function that means that for each new system or new configuration, a new optimal $\hat{G}(\mathbf{k})$ has to be computed. Besides, $\hat{G}(\mathbf{k})$ depends on the size and charge shape of the system. Moreover, in the PPPM method, the accuracy of the calculations depends on the grid size, that means, smaller grid size more accurate is the computation, however, on the other hand, increasing the number of grid points increases the computation complexity of the algorithm. Also, the accuracy is influenced by the force calculations in each grid point using the numerical differentiation of the potential.

10.7.6 Particle-Mesh Ewald

The Particle-Mesh Ewald (PME) method (Darden et al. 1993) decomposes the potential energy function into two terms, namely the standard direct and reciprocal sums of the Ewald approach. Besides, PME uses the standard Gaussian charge distributions. In Eq. (10.17), the direct summation is calculated explicitly using the cutoff radius, and the reciprocal sum is computed using the Fast Fourier Transformation method with convolutions on the grid points. In PME the charges are interpolated to the grid points. The advantage of the PME method compare to the PPPM method is that in the PME approach the forces are calculated analytically by differentiating the energy function, and hence additionally reducing the memory requirements significantly. The PME is considered a highly accurate method, achieving an accuracy of 10^{-6} on the relative force error with less computational efforts compared to other approaches.

In the PME, the time complexity of the algorithm is $\mathcal{O}(N \log N)$ that indicates the time complexity of the calculation of the reciprocal sum using three-dimensional FFT. In particular, the reciprocal sum, U_m, is calculated using Eq. (10.24) with $S(\mathbf{m}) \approx \tilde{S}(\mathbf{m})$, where $\tilde{S}(\mathbf{m})$ is given as (Toukmaji and Board 1996)

$$\begin{aligned}\tilde{S}(\mathbf{m}) &= \sum_{k_1,k_2,k_3} Q(k_1,k_2,k_3) \exp\left(2\pi i \left(\frac{m_1 k_1}{K_1} + \frac{m_2 k_2}{K_2} + \frac{m_3 k_3}{K_3}\right)\right) \\ &\equiv \hat{F}_Q(m_1,m_2,m_3)\end{aligned} \tag{10.28}$$

where $\hat{F}_Q(m_1, m_2, m_3)$ is the three-dimensional FFT of the charge matrix Q. Here, the matrix Q is obtained using the interpolation of the point-charges uniformly in a three-dimensional grid with dimensions $(K_1,\ K_2,\ K_3)$ of the simulation cell.

Combining Eqs. (10.24) and (10.28), we obtain

$$\tilde{U}_m = \frac{k_e}{2\pi V} \sum_{m \neq 0} \frac{\exp\left(-(\pi m/\alpha)^2\right)}{m^2} \mid \hat{F}_Q(\mathbf{m}) \mid^2 \tag{10.29}$$

which can be written as convolution (Toukmaji and Board 1996)

$$\tilde{U}_m = \frac{k_e}{2} \sum_{k_1=0}^{K_1-1} \sum_{k_2=0}^{K_2-1} \sum_{k_3=0}^{K_3-1} Q(k_1,k_2,k_3)(\hat{U}_m * Q)(k_1,k_2,k_3) \tag{10.30}$$

where $(\hat{U}_m * Q)$ denotes the convolution of the reciprocal-space potential energy function $\hat{U}_m$ with real-space charge matrix Q.

10.7.7 Multipole Ewald Summation Methods

The primary goal of the multipole-based Ewald methods is to decompose the potential energy functions into the term representing the short-range term, which is computed directly, and the long-range term, which is approximated using multipole expansions. Let us consider first the electrostatic potential $V_{\text{Elec}}(\mathbf{r}_1, \cdots, \mathbf{r}_N)$:

$$V_{\text{Elec}}(\mathbf{r}_1, \cdots, \mathbf{r}_N) = \frac{1}{2}\sum_{i=1}^{N}\sum_{j=1\neq i}^{N} k_e \frac{q_i q_j}{\varepsilon r_{ij}} = \frac{1}{2}\sum_{i=1}^{N} q_i \Phi(\mathbf{r_i}) \tag{10.31}$$

where

$$\Phi(\mathbf{r_i}) = \frac{k_e}{\varepsilon}\sum_{j=1\neq i}^{N} \frac{q_j}{r_{ij}} \tag{10.32}$$

denotes the electric potential created by all other charges at the position where the charge i is placed.

Consider a system of N charges distributed in a three-dimension grid represented by spherical coordinates $(\rho_i, \alpha_i, \beta_i)$ inside a sphere of radius a $(|\ \rho_i\ | < a)$. In spherical coordinates, every point in three-dimension outside the sphere is represented as

$$\mathbf{r} \equiv (r, \theta, \phi), \qquad r > a$$

An expansion at the point $\mathbf{r}_i$ around the origin can be written in terms of infinite multipoles as (Toukmaji and Board 1996; Frenkel and Smit 2001; Schlick 2010):

$$\Phi(\mathbf{r_i}) = \sum_{n=0}^{\infty}\sum_{m=-n}^{n} \frac{M_n^m}{r_i^{n+1}} Y_n^m(\theta, \phi) \tag{10.33}$$

where the functions $Y_n^m(\theta, \phi)$ is the *spherical harmonic polynomial* or *multipoles* and the M_n^m are the *moments of the expansion*. These are defined as the following, based on the literature (Schlick 2010):

$$Y_n^m(\theta, \phi) = \begin{cases} r^{n+1} A_n^m \dfrac{\partial^n}{\partial z^n}\left(\dfrac{1}{r}\right), & m = 0 \\ r^{n+1} A_n^m \left(\dfrac{\partial}{\partial x} + i\dfrac{\partial}{\partial y}\right)^m \left(\dfrac{\partial}{\partial z}\right)^{n-m} \left(\dfrac{1}{r}\right), & m = 1, 2, \cdots \\ r^{n+1} A_n^{-m} \left(\dfrac{\partial}{\partial x} - i\dfrac{\partial}{\partial y}\right)^{-m} \left(\dfrac{\partial}{\partial z}\right)^{n+m} \left(\dfrac{1}{r}\right), & m = -1, -2, \cdots \end{cases} \tag{10.34}$$

and

$$M_n^m = \sum_{j=1\neq i}^{N} \frac{k_e q_j}{\varepsilon} r_j^n Y_n^{-m}(\theta_j, \phi_j) \tag{10.35}$$

with

$$A_n^m = \frac{(-1)^n}{\sqrt{(n-m)!(n+m)!}}$$

The relationship between the spherical harmonic polynomial and the so-called *associated Lengendre polynomials of degree n*, P_n^m, is given as

$$Y_n^m(\theta, \phi) = \sqrt{\frac{(n- \mid m \mid)!}{(n+ \mid m \mid)!}} P_n^{|m|}(\cos\theta)\exp(im\phi) \tag{10.36}$$

where

$$P_n^m(x) = (-1)^m(1-x^2)^{m/2}\frac{d^m}{dx^m}P_n(x) \tag{10.37}$$

The expansion power p is a criteria for achieving a certain accuracy of the calculations, defined as

$$\mid \sum_{n=p+1}^{\infty}\sum_{m=-n}^{n}\frac{M_n^m(r_i)}{r_i^{n+1}}Y_n^m(\theta, \phi) \mid \leq \frac{\sum_{i=1}^{N} \mid q_i \mid}{r-a}\left(\frac{a}{r}\right)^{p+1} = \epsilon \tag{10.38}$$

To reduce ϵ, we can either increase p or decrease the radius a.

This method is known as the *fast multipole algorithm* developed in Greengard and Rokhlin (1987). Other variants of multipole method have also been developed, such as the *reduced cell multipole method* (Ding et al. 1992), the *Particle-Particle Particle-Mesh/Multipole expansion method* (Shimada et al. 1993, 1994) and the *Macroscopic Multipole method* (Toukmaji and Board 1996). Today's improvements on the computer architecture encourage the development of fast algorithms to compute the electrostatic interactions of macromolecules using the methods mentioned above, as discussed in Hardy et al. (2011) about the fast calculations of electrostatics in the Graphical Processing Unit (GPU).

10.7.8 Reaction Field Method

According to this approach, the molecule is surrounded by a sphere with radius equal to the cutoff distance, R_c. Then, the electrostatic interactions with all molecules inside the sphere are calculated using explicit models. The region outside the cavity will be assumed a continuous dielectric medium with a dielectric constant equal to $\varepsilon_{\mathrm{RF}}$. The outside region additionally will be considered polarized due to the electric field created by molecules inside the cavity. This polarization will create an electric field, the so-called *reaction field*, E_i, inside the cavity given as (Fukuda and Nakamura 2012):

$$\mathbf{E}_i = k_e\frac{2(\varepsilon_{\mathrm{RF}}-1)}{2\varepsilon_{\mathrm{RF}}+1}\frac{1}{R_c^3}\sum_{j\ (r_{ij}\leq R_c)}\boldsymbol{\mu}_j \tag{10.39}$$

Here, $\boldsymbol{\mu}_j$ are the dipoles of the neighbouring molecules within a cutoff distance R_c of molecule i, which is given as

$$\boldsymbol{\mu}_j = \sum_{\nu \in j} q_\nu \mathbf{r}_\nu$$

where the sum is over all atoms ν of the molecule j and $\mathbf{r}_\nu$ is the vector position of the atom ν.

Now, the interaction energy between the molecule i and the reaction field is

$$E_r = \mathbf{E}_i \cdot \boldsymbol{\mu}_i = k_e \frac{2(\varepsilon_{\mathrm{RF}} - 1)}{2\varepsilon_{\mathrm{RF}} + 1} \frac{1}{R_c^3} \sum_{j\ (r_{ij} \leq R_c)} \boldsymbol{\mu}_i \cdot \boldsymbol{\mu}_j \tag{10.40}$$

Following Fukuda and Nakamura (2012), we assume that the total charge inside the cavity is zero and that each molecule i is too small such that the set of atoms inside the cavity of a molecule of the system equals the set of atoms inside the cavity centered at each atom of that molecule. Using these assumptions, we can write that the reaction field interaction energy between any pair of the atoms is:

$$V_{\mathrm{RF}}^{\mathrm{Corr}}(r_{ij}) = k_e \frac{q_i q_j r_{ij}^2}{R_c^3} \frac{\varepsilon_{\mathrm{RF}} - 1}{2\varepsilon_{\mathrm{RF}} + 1}, \quad \text{for } r_{ij} \geq R_c \tag{10.41}$$

This correction is added to the short-range potential energy of the molecule-molecule interaction accounting for long range corrections as follows:

$$V_{\mathrm{RF}}(r_{ij}) = k_e \frac{q_i q_j}{r_{ij}} \left(1 + \frac{r_{ij}^3}{R_c^3} \frac{\varepsilon_{\mathrm{RF}} - 1}{2\varepsilon_{\mathrm{RF}} + 1} \right) \tag{10.42}$$

There are a few drawbacks arising from the assumptions made by the approach (Leach 2001). The first problem is related to the discontinuity of the potential energy function and force if the number of molecules within the cutoff range varies during the simulation run. This problem, practically, can be avoided by using the switching functions at the border between the reaction sphere and surrounding dielectric media:

$$V_{\mathrm{RF}}(r_{ij}) = k_e \frac{q_i q_j}{r_{ij}} \left(1 + \frac{r_{ij}^3}{R_c^3} \frac{\varepsilon_{\mathrm{RF}} - 1}{2\varepsilon_{\mathrm{RF}} + 1} \right) - k_e \frac{q_i q_j}{R_c} \frac{3\varepsilon_{\mathrm{RF}}}{2\varepsilon_{\mathrm{RF}} + 1} \tag{10.43}$$

which is such that $V_{\mathrm{RF}}(R_c) = 0$.

Secondly, in this approach assuming that a constant dielectric constant characterizes the surrounding dielectric medium is a good approximation if this media is a homogeneous fluid; otherwise, extra efforts must be taken to estimate the dielectric constant of the surrounding media using, for example, mean field approximation.

10.8 Equilibration

A trajectory during MD simulations is a sequence of configurations for a given system saved at regular time steps, let say every N_{steps}. For example, for a system of N particles, every configuration of system is a vector of $3N$ dimensions

$$\mathbf{r}(t) \equiv (\mathbf{r}_1(t), \mathbf{r}_2(t), \cdots, \mathbf{r}_N(t))^T$$

where $\mathbf{r}_i(t) = (x_i(t), y_i(t), z_i(t))$ are the coordinates of the i-th atom. If we consider a path through the configurational phase space, then $\mathbf{r}(t)$ is the vector of coordinates of all atoms at the time step t in this path. The path $\mathbf{r}(t)$ is called a *trajectory*.

In general, the values of any physical property, such as temperature, pressure, energy, and so on, fluctuate as we move along the trajectory. The average value of a physical property, $\mathcal{P}(\mathbf{r})$, overall the configurations visited by the system during a trajectory with T steps is

$$\langle \mathcal{P} \rangle_T = \sum_{t=1}^{T} \mathcal{P}(\mathbf{r}(t))$$

Assuming that the trajectory is ergodic, then the time averages are equal to ensemble averages, that is:

$$\langle \mathcal{P} \rangle = \lim_{T \to \infty} \langle \mathcal{P} \rangle_T$$

In practice, MD simulations run for a finite time, typically, of order hundred of nanoseconds, hence the average over configurations provides only an estimate of $\langle \mathcal{P} \rangle$.

As an example, Fig. 10.5 shows the running averages versus T along with a trajectory. The magnitude of the fluctuations along the average $\langle \mathcal{P} \rangle$ equals the statistical uncertainty in the average. For this, we can divide the trajectory into blocks of the length L. Assuming that L are long enough, then averages on each block can be considered statistically independent. Let us denote with $\langle \mathcal{P} \rangle_m$ the average of $\mathcal{P}(\mathbf{r}(t))$ over the configurations of the block m, then

$$\langle \mathcal{P} \rangle_T = \frac{1}{M} \sum_{m=1}^{M} \langle \mathcal{P} \rangle_m$$

where $M = T/L$ is the total number of blocks.

The standard deviation gives the statistical error of the average value as

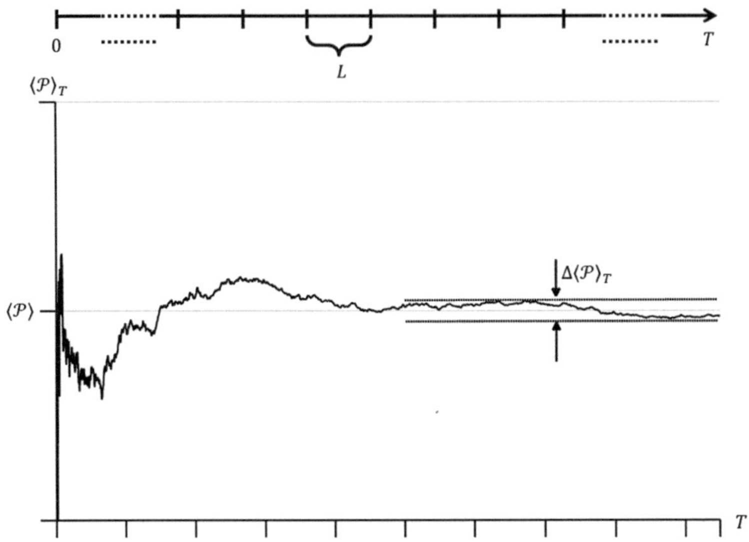

Fig. 10.5 Running averages along a MD trajectory

$$\Delta\langle\mathcal{P}\rangle_T = \frac{1}{M}\sqrt{\sum_{m=1}^{M}\left(\langle\mathcal{P}\rangle_m - \langle\mathcal{P}\rangle_T\right)^2}$$

which for $T \to \infty$ tends to zero.

The prove of the *ergodic hypothesis* is difficult for any system. However, there can be a simple evaluation of the ergodic hypothesis using MD simulation results from different independent trajectories. In that case, averages of a macroscopic property, for example, total energy or pressure of an ergodic system, over two different independent trajectories should be equal. For that criteria to be satisfied, averages over a trajectory should be identical to the statistical path over conformation space.

It can be argued that correlations in low-frequency atomic displacements of order one nanosecond are not sampled efficiently due in part to the fact that trajectory remains constrained only in some region of the phase space, especially for complex molecular systems. Practically, the following quantity is used as an estimation of the simulation length needed to guarantee that the ergodic hypothesis is satisfied:

$$E(t) = \frac{1}{N_{\text{bins}}}\sum_{i=1}^{N_{\text{bins}}}\left(\mathcal{P}_i^{(1)}(t) - \mathcal{P}_i^{(2)}(t)\right)^2 \tag{10.44}$$

where $\mathcal{P}_i^{(j)}$ is the value of the quantity measured from the j-th trajectory at the bin i and N_{bins} is the total number of histogram bins. In general, the quantity $\mathcal{P}$ can be any macroscopic property of the system, for instance, root mean square deviation, energy, and so on, and E is presented as a function of the simulation time t. If the system is ergodic, then E should decay to zero as $1/D\tau$, where D is the generalized diffusion constant, and τ is a timescale for self-averaging in the simulation. The decay of the ergodic measure, $E(t)$, to zero in the limit of the long times is considered a necessary condition for the system's average properties to correspond to equilibrium thermodynamic averages.

Chapter 11
Symplectic and Time Reversible Integrator

In this chapter, we will discuss numerical integrator algorithms used for solving differential equations used in molecular dynamics simulations. In particular, we will propose different numerical integrator algorithms, which satisfy time reversibility or symplectic properties.

11.1 Flow-Maps

A discrete flow-map is an infinitesimal canonical transformation when the time t advances from t to $t + \Delta t$ by a small time step Δt:

$$\Phi_{\Delta t} \, : \, \Re^{2d} \longmapsto \Re^{2d}$$

where $\Re^{2d}$ is an $2d$-dimensional manifold with coordinates

$$(\mathbf{q}, \, \mathbf{p}) = (q_1, q_2, \cdots, q_d, p_1, p_2, \cdots, p_d)$$

such that

$$z(t + \Delta t) = \Phi_{\Delta t}(z(t))$$

where $z(t) = (\mathbf{q}(t), \, \mathbf{p}(t))$ is the state vector at time t, and $\mathbf{q}$ and $\mathbf{p}$ are, respectively, the generalized coordinates and their conjugated momenta. Based on the Chap. 1, we can write for an infinitesimal canonical transformation that

$$z(t + \Delta t) = z(t) + \Delta t \mathbf{J} \frac{\partial G}{\partial z}$$

H. Kamberaj, *Molecular Dynamics Simulations in Statistical Physics: Theory and Applications*, Scientific Computation,
https://doi.org/10.1007/978-3-030-35702-3_11

where $\mathbf{J}$ is the $2d \times 2d$ zero-one matrix defined in Chap. 1 (see Eq. (1.109)), and $G(\mathbf{q}, \mathbf{p})$ is the generating function of the infinitesimal canonical transformation such that in the case of the Hamiltonian system with Hamiltonian function $H(\mathbf{q}, \mathbf{p})$ is given by:

$$\mathbf{G} = \mathbf{J}\frac{\partial G}{\partial z} = \mathbf{J}\nabla_z H(z) = \left(\frac{\partial H}{\partial \mathbf{p}}, -\frac{\partial H}{\partial \mathbf{q}}\right)$$

Here, $\mathbf{G}$ is the metric tensor.

11.1.1 Symplectic Maps

The term symplectic was first used mathematically by Hermann Weyl and is taken from the Greek word meaning "twining or plaiting together". Symplectic systems consist of a pair of d-dimensional variables, generally position $\mathbf{q}$ and momentum $\mathbf{p}$, "intertwined" by the symplectic two form,

$$\omega = d\mathbf{q} \wedge d\mathbf{p} . \tag{11.1}$$

This is an antisymmetric, bilinear form acting on a pair of tangent vectors to compute the sum of areas of the parallelograms formed by projecting the vectors onto planes defined by the pairs (q_i, p_i), $i = 1, \cdots, d$ giving,

$$\omega(v, w) = \sum_{i=1}^{d} \left(v_{p_i} w_{q_i} - v_{q_i} w_{p_i}\right) . \tag{11.2}$$

A discrete flow-map $\Phi_{\Delta t}$ is symplectic if it preserves the symplectic form (Arnold 1988). If we write the Jacobian matrix of the infinitesimal time transformation as (see Chap. 1):

$$\mathbf{M} = \mathbf{I} + \Delta t \mathbf{J}\frac{\partial^2 G}{\partial z \partial z} \tag{11.3}$$

then the symplectic condition becomes (see also Eq. (1.142)), if we ignore the second order terms,

$$\mathbf{M}\mathbf{J}\mathbf{M}^T = \mathbf{J} \tag{11.4}$$

or

$$\mathbf{M}^T\mathbf{J}^{-1}\mathbf{M} = \mathbf{J}^{-1} \tag{11.5}$$

Here, $\mathbf{I}$ is the $d \times d$ identity matrix and $\mathbf{M}^T$ is the transpose of $\mathbf{M}$:

$$\mathbf{M}^T = \mathbf{I} - \Delta t \frac{\partial^2 G}{\partial z \partial z} \mathbf{J}$$

11.1.2 Symplecticness of Hamiltonian Flow-Maps

Consider now the continuous flow-map $\Phi_{t,H}$ of a Hamiltonian system H. To prove that $\Phi_{t,H}$ is symplectic we can use the matrix form for the Hamiltonian equations of motion,

$$\dot{z} = \mathbf{J} \nabla_z H(z) \, , \tag{11.6}$$

where $\mathbf{J}$ is the usual invertible skew-symmetric matrix defined in Chap. 1 ($\mathbf{J}^T = -\mathbf{J}$). We define

$$\mathbf{M}(t) = \frac{\partial}{\partial z} \Phi_{t,H} \tag{11.7}$$

Note that for the case of discrete flow-maps, $\mathbf{M}$ given by Eq. (11.3) is a Taylor expansion of $\mathbf{M}(t)$ given by Eq. (11.7) around Δt, where only the terms up to the first order retain.

Since $\mathbf{M}(0)$ is defined as a symplectic map, for which Eqs. (11.4) and (11.5) hold, we have to show that

$$\frac{d}{dt} \left[\mathbf{M}(t)^T \mathbf{J}^{-1} \mathbf{M}(t) \right] = 0 \, . \tag{11.8}$$

to prove that $\Phi_{t,H}$ is symplectic flow-map.

For that we can write:

$$\begin{aligned}
\frac{d}{dt} \left[\mathbf{M}^T \mathbf{J}^{-1} \mathbf{M} \right] &= \mathbf{M}^T \mathbf{J}^{-1} \frac{d}{dt} \mathbf{M} + \left(\frac{d}{dt} \mathbf{M} \right)^T \mathbf{J}^{-1} \mathbf{M} \\
&= \mathbf{M}^T \mathbf{J}^{-1} \left(\mathbf{J} \left(\nabla_z \nabla_z H(z) \right) \mathbf{M} \right) + \left(\mathbf{M}^T \left(\nabla_z \nabla_z H(z) \right) \mathbf{J}^T \right) \mathbf{J}^{-1} \mathbf{M} \\
&= \mathbf{M}^T \left(\nabla_z \nabla_z H(z) \right) \mathbf{M} - \mathbf{M}^T \left(\nabla_z \nabla_z H(z) \right) \mathbf{M} = 0 \, .
\end{aligned}$$

If a flow-map is symplectic, it possesses certain integral invariants which relate to the evolution of subsets of phase-space. One such integral invariant is the preservation of phase-space area for systems with one degree of freedom, $d = 1$, and volume for $d > 1$, which also follows from Liouville's theorem (Arnold 1988). Since the existence of integral invariants, such as this restricts the possible solutions for a Hamiltonian system, can be an important feature for numerical integrator algorithms, in particular, for long time results.

11.1.3 Phase-Space Area Preservation for d = 1

A one degree of freedom symplectic map, $\Phi_t : \Re^2 \longmapsto \Re^2$, has a Jacobian,

$$\mathbf{M}(t) = \frac{\partial \Phi_t}{\partial z} = \begin{bmatrix} a & b \\ c & d \end{bmatrix} \tag{11.9}$$

for some $a, b, c, d \in \Re$. Substituting Eq. (11.9) into Eq. (11.5) yields,

$$ad - bc = 1 ,$$

indicating that

$$\det[\mathbf{M}] = 1$$

Consider the motion of the volume in a two-dimensional phase space as illustrated in Fig. 1.6 (see Chap. 1). If we let Γ be a bounded subset of phase-space and $\hat{\Gamma} = \Phi_t(\Gamma)$ its image under the flow-map Φ_t, then the area $\alpha(\Gamma)$ is given by,

$$\alpha(\Gamma) = \int_{\Gamma} d\mathbf{q}d\mathbf{p}$$

Similarly, the area $\alpha(\hat{\Gamma})$ is given by,

$$\begin{aligned}\alpha(\hat{\Gamma}) &= \int_{\hat{\Gamma}} d\hat{\mathbf{q}}d\hat{\mathbf{p}} \\ &= \int_{\Gamma} \det[\nabla_z \Phi_t(z)]\, d\mathbf{q}d\mathbf{p} = \int_{\Gamma} \det[\mathbf{M}]\, d\mathbf{q}d\mathbf{p} \\ &= \int_{\Gamma} d\mathbf{q}d\mathbf{p} = \alpha(\Gamma)\end{aligned}$$

and hence a one degree of freedom symplectic map preserves the area of phase-space. The proof of the conservation of phase-space volume for $d > 1$ can be found in references such as Arnold (1988).

11.1.4 Time-Reversal Symmetry

Newton's equations of motion possess the geometric property of time-reversibility, which manifests itself as the invariant of a Hamiltonian $H(q, p)$ under the reflection symmetry $p \longmapsto -p$. The equations of motion for this Hamiltonian are,

$$\dot{q} = \nabla_p H(q, p),$$
$$\dot{p} = -\nabla_q H(q, p)\,. \tag{11.10}$$

If we assume $(q(t), p(t))$ is a solution of Eq. (11.10) and define $(\hat{q}(t), \hat{p}(t)) = (q(-t), -p(-t))$, we get

$$\frac{d}{dt}\hat{q}(t) = -\dot{q}(-t) = -\nabla_p H(q(-t), p(-t)) = \nabla_p H(\hat{q}(t), \hat{p}(t)), \tag{11.11}$$

$$\frac{d}{dt}\hat{p}(t) = \dot{p}(-t) = -\nabla_q H(q(-t), p(-t)) = \nabla_q H(\hat{q}(t), \hat{p}(t)), \tag{11.12}$$

since $H(q, p)$ is even in p for Newtonian mechanics, giving $\nabla_q H$ even in p and $\nabla_p H$ odd in p. This shows that $(\hat{q}(t); \hat{p}(t))$ is a solution of Eq. (11.10). This invariant implies that for every solution of the Hamiltonian system there is another one whose trajectory is in the opposite direction with negated momentum.

This time-reversal invariant can be written for $z = (q, p)$

$$H(z) = H(\mathbf{J}z) \tag{11.13}$$

For the vector field of a Hamiltonian of the form Eq. (11.6) this gives

$$f(z) = -\mathbf{J}f(\mathbf{J}z) \tag{11.14}$$

where $f(z) = \mathbf{J}\nabla_z H(z)$, since for any flow map $\Phi_{-t,H} = \left[\Phi_{t,H}\right]^{-1}$. A mapping Ψ with the property $\Psi_{-t,H} = \left[\Psi_{t,H}\right]^{-1}$ is said to be symmetric or self-adjoint.

11.2 Symplectic and Hamiltonian Splitting Methods

In general, the Hamiltonian systems of interest may not have an analytical solution, and this has led to the development of numerical integrator algorithms, which solve the equations of motion by taking discrete time steps until the required integration time, T, is reached.

That was first seen in simple schemes such as Euler's method, but both the mathematician De Vogalaere and the physicist Ruth had postulated that if the numerical integrator possessed some of the properties of Hamiltonian system's flow-map then simulations would display improved behavior. This idea has led to the development and classification of Geometric Integrator Algorithms, where their use preserves geometrical properties of the original system. For Hamiltonian systems, the symplectic property is perhaps the most important geometrically and can lead to efficient explicit Hamiltonian splitting methods as discussed by Leimkuhler and Reich (2004).

11.2.1 Hamiltonian Splitting Methods

Symplectic numerical integrator algorithms are desirable for the approximation of Hamiltonian flow-maps but, for complicated systems, can lead to implicit methods which are difficult to solve. From the definition of symplecticness we see that the composition of symplectic maps is again symplectic and this leads to the idea of splitting the Hamiltonian. This can be achieved if it is possible to split the Hamiltonian H into the sum of $k \geq 2$ Hamiltonians $H_i,\ i = 1, \cdots, k$,

$$H(z) = \sum_{i=1}^{k} H_i(z)\ , \tag{11.15}$$

where each Hamiltonian equation

$$\dot{z} = \mathbf{J}\nabla_z H_i(z)\ , \tag{11.16}$$

can be solved explicitly. From this the composition method,

$$\Phi_{\Delta t} = \phi_{\Delta t, H_1} \circ \phi_{\Delta t, H_2} \circ \cdots \circ \phi_{\Delta t, H_k}\ , \tag{11.17}$$

is a first-order symplectic integrator.

For the simulation of systems with time-reversal symmetry, in addition to the symplectic property, a symmetric method is required but methods composed in the above manner are generally not symmetric. To achieve this the same Hamiltonian splitting can be utilized but composed in a symmetric manner as follows,

$$\hat{\Phi}_{\Delta t} = \phi_{\Delta t/2, H_1} \circ \phi_{\Delta t/2, H_2} \circ \cdots \circ \phi_{\Delta t, H_k} \circ \cdots \circ \phi_{\Delta t/2, H_2} \circ \phi_{\Delta t/2, H_1}\ . \tag{11.18}$$

Using this relation, we can derive

$$\left[\hat{\Phi}_{-\Delta t}\right]^{-1} = \left[\phi_{-\Delta t/2, H_1}\right]^{-1} \circ \cdots \circ \left[\phi_{-\Delta t, H_k}\right]^{-1} \circ \cdots \circ \left[\phi_{-\Delta t/2, H_1}\right]^{-1}\ . \tag{11.19}$$

If each $\phi_{\Delta t, H_i}$ is symmetric then we get:

$$\left[\hat{\Phi}_{-\Delta t}\right]^{-1} = \hat{\Phi}_{\Delta t}\ .$$

This yields a symplectic, time-reversible mapping which is suitable for Hamiltonian systems based on Newton's equations. In addition it can be shown (Leimkuhler and Reich 2004) that symmetric methods necessarily have order two.

If the condition that each Hamiltonian vector field can be solved explicitly is not satisfied, then the splitting methods can still be employed. Here, if the majority of solutions are explicit, splitting leads to a reduced number of simplified vector fields which need to be solved implicitly.

11.3 Liouville Formalism and Trotter Formula

For the systems under investigation, also discussed in Chap. 6, we can apply the Liouville formalism of the Hamiltonian dynamics to find the time derivative of a set

$$\mathbf{f} \equiv (\mathbf{p}, \mathbf{q})$$

of configuration and momentum variables along the trajectory (Allen and Tildesley 1989)

$$\frac{d\mathbf{f}}{dt} = \sum_j \left(\dot{q}_j \frac{\partial \mathbf{f}}{\partial q_j} + \dot{p}_j \frac{\partial \mathbf{f}}{\partial p_j} \right), \tag{11.20}$$

where the summation is over all degrees of freedom in the system and

$$\dot{q}_j = \frac{\partial H}{\partial p_j}, \tag{11.21}$$

$$\dot{p}_j = -\frac{\partial H}{\partial q_j} \tag{11.22}$$

are the Hamiltonian equations of motion (Goldstein 2002). The Liouvillean operator, iL, is defined as (Allen and Tildesley 1989):

$$iL = \sum_j \left(\dot{q}_j \frac{\partial}{\partial q_j} + \dot{p}_j \frac{\partial}{\partial p_j} \right).$$

The exponential operator $O(t) \equiv e^{iLt}$ is the classical propagator, which evolves the system phase space point from the initial state $(q(0), p(0))$ to the state $(q(t), p(t))$ by acting on $\mathbf{f}$ as (Allen and Tildesley 1989) $\mathbf{f}(t) = O(t)\mathbf{f}(0)$. It satisfies the time-symmetry property $O(\Delta t)O(-\Delta t) = 1$, therefore the flow map is time-reversible. This formalism is completely general and it can also be applied to non-Hamiltonian systems sometimes leading to time-reversible numerical integrator algorithms as described in Martyna et al. (1996). In this case, the function $\mathbf{f}$ can be the state vector $(\eta_k, \pi_k, \mathbf{p}, \mathbf{q})$ for the NVT or $(\eta_k, \pi_k, \varepsilon, \pi_\epsilon, \mathbf{p}, \mathbf{q})$ $(k = 1, \cdots, M)$ for NPT ensembles as presented here.

In general, the action of evolution operator, $O(t)$, on the state vector cannot be solved exactly. Therefore, a short time approximation to the true operator, accurate at time $\Delta t = t/n$ is applied n times in succession to evolve the system considering small time steps Δt used in MD simulations. Thus, if the Hamiltonian of system can be written as:

$$H = \sum_k H_k$$

then using the Trotter formula (Trotter 1959; Raedt and Raedt 1983), we can write

$$O(t) = \left(\prod_k e^{iL_k \Delta t}\right)^n + \mathcal{O}\left(\Delta t^{p+1}/n^p\right), \tag{11.23}$$

which has an overall accuracy of order $\Delta t^{p+1}/n^p$ for a pth order factorisation where iL_k is the Liouville operator of the associated variables with the Hamiltonian function H_k. For example for $p = 2$ we can write

$$H = H_1 + H_2$$

and the Poisson bracket

$$\{H_1, H_2\} = \frac{\partial H_1}{\partial q}\frac{\partial H_2}{\partial p} - \frac{\partial H_1}{\partial p}\frac{\partial H_2}{\partial q} \neq 0$$

The propagator operator can be written as:

$$\exp(iL\Delta t) = \exp(iL_1\Delta t/2)\exp(iL_2\Delta t)\exp(iL_1\Delta t/2) + \mathcal{O}(\Delta t^3), \tag{11.24}$$

which is known as symmetric Trotter factorisation or Strang splitting (Strang 1968) and it is correct to the second order. The Strang splitting scheme can be applied for each set of two operators, in order to obtain time reversible and explicit integrator algorithms, in the case of dynamical systems where the Hamiltonian can be split into terms with a non-vanishing Poisson bracket, H_k $(k = 1, 2 \cdots p)$. Then the time evolution operator can be written

$$\begin{aligned} e^{iL\Delta t} &= e^{i\frac{\Delta t}{2}L_1} \cdots e^{i\frac{\Delta t}{2}L_{p-1}} e^{i\Delta t L_p} e^{i\frac{\Delta t}{2}L_{p-1}} \cdots e^{i\frac{\Delta t}{2}L_1} + \mathcal{O}\left(\Delta t^3\right) \\ &\equiv \phi_{\Delta t} + \mathcal{O}(\Delta t^3), \end{aligned} \tag{11.25}$$

which can be considered as a Strang splitting of the pth order of factorisation. We see that this splitting is also correct to the second order since the factorisation of each set of two operators is correct to the second order (see equation Eq. (11.24)). It has been shown elsewhere (Kamberaj et al. 2005) that the limited accuracy in the numerical integration will not destroy the time reversibility property of the flow map ($\phi_{\Delta t}\phi_{-\Delta t} = 1$). It is worth noting that similar factorisation schemes have also been proposed, e.g. see Martyna et al. (1996). Higher order integrator algorithms can also be generated using Yoshida-Suzuki integration (Yoshida 1990; Suzuki 1991) from low order integrator algorithms. The splitting is easy to implement, and with low computation cost, furthermore, it resulted in the integrator algorithms with long-time stability in terms of energy conservation.

The explicit integration schemes can be straightforward obtained by using the exponential expansion

$$\exp\left(a\nabla_x\right) f(x) = f(x + a), \tag{11.26}$$

where a does not depend on x. For extended systems two additional analytical forms for the exponential operators are obtained with a being a scalar or a matrix, $\mathbf{a}$ containing $\mathbf{x}$. In these cases the following expressions are applied (Martyna et al. 1996):

$$\exp\left(a\mathbf{x}\cdot\nabla_{\mathbf{x}}\right) f(\mathbf{x}) = f\left(e^{a}\mathbf{x}\right), \quad \exp\left([\mathbf{a}\mathbf{x}]\cdot\nabla_{\mathbf{x}}\right) f(\mathbf{x}) = f\left(e^{\mathbf{a}}\mathbf{x}\right). \tag{11.27}$$

The exponential matrix $e^{\mathbf{a}}$ can be obtained by diagonalisation of $\mathbf{a}$ as:

$$\exp\left(\mathbf{a}\right) = \mathbf{CDC}^{-1}, \tag{11.28}$$

where $\mathbf{C}$ is the associated matrix of eigenvectors of the matrix $\mathbf{a}$ and $\mathbf{D}$ is a diagonal matrix with diagonal element equal to exponential of the eigenvalues of the matrix $\mathbf{a}$. Here, for a 3×3 matrix the eigenvalues and eigenvectors are determined analytically in order to maintain the second order accuracy of the overall integration scheme and to decrease the computational time needed.

11.4 Microcanonical Ensemble

First we will discuss the NVE ensemble described by the following equations of motion (as discussed in Chap. 6):

$$\begin{aligned}\dot{\mathbf{p}}_i &= -\nabla U(\mathbf{r}),\\ \dot{\mathbf{r}}_i &= \mathbf{p}_i/m_i, \quad i = 1,\cdots,N\,.\end{aligned} \tag{11.29}$$

The Liouville operator $iL^{(1)}_{NVE}$ is written as

$$iL^{(1)}_{\mathrm{NVE}} = iL_1 + iL_2, \tag{11.30}$$

where

$$\begin{aligned}iL_1 &= \sum_{j=1}^{N}\left(\frac{\mathbf{F}_j}{m_j}\right)\cdot\nabla_{\mathbf{v}_j},\\ iL_2 &= \sum_{j=1}^{N}\mathbf{v}_j\cdot\nabla_{\mathbf{q}_j},\end{aligned} \tag{11.31}$$

where $\mathbf{v}_j$ is the velocity of the jth particle ($=\dfrac{\mathbf{p}_j}{m_j}$) and N is the number of particles of the system. In the limit of the short time-step (Δt), the Strang splitting scheme (Strang 1968) can be used, in the form presented here by equation Eq. (11.25), to generate the classical propagator (Kamberaj et al. 2005)

$$\begin{aligned}\exp\left(iL_{NVE}^{(1)}\Delta t\right) &= \exp\left(i\Delta t L_1/2\right)\\&\times \exp\left(i\Delta t L_2\right)\\&\times \exp\left(i\Delta t L_1/2\right) + \mathcal{O}\left(\Delta t^3\right)\\&\equiv \phi_{\Delta t} + \mathcal{O}\left(\Delta t^3\right).\end{aligned} \tag{11.32}$$

Using this equation, the time reversible numerical integrator can be derived since the flow map $\phi_{\Delta t}$ has the time symmetry property $\phi_{\Delta t}\phi_{-\Delta t} = 1$. Acting with the operator $\phi_{\Delta t}$ onto the state vector at time $t = 0$ we can update coordinates and momenta at a later time Δt. Since the forces $\mathbf{F}_j$ depend only on the positions $\mathbf{q}$, the operator $\exp(iL_1\Delta t/2)$ becomes a translation operator on the momenta. Similarly, $\exp(iL_2\Delta t)$ is a translation operator on the positions. The resulting algorithm for translation motion is completely equivalent to the well-known velocity Verlet (Allen and Tildesley 1989).

The full integration procedure of the translation motion is summarized in following (Kamberaj et al. 2005)

Algorithm 1 Numerical integrator

▷ Step 1: Update velocities half-step and positions a full step
For $i = 1$ *to* N

$$\mathbf{v}_i\left(t + \frac{\Delta t}{2}\right) = \mathbf{v}_i(t) + \frac{\Delta t}{2}\frac{\mathbf{F}_i(t)}{m_i},$$

$$\mathbf{q}_i(t + \Delta t) = \mathbf{q}_i(t) + \Delta t\mathbf{v}_i\left(t + \frac{\Delta t}{2}\right),$$

▷ Step 2: Calculate forces and update velocity a full step
For $i = 1$ *to* N

$$\mathbf{v}_i(t + \Delta t) = \mathbf{v}_i\left(t + \frac{\Delta t}{2}\right) + \frac{\Delta t}{2}\frac{\mathbf{F}_i(t + \Delta t)}{m_i}. \tag{11.33}$$

It is worth noting that the algorithm for translational motion, derived here using Liouville formalism, is equivalent to the velocity-Verlet algorithm and therefore it is symplectic.

11.5 Canonical Ensemble

The integration scheme for the NVT ensemble is also formulated using the approach described above. First we write the Liouville operator for equations of motion, equation Eq. (6.29) (Chap. 6), as

$$iL_{\text{NVT}}^{(1)} = iL_{\text{NVE}}^{(1)} + iL_{NHC}^{(1)}, \tag{11.34}$$

where

$$\begin{aligned} iL_{NHC}^{(1)} = \sum_{\alpha} \Bigg(& \sum_{j=1}^{N_\alpha} [-\pi_{\eta_1}^{(\alpha)} \mathbf{v}_{j,\alpha}] \cdot \nabla_{\mathbf{v}_{j,\alpha}} \\ & + \sum_{k=1}^{M} \pi_{\eta_k}^{(\alpha)} \frac{\partial}{\partial \eta_k^{(\alpha)}} \\ & + \sum_{k=1}^{M-1} \left[\frac{G_k^{(\alpha)}}{Q_{p_k}^{(\alpha)}} - \pi_{\eta_{k+1}}^{(\alpha)} \pi_{\eta_k}^{(\alpha)} \right] \frac{\partial}{\partial \pi_{\eta_k}^{(\alpha)}} + \frac{G_M^{(\alpha)}}{Q_{p_M}^{(\alpha)}} \frac{\partial}{\partial \pi_{\eta_M}^{(\alpha)}} \Bigg), \end{aligned} \tag{11.35}$$

where the index α run over all groups of the atoms in the system and $G_k^{(\alpha)}$ denotes the thermostat forces given by:

$$\begin{aligned} G_1^{(\alpha)} &= \sum_{i=1}^{N_\alpha} \frac{\mathbf{p}_{i,\alpha}^2}{m_{i,\alpha}} - g_N^{(\alpha)} k_B T \\ G_k^{(\alpha)} &= Q_{p_k}^{(\alpha)} \left(\pi_{\eta_k}^{(\alpha)}\right)^2 - k_B T, \quad k = 2, \cdots, M \end{aligned} \tag{11.36}$$

where $g_N^{(\alpha)}$ is the number of degrees of freedom associated with group α of atoms.

Similarly to the NVE case, the Strang splitting scheme given by equation Eq. (11.25) can be used to obtain the evolution operator. Many choices for this splitting are possible, such as that given in Martyna et al. (1996) and Kamberaj et al. (2005). The numerical stability of the splitting may depend on how far from the equilibrium temperature is the initial temperature of the system and the homogeneity of the system. In addition, the choice of the integration scheme for the thermostat degrees of freedom may influence the splitting. We will use the following splitting:

$$\begin{aligned} \exp(i\Delta t L_{\text{NVT}}^{(1)}) = & \exp\left(i\frac{\Delta t}{2} L_1\right) \exp\left(\frac{\Delta t}{2} \sum_{\alpha} \sum_{j=1}^{N_\alpha} \left[-\pi_{\eta_1}^{(\alpha)} \mathbf{v}_{j,\alpha}\right] \cdot \nabla_{\mathbf{v}_{j,\alpha}}\right) \\ & \times \exp\left(\frac{\Delta t}{2} \sum_{\alpha} \frac{G_M^{(\alpha)}}{Q_{p_M}^{(\alpha)}} \frac{\partial}{\partial \pi_{\eta_M}^{(\alpha)}}\right) \end{aligned}$$

$$
\begin{aligned}
&\times \exp\left(\frac{\Delta t}{2}\sum_{\alpha}\sum_{k=1}^{M-1}\left[\frac{G_k^{(\alpha)}}{Q_{p_k}^{(\alpha)}} - \pi_{\eta_{k+1}}^{(\alpha)}\pi_{\eta_k}^{(\alpha)}\right]\frac{\partial}{\partial \pi_{\eta_k}^{(\alpha)}}\right)\\
&\times \exp\left(\Delta t \sum_{k=1}^{M}\pi_{\eta_k}^{(\alpha)}\frac{\partial}{\partial \eta_k^{(\alpha)}}\right)\\
&\times \exp\left(\frac{\Delta t}{2}\sum_{\alpha}\sum_{k=1}^{M-1}\left[\frac{G_k^{(\alpha)}}{Q_{p_k}^{(\alpha)}} - \pi_{\eta_{k+1}}^{(\alpha)}\pi_{\eta_k}^{(\alpha)}\right]\frac{\partial}{\partial \pi_{\eta_k}^{(\alpha)}}\right)\\
&\times \exp\left(\frac{\Delta t}{2}\sum_{\alpha}\frac{G_M^{(\alpha)}}{Q_{p_M}^{(\alpha)}}\frac{\partial}{\partial \pi_{\eta_M}^{(\alpha)}}\right)\\
&\times \exp\left(i\Delta t L_2\right)\\
&\times \exp\left(\frac{\Delta t}{2}\sum_{\alpha}\sum_{j=1}^{N_\alpha}\left[-\pi_{\eta_1}^{(\alpha)}\mathbf{v}_{j,\alpha}\right]\cdot\nabla_{\mathbf{v}_{j,\alpha}}\right)\exp\left(\frac{i\Delta t}{2}L_1\right)\\
&+ \mathcal{O}(\Delta t^3) \equiv \phi_{\Delta t} + \mathcal{O}(\Delta t^3),
\end{aligned}
\tag{11.37}
$$

which is correct to the second order. The approach described above for the constant NVE case is actually straightforward to implement here. Applying the approximate flow map $\phi_{\Delta t}$ on the state vector $(\eta_k, \pi_k, \mathbf{v}, \mathbf{q})$ $(k = 1, \cdots, M)$. First, the operator $\exp\left(i\frac{\Delta t}{2}L_1\right)$ acts to update the state vector for each chain thermostat α. It can be seen that the particle velocities can be updated at the half-time step using the velocity Verlet scheme described above, while other components remain unchanged. Then, the operator

$$
\exp\left(\frac{\Delta t}{2}\sum_{\alpha}\sum_{j=1}^{N_\alpha}[-\pi_{\eta_1}^{(\alpha)}\mathbf{v}_{j,\alpha}]\cdot\nabla_{\mathbf{v}_{j,\alpha}}\right)
$$

scales the output velocities according to Nosé-Hoover thermostats. At this stage, η_k, π_k, and $\mathbf{q}$, have not changed.

Next, the coordinates of the particles $\mathbf{q}$, using the velocity Verlet scheme is updated a full-time step. At this step, the extended coordinates and velocities of each thermostat are also updated using the updated velocities as input to compute the thermostat forces. In our algorithm the operator

$$
\exp\left(t\left[\frac{G_k}{Q_{p_k}} - \pi_{k+1}\pi_k\right]\frac{\partial}{\partial \pi_k}\right)
$$

is factorized as proposed by Martyna et al. (1996)

$$\exp\left(t\left[\frac{G_k}{Q_{p_k}} - \pi_{k+1}\pi_k\right]\frac{\partial}{\partial \pi_k}\right) = \exp\left(-\frac{t}{2}\pi_{k+1}\pi_k\frac{\partial}{\partial \pi_k}\right)$$
$$\exp\left(tG_k\frac{\partial}{\partial \pi_k}\right)$$
$$\exp\left(-\frac{t}{2}\pi_{k+1}\pi_k\frac{\partial}{\partial \pi_k}\right), \tag{11.38}$$

where the index α and η are omitted for simplicity in notation. The action of this operator on π_k yields

$$\pi_k \to \pi_k \exp\left(-t\pi_{k+1}\right) + tG_k \exp\left(-\frac{t}{2}\pi_{k+1}\right)\frac{\sinh\left(\frac{t}{2}\pi_{k+1}\right)}{\frac{t}{2}\pi_{k+1}}. \tag{11.39}$$

Here, the first step of the integration finishes. The rest of the evolution operator is applied in the same way after the forces acting on each particle are calculated at new particle positions, which represents the second step of the integration. The action of the full operator to the state vector $(\eta, \pi, \mathbf{v}, \mathbf{q})$ associated with chain thermostat α is summarized as following:

Algorithm 2 Numerical integrator for the NVT ensemble

▷ Step 1: Update the velocities half-step and positions full time step
$UPDATE_{NHCP}\,(t \to t + \Delta t/2)$
For $i \leftarrow 1$ *to* N

$$\mathbf{v}_i\left(t + \frac{\Delta t}{2}\right) \leftarrow \mathbf{v}_i(t) + \frac{\Delta t}{2}\frac{\mathbf{F}_i(t)}{m_i},$$

$$\mathbf{q}_i\,(t + \Delta t) \leftarrow \mathbf{q}_i(t) + \Delta t\mathbf{v}_i\left(t + \frac{\Delta t}{2}\right),$$

▷ Step 2: Calculate $\mathbf{F}(t + \Delta t)$
For $i \leftarrow 1$ *to* N

$$\mathbf{v}_i\,(t + \Delta t) \leftarrow \mathbf{v}_i\left(t + \frac{\Delta t}{2}\right) + \frac{\Delta t}{2}\frac{\mathbf{F}_i\,(t + \Delta t)}{m_i} \tag{11.40}$$

$UPDATE_{NHCP}\,(t + \Delta t/2 \to t + \Delta t)$

---

where $UPDATE_{NHCP}\,(\delta t)$ includes the following steps:

Algorithm 3 UPDATE$_{NHCP}$ (δt)

For $m \leftarrow 1$ to n_c

$$v_{i,\alpha}(t) \leftarrow v_{i,\alpha}(t)\exp\left(-\frac{\delta t}{2}\pi_{\eta_1}^{(\alpha)}(t)\right)$$

$$G_1^{(\alpha)}\left(t+\frac{\delta t}{2}\right) \leftarrow (m_{i,\alpha}\left(v_{i,\alpha}(t)\right)^2 - k_B T)/Q_1^{(\alpha)}$$

$$\pi_{\eta M}^{(\alpha)}\left(t+\frac{\delta t}{2}\right) \leftarrow \pi_{\eta M}^{(\alpha)}(t) + \frac{\delta t}{2}G_M^{(\alpha)}(t)$$

$$\pi_{M-k}^{(\alpha)}\left(t+\frac{\delta t}{2}\right) \leftarrow \exp\left(-\frac{\delta t}{4}\pi_{M-k+1}^{(\alpha)}(t)\right)$$
$$\times\left[\exp\left(-\frac{\delta t}{4}\pi_{M-k+1}^{(\alpha)}(t)\right)\pi_{M-k}^{(\alpha)}(t)\right.$$
$$\left.+\frac{\delta t}{2}G_{M-k}^{(\alpha)}(t)\frac{\sinh\left(\frac{\delta t}{4}\pi_{M-k+1}^{(\alpha)}(t)\right)}{\frac{\delta t}{4}\pi_{M-k+1}^{(\alpha)}(t)}\right] \quad (k=1,\cdots M-1),$$

$$\eta_k^{(\alpha)}(t+\delta t) \leftarrow \eta_k^{(\alpha)}(t) + \delta t\pi_k^{(\alpha)}\left(t+\frac{\delta t}{2}\right) \quad (k=1,\cdots,M),$$

$$G_k^{(\alpha)}\left(t+\frac{\delta t}{2}\right) \leftarrow (Q_{k-1}^{(\alpha)}\left(\pi_{k-1}^{(\alpha)}\left(t+\frac{\delta t}{2}\right)\right)^2 - k_B T)/Q_k^{(\alpha)}$$
$$(k=2,\cdots,M),$$

$$\pi_{\eta_k}^{(\alpha)}(t+\delta t) \leftarrow \exp\left(-\frac{\delta t}{4}\pi_{\eta_{k+1}}^{(\alpha)}\left(t+\frac{\delta t}{2}\right)\right)$$
$$\times\left[\exp\left(-\frac{\delta t}{4}\pi_{\eta_{k+1}}^{(\alpha)}\left(t+\frac{\delta t}{2}\right)\right)\right.$$
$$\left.\times\,\pi_{\eta_k}^{(\alpha)}\left(t+\frac{\delta t}{2}\right) + \frac{\delta t}{2}G_k^{(\alpha)}\left(t+\frac{\delta t}{2}\right)\frac{\sinh\left(\frac{\delta t}{4}\pi_{\eta_{k+1}}^{(\alpha)}\left(t+\frac{\delta t}{2}\right)\right)}{\frac{\delta t}{4}\pi_{\eta_{k+1}}^{(\alpha)}\left(t+\frac{\delta t}{2}\right)}\right]$$
$$(k=1,\cdots,M-1),$$

$$\pi_{\eta M}^{(\alpha)}(t+\delta t) \leftarrow \pi_{\eta M}^{(\alpha)}\left(t+\frac{\delta t}{2}\right) + \frac{\delta t}{2}G_M^{(\alpha)}\left(t+\frac{\delta t}{2}\right),$$

$$v_{i,\alpha}(t) \leftarrow v_{i,\alpha}(t)\exp\left(-\frac{\delta t}{2}\pi_{\eta_1}^{(\alpha)}(t+\delta t)\right). \tag{11.41}$$

Here, $\delta t = (\Delta t/2)/n_c$ where n_c is the number of multiple steps. The term

$$\frac{\sinh(x)}{x}$$

that appears in the non-factorized result has a singularity for $x = 0$, which can be removed by expanding it in a Maclaurin series. Typically, only the first eight terms are suggested (Martyna et al. 1996; Kamberaj et al. 2005). It has to be noted that higher order integrator algorithms can be constructed by applying the Yoshida-Suzuki factorization scheme (Yoshida 1990; Suzuki 1991). Following Martyna et

Table 11.1 The values of the (n_{ys}, w_j) parameters involved in the Yoshida-Suzuki multiple time steps integrator algorithms (Yoshida 1990; Suzuki 1991; Martyna et al. 1996)

n_{ys}	w_j
3	$w_1 = w_3 = 1/(2 - 2^{1/3})$
	$w_2 = 1 - 2w_1$
5	$w_1 = w_2 = w_4 = w_5 = 1/(4 - 4^{1/3})$
	$w_3 = 1 - 2w_1$
7	$w_1 = w_7 = -1.17767998417887$
	$w_2 = w_6 = 0.235573213359357$
	$w_3 = w_5 = 0.78451361047756$
	$w_4 = 1 - 2(w_1 + w_2 + w_3)$

al. (1996), we have used a multiple time step, denoted $n_c \times n_{ys}$ where (n_c, n_{ys}) is the number of inner steps. In this case, $UPDATE_{NHCP}$ $(t \to t + \delta t/2)$, equation Eq. (11.41), is performed in $n_c \times n_{ys}$ steps where $\delta t \to w_j(\Delta t/2)/n_c$. Some of the values of the (n_{ys}, w_j) parameters are given in Table 11.1 as reported in Yoshida (1990), Suzuki (1991) and Martyna et al. (1996).

The dynamics generated by equations Eq. (6.29) (Chap. 6) in the NVT the ensemble is not Hamiltonian and hence we can not speak of symplectic integrator algorithms for the t-flows defined by equations Eq. (11.40). However, the algorithms are time reversible (for the same reason as the NVE case, $\phi_{\Delta t}\phi_{-\Delta t} = 1$) and second-order similar to the constant NVE case, since the Trotter factorisation scheme, equation Eq. (11.37), is again correct to the second order. And as we will show in the next section these integrator algorithms for the non-microcanonical ensembles are also stable for long time trajectories, as are the symplectic integrator algorithms for the NVE ensemble. The algorithm is explicit and can also be implemented in two steps.

11.6 Isothermal-Isobaric Ensemble

Similarly to the NVT ensemble, the integration scheme for the NPT ensemble is also formulated using the Liouville formalism.

The Liouville operator for the equations of motion, Eq. (6.85) (in Chap. 6), is

$$iL^{(1)}_{\text{NPT}} = iL^{(1)}_{\text{NVT}} + iL^{(1)}_{NHB}, \tag{11.42}$$

where $iL^{(1)}_{NHB}$ is given by

$$iL^{(1)}_{NHB} = \sum_{j=1}^{N} [-\left(1 + \frac{d}{g_f}\right)\pi_\epsilon \mathbf{v}_j] \cdot \nabla_{\mathbf{v}_j} + \left[\pi_\epsilon \mathbf{q}_j\right] \cdot \nabla_{\mathbf{q}_j}$$

$$
\begin{aligned}
&+\left(\frac{G_\epsilon}{W}-\pi_\epsilon\pi_{\eta_1}\right)\frac{\partial}{\partial\pi_\epsilon}+\pi_\epsilon\frac{\partial}{\partial\epsilon}\\
&+\sum_{k=1}^{M}\pi_{\eta_k}\frac{\partial}{\partial\eta_k}\\
&+\sum_{k=1}^{M-1}\left[\frac{G_{\eta_k}}{Q_{b_k}}\frac{\partial}{\partial\pi_{\eta_k}}-\pi_{\eta_{k+1}}\pi_{\eta_k}\frac{\partial}{\partial\pi_{\eta_k}}\right]+\frac{G_{\eta_M}}{Q_{b_M}}\frac{\partial}{\partial\pi_{\eta_M}}.
\end{aligned}
\tag{11.43}
$$

The chain thermostat part of the operator retains its previous definition. Similarly, other choices for this splitting are possible, such as that given in Martyna et al. (1996) and Kamberaj et al. (2005). We will use the following splitting to approximate evolution operator:

$$
\begin{aligned}
\exp\left(i\Delta t L_{NPT}^{(1)}\right) &= \exp\left(\frac{\Delta t}{2}\frac{G_\epsilon}{W}\frac{\partial}{\partial\pi_\epsilon}\right)\\
&\times\exp\left(\frac{\Delta t}{2}\pi_{\eta_1}\pi_\epsilon\frac{\partial}{\partial\pi_\epsilon}\right)\\
&\times\exp\left(\frac{\Delta t}{2}\frac{G_{\eta_M}}{Q_{b_M}}\frac{\partial}{\partial\pi_{\eta_M}}\right)\\
&\times\exp\left(\frac{\Delta t}{2}\sum_{k=1}^{M-1}\left[\frac{G_{\eta_k}}{Q_{b_k}}-\pi_{\eta_{k+1}}\pi_{\eta_k}\right]\frac{\partial}{\partial\pi_{\eta_k}}\right)\\
&\times\exp\left(\Delta t\sum_{k=1}^{M}\pi_{\eta_k}\frac{\partial}{\partial\eta_k}\right)\\
&\times\exp\left(i\frac{\Delta t}{2}L_1\right)\\
&\times\exp\left(\frac{\Delta t}{2}\sum_\alpha\sum_{j=1}^{N_\alpha}\left[-\left(1+\frac{d}{g_f}\right)\pi_\epsilon-\pi_{\eta_1}^{(\alpha)}\right]\mathbf{v}_{j,\alpha}\cdot\nabla_{\mathbf{v}_{j,\alpha}}\right)\\
&\times\exp\left(\frac{\Delta t}{2}\sum_\alpha\frac{G_M^{(\alpha)}}{Q_{pM}^{(\alpha)}}\frac{\partial}{\partial\pi_M^{(\alpha)}}\right)\\
&\times\exp\left(\frac{\Delta t}{2}\sum_\alpha\sum_{k=1}^{M-1}\left[\frac{G_k^{(\alpha)}}{Q_{pk}^{(\alpha)}}-\pi_{\eta_{k+1}}^{(\alpha)}\pi_k^{(\alpha)}\right]\frac{\partial}{\partial\pi_{\eta_k}^{(\alpha)}}\right)
\end{aligned}
$$

$$\times \exp\left(\Delta t \sum_{k=1}^{M} \pi_k^{(\alpha)} \frac{\partial}{\partial \eta_k^{(\alpha)}}\right)$$

$$\times \exp\left(\frac{\Delta t}{2} \sum_{\alpha} \sum_{k=1}^{M-1} \left[\frac{G_k^{(\alpha)}}{Q_{p_k}^{(\alpha)}} - \pi_{\eta_{k+1}}^{(\alpha)} \pi_k^{(\alpha)}\right] \frac{\partial}{\partial \pi_{\eta_k}^{(\alpha)}}\right)$$

$$\times \exp\left(\frac{\Delta t}{2} \sum_{\alpha} \frac{G_M^{(\alpha)}}{Q_{p_M}^{(\alpha)}} \frac{\partial}{\partial \pi_M^{(\alpha)}}\right)$$

$$\times \exp\left(\Delta t \sum_{j=1}^{N} \left[\mathbf{v}_j - \pi_\epsilon \mathbf{q}_j\right] \cdot \nabla_{\mathbf{q}_j}\right)$$

$$\times \exp\left(\frac{\Delta t}{2} \sum_{\alpha} \sum_{j=1}^{N_\alpha} \left[-\left(1 + \frac{d}{g_f}\right)\pi_\epsilon - \pi_{\eta_1}^{(\alpha)}\right] \mathbf{v}_{j,\alpha} \cdot \nabla_{\mathbf{v}_{j,\alpha}}\right)$$

$$\times \exp\left(i\frac{\Delta t}{2} L_1\right)$$

$$\times \exp\left(\frac{\Delta t}{2} \sum_{k=1}^{M-1} \left[\frac{G_{\eta_k}}{Q_{b_k}} - \pi_{\eta_{k+1}} \pi_{\eta_k}\right] \frac{\partial}{\partial \pi_{\eta_k}}\right)$$

$$\times \exp\left(\frac{\Delta t}{2} \frac{G_{\eta_M}}{Q_{b_M}} \frac{\partial}{\partial \pi_{\eta_M}}\right)$$

$$\times \exp\left(\Delta t \pi_\epsilon \frac{\partial}{\partial \epsilon}\right)$$

$$\times \exp\left(\frac{\Delta t}{2} \pi_{\eta_1} \pi_\epsilon \frac{\partial}{\partial \pi_\epsilon}\right)$$

$$\times \exp\left(\frac{\Delta t}{2} \frac{G_\epsilon}{W} \frac{\partial}{\partial \pi_\epsilon}\right) + \mathcal{O}\left(\Delta t^3\right) \equiv \phi_{\Delta t} + \mathcal{O}\left(\Delta t^3\right). \quad (11.44)$$

The full approximate propagator, $\phi_{\Delta t}$ (Eq. 11.48), is applied to the full phase space by first acting with operator

$$\exp\left(\frac{\Delta t}{2} \pi_{\eta_1} \pi_\epsilon \frac{\partial}{\partial \pi_\epsilon}\right) \exp\left(\frac{\Delta t}{2} \frac{G_\epsilon}{W} \frac{\partial}{\partial \pi_\epsilon}\right)$$

to update the state vector $(s_k, \xi_k, \varepsilon, \pi_\epsilon, \mathbf{q}, \mathbf{v})$ $(k = 1, \cdots, M)$. As one can see, it updates only π_ϵ. Then the variables of the chain thermostat coupled to barostat are updated precisely in the same way as the chains of thermostats variables associated with particles, already discussed in the previous case for the NVT ensemble. The rest of the operator acts similarly to the NVT ensemble.

To update the coordinates a non-factorized operator was employed as was suggested by Martyna et al. (1996) in order to maintain the second order accuracy of factorization given by equation Eq. (11.48). The action of the operator

$$\exp\left(\Delta t\,[\mathbf{v}+\pi_\epsilon \mathbf{q}]\cdot\nabla_{\mathbf{q}}\right)$$

can be simplified using the same procedure presented above by equations Eqs. (11.38) and (11.39) as (Martyna et al. 1996):

$$\exp\left(\Delta t \pi_\epsilon \mathbf{q}\cdot\nabla_{\mathbf{q}}\right)\mathbf{q}(0) = \exp\left(\Delta t \pi_\epsilon\right)\mathbf{q}(0) + \Delta t \exp\left(\frac{\Delta t}{2}\pi_\epsilon\right)\frac{\sinh\left(\frac{\Delta t}{2}\pi_\epsilon\right)}{\frac{\Delta t}{2}\pi_\epsilon}\mathbf{v}(0). \tag{11.45}$$

Similar to the case of the NVT ensemble, the term

$$\frac{\sinh\left(\frac{\Delta t}{2}\pi_\epsilon\right)}{\frac{\Delta t}{2}\pi_\epsilon}$$

that appears in the non-factorized result has a singularity for $\pi_\epsilon = 0$, therefore it is expanded in a Maclaurin series. We also considered only the first eight terms as suggested in Martyna et al. (1996), which as will be shown here in the following section do not decrease the overall accuracy of the integrator.

The explicit integration method can be written as shown by Algorithm 4.

In Eq. (11.46), $UPDATE_{NHCP}\,(t \rightarrow t+\Delta t)$ updates the chains of thermostats coupled to translational degrees of freedom associated with α-species and is described by equations Eq. (11.41). The chain thermostat variables coupled to barostat are updated using $UPDATE_{NHCB}\,(t \rightarrow t+\Delta t)$ which is completely described by Eq. (11.41) by substituting the variables of chain thermostat coupled to particles with these coupled to barostat (see also discussion in Kamberaj et al. 2005).

For convenience we have written the steps to update the chain thermostat variables coupled to barostat in the table presenting Algorithm 5.

Here, $\delta t = (\Delta t/2)/n_c$ and n_c is the number of multiple steps. As above, the term

$$\frac{\sinh(x)}{x}$$

has a singularity for $x = 0$, which can be removed by expanding it in a Maclaurin series. Moreover, higher order integrator algorithms can also be constructed by applying the Yoshida-Suzuki factorisation scheme (Yoshida 1990; Suzuki 1991).

Algorithm 4 Numerical integrator for the NPT ensemble

▷ Step 1: Update velocities half-step and positions a full time step

$UPDATE_{NHCP}\ (t \rightarrow t + \Delta t/2)$

$UPDATE_{NHCB}\ (t \rightarrow t + \Delta t/2)$

$\pi_\epsilon\left(t + \frac{\Delta t}{2}\right) \leftarrow \pi_\epsilon(t) + \frac{\Delta t}{2} G_\epsilon(t)$

$\varepsilon\,(t + \Delta t) \leftarrow \varepsilon(t) + \Delta t \pi_\epsilon\left(t + \frac{\Delta t}{2}\right)$

For $i \leftarrow 1\ to\ N$

$$
\begin{aligned}
&\mathbf{v}_i\,(t) \leftarrow \mathbf{v}_i(t) + \frac{\Delta t}{2}\frac{\mathbf{F}_i(t)}{m_i} \\
&\mathbf{v}_i\left(t + \frac{\Delta t}{2}\right) \leftarrow \exp\left(-\frac{\Delta t}{2}\left(1 + \frac{d}{g_f}\right)\pi_\epsilon\left(t + \frac{\Delta t}{2}\right)\right)\mathbf{v}_i\,(t) \\
&\mathbf{q}_i\,(t + \Delta t) \leftarrow \exp\left(\Delta t \pi_\epsilon\left(t + \frac{\Delta t}{2}\right)\right)\mathbf{q}_i\,(t) \\
&\quad + \Delta t \exp\left(\frac{\Delta t}{2}\pi_\epsilon\left(t + \frac{\Delta t}{2}\right)\right) \\
&\quad \times \frac{\sinh\left(\frac{\Delta t}{2}\pi_\epsilon\left(t + \frac{\Delta t}{2}\right)\right)}{\frac{\Delta t}{2}\pi_\epsilon\left(t + \frac{\Delta t}{2}\right)}\mathbf{v}_i\left(t + \frac{\Delta t}{2}\right)
\end{aligned}
$$

▷ Step 2: Calculate $\mathbf{F}(t + \Delta t)$

For $i \leftarrow 1\ to\ N$

$$
\begin{aligned}
&\mathbf{v}_i\left(t + \frac{\Delta t}{2}\right) \leftarrow \exp\left(-\frac{\Delta t}{2}\left(1 + \frac{d}{g_f}\right)\pi_\epsilon\left(t + \frac{\Delta t}{2}\right)\right)\mathbf{v}_i\left(t + \frac{\Delta t}{2}\right) \\
&\mathbf{v}_i\,(t + \Delta t) \leftarrow \mathbf{v}_i\left(t + \frac{\Delta t}{2}\right) + \frac{\Delta t}{2}\frac{\mathbf{F}_i\,(t + \Delta t)}{m_i} \\
&\pi_\epsilon\,(t + \Delta t) \leftarrow \pi_\epsilon\left(t + \frac{\Delta t}{2}\right) + \frac{\Delta t}{2} G_\epsilon\,(t + \Delta t) \\
&UPDATE_{NHCB}\,(t + \Delta t/2 \rightarrow t + \Delta t) \\
&UPDATE_{NHCP}\,(t + \Delta t/2 \rightarrow t + \Delta t)
\end{aligned}
\tag{11.46}
$$

Similarly to the NVT ensemble, the dynamics generated by equations in the NPT ensemble are not Hamiltonian, therefore, the algorithm is not symplectic, but it is time-reversible ($\phi_{\Delta t}\phi_{-\Delta t} = 1$) and correct to the second order as the Trotter factorization scheme applied here (see Eq. (11.44)) is accurate to the second order. The integration procedure implements in two steps as in the previous cases.

It is worth noting that other factorization schemes can also be proposed. In particular, in NPT ensemble MD simulations when the system is far from the

Algorithm 5 UPDATE$_{NHCB}$ (δt)

For $m \leftarrow 1$ *to* n_c

$$\pi_\epsilon(t) \leftarrow \pi_\epsilon(t)\exp\left(-\frac{\delta t}{2}\pi_{\eta_1}(t)\right)$$

$$G_{\eta_1}\left(t+\frac{\delta t}{2}\right) \leftarrow (W(\pi_\epsilon(t))^2 - k_B T)/Q_{b_1}$$

$$\pi_{\eta M}\left(t+\frac{\delta t}{2}\right) \leftarrow \pi_{\eta M}(t) + \frac{\delta t}{2}G_{\eta M}(t)$$

$$\pi_{\eta M-k}\left(t+\frac{\delta t}{2}\right) \leftarrow \exp\left(-\frac{\delta t}{4}\pi_{\eta M-k+1}(t)\right)$$

$$\times\left[\exp\left(-\frac{\delta t}{4}\pi_{\eta M-k+1}(t)\right)\pi_{\eta M-k}(t)\right.$$

$$\left.+\frac{\delta t}{2}G_{\eta M-k}(t)\frac{\sinh\left(\frac{\delta t}{4}\pi_{\eta M-k+1}(t)\right)}{\frac{\delta t}{4}\pi_{\eta M-k+1}(t)}\right] \quad (k=1,\cdots M-1),$$

$$\eta_k(t+\delta t) \leftarrow \eta_k(t) + \delta t \pi_{\eta_k}\left(t+\frac{\delta t}{2}\right) \quad (k=1,\cdots,M),$$

$$G_{\eta_k}\left(t+\frac{\delta t}{2}\right) \leftarrow \left(Q_{b_{k-1}}\left(\pi_{\eta_{k-1}}\left(t+\frac{\delta t}{2}\right)\right)^2 - k_B T\right)/Q_{\eta_k}$$

$$(k=2,\cdots,M),$$

$$\pi_{\eta_k}(t+\delta t) \leftarrow \exp\left(-\frac{\delta t}{4}\pi_{\eta_{k+1}}\left(t+\frac{\delta t}{2}\right)\right)$$

$$\times\left[\exp\left(-\frac{\delta t}{4}\pi_{\eta_{k+1}}\left(t+\frac{\delta t}{2}\right)\right)\right.$$

$$\left.\times\,\pi_{\eta_k}\left(t+\frac{\delta t}{2}\right) + \frac{\delta t}{2}G_{\eta_k}\left(t+\frac{\delta t}{2}\right)\frac{\sinh\left(\frac{\delta t}{4}\pi_{\eta_{k+1}}\left(t+\frac{\delta t}{2}\right)\right)}{\frac{\delta t}{4}\pi_{\eta_{k+1}}\left(t+\frac{\delta t}{2}\right)}\right]$$

$$(k=1,\cdots,M-1),$$

$$\pi_{\eta M}(t+\delta t) \leftarrow \pi_{\eta M}\left(t+\frac{\delta t}{2}\right) + \frac{\delta t}{2}G_{\eta M}\left(t+\frac{\delta t}{2}\right)$$

$$\pi_\epsilon(t) \leftarrow \pi_\epsilon(t)\exp\left(-\frac{\delta t}{2}\pi_{\eta_1}(t+\delta t)\right). \tag{11.47}$$

equilibrium pressure and high fluctuations of the simulation box are expected. In this case, we propose in the following another factorization scheme:

$$
\begin{aligned}
\exp\left(i\Delta t L_{NPT}^{(1)}\right) &= \exp\left(i\frac{\Delta t}{2}L_1\right) \\
&\times \exp\left(\frac{\Delta t}{2}\sum_{\alpha}\sum_{j=1}^{N_\alpha}\left[-\pi_{\eta_1}^{(\alpha)}\right]\mathbf{v}_{j,\alpha}\cdot\nabla_{\mathbf{v}_{j,\alpha}}\right) \\
&\times \exp\left(\frac{\Delta t}{2}\sum_{\alpha}\frac{G_M^{(\alpha)}}{Q_{p_M}^{(\alpha)}}\frac{\partial}{\partial \pi_M^{(\alpha)}}\right) \\
&\times \exp\left(\frac{\Delta t}{2}\sum_{\alpha}\sum_{k=1}^{M-1}\left[\frac{G_k^{(\alpha)}}{Q_{p_k}^{(\alpha)}} - \pi_{\eta_{k+1}}^{(\alpha)}\pi_k^{(\alpha)}\right]\frac{\partial}{\partial \pi_{\eta_k}^{(\alpha)}}\right) \\
&\times \exp\left(\Delta t\sum_{k=1}^{M}\pi_k^{(\alpha)}\frac{\partial}{\partial \eta_k^{(\alpha)}}\right) \\
&\times \exp\left(\frac{\Delta t}{2}\frac{G_\epsilon}{W}\frac{\partial}{\partial \pi_\epsilon}\right) \\
&\times \exp\left(\frac{\Delta t}{2}\pi_{\eta_1}\pi_\epsilon\frac{\partial}{\partial \pi_\epsilon}\right) \\
&\times \exp\left(\frac{\Delta t}{2}\sum_{\alpha}\sum_{j=1}^{N_\alpha}\left[-\left(1+\frac{d}{g_f}\right)\pi_\epsilon\right]\mathbf{v}_{j,\alpha}\cdot\nabla_{\mathbf{v}_{j,\alpha}}\right) \\
&\times \exp\left(\frac{\Delta t}{2}\frac{G_{\eta_M}}{Q_{b_M}}\frac{\partial}{\partial \pi_{\eta_M}}\right) \\
&\times \exp\left(\frac{\Delta t}{2}\sum_{k=1}^{M-1}\left[\frac{G_{\eta_k}}{Q_{b_k}} - \pi_{\eta_{k+1}}\pi_{\eta_k}\right]\frac{\partial}{\partial \pi_{\eta_k}}\right) \\
&\times \exp\left(\Delta t\sum_{k=1}^{M}\pi_{\eta_k}\frac{\partial}{\partial \eta_k}\right) \\
&\times \exp\left(\Delta t\sum_{j=1}^{N}\left[\mathbf{v}_j - \pi_\epsilon\mathbf{q}_j\right]\cdot\nabla_{\mathbf{q}_j}\right) \\
&\times \exp\left(\Delta t\pi_\epsilon\frac{\partial}{\partial \epsilon}\right) \\
&\times \exp\left(\frac{\Delta t}{2}\sum_{k=1}^{M-1}\left[\frac{G_{\eta_k}}{Q_{b_k}} - \pi_{\eta_{k+1}}\pi_{\eta_k}\right]\frac{\partial}{\partial \pi_{\eta_k}}\right)
\end{aligned}
$$

$$
\begin{aligned}
&\times \exp\left(\frac{\Delta t}{2}\frac{G_{\eta M}}{Q_{b_M}}\frac{\partial}{\partial \pi_{\eta M}}\right) \\
&\times \exp\left(\frac{\Delta t}{2}\sum_{\alpha}\sum_{j=1}^{N_\alpha}\left[-\left(1+\frac{d}{g_f}\right)\pi_\epsilon\right]\mathbf{v}_{j,\alpha}\cdot\nabla_{\mathbf{v}_{j,\alpha}}\right) \\
&\times \exp\left(\frac{\Delta t}{2}\pi_{\eta_1}\pi_\epsilon\frac{\partial}{\partial \pi_\epsilon}\right) \\
&\times \exp\left(\frac{\Delta t}{2}\frac{G_\epsilon}{W}\frac{\partial}{\partial \pi_\epsilon}\right) \\
&\times \exp\left(\frac{\Delta t}{2}\sum_{\alpha}\sum_{k=1}^{M-1}\left[\frac{G_k^{(\alpha)}}{Q_{p_k}^{(\alpha)}}-\pi_{\eta_{k+1}}^{(\alpha)}\pi_k^{(\alpha)}\right]\frac{\partial}{\partial \pi_{\eta_k}^{(\alpha)}}\right) \\
&\times \exp\left(\frac{\Delta t}{2}\sum_{\alpha}\frac{G_M^{(\alpha)}}{Q_{p_M}^{(\alpha)}}\frac{\partial}{\partial \pi_M^{(\alpha)}}\right) \\
&\times \exp\left(\frac{\Delta t}{2}\sum_{\alpha}\sum_{j=1}^{N_\alpha}\left[-\pi_{\eta_1}^{(\alpha)}\right]\mathbf{v}_{j,\alpha}\cdot\nabla_{\mathbf{v}_{j,\alpha}}\right) \\
&\times \exp\left(i\frac{\Delta t}{2}L_1\right) \\
&+\mathcal{O}\left(\Delta t^3\right)\equiv\phi_{\Delta t}+\mathcal{O}\left(\Delta t^3\right).
\end{aligned}
\tag{11.48}
$$

11.7 Multiple Time Step Integrator

MD simulations of macromolecular systems, such as proteins characterized by multiple time scales, show some disadvantage because of the small time steps used to ensure the stability of numerical integration of the fast motions. Hence, the slow conformation transitions are observed only after many time steps, which practically requires a large number of force computations. However, developed methods such as the Reference System Propagator Algorithm (RESPA) reduce computational efforts for simulations of such system (Tuckerman et al. 1990, 1991; Tuckerman and Berne 1991a,b). The time-reversible approaches to the RESPA methods have also been developed, named r-RESPA, which have shown to be very stable concerning the order and stability of numerical integrator algorithms (Tuckerman et al. 1992). In the following discussion, the r-RESPA is introduced using the Trotter factorization of the classical Liouville propagation operator (Creutz and Goksch 1989; Raedt and Raedt 1983; Takahashi and Imada 1984).

Following the discussion in literature (Tuckerman et al. 1992) (see also Allen and Tildesley 1989), for a system with f degrees of freedom the Liouville operator, L, is defined as

$$iL = \{\cdots, H\} = \sum_{j=1}^{f} \left[\dot{x}_j \frac{\partial}{\partial x_j} + \dot{p}_j \frac{\partial}{\partial p_j} \right] \tag{11.49}$$

Eq. (11.49) uses the Cartesian coordinates with $(x_j, p_j) \equiv \Gamma$ being the position and conjugate momenta of the system, $\dot{p}_j$ gives the force along the jth direction, and $\{\cdots\}$ represents the Poisson bracket of the system. L is a linear Hermitian operator of square integrable function on the phase space of Γ. The dependence on L of the time propagation operator is as follows:

$$U(t) = \exp(iLt)$$

which is a unitary: $U(-t) = U^{-1}(t)$. The positions and their conjugate momenta state point of the system at a given time t is defined as $\Gamma(t) = U(t)\Gamma(0)$, which allows determining one time step propagation as the following:

$$\Gamma(\Delta t) = \exp(iL\Delta t)\,\Gamma(0)$$

where $\Delta t = t/P$ is the size of a time step. Here, t represents the total evolution time and P is the number of integration points.

The Liouville operator splits into n different terms as:

$$iL = \sum_{k=1}^{n} iL_k$$

and use the Trotter factorization scheme (Trotter 1959), then the propagator becomes

$$U(t) = \left\{ \left[\sum_{k=1}^{n-1} U_k(\Delta t/2) \right] U_n(\Delta t) \times \left[\sum_{k=1}^{n-1} U_{n-k}(\Delta t/2) \right] \right\}^P + O(t^3/P^2) \tag{11.50}$$

where $U_k(h) = \exp(iL_k h)$. Denoting

$$G(\Delta t) = \left[\sum_{k=1}^{n-1} U_k(\Delta t/2) \right] U_n(\Delta t)$$

$$\times \left[\sum_{k=1}^{n-1} U_{n-k}(\Delta t/2) \right]$$

As shown in Tuckerman et al. (1992), $G(\Delta t)G(-\Delta t) = 1$, therefore, $G(\Delta t)$ generates time-reversible dynamics.

The multiple time step integration algorithm uses splitting the system into the fast and slow degrees of freedom. Equivalently, decomposing the forces entering into the equations of motion into long-range forces, $F_l(\mathbf{r})$ and short-range forces $F_s(\mathbf{r})$ (Tuckerman et al. 1992):

$$F(\mathbf{r}) = F_s(\mathbf{r}) + F_l(\mathbf{r})$$

The short-range forces in the system associated with the slow degrees of freedom, and thus, they define the multiple timesteps of the numerical integration, namely δt. On the other hand, the long-range forces relate to the fast degrees of freedom, and thus, they define the most extended time step of the numerical integration, namely Δt. The relationship is as

$$\delta t = \frac{\Delta t}{N_{MTS}} \tag{11.51}$$

where N_{MTS} is the number of multiple steps. The short-range forces are evaluated every timestep δt, and on the other hand, the long-range forces are evaluated after every N_{MTS} time steps (i.e., every time step Δt). Hence, the degrees of freedom are advanced using Δt as a time step. Therefore, the r-RESPA implementation procedure decreases the number of calls for evaluations of the forces, which reduces, in turn, the overall computational time.

The main idea of the r-RESPA implementation (Tuckerman et al. 1992; Tuckerman and Martyna 2000; Minary et al. 2004) is on determining a reference system force $F_s(\mathbf{r})$ for short range interactions. Then, Eq. (11.49) can be written in the following form:

$$iL = \sum_{j=1}^{f} \left(\dot{x}_j \frac{\partial}{\partial x_j} + F_s(x_j) \frac{\partial}{\partial p_j} + F_l(x_j) \frac{\partial}{\partial p_j} \right) \tag{11.52}$$

$$= iL_s + \sum_{j=1}^{f} F_l(x_j) \frac{\partial}{\partial p_j}$$

and the propagator operator is factorized as

$$G(\Delta t) = \prod_{j=1}^{f} \exp\left(\frac{\Delta t}{2} F_l(x_j) \frac{\partial}{\partial p_j} \right) \tag{11.53}$$

$$\times \exp(iL_s \Delta t)$$
$$\times \prod_{j=1}^{f} \exp\left(\frac{\Delta t}{2} F_l(x_j) \frac{\partial}{\partial p_j}\right)$$

where the operator $\exp(iL_s \Delta t)$ propagates the state vector using the short-range forces with a smaller time step δt (see Eq. 11.51). Here, this operator factorizes using the Trotter formula as follows (Tuckerman et al. 1992):

$$\exp(iL_s \Delta t) = \left[\prod_{j=1}^{f} \exp\left(\frac{\delta t}{2} F_s(x_j) \frac{\partial}{\partial p_j}\right) \times \prod_{j=1}^{f} \exp\left(\delta t F_s(x_j) \dot{x}_j \frac{\partial}{\partial x_j}\right) \times \prod_{j=1}^{f} \exp\left(\frac{\delta t}{2} F_s(x_j) \frac{\partial}{\partial p_j}\right)\right]^{N_{MTS}} \tag{11.54}$$

Here, N_{MTS} is usually chosen *a priory* to guarantee the stability of numerical integrator (Tuckerman et al. 1992). Usually, when the operator $G(\Delta t)$ is applied to an initial state $(\mathbf{r}(0), \mathbf{p}(0))$, it gives a solution for both position and velocity similar to Verlet numerical integrator (Tuckerman et al. 1992).

A Lennard-Jones type of fluid has only the translational relaxation time characteristic; therefore, the numerical integration time step can easily be chosen (Tuckerman et al. 1992; Tuckerman and Parrinello 1994). However, for macromolecular systems, indeed there exists more than one time-scale, such as those characterizing the intra-molecular motion (e.g., bond stretching, angle bending, and dihedral angle). Besides, the inter-molecular motion (e.g., van der Waals and electrostatic interactions) is of the typical timescale of one or more orders in magnitude slower than intra-molecular movement. Therefore, the system is characterized by stiff nonlinear differential equations, requiring small enough time step to observe even the fast motion, if treated using one time-scale.

High-frequency oscillators interacting with a bath of slow motion (Tuckerman et al. 1990) or systems consisting of particles with different masses (e.g., large mass particles, namely slow degrees of freedom, and small mass particles, namely fast particles (Tuckerman et al. 1991)) represent other classes of systems characterized by multiple time scale motions.

The approach treats the systems coupled to a Nosé heat bath (Nosé 1984a,b; Hoover 1985b), too, used to keep temperature and pressure fixed during MD simulations. The heat bath represents extra fast degrees of freedom of the system, treated using multiple time stepping algorithms (Tuckerman et al. 1992), typically, two or more time steps. The method is used in all molecular dynamics simulation runs throughout this book.

It is important to note that the approach is limited by the so-called resonance phenomena, which restricts the use of time steps higher than $\Delta t < 8$ fs by r-RESPA in MD simulations of macromolecular systems (Bishop et al. 1997; Schlick et al. 1998; Ma et al. 2003). Note that not just time-reversible integrator algorithms, but also multiple time step symplectic integrator algorithms (Skeel et al. 1997) show numerical instability limiting the use of large time steps (Wolfram 2002). According to Schlick et al. (1998), the resonance phenomena is the result of using the perturbation techniques to derive the numerical integrator algorithms. These problems overcome using numerical integration algorithms that allow increasing the time steps in molecular dynamics simulations, such as the non-symplectic Langevin Molly (LM) (Izaguirre et al. 2001) and the so-called LN integration algorithms (Barth and Schlick 1998a,b). These methods allow using more substantial time steps in MD simulations using stochastic approaches to increase the numerical stability of integration. An improved version of r-RESPA integration algorithm, namely the Targeted Mollified Impulse algorithm (Ma and Izaguirre 2003), which includes the Langevin dynamics to improve the accuracy of multiple time stepping integrator.

The resonance-free numerical integration algorithms (Minary et al. 2004) allow for using time steps of the order up to 100 fs or even larger depending on the time length correlations studied. These algorithms use non-Hamiltonian dynamics to sample a canonical distribution of physical configuration space (Minary et al. 2004)

$$(q_1, q_2, \cdots, q_{3N}) \equiv ((x_1, y_1, z_1), \cdots, (x_N, y_N, z_N))$$

Chapter 12
Generalized Ensemble Molecular Dynamics Methods

Generalized ensemble molecular dynamics simulation methods can be used to improve the sampling of lower energy configurations. In this class of methods the following approaches have been widely used in simulations of macromolecular systems (Hansmann and Okamoto 1999): multicanonical sampling (Berg and Neuhaus 1991, 1992), the broad histogram method (de Oliveira et al. 1996; de Oliveira 1998), Wang-Landau algorithm (Wang and Landau 2001a), Tsallis weights methods (Tsallis 1988), and parallel tempering or replica exchange method (Penna 1995; Hukushima and Nemoto 1996; Geyer 1992).

All of the above mentioned generalized-ensemble approaches have the same starting point, that is, the replacement of canonical Boltzmann-like weights at temperature T with non-Boltzmann weights, which allows the system is escaping from the local minimum states.

In this chapter, we will discuss the choice of different weights for those methods that are most often used in molecular dynamics simulations according to Kamberaj (2019).

12.1 Multicanonical Sampling Method

In the multicanonical ensemble (MUCA), the states are multiplied by a non-Boltzmann multicanonical factor, $W_{\rm mu}(E)$, generating in this way a uniform probability distribution of the energy, $P_{\rm mu}(E)$ (Berg and Neuhaus 1991, 1992):

$$P_{\rm mu}(E) \propto \Omega(E) W_{\rm mu}(E) \equiv \text{constant} \tag{12.1}$$

Therefore, the multicanonical ensemble represents a free random walks in the potential energy space, since it is characterized by a flat distribution, and hence

H. Kamberaj, *Molecular Dynamics Simulations in Statistical Physics: Theory and Applications*, Scientific Computation,
https://doi.org/10.1007/978-3-030-35702-3_12

the system is able to escape faster any local energy minimum state. This yields an enhancement of the sampling the configuration phase space in an MD simulation. From Eq. (12.1), the non-Boltzmann weights are

$$W_{\mathrm{mu}}(E) \equiv \exp\left(-\beta E_{\mathrm{mu}}(E, \beta_0)\right) \propto \frac{1}{\Omega(E)} \tag{12.2}$$

where $E_{\mathrm{mu}}(E, \beta_0)$ is the multicanonical potential energy function given by

$$E_{\mathrm{mu}}(E, \beta_0) = k_B T_0 \ln \Omega(E) = \frac{1}{k_B \beta_0} S(E) \tag{12.3}$$

where $S(E) = k_B \ln \Omega(E)$ is the entropy function of the microcanonical ensemble and β_0 is the multicanonical inverse temperature.

In general, the non-Boltzmann's weights $W_{\mathrm{mu}}(E)$ are computed using short MD simulation runs (Berg and Neuhaus 1991, 1992), which is a limitation of the standard multicanonical ensemble approach.

The implementation of the MUCA in a MD simulation includes modification of the equations of motion with new forces, $\tilde{F}_i$ (for $i = 1, 2, \cdots, g_f$), as the following (Bartels and Karplus 1998; Hansmann et al. 1996; Nakajima et al. 1997):

$$\tilde{F}_i = -\frac{\partial E_{\mathrm{mu}}(E, \beta_0)}{\partial q_i} = \frac{\partial E_{\mathrm{mu}}(E, \beta_0)}{\partial E} F_i \tag{12.4}$$

In Eq. (12.4), F_i is the Newton force along the i degree of freedom. Equation (6.29) (see Chap. 6) are modified to describe the dynamics of a system in the multicanonical ensemble as:

$$\begin{aligned}
\frac{dq_i}{dt} &= \frac{p_i}{m_i} \\
\frac{dp_i}{dt} &= \frac{\beta(E)}{\beta_0} F_i - \frac{\pi_{\eta_1}}{Q_1} p_i \\
\frac{d\pi_{\eta_1}}{dt} &= \frac{p_i^2}{m_i} - k_B T - \frac{\pi_{\eta_2}}{Q_2} \pi_{\eta_1} \\
\frac{d\pi_{\eta_k}}{dt} &= \frac{\pi_{\eta_{k-1}}^2}{Q_{k-1}} - k_B T - \frac{\pi_{\eta_{k+1}}}{Q_{k+1}} \pi_{\eta_k} , \quad k = 2, \cdots, M-1 , \\
\frac{d\pi_{\eta_M}}{dt} &= \frac{\pi_{\eta_{M-1}}^2}{Q_{M-1}} - k_B T , \\
\frac{d\eta_k}{dt} &= \frac{\pi_{\eta_k}}{Q_k} , \quad k = 1, \cdots, M ,
\end{aligned} \tag{12.5}$$

where β denotes the simulation inverse temperature:

$$\beta(E_0) = \frac{1}{k_B}\left(\frac{\partial S(E)}{\partial E}\right)_{E_0}$$

Usually, the multicanonical weighting factor is computed using short MD simulation runs at some higher temperature T_0 using a canonical ensemble (Berg and Celik 1992; Okamoto and Hansmann 1995), introduced in Chap. 6. From these simulation runs, the multicanonical weighting factor is:

$$\begin{cases} E_{mu}^{(1)}(E, \beta_0) = E \\ W_{mu}^{(1)}(E, \beta_0) = W_B(E, \beta_0) = \exp(-\beta_0 E) \end{cases}$$

From the canonical ensemble MD simulation runs at the temperature T_0, a maximum value of energy E_{max} is estimated as an average of potential energy function:

$$E_{max} = \langle E \rangle_{T_0}$$

For $E \le E_{max}$, a uniform energy distribution is obtained, and for $E > E_{max}$, the system samples the canonical ensemble distribution at T_0.

In the following we are explaining the basis of the algorithm. The probability distribution weighting factor of the phase space at any given MD iteration time step t is determined as:

$$W^{(t)}(E, \beta_0) = \exp\left(-\beta_0 E^{(t)}(E, \beta_0)\right)$$

During the MD run a histogram $N^{(t)}(E)$ is accumulated of the potential energy distribution $P_{mu}^{(t)}(E)$. Denoting by $E_{min}^{(t)}$ the minimum energy value obtained until the t time step, at the $(t + 1)$ time step the multicanonical potential energy is determined as

$$E_{mu}^{(t+1)}(E, \beta_0) = \begin{cases} E, \quad E \ge E_{max} \\ E_{mu}^{(t)}(E, \beta_0) + \dfrac{1}{\beta_0} \ln\left(N^{(t)}(E)\right) - c^{(t)}, \\ \quad E_{min}^{(t)} \le E < E_{max} \\ \dfrac{\beta^{(t+1)}(E_{min}^{(t)})}{\beta_0}\left(E - E_{min}^{(t)}\right) \\ +E_{mu}^{(t+1)}(E_{min}^{(t)}, \beta_0), \quad E < E_{min}^{(t)} \end{cases} \tag{12.6}$$

Here, $c^{(t)}$ guarantees the continuity of energy function at $E = E_{max}$, defined as

$$c^{(t)} = \frac{1}{\beta_0} \ln\left(N^{(t)}(E_{max})\right)$$

The MD simulation continues until an approximately uniform probability distribution of the potential energy function is obtained. This is established by requiring that the values of energy for all $E < E_{max}$ to be of the same order of magnitude. If the convergence is achieved, $E_{min}^{(t)}$ equals the global minimum potential energy. Note that during MD simulation, a polynomial or sometimes a cubic spline function (Sugita and Okamoto 2000) is used to fit the histograms each MD simulation time step (Nakajima et al. 1997).

After the optimal weighting factor is determined, long MD simulation run in the multicanonical ensemble is performed. It is important noting that to obtain an ensemble average of any physical quantity, $\mathcal{A}$, the Weighted Histogram Analysis Method (WHAM) (Gallicchio et al. 2005) employs.

12.2 Tsallis Statistics Molecular Dynamics Method

Tsallis statistics molecular dynamics (TSMD) method (Tsallis 1988) is obtained using the principle of maximum generalized entropy. In particular, this principle is employed to obtain the so-called generalized statistical mechanics formalism with the probability weights of the each point in the phase space determined as (Tsallis 1988)

$$W_T(E,\beta) = \exp_q\left(-\beta(E-E_0)\right) = \begin{cases} \exp\left(-\beta\left(E-E_0\right)\right), & q=1 \\ \left[1-(1-q)\beta(E-E_0)\right]^{\frac{1}{1-q}}, & q\neq 1 \wedge (1-q)\beta(E-E_0)<1 \\ 0, & q\neq 1 \wedge (1-q)\beta(E-E_0)\geq 1 \end{cases} \tag{12.7}$$

In Eq. (12.7), q denotes an adjustable real parameter and E_0 is the ground state energy. Note that $W_T(E,\beta) > 0$ and for $q \to 1$, $W_T(E,\beta)$ equals the Boltzmann's weight. Besides, for $q > 1$ the Tsallis probability distribution is characterized by longer tails. The weighting factor $W_T(E,\beta)$ of generalized Tsallis distribution allows the excursion towards regions with higher energy by decreasing the magnitude of the force close to these regions. This yields an increase of the barrier crossing rate and hence allows the system escaping the local minimum energy states (Andricioaei and Straub 1997; Sugita and Okamoto 1999; Kamberaj and van der Vaart 2007; Kim and Straub 2009; Karolak and van der Vaart 2012). This is also illustrated in Fig. 12.1 for an arbitrary potential energy function given as $U(x) = 0.1521(x^4-2x^2+1)$ (Fig. 12.1a, b), and for Lennard-Jones potential energy function with $\varepsilon = 0.1521$ kcal/mol and $r_{min} = 1.7682$ Å (Fig. 12.1c, d). Here, $\beta = 1$ and Tsallis parameter is taking the following values: $q = 1.0,\ 1.5,\ 2.0,\ 3.0$. The plots are for both the potential energy function and its gradient. The graphs

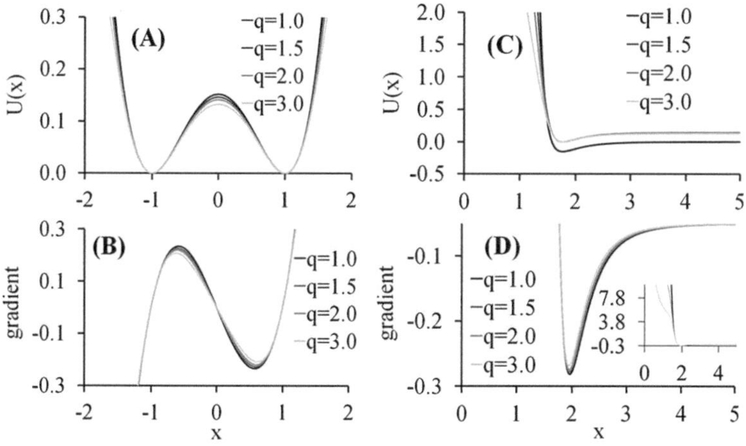

Fig. 12.1 The potential energy function and its gradient using Tsallis distribution function for $q = 1.0, 1.5, 2.0, 3.0$. (**a**)–(**b**) For an arbitrary potential energy function given as $U(x) = 0.1521(x^4 - 2x^2 + 1)$ (Barth et al. 2003); (**c**)–(**d**) For Lennard-Jones potential energy function with $\varepsilon = 0.1521$ kcal/mol and $r_{min} = 1.7682$ Å. Here, $\beta = 1$

indicate both the decrease in the barrier heights and magnitude of the force close to the barrier regions.

In the Tsallis statistical ensemble each state point of the phase space weighs by a factor, $W_{\mathrm{T}}(E, \beta)$ (Sugita and Okamoto 1999):

$$P_{\mathrm{T}}(E, \beta) \propto \Omega(E) W_{\mathrm{T}}(E, \beta) \tag{12.8}$$

where the Tsallis weights are as the following (Sugita and Okamoto 1999)

$$W_T(E, \beta) = \exp(-\beta U_{\mathrm{eff}})$$

Here, U_{eff} is the so-called effective potential defined as

$$U_{\mathrm{eff}}(E, \beta) = \frac{1}{\beta(q-1)} \ln(1 + \beta(q-1)(E - E_0)) \tag{12.9}$$

In the generalized ensemble, MD simulation runs using the new potential function U_{eff}, replacing the force field potential energy function E. Then, new forces drive the Newton's equations of motion given as (Sugita and Okamoto 1999)

$$\tilde{\mathbf{F}}_i = -\frac{\partial U_{\mathrm{eff}}(E, \beta)}{\partial \mathbf{q}_i} = \frac{1}{1 + \beta(q-1)(E - E_0)} \mathbf{F}_i \tag{12.10}$$

where $\mathbf{F}_i$ is the Newton force on particle i ($i = 1, 2, \cdots, N$). The equations of motion governing the generalized canonical ensemble according to Tsallis statistics are as the following:

$$
\begin{aligned}
\frac{dq_i}{dt} &= \frac{p_i}{m_i} \\
\frac{dp_i}{dt} &= \frac{1}{1+\beta(q-1)(E-E_0)} F_i - \frac{\pi_{\eta 1}}{Q_1} p_i \\
\frac{d\pi_{\eta 1}}{dt} &= \frac{p_i^2}{m_i} - k_B T - \frac{\pi_{\eta 2}}{Q_2} \pi_{\eta 1} \\
\frac{d\pi_{\eta k}}{dt} &= \frac{\pi_{\eta k-1}^2}{Q_{k-1}} - k_B T - \frac{\pi_{\eta k+1}}{Q_{k+1}} \pi_{\eta k}\ , \ k = 2, \cdots, M-1\ , \\
\frac{d\pi_{\eta M}}{dt} &= \frac{\pi_{\eta M-1}^2}{Q_{M-1}} - k_B T\ , \\
\frac{d\eta_k}{dt} &= \frac{\pi_{\eta k}}{Q_k}\ , \ k = 1, \cdots, M\ .
\end{aligned}
\tag{12.11}
$$

There exist several applications of the TSMD method, such as simulation of atomic clusters (Andricioaei and Straub 1996a, 1997), protein folding (Hansmann and Okamoto 1997; Pak and Wang 1999; Fukuda and Nakamura 2002; Jang et al. 2008; Kamberaj and van der Vaart 2007), molecular docking (Pak and Wang 2000), and replica exchange approach (Whitfield et al. 2002; Jang et al. 2003; Kim and Straub 2009; Kamberaj and van der Vaart 2007).

12.3 Swarm Particle-Like Molecular Dynamics Method

Recently, Kamberaj (2015) introduced a new approach based on the swarm particle social intelligence, which is tested to improve the conformation sampling (Kamberaj 2015, 2018). In this approach, in additional to the Newton force, generated from the employed force field, a new random force is applied on each particle (Kamberaj 2015), similar to Langevin dynamics (Schlick 2010). Then, the MD equations of motion given by Eq. (6.29) (see Chap. 6) modify as following (Kamberaj 2015, 2018):

$$
\begin{aligned}
\frac{dq_i}{dt} &= \frac{p_i}{m_i} \\
\frac{dq_i^{\text{Lbest}}}{dt} &= \frac{p_i^{\text{Lbest}}}{m_i} \delta\left(U(\mathbf{q}) < U\left(\mathbf{q}^{\text{Lbest}}\right)\right),
\end{aligned}
\tag{12.12}
$$

$$
\begin{aligned}
\frac{dq_i^{\text{Gbest}}}{dt} &= \frac{p_i^{\text{Gbest}}}{m_i}\delta\left(U(\mathbf{q}) < U\left(\mathbf{q}^{\text{Gbest}}\right)\right), \\
\frac{dp_i}{dt} &= F_i - \frac{\pi_{\eta_1}}{Q_1}p_i \\
&\quad + \sum_{j=1}^{m} P_{ij}\left(\gamma_1 u_1(c_j^{\text{Lbest}} - c_j) + \gamma_2 u_2(c_j^{\text{Gbest}} - c_j)\right) \\
\frac{dp_i^{\text{Lbest}}}{dt} &= -\gamma_1 u_1(q_i^{\text{Lbest}} - q_i) \\
\frac{dp_i^{\text{Gbest}}}{dt} &= -\gamma_2 u_2(q_i^{\text{Gbest}} - q_i) \\
\frac{d\pi_{\eta_1}}{dt} &= \frac{p_i^2}{m_i} - k_B T - \frac{\pi_{\eta_2}}{Q_2}\pi_{\eta_1} \\
\frac{d\pi_{\eta_k}}{dt} &= \frac{\pi_{\eta_{k-1}}^2}{Q_{k-1}} - k_B T - \frac{\pi_{\eta_{k+1}}}{Q_{k+1}}\pi_{\eta_k}\ ,\ k = 2, \cdots, M-1\ , \\
\frac{d\pi_{\eta_M}}{dt} &= \frac{\pi_{\eta_{M-1}}^2}{Q_{M-1}} - k_B T\ , \\
\frac{d\eta_k}{dt} &= \frac{\pi_{\eta_k}}{Q_k}\ ,\ k = 1, \cdots, M
\end{aligned}
$$

These equations (see Eq. (12.12)) determine an augmented dynamical system. The vector $\mathbf{c} = (c_1, c_2, \cdots, c_m)^T$ denotes the essential degrees of freedom in the system, determined by the so-called collective coordinates. Projection operator, **P**, transforms the real coordinates **q** to the collective coordinates **c** according to:

$$
c_j = \sum_{i=1}^{f} P_{ij} q_i
$$

In Eq. (12.12), $\{c_j^{\text{Lbest}}\}_{j=1}^m$ and $\{c_j^{\text{Gbest}}\}_{j=1}^m$, updated every time step, are defined as (Kamberaj 2018):

$$
\begin{aligned}
c_j^{\text{Lbest}} &= \sum_{i=1}^{f} P_{ij} q_i^{\text{Lbest}} \\
c_j^{\text{Gbest}} &= \sum_{i=1}^{f} P_{ij} q_i^{\text{Gbest}}
\end{aligned}
$$

Here, $\mathbf{q}^{\mathrm{Lbest}}$ is configuration vector with the lowest value of the potential energy of the system and $\mathbf{q}^{\mathrm{Gbest}}$ is configuration vector of the final state of the system.

In Eq. (12.12), u_i $(i = 1, 2)$ denotes the uniformly distributed random numbers in $(0, 1)$, and γ_1 and γ_2 are adjustable fixed parameters.

In Eq. (12.12), δ function is given as:

$$\delta\left(U(\mathbf{q}) < U\left(\mathbf{q}^{\mathrm{Lbest}}\right)\right) = \begin{cases} 1, & \text{if } U(\mathbf{q}) < U\left(\mathbf{q}^{\mathrm{Lbest}}\right) \\ 0, & \text{otherwise} \end{cases}$$

and

$$\delta\left(U(\mathbf{q}) < U\left(\mathbf{q}^{\mathrm{Gbest}}\right)\right) = \begin{cases} 1, & \text{if } U(\mathbf{q}) < U\left(\mathbf{q}^{\mathrm{Gbest}}\right) \\ 0, & \text{otherwise} \end{cases}$$

The augmented dynamical system, determined by Eq. (12.12), sample an equilibrium canonical distribution with conserved total energy given by:

$$\begin{aligned} E_{\mathrm{ext}} = & \sum_{i=1}^{g_f} \frac{p_i^2}{2m_i} \\ & \underbrace{+ \left(\sum_{i=1}^{g_f} \frac{\left(p_i^{\mathrm{Lbest}}\right)^2}{2m_i}\right) \delta\left(U(\mathbf{q}) < U\left(\mathbf{q}^{\mathrm{Lbest}}\right)\right) + \left(\sum_{i=1}^{g_f} \frac{\left(p_i^{\mathrm{Gbest}}\right)^2}{2m_i}\right) \delta\left(U(\mathbf{q}) < U\left(\mathbf{q}^{\mathrm{Gbest}}\right)\right)}_{E_{\mathrm{tot,kin}}} \\ & \underbrace{+ U(\mathbf{q}) + \frac{1}{2} \sum_{j=1}^{g_f} \left[u_1\gamma_1 \left(q_j^{\mathrm{Lbest}} - q_j\right)^2 + u_2\gamma_2 \left(q_j^{\mathrm{Gbest}} - q_j\right)^2\right]}_{U_{\mathrm{bias}}} \\ & \underbrace{+ \sum_{i=1}^{g_f} \sum_{k=1}^{M} \left(\frac{Q_{i,k}\pi_{i,\eta_k}^2}{2} + k_B T \eta_{i,k}\right)}_{E_{\mathrm{thermo}}} \end{aligned} \tag{12.13}$$

where g_f is the total number of degrees of freedom of the system, namely $g_f = 3N$. These equations represent an extended phase space of the augmented dynamical system with real variables:

$$\left((q_i, p_i), \left(q_i^{\mathrm{Lbest}}, p_i^{\mathrm{Lbest}}\right), \left(q_i^{\mathrm{Gbest}}, p_i^{\mathrm{Gbest}}\right)\right), \quad i = 1, 2, \cdots, g_f$$

and thermostats variables:

$$\left(\eta_{i,k}, \pi_{i,\eta_k}\right), \quad i = 1, 2, \cdots, g_f; \; k = 1, 2, \cdots, M$$

In Eq. (12.13), $E_{\text{tot,kin}}$ is the total kinetic energy of augmented system, U_{bias} is the total potential energy including bias term, and E_{thermo} is the thermostat energy. The augmented dynamical system uses the WHAM to recover the canonical equilibrium distribution of the real system (Kamberaj 2015, 2018). The augmented dynamical system is shown to sample metastable, rare transition events, and to enhance the conformation sampling (Kamberaj 2015, 2018). In Eq. (12.12), the first bias term steers the system towards the state with the lowest energy, which has been visited at any instant time t and hence enhancing the local basin sampling. Besides, the second bias term indicates the "information" about configuration with the lowest energy ever visited, and hence enhancing the barrier crossing rate.

Example 1 (Two-Dimensional Surface Model) To test the phase space sampling efficiency of the SPMD algorithm, we considered the following potential energy function model (Kamberaj 2018):

$$\begin{aligned} U(x, y) = &-4 \exp\left(-\frac{1}{4}(x+4)^2 - y^2\right) \\ &-4 \exp\left(-\frac{1}{4}(x-4)^2 - y^2\right) \\ &+\frac{1}{5625}\left(\frac{17}{400}x^6 + \frac{1}{2}(y-2)^6\right) \\ &+5 \exp\left(-4x^2 - \frac{1}{100}(y+1)^4\right) \\ &+5 \exp\left(-\frac{81}{10000}x^4 - 4y^2\right) \\ &-2 \exp\left(-\frac{81}{4}\left((x+3)^2 + \left(y - \frac{24}{5}\right)^2\right)\right) \\ &-2 \exp\left(-\frac{81}{4}\left(\left(x+\frac{1}{2}\right)^2 + \left(y - \frac{16}{5}\right)^2\right)\right) \end{aligned} \tag{12.14}$$

This potential function has two minimum stable states, namely the state A at position $(-4.0,\ 0.0)$ and state B at the position $(4.0,\ 0.0)$. Besides, $U(x, y)$ has two metastable states, namely the metastable state I at the position $(-3.0,\ 4.8)$ and the metastable state II at the position $(-0.5,\ 3.2)$. The two stable states A and B are connected via a channel that passes through the metastable states I and II (Rogal and Bolhuis 2008).

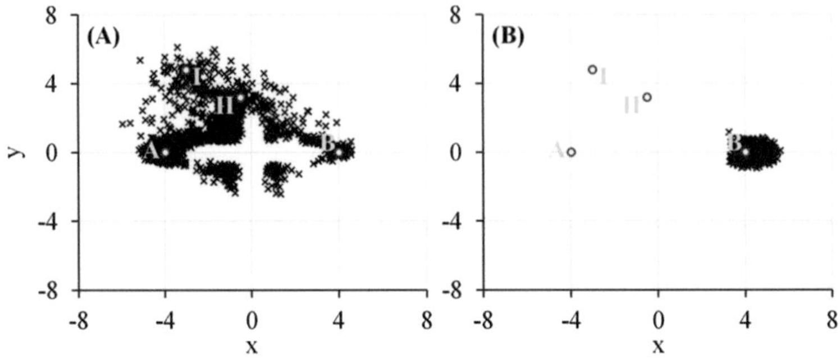

Fig. 12.2 A two-dimensional scatter plot of x-coordinate versus y-coordinate for the model potential energy function $U(x, y)$, which has four minimum states, namely the stable states A and B and the metastable states I and II. (**a**) For SPMD simulation run with $\gamma_1 = 0.01\beta$ and $\gamma_2 = 0.1\beta$; (**b**) For standard MD simulation run. A Nosé-Hoover chain of thermostats, with length $M = 3$, was coupled to the system. The mass of particle moving in that two-dimension surface is $m = 1$ amu. The target temperature is $T = 300$ K

Here, we assumed a particle with mass $m = 1$ amu moves on this two-dimensional potential energy surface, which is also coupled to a heat bath taken as a Nosé-Hoover chain of thermostats with length $M = 3$ and relaxation time $\tau = 0.01$ ps. The temperature of the bath was $T = 300$ K. Two MD simulations run, the first using the standard MD method and the second using SPMD method. In SPMD method, we fixed the parameters $\gamma_1 = 0.01\beta$ and $\gamma_2 = 0.1\beta$. Both simulations started at the particle position taken randomly in a square with minimum and maximum coordinates along the x- and y-axes as $(-8.0, +8.0)$. The initial velocities were sampled according to the Maxwell-Boltzmann distribution at $T = 300$ K, the center of mass was constraint to zero. Total simulation run for each simulation was 20 ns. The results are shown graphically in Fig. 12.2. In the SPMD (single) simulation run, every time that one of the states was visited the global best position was randomly chosen between three other minimum states with an equal probability. While, the local best position are updated every time step to the local best value of the potential energy function. Our data show that in the SPMD simulation all the minimum states are visited, and more interestingly, the path connecting the stable states A and B passes through metastable states I and II. On the other hand, in standard MD simulation, which randomly started at a position close to stable state B, only the region close to that state was sampled. These results indicate that the SPMD approach is an efficient algorithm for improving the sampling in the configuration phase space. Besides, there exists an efficient set up for enhancing the transition pathways sampling, if the positions of the minimum states are know.

12.4 Replica Exchange Method

The generalized distributions for sampling the conformation phase space are also generated using the so-called temperature *replica exchange method* (REM) (Wang and Swendsen 1986; Neal 1996; Sugita and Okamoto 1999; Falcioni and Deem 1999; Earl and Deem 2005). REM solves the problems of quasi-ergodicity in simulations of (bio)molecular systems using the so-called replicas, representing the system, simulated independently at different temperatures (Wang and Swendsen 1986). For a system of N atoms with masses m_i, position vector $\mathbf{r}_i = (x_i, y_i, z_i)$, and conjugated momentum $\mathbf{p}_i = \left(p_{xi}, p_{yi}, p_{zi}\right)$, the standard REM of a generalized ensemble corresponds to $\mathcal{L}$ independent replications of the original system coupled to $\mathcal{L}$ chain of thermostats at different temperatures (Nosé 1984c; Hoover 1985a; Tuckerman et al. 1992) with equations of motion given here as the following for each replica α:

$$\begin{aligned}
\frac{dq_{i,\alpha}}{dt} &= \frac{p_{i,\alpha}}{m_i} \\
\frac{dp_{i,\alpha}}{dt} &= F_{i,\alpha} - \frac{\pi_{\eta 1}^{(\alpha)}}{Q_1^{(\alpha)}} p_{i,\alpha} \\
\frac{d\pi_{\eta 1}^{(\alpha)}}{dt} &= \frac{p_{i,\alpha}^2}{m_i} - k_B T_\alpha - \frac{\pi_{\eta 2}^{(\alpha)}}{Q_2^{(\alpha)}} \pi_{\eta 1}^{(\alpha)} \\
\frac{d\pi_{\eta k}^{(\alpha)}}{dt} &= \frac{(\pi_{\eta k-1}^{(\alpha)})^2}{Q_{k-1}^{(\alpha)}} - k_B T_\alpha - \frac{\pi_{\eta k+1}^{(\alpha)}}{Q_{k+1}^{(\alpha)}} \pi_{\eta k}^{(\alpha)}\ , \quad k = 2, \cdots, M-1\ , \\
\frac{d\pi_{\eta M}^{(\alpha)}}{dt} &= \frac{(\pi_{\eta M-1}^{(\alpha)})^2}{Q_{M-1}^{(\alpha)}} - k_B T_\alpha\ , \\
\frac{d\eta_k^{(\alpha)}}{dt} &= \frac{\pi_{\eta k}^{(\alpha)}}{Q_k^{(\alpha)}}\ , \quad k = 1, \cdots, M
\end{aligned} \tag{12.15}$$

Two replicas between the neighboring thermostats (e.g., i and j) swap at regular interval of times with a probability P_{acc}, which preserves the detailed balance criteria (Wang and Swendsen 1986; Sugita and Okamoto 1999):

$$P_{\mathrm{acc}} = \min\left\{1, \exp\left(-(\beta_j - \beta_i)(E_i - E_j)\right)\right\} \tag{12.16}$$

Here, E_i and E_j are, respectively, the total energies of replicas i and j. In REM, high-temperature replicas are able to sample more states characterized by high energies and hence cross more often high energy barrier. On the other hand, low-temperature replicas sample more often potential energy basins and have less tendency to cross high energy barriers. The number of replicas in a standard REM scales as the square root of system's degrees of freedom (Fukunishi et al. 2002). Therefore, for systems with large number of degrees of freedom, the number of the replicas increases significantly, requiring longer MD simulation runs, necessary to optimize the rate of round trips between the two extreme temperatures.

Using implicit or combination of explicit/implicit solvent models allows decreasing the number of degrees of freedom coupled to thermostats and hence improving the optimization of the REM algorithm (Bashford and Case 2000; Zhou and Berne 2002; Zhou 2003; Garcia and Onuchic 2003; Okur et al. 2006). Other approaches include the coupling of solvent and solute to separate thermostats (Cheng et al. 2005). However, for the macromolecular systems, that reductions in system size may not accurately describe the structure and dynamics of the system (Zhou and Berne 2002; Zhou 2003; Garcia and Onuchic 2003; Liu et al. 2005). Temperature scaling of the solvent-solvent and solvent-protein interactions in REM reduced the number of replicas (Liu et al. 2005). Also, using the Tsallis biasing potential added to each replica improved further the sampling of the REM (Kamberaj and van der Vaart 2007).

Optimization of the temperature distribution among the replicas was subject of several studies, including the protein folding/unfolding transitions (Sugita and Okamoto 1999; Predescu et al. 2004, 2005; Berg 2004; Kone and Kofke 2005; **?**; Rathore et al. 2005; Gront and Kolinski 2007; Trebst et al. 2004; Katzgraber et al. 2006; Trebst et al. 2006; Nadler and Hansmann 2007; Escobedo and Martinez-Veracoechea 2007; Sabo et al. 2008; Li et al. 2007). Further modifications of the REM algorithm (Kamberaj and van der Vaart 2009) aim to obtain a flat generalized probability distribution function in temperature space using the Wang-Landau algorithm (Wang and Landau 2001b,c). It is important to note that a WHAM is used for analyzing the data from all replicas as discussed in Chodera et al. (2007).

12.5 Swarm Particle-Like Replica Exchange Method

A combination of the replica exchange method with the swarm particle-like molecular dynamics (REM:SPMD) showed to improve conformation sampling when applied to Lennard-Jones atomic cluster systems (Kamberaj 2015) and protein folding problems (Kamberaj 2018). Here, the equations of motion of the REM:SPMD algorithm are as the following (Kamberaj 2018):

$$
\begin{aligned}
\frac{dq_{i,\alpha}}{dt} &= \frac{p_{i,\alpha}}{m_i} \\
\frac{dq_{i,\alpha}^{\mathrm{Lbest}}}{dt} &= \frac{p_{i,\alpha}^{\mathrm{Lbest}}}{m_i}\delta\left(U(\mathbf{q}_\alpha) < U\left(\mathbf{q}_\alpha^{\mathrm{Lbest}}\right)\right),
\end{aligned}
\tag{12.17}
$$

$$\frac{dq_{i,\alpha}^{\text{Gbest}}}{dt} = \frac{p_{i,\alpha}^{\text{Gbest}}}{m_i}\delta\left(U(\mathbf{q}_\alpha) < U\left(\mathbf{q}_\alpha^{\text{Gbest}}\right)\right),$$

$$\frac{dp_{i,\alpha}}{dt} = F_{i,\alpha} - \frac{\pi_{\eta_1}^{(\alpha)}}{Q_1^{(\alpha)}} p_{i,\alpha}$$

$$+ \sum_{j=1}^{m} P_{ij}\left(\gamma_1 u_1 (c_j^{\alpha,\text{Lbest}} - c_j^{(\alpha)}) + \gamma_2 u_2 (c_j^{\text{Gbest}} - c_j^{(\alpha)})\right)$$

$$\frac{dp_{i,\alpha}^{\text{Lbest}}}{dt} = -\gamma_1 u_1 (q_{i,\alpha}^{\text{Lbest}} - q_{i,\alpha})$$

$$\frac{dp_{i,\alpha}^{\text{Gbest}}}{dt} = -\gamma_2 u_2 (q_{i,\alpha}^{\text{Gbest}} - q_{i,\alpha})$$

$$\frac{d\pi_{\eta_1}^{(\alpha)}}{dt} = \frac{p_{i,\alpha}^2}{m_i} - k_B T_\alpha - \frac{\pi_{\eta_2}^{(\alpha)}}{Q_2^{(\alpha)}} \pi_{\eta_1}^{(\alpha)}$$

$$\frac{d\pi_{\eta_k}^{(\alpha)}}{dt} = \frac{(\pi_{\eta_{k-1}}^{(\alpha)})^2}{Q_{k-1}^{(\alpha)}} - k_B T_\alpha - \frac{\pi_{\eta_{k+1}}^{(\alpha)}}{Q_{k+1}^{(\alpha)}} \pi_{\eta_k}^{(\alpha)}, \quad k = 2, \cdots, M-1,$$

$$\frac{d\pi_{\eta_M}^{(\alpha)}}{dt} = \frac{(\pi_{\eta_{M-1}}^{(\alpha)})^2}{Q_{M-1}^{(\alpha)}} - k_B T_\alpha,$$

$$\frac{d\eta_k^{(\alpha)}}{dt} = \frac{\pi_{\eta_k}^{(\alpha)}}{Q_k^{(\alpha)}}, \quad k = 1, \cdots, M$$

In Eq. (12.17), all the variables have the same meaning as in Eqs. (12.12) and (12.15) for the replica α. $\{c_j^{\text{Gbest}}\}_{j=1}^m$ denotes the *global best* coordinates $\mathbf{q}^{\text{Gbest}}$ representing configuration with the lowest energy among all replicas through the projection operator $\mathbf{P}$:

$$c_j^{\text{Gbest}} = \sum_{i=1}^{f} P_{ij} q_i^{\text{Gbest}}$$

Eq. (12.17) preserves the detailed balance condition (Kamberaj 2018), assuming that a Markovian chain of states is generated using a REM:SPMD simulation run. In that case, the probability of obtaining a trajectory in the configuration space of the replica k can be written as:

$$P_k(\mathbf{X}_T^k) = \exp\left(-\beta_k E(\mathbf{x}_{k,0})\right) \prod_{t=0}^{T-1} \pi(\mathbf{x}_{k,t} \to \mathbf{x}_{k,t+1}) \tag{12.18}$$

β_k denotes the inverse temperature of the thermostat k and $E(\mathbf{x}_{k,t})$ the total energy obtained for the configuration $\mathbf{x}_{k,t}$. Here, $\mathbf{X}_T^k$ vector represents T snapshots of the system (i.e., a trajectory) from replica k:

$$\mathbf{X}_T^k = \left\{\mathbf{x}_{k,0} \to \mathbf{x}_{k,1} \to \cdots \to \mathbf{x}_{k,T-1}\right\}$$

The initial configurations of the replicas are obtained according to a canonical distribution using an unbiased energy of the system as given by the force field for replica k $E(\mathbf{x}_{k,0})$:

$$\rho_{init}(\mathbf{x}_{k,0}) = \exp\left(-\beta_k E(\mathbf{x}_{k,0})\right)$$

In Eq. (12.18), $\pi(\mathbf{x}_{k,t} \to \mathbf{x}_{k,t+1})$ is the time step propagation probability, which depends on the details of deterministic or stochastic dynamics. In general, the Markovian transition probability $\pi(\mathbf{x}_{k,t} \to \mathbf{x}_{k,t+1})$ can have any distribution that conserves the Boltzmann distribution. Here, $\pi(\mathbf{x}_{k,t} \to \mathbf{x}_{k,t+1})$ represents the action characterized by augmented system given in Eq. (12.17), which produces a Boltzmann distribution in the extended phase space of variables. In the general case of the Newtonian dynamics, we can write:

$$p(\mathbf{x}_t \to \mathbf{x}_{t+1}) = \delta\left(\mathbf{x}_{t+1} - \Phi_{\Delta t}(\mathbf{x}_t)\right)$$

where δ is the delta function and $\Phi_{\Delta t}(\mathbf{x}_t)$ is the discrete flow map of one time step Δt propagation operator. In this case, a trajectory can be generated using an initial state sampled from some canonical distribution and then propagating in time using usual Hamiltonian dynamics. Note that for Hamiltonian dynamics is easy to find a time-reversible discrete flow map. On the other hand, when dynamics are governed by Eq. (12.17), the structure is not symplectic, but still, it is time reversible.

The WHAM is used to analyze the data from all replicas in the case of REM:SPMD simulations introduced in Kamberaj (2015, 2018).

12.6 Weighted Histogram Analysis Method

To analyze the data from replica exchange molecular dynamics simulation, often, the WHAM is employed. WHAM is considered an efficient approach of data processing since it combines all the data from replicas and thus the predictions include all statistical fluctuations. In WHAM, K copies of the same system (namely the replicas) are in equilibrium with L thermostats at inverse temperature β_ℓ ($\ell = 1, 2, \cdots, L$). Besides, to the unbiased potential energy, $U_\ell(\mathbf{q}_k)$ ($\ell = 1, 2, \cdots, L$; $k = 1, 2, \cdots, K$) of each replica a biasing potential energy term is added $\Delta U_\ell(\mathbf{q}_k)$. A histogram of M bins counts the unbiased potential energy values combining all of the replicas, with U_m ($m = 1, 2, \cdots, M$) being the energy at the center of the bin. In WHAM, for each replica k and histogram unbiased potential energy bin m, we count the number of independent configurations, namely H_{km}. The energy probability distribution of the system counted at the bin m visiting the thermostat ℓ is as the following (Gallicchio et al. 2005):

$$P_{\ell m} = Z_{\ell}^{-1} C_{\ell m} \Omega_m e^{-\beta_0 U_m} \tag{12.19}$$

where $\Omega_m = \Omega(U_m)$ is the density of states at the energy U_m, associated with bin m, and the constant $C_{\ell m}$ is a bias term that determines both the effect of temperature and biasing potential in probability distribution as:

$$C_{\ell m} = \exp\left(-\left(\beta_\ell - \beta_0\right) U_m\right) \times \exp\left(-\beta_\ell \Delta U_\ell\right) \tag{12.20}$$

In Eq. (12.19), $\Omega_m e^{-\beta_0 U_m}$ gives the unbiased probability of observing the system at the bin m at the target temperature and Z_ℓ is the partition function at β_ℓ. Note that the normalization condition is:

$$\sum_{m=1}^{M} P_{\ell m} = 1$$

Combination of Eqs. (12.19) and (12.20) gives

$$P_{\ell m} = \Omega_m e^{-\beta_\ell (U_m^{\text{bias}} - F_\ell)} \tag{12.21}$$

where U_m^{bias} is the bias potential energy value at the center of bin m and F_ℓ is the so-called Helmholtz free energy:

$$F_\ell = -(1/\beta_\ell) \ln Z_\ell$$

Here, F_ℓ is evaluated using an iterative procedure. For that, let $n_{k\ell}$ be the number of configurations saved from replica k visiting thermostat ℓ, then the accumulated probability density for energy bin m is:

$$P_m = \Omega_m \sum_{k=1}^{K} \sum_{\ell=1}^{L} \frac{n_{k\ell}}{N_k} e^{-\beta_\ell (U_m^{\text{bias}} - F_\ell)} \tag{12.22}$$

where N_k is the total number of configurations saved from the replica k. Besides, P_m can also be approximated as (Chodera et al. 2007):

$$P_m \approx \sum_{k=1}^{K} \frac{H_{km}}{N_k} \tag{12.23}$$

Using Eqs. (12.22) and (12.23), we get

$$\Omega_m = \frac{\sum_{k=1}^{K} H_{km}}{\sum_{k=1}^{K}\sum_{\ell=1}^{L} n_{k\ell} e^{-\beta_\ell (U_m^{\text{bias}} - F_\ell)}} \tag{12.24}$$

$$F_\ell = -\frac{1}{\beta_\ell} \ln \sum_{m=1}^{M} \Omega_m e^{-\beta_\ell U_m^{\text{bias}}}$$

Accounting for possible correlations between configurations saved from simulations, a histogram bin statistical inefficiency for each energy bin m from replica k, g_{km}, is evaluated (Chodera et al. 2007), which determines the effective number of snapshots from replica k with unbiased potential energy counted at the bin m, H_{km}^{eff}, and the effective number of snapshots from replica k in equilibrium with thermostat ℓ, $n_{k\ell}^{\text{eff}}$:

$$H_{km}^{\text{eff}} = \frac{H_{km}}{g_{km}}; \quad n_{k\ell}^{\text{eff}} = \frac{n_{k\ell}}{g_{km}}$$

Then, the calculated density of states $\hat{\Omega}_m$ is:

$$\hat{\Omega}_m = \frac{\sum_{k=1}^{K} H_{km}^{\text{eff}}}{\sum_{k=1}^{K}\sum_{\ell=1}^{L} n_{k\ell}^{\text{eff}} e^{-\beta_\ell (U_m^{\text{bias}} - F_\ell)}} \tag{12.25}$$

$$F_\ell = -\frac{1}{\beta_\ell} \ln \sum_{m=1}^{M} \hat{\Omega}_m e^{-\beta_\ell U_m^{\text{bias}}}$$

From Eq. (12.25), $\hat{\Omega}_m$ depends on F_ℓ, and, in turn, F_ℓ depends on $\hat{\Omega}_m$. Therefore, F_ℓ and $\hat{\Omega}_m$ are usually determined using an iterative procedure from Eqs. 12.25; starting from some arbitrary choice $F_\ell = 0$ ($\ell = 1, 2, \cdots, L$), iteration continues until a desired value of error is established. The statistical error $\sigma^2_{\hat{\Omega}_m}$ of $\hat{\Omega}_m$ is as follows (Chodera et al. 2007):

$$\sigma^2_{\hat{\Omega}_m} = \frac{\hat{\Omega}_m}{\sum_{k=1}^{K}\sum_{\ell=1}^{L} n_{k\ell}^{\text{eff}} e^{-\beta_\ell (U_m^{\text{bias}} - F_\ell)}} \tag{12.26}$$

The predicted average value of some physical quantity A of the system at the reference inverse temperature β_0 is calculated by summing the weighted values from all configurations:

$$\hat{A}(\beta_0) = \frac{\sum_{k=1}^{K} \sum_{n=1}^{N_k} W_{kn}(\beta_0) A_{kn}}{\sum_{k=1}^{K} \sum_{n=1}^{N_k} W_{kn}(\beta_0)} \tag{12.27}$$

In Eq. (12.27), $W_{kn}(\beta_0)$ are the weights given by

$$W_{kn}(\beta_0) = \sum_{m=1}^{M} \frac{\hat{\Omega}_m}{H_{km}} e^{-\beta_0 U_m}$$

The chain rule of error propagation is used to obtain the statistical error of $\hat{A}(\beta_0)$ (Chodera et al. 2007):

$$\sigma_{\hat{A}}^2 = \left(\frac{\langle X \rangle}{\langle Y \rangle} \right)^2 \left(\frac{\sigma_X^2}{(\langle X \rangle)^2} + \frac{\sigma_Y^2}{(\langle Y \rangle)^2} - 2 \frac{\sigma_{XY}^2}{\langle X \rangle \langle Y \rangle} \right) \tag{12.28}$$

where

$$\langle X \rangle = \frac{1}{N_k} \sum_{n=1}^{N_k} W_n(\beta_0) A_n \tag{12.29}$$

$$\langle Y \rangle = \frac{1}{N_k} \sum_{n=1}^{N_k} W_n(\beta_0) \tag{12.30}$$

$$\sigma_X^2 = \frac{g_X}{N_k(N_k - 1)} \sum_{n=1}^{N_k} (W_n(\beta_0) A_n - \langle X \rangle)^2 \tag{12.31}$$

$$\sigma_Y^2 = \frac{g_Y}{N_k(N_k - 1)} \sum_{n=1}^{N_k} (W_n(\beta_0) - \langle Y \rangle)^2 \tag{12.32}$$

$$\sigma_{XY}^2 = \frac{g_{XY}}{N_k(N_k - 1)} \sum_{n=1}^{N_k} (W_n(\beta_0) A_n - \langle X \rangle) \times (W_n(\beta_0) - \langle Y \rangle) \tag{12.33}$$

Here, $g_{X(Y,XY)}$ are the statistical inefficiencies determined from (auto)correlation functions of replica exchange simulations.

If $\Delta U_\ell = 0$ ($\ell = 1, 2, \cdots, L$), the standard WHAM of replica exchange simulations is obtained, discussed already in the literature (Chodera et al. 2007).

References

Aalten, D., Amadei, A., Linssen, A., Eijsink, V., Vriend, G., Berendsen, H.: The essential dynamics of thermolysin-conformation of the hinge-bending motion and comparison of simulations in vacuum and water. Proteins **22**, 45–54 (1993)

Abarbanel, H.D.I.: Analysis of Observed Chaotic Data. Springer, New York (1996)

Abarbanel, H.D.I., Kennel, M.B.: Local false nearest neighbors and dynamical dimensions from observed chaotic data. Phys. Rev. E **47**(5), 3057–3068 (1993)

Abraham, F.F.: Computational statistical mechanics, methodology, applications, and supercomputing. Adv. Phys. **35**, 1 (1986)

Abraham, F.F., Rudge, W.E., Auerbach, D.J., Koch, S.W.: Molecular-dynamic simulations of the incommensurate phase of krypton on graphite using more than 100,000 atoms. Phys. Rev. Lett. **52**, 445 (1984)

Albers, J., Deutch, J.M., Oppenheim, I.: Generalized Langevin equations. J. Chem. Phys. **54**(8), 3541–3546 (1971)

Alder, B.J., Hoover, W.G., Young, D.A.: Studies in molecular dynamics. V. High-Density equation of state and entropy for hard disks and spheres. J. Chem. Phys. **49**, 3688–3696 (1968)

Alder, B.J., Wainwright, T.E.: Studies in molecular dynamics. I. General method. J. Chem. Phys. **31**(2), 459 (1959)

Alfrey, T. Jr., Berg, P.W., Morawetz, H.: The counter ion distribution in solutions of rod-shaped polyelectrolytes. J. Polym. Sci. **7**, 543–547 (1951)

Allen, M.P., Tildesley, D.J.: Computer Simulation of Liquids. Oxford University Press, New York (1989)

Allinger, N.L.: Conformational analysis 130. MM2. a hydrocarbon force field utilizing V_1 and V_2 torsional terms. J. Am. Chem. Soc. **99**, 8127–8134 (1977)

Allinger, N.L., Li, F., Yan, L., Tai, J.C.: Molecular mechanics. the MM3 force field for alkenes. J. Comput. Chem. **11**, 848–867 (1990a)

Allinger, N.L., Li, F., Yan, L., Tai, J.C.: Molecular mechanics (MM3) calculations on conjugated hydrocarbons. J. Comput. Chem. **11**, 868–895 (1990b)

Allinger, N.L., Chen, K., Katzenelenbogen, J.A., Wilson, S.R., Anstead, G. M.: Hyperconjugative effects on carbon-carbon bond lengths in molecular mechanics (MM4). J. Comput. Chem. **17**, 747–755 (1996a)

Allinger, N.L., Chen, K., Lii, J-H.: An improved force field (MM4) for saturated hydrocarbons. J. Comput. Chem. **17**, 642–668 (1996b)

Amadei, A., Linssen, A.B.M., Berendsen, H.J.C.: Essential dynamics of proteins. Proteins Struct. Funct. Genet. **17**, 412–425 (1993)

H. Kamberaj, *Molecular Dynamics Simulations in Statistical Physics: Theory and Applications*, Scientific Computation,
https://doi.org/10.1007/978-3-030-35702-3

Amadei, A., de Groot, B., Ceruso, M., Paci, M., Di Nola, A., Berendsen, H.: A kinetic model for the internal motions of proteins: diffusion between multiple harmonic wells. Proteins **35**(3), 283–292 (1999a)

Amadei, A., de Groot, B.L., Ceruso, M.A., Paci, M., Di Nola, A., Berendsen, H.J.C.: A kinetic model for the internal motions of proteins: diffusion between multiple harmonic wells. Proteins Struct. Funct. Genet. **35**, 283 (1999b)

Andersen, H.C.: Molecular dynamics simulations at constant pressure and/or temperature. J. Chem. Phys. **72**, 2384–2393 (1980)

Anderson, J.A.: An Introduction to Neural Networks. MIT Press, Cambridge (1995)

Andricioaei, I., Straub, J.E.: Generalized simulated annealing algorithms using Tsallis statistics: application to conformational optimization of a tetrapeptide. Phys. Rev. E **53**, R3055 (1996a)

Andricioaei, I., Straub, J.E.: Finding the needle in the haystack: algorithms for conformational optimisation. Comput. Phys. **10**, 449–454 (1996b)

Andricioaei, I., Straub, J.E.: Global optimisation using bad derivatives: derivative-free method for molecular energy minimisation. J. Comput. Chem. **19**(13), 1445–1455 (1998)

Andricioaei, I., Straub, J.E.: On Monte Carlo and molecular dynamics methods inspired by Tsallis statistics: methodology, optimization, and application to atomic clusters. J. Chem. Phys. **107**, 9117–9124 (1997)

Anfinsen, C.B., Scheraga, H.A.: Experimental and theoretical aspects of protein folding. Adv. Protein Chem. **29**, 205–300 (1975)

Arkhipov, A., Yin, Y., Schulten, K.: Four-scale description of membrane sculpting by BARdomains. Biophys. J. **95**, 2806 (2008)

Arnold, V.I.: Mathematical Methods of Classical Mechanics. Springer, Berlin (1978)

Arnold, V.I.: Mathematical Methods of Classical Mechanics, 2nd edn. Springer, New York (1988)

Ashbaugh, H.S., Kaler, E.W., Paulaitis, M.E.: A universal surface area correlation for molecular hydrophobic phenomena. J. Am. Chem. Soc. **121**, 9243–9244 (1999)

Bahar, I., Jernigan, R.L.: Inter-residue potentials in globular proteins and the dominance of highly specific hydrophilic interactions at close separation. J. Mol. Biol. **266**, 195–214 (1997)

Baker, D.: Metastable states and folding free energy barriers. Nat. Struct. Biol. **5**, 1021–1024 (1998)

Baker, D., Agard, D.A.: Kinetics versus thermodynamics in protein folding. Biochemistry **33**, 7505–7509 (1994)

Balucani, U., Zoppi, M.: Dynamics of the Liquid State. Clarendon, Oxford (1994)

Bandt, C., Pompe, B.: Permutation entropy: a natural complexity measure for time series. Phys. Rev. Lett. **88**, 174102 (2002)

Bartels, C., Karplus, M.: Probability distributions for complex systems: adaptive umbrella sampling of the potential energy. J. Phys. Chem. B **102**, 865–880 (1998)

Barth, E., Schlick, T.: Overcoming stability limitations in biomolecular dynamics. I. Combining force splitting via extrapolation with Langevin dynamics in LN. J. Chem. Phys. **109**(5), 1617–1632 (1998a)

Barth, E., Schlick, T.: Extrapolation versus impulse in multiple-timestepping schemes. II. Linear analysis and applications to Newtonian and Langevin dynamics. J. Chem. Phys. **109**(5), 1633–1642 (1998b)

Barth, E., Schlick, T.: Overcoming stability limitations in biomolecular dynamics: I. Combining force splitting via extrapolation with Langevin dynamics in LN. J. Chem. Phys. **109**, 1617–1632 (1998c)

Barth, E.J., Laird, B.B., Leimkuhler, B.J.: Generating generalised distributions from dynamical simulation. J. Chem. Phys. **118**, 5759–5768 (2003)

Bashford, D., Case, D.A.: Generalized Born model of macromolecular salvation effects. Annu. Rev. Phys. Chem. **51**, 129–152 (2000)

Beglov, D., Roux, B.: Dominant solvations effects from the primary shell of hydration: approximation for molecular dynamics simulations. Biopolymers **35**, 171–178 (1994a)

Beglov, D., Roux, B.: Finite representation of an infinite bulk system: solvent boundary potential for computer simulations. J. Chem. Phys. **100**, 9050–9063 (1994b)

Beglov, D., Roux, B.: Numerical solutions of the hypernetted chain equation for a solute of arbitrary geometry in three dimensions. J. Chem. Phys. **103**, 360–364 (1995)

Ben-Naim, A.: Standard thermodynamics of transfer. Uses and misuses. J. Phys. Chem. **82**, 792–803 (1978)

Benguria, R., Kac, M.: Quantum langevin equation. Phys. Rev. Lett. **46**, 1 (1981)

Bereau, T., DiStasio, R.A. Jr., Tkatchenko, A., von Lilienfeld, O.A.: Non-covalent interactions across organic and biological subsets of chemical space: physics-based potentials parametrized from machine learning. J. Chem. Phys. **148**, 241706–241714 (2018)

Berendsen, H.J.C., Postma, J.P.M., van Gunsteren, W.F., DiNola, A., Haak, J.R.: Molecular dynamics with coupling to an external bath. J. Chem. Phys. **81**, 3684–3690 (1984)

Berg, B.A.: Markov Chain Monte Carlo Simulations and their Statistical Analysis. World Scientific, Singapore (2004)

Berg, B.A., Celik, T.: New approach to spin-glass simulations. Phys. Rev. Lett. **69**, 2292–2295 (1992)

Berg, B.A., Neuhaus, T.: Multicanonical algorithms for first order phase transitions. Phys. Lett. **B267**, 249–253 (1991)

Berg, B.A., Neuhaus, T.: Multicanonical ensemble: A new approach to simulate first-order phase transitions. Phys. Rev. Lett. **68**, 9–12 (1992)

Bernal, J.D.: The Bakerian lecture, 1962: the structure of liquids. Proc. R. Soc. **280**(1382), 299–322 (1964)

Beveridge, D.L., DiCapua, F.M.: Free energy via molecular simulation. Ann. Rev. Biophys. **18**, 431–492 (1989)

Bishop, T.C., Skeel, R.D., Schulten, K.: Difficulties with multiple time stepping and fast multipole algorithm in molecular dynamics. J. Comput. Chem. **18**(14), 1785–1791 (1997)

Bolhuis, P.G., Chandler, D., Dellago, C., Geissler, P.L.: Transition path sampling: throwing ropes over rough mountain passes, in the dark. Annu. Rev. Phys. Chem. **53**, 291–318 (2002)

Bonanno, C., Mega, M.: Toward a dynamical model for prime numbers. Chaos Solitons Fractals **20**, 107–118 (2004)

Bond, S.D., Laird, B.B., Leimkuhler, B.J.: The Nosé-Poincaré method for constant temperature molecular dynamics. J. Comput. Phys. **151**, 114 (1999)

Born, M.: Volumen und Hydratationswärme der ionen. Z. Phys. **1**, 45–48 (1920)

Bren, M., Florián, J., Mavri, J., Bren, U.: Do all pieces make a whole? Thiele cumulants and the free energy decomposition. Theor. Chem. Acc. **117**, 535–540 (2007)

Bren, U., Martínek, V., Florián, J.: Decomposition of the solvation free energis of Deoxyribonucleoside Triphosphates using the free energy peturbation method. J. Phys. Chem. B **110**, 12782–12788 (2006)

Brooks, B.R., Brooks, C.L., MacKerell, A.D., Nilsson, L., Petrella, R.J., Roux, B., Won, Y., Archontis, G., Bartels, C., Boresch, S., Caflisch, A., Caves, L., Cui, Q., Dinner, A.R., Feig, M., Fischer, S., Gao, J., Hodoscek, M., Im, W., Kuczera, K., Lazaridis, T., Ma, J., Ovchinnikov, V., Paci, E., Pastor, R.W., Post, C.B., Pu, J.Z., Schaefer, M., Tidor, B., Venable, R.M., Woodcock, H.L., Wu, X., Yang, W., York, D.M., Karplus, M.: CHARMM: the biomolecular simulation program. J. Comput. Chem. **30**(10), 1545–1614 (2009)

Brooks, B.B., Janezic, D., Karplus, M.: Hamonic-analysis of large systems. 1. Methodology. J. Comput. Chem. **16**, 1522–1542 (1995)

Brünger, A., Brooks III, C.L., Karplus, M.: Stochastic boundary conditions for molecular dynamics simulations of ST2 water. Chem. Phys. Lett. **105**, 495–500 (1982)

Cagin, T., Pettitt, B.M.: Molecular dynamics with a variable number of molecules. Mol. Phys. **72**, 169 (1991a)

Cagin, T., Pettitt, B.M.: Grand molecular dynamics: A method for open systems. Mol. Simul. **6**, 5 (1991b)

Callen, H.: Introduction to Thermodynamics and Thermostatics. Wiley, New York (1985)

Camsari, K.Y., Faria, R., Sutton, B.M., Datta, S.: Stochastic p-bits for invertible logic. Phys. Rev. X **7**, 031014 (2017)

Camsari, K.Y., Sutton, B.M., Datta, S.: P-bits for probabilistic spin logic. Appl. Phys. Rev. **6**, 011305 (2019)

Canutescu, A.A., Shelenkov, A.A., Dunbrack, R.L. Jr.: A graph-theory algorithm for rapid protein side-chain prediction. Protein Sci. **12**, 2001–2014 (2003)

Carlsson, G.: Topology and data. Bull. Am. Math. Soc. **46**, 255 (2009)

Case, D.A.: Normal mode analysis of protein dynamics. Curr. Opin. Struc. Biol. **4**, 290–385 (1994)

Cellucci, C.J., Albano, A.M., Rapp, P.E.: Comparative study of embedding methods. Phys. Rev. E **67**, 066210–066213 (2003)

Cellucci, C.J., Albano, A.M., Rapp, P.E.: Statistical validation of mutual information calculations: comparison of alternative numerical algorithms. Phys. Rev. E **71**, 066208–066214 (2005)

Chandrasekhar, S.: Brownian motion, dynamical friction, and stellar dynamics. Mod. Mod. Phys. **21**, 383 (1949)

Chapman, D.L.: A contribution to the theory of electrocapillarity. Phil. Mag. **25**, 475–481 (1913)

Cheatham III, T.E., Miller, J.L., Fox, T., Darden, T.A., Kollman, P.A.: Molecular dynamics simulations on solvated biomolecular systems: the particle mesh ewald method leads to stable trajectories of DNA, RNA and proteins. J. Am. Chem. Soc. **117**, 4193–4194 (1995)

Chen, J., Brooks III, C.L.: Implicit modelling of non polar solvation for simulating protein folding and conformational transitions. Phys. Chem. Chem. Phys. **10**, 471–481 (2008)

Chen, R., Liu, X., Jin, S., Lin, J., Liu, J.: Machine learning for drug-target interaction prediction. Molecules **23**, 2208–2215 (2018)

Cheng, X., Cui, G., Hornak, V., Simmerling, C.: Modified replica exchange simulation methods for local structure refinement. J. Phys. Chem. B **109**, 8220–8230 (2005)

Chodera, J.D., Swope, W.C., Pitera, J.W., Seok, C., Dill, K.A.: Use of the weighted histogram analysis method for the analysis of simulated and parallel tempering simulations. J. Chem. Theory Comput. **3**, 26–41 (2007)

Chothia, C.: Hydrophobic binding and accessible surface area in proteins. Nature **248**, 338–339 (1974)

Ciccotti, G., Vanden-Eijnden, E.: The trees and the forest. Aims and objectives of molecular dynamics simulations. Eur. Phys. J. Special Topics **224**, 2515–2518 (2015)

Clarage, J.B., Romo, T., Andrews, B.K., Pettitt, B.M., Phillips, G.N. Jr.: A sampling problem in molecular dynamics simulations of macromolecules. Proc. Natl. Acad. Sci. USA **92**, 3288–3292 (1995)

Coffey, W.T., Kalmykov, Y.P., Waldron, J.T.: The Langevin Equation, vol. 14. World Scientific Series in Contemporary Chemical Physics, 2nd Edition, World Scientific, (1996)

Collins, C.R., Gordon, G.J., von Lilienfeld, O.A., Yaron, D.J.: Constant size descriptors for accurate machine learning models of molecular properties. J. Chem. Phys. **148**, 241718–241711 (2018)

Connolly, M.L.: Solvent-accessible surfaces of proteins and nucleic acids. Science **221**, 709 (1983a)

Connolly, M.L.: Analytical molecular surface calculation. J. Appl. Cryst. **16**, 548–558 (1983b)

Connolly, M.L.: Molecular surface triangulation. J. Appl. Cryst. **18**, 499 (1985)

Cornell, W.D., Cieplak, P., Bayly, C.I., Gould, I.R., Merz, K.M. Jr., Ferguson, D.M., Spellmeyer, D.C., Fox, T., Caldwell, J.W., Kollman, P.A.: A second generation force field for the simulation of proteins, nucleic acids and organic molecules. J. Am. Chem. Soc. **117**, 5179–5197 (1995)

Cramer, C.J., Truhlar, D.G.: Implicit solvation models: equilibria, structure, spectra, and dynamics. Chem. Rev. **99**(8), 2161–2200 (1999)

Creutz, M., Goksch, A.: Higher-order hybrid Monte-Carlo algorithms. Phys. Rev. Lett. **63**, 9–12 (1989)

Dama, J.F., Sinitskiy, A.V., McCullagh, M., Weare, J., Roux, B., Dinner, A.R., Voth, G.A.: J. Chem. Theory Comput. **9**, 2466 (2013)

Darden, T., York, D., Perdersen, L.: Particle-mesh Ewald: an $n \log n$ method for Ewald sums in large systems. J. Chem. Phys. **98**, 10089–10092 (1993)

Darden, T.A., Perera, L., Li, L., Pedersen, L.: New tricks for modellers from the crystallography toolkit: the particle mesh Ewald algorithm and its use in nucleic acid simulations. Struct. Folding Des. **7**, R55–R60 (1999)
de Oliveira, P.M.C.: Broad histogram simulation: microcanonical ising dynamics. Int. J. Mod. Phys. C **9**, 497–503 (1998)
de Oliveira, P.M.C., Penna, T.J.P., Herrmann, H.J.: Broad histogram method. Braz. J. Phys. **26**, 677 (1996)
Debye, P., Hückel, E.: Zur theorie der elektrolyte. Phys. Zeitschr. **24**, 185–206 (1923)
Decherchi, S., Berteotti, A., Bottegoni, G., Rocchia, W., Cavalli, A.: The ligand binding mechanism to purine nucleoside phosphorylase elucidated via molecular dynamics and machine learning. Nat. Commun. **6**(6155), 1–10 (2015)
Derjaguin, B., Landau, L.: A theory of the stability of strongly charged lyophobic sold and the coalescence of strongly charged particles in electrolytic solution. Acta Phys. Chim. USSR **14**, 633–662 (1941)
Derreumaux, P., Schlick, T.: Long-time integration for peptides by the dynamics driver approach. Proteins Struct. Funct. Genet. **21**, 282–302 (1995)
Deserno, M., Holm, C.: How to mesh up Ewald sums. I A theoretical and numerical comparison of various particle mesh routines. J. Chem. Phys. **109**, 7678–7693 (1998a)
Deserno, M., Holm, C.: How to mesh up Ewald sums. II An accurate error estimate for the particle-particle-particle-mesh algorithm. J. Chem. Phys. **109**, 7694–7701 (1998b)
Ding, H., Karasawa, N., Goddard III, W.A.: The reduced cell multipole method for Coulomb interactions in periodic systems with million-atom unit cell. Chem. Phys. Lett. **196**, 6–10 (1992)
Dinner, A.R., Karplus, M.: A metastable state in folding simulations of a protein model. Nat. Struct. Biol. **5**, 236–241 (1998)
Doniach, S., Eastman, P.: Protein dynamics simulations from nanoseconds to microseconds. Curr. Opin. Struct. Biol. **9**, 157–163 (1999)
Dror, R.O., Arlow, D.H., Maragakis, P., Mildorf, T.J., Pan, A.C., Xu, H., Borhani, D.W.: Activation mechanism of the β2-adrenergic receptor. Proc. Natl. Acad. Sci. USA **108**, 18684–18689 (2011)
Earl, D.J., Deem, M.W.: Parallel tempering: theory, applications, and new perspectives. Phys. Chem. Chem. Phys. **7**, 3910 (2005)
Edberg, R., Evans, D.J., Moriss, G.P.: Constrained molecular dynamics: simulations of liquid alkanes with a new algorithm. J. Chem. Phys. **84**(12), 6933 (1986)
Edelsbrunner, H., Harer, J.: Computational Topology: An Introduction. Amer. Math. Soc., (2010)
Einstein, A.: Investigations on the Theory of the Brownian Movement. (Edited by Fürth), Methuen and Co. Ltd., London (1926)
Eisenberg, D., McLachlan, A.D.: Solvation energy in protein folding and binding. Nature **319**, 199–203 (1986)
Ernst, J.A., Clubb, R.T., Zhou, H.Z., Gronenborn, A.M., Clore, G.M.: Demonstration of positionally disordered water within a protein hydrophobic cavity by NMR. Science **267**, 1813–1816 (1995)
Escobedo, F.A., Martinez-Veracoechea, F.J.: Optimized expanded ensembles for simulations involving molecular insertions and deletions. I. Closed systems. J. Chem. Phys. **127**, 174103 (2007)
Evans, D.J., Hoover, W.G., Failor, B.H., Moran, B., Ladd, A.J.C.: Nonequilibrium molecular dynamics via Gauss's principle of least constraint. Phys. Rev. A **28**(2), 1016 (1983)
Ewald, P.: Die Berechnung optischer und elektrostatischer Gitter potenciale. Annalen der Physik **64**, 253–287 (1921)
Faber, F.A., Christensen, A.S., Huang, B., von Lilienfeld, O.A.: Alchemical and structural distribution based representation for universal quantum machine learning. J. Chem. Phys. **148**, 241717–12 (2018)
Falcioni, M., Deem, M.W.: A biased Monte Carlo scheme for zeolite structure solution. J. Chem. Phys. **110**(3), 1754 (1999)
Feynman, R.P.: Simulating physics with computers. Int. J. Theor. Phys. **21**, 467–488 (1982)

Fogolari, F., Briggs, J.M.: On variational approach to the Poisson-Boltzmann free energies. Chem. Phys. Lett. **281**, 135–139 (1997)

Fogolari, F., Zuccato, P., Esposito, G., Viglino, P.: Biomolecular electrostatics with the linearised Poisson-Boltzmann equation. Biophys. J. **76**, 1–16 (1999)

Fogolari, F., Brigo, A., Molinari, H.: The Poisson-Boltzmann equation for biomolecular electrostatics: a tool for structural biology. J. Mol. Recognit. **15**, 377–392 (2002)

Foloppe, N., MacKerell, A.: All-atom empirical force field for nucleic acids: I. Parameter optimization based on small molecule and condensed phase macromolecular target data. J. Comput. Chem. **21**, 86–104 (2000)

Ford, G.W., Kac, M.: On the quantum Langevin equation. J. Stat. Phys. **46**, 803–810 (1987)

Ford, G.W., Lewis, J.T., O'Connell, R.F.: Quantum Langevin equation. Phys. Rev. A **37**, 4419 (1988)

Fowler, R.H., Guggenheimer, E.A.: Statistical Thermodynamics. Cambridge University Press, Cambridge (1939)

Frauenfelder, H., Sligar, S.G., Wolynes, P.G.: The energy landscape and motions of proteins. Science **254**, 1598–1603 (1991)

Frenkel, D., Smit, B.: Understanding Molecular Simulation from Algorithms to Applications. Academic, San Diego (2001). ISBN 9780122673511

Friedrichs, M.S., Eastman, P., Vaidyanathan, V., Houston, M., Legrand, S., Beberg, A.L., Ensign, D.L., Bruns, C.M., Pande, V.S.: Accelerating molecular dynamics simulation on graphics processing units. J. Comput. Chem. **30**, 864–872 (2009)

Fukuda, I., Nakamura, H.: Tsallis dynamics using the Nosé-Hoover approach. Phys. Rev. E **65**, 026105 (2002)

Fukuda, I., Nakamura, H.: Non-Ewald methods: theory and applications to molecular systems. Biophys. Rev. **4**, 161–170 (2012)

Fukunishi, H., Watanabe, O., Takada, S.: On the Hamiltonian replica exchange method for efficient sampling of biomolecular systems: aplacation to protein structure prediction. J. Chem. Phys. **116**, 9058–9067 (2002)

Gallicchio, E., Levy, R.M.: AGBNP: an analytic implicit solvent model suitable for molecular dynamics simulations and high-resolution modeling. J. Comput. Chem. **25**, 479–499 (2004)

Gallicchio, E., Andrec, M., Felts, A.K., Levy, R.M.: Temperature weighted histogram analysis method, replica exchange, and transition paths. J. Phys. Chem. B **109**, 6722–6731 (2005)

Gallo, P., Rovere, M., Ricci, M.A., Hartnig, C., Spohr, E.: Non-exponential kinetic behaviour of confined water. Europhys. Lett. **49**(2), 183 (2000)

Garcia, A.E.: Large-amplitude nonlinear motions in proteins. Phys. Rev. Lett. **68**, 2696 (1992)

Garcia, A.E., Onuchic, J.N.: Folding a protein in a computer: an atomic description of the folding/unfolding of protein A. Proc. Natl. Acad. Sci. USA **100**, 13898–13903 (2003)

Gastegger, M., Schwiedrzik, L., Bittermann, M., Berzsenyi, F., Marquetand, P.: wACSF-Weighted atom-centered symmetry functions as descriptors in machine learning potentials. J. Chem. Phys. **148**, 241709–241711 (2018)

Gay, J.G., Berne, B.J.: Modification of the overlap potential to mimic a linear site-site potential. J. Chem. Phys. **74**, 3316 (1981)

Geyer, G.J.: Practical Markov chain Monte Carlo. Stat. Sci. **7**, 473–483 (1992)

Gibbs, J.W.: Elementary Principles in Statistical Mechanics. Yale University Press, New Haven (1902)

Gilson, M.K., Honig, B.: The dielectric constant of a folded protein. Biopolymers **25**, 2097–2119 (1986)

Gilson, M.K., Honig, B.H.: Calculation of the total electrostatic energy of a macromolecular system: solvation energies, binding energies, and conformational analysis. Proteins **4**, 7–18 (1988)

Go, N.: A theorem on amplitudes of thermal atomic fluctuations in large molecules assuming specific conformations calculated by normal mode analysis. Biophys. Chem. **35**, 105–112 (1990)

Goh, G.B., Siegel, C., Vishnu, A., Hodas, N., Baker, N.: How much chemistry does a deep neural network need to know to make accurate predictions? In: 2018 IEEE Winter Conference on Applications of Computer Vision (WACV), pp. 1340–1349 (2018)
Gohlke, H., Thorpe, M.F.: A natural coarse graining for simulating large biomolecular motion. Biophys. J. **91**, 2115–2120 (2006)
Gohlke, H., Kiel, C., Case, D.A.: Insights into protein-protein binding by binding free energy calculation and free energy decomposition for Ras-Raf and Ras-RalGDS complexes. J. Mol. Biol. **330**(4), 891–913 (2003)
Goldberg, M.E.: The second translation of the genetic message: protein folding and assembly. TIBS **10**, 388–391 (1985)
Goldstein, H.: Classical Mechanics, 2nd edn. Addison-Wesley, San Francisco (2002)
Gopal, S.M., Mukherjee, S., Cheng, Y.M., Feig, M.: PRIMO/PRIMONA: A coarse-grained model for proteins and nucleic acids that preserves near-atomistic accuracy. Proteins **78**, 1266–1281 (2010)
Gourévitch, B., Eggermont, J.: Evaluating information transfer between auditory cortical neurons. J. Neurophysiol. **97**, 2533–2543 (2007)
Gouy, M.: Sur la constitution de la charge électrique a la surface d'un électrolyte. J. Phys. **9**, 457–468 (1910)
Granger, J.: Investigating causal relations by econometric models and crossspectral methods. Acta Physica Polonica B **37**, 424–438 (1969)
Grassberger, P.: Finite sample corrections to entropy and dimension estimates. Phys. Lett. A **128**, 369–373 (1988)
Grassberger, P., Procaccia, I.: Measuring the strangeness of strange attractors. Physica D **9**, 189 (1983)
Greengard, L., Rokhlin, V.: A fast algorithm for particle simulations. J. Comput. Phys. **73**, 325–348 (1987)
Grimme, S.: A general quantum mechanically derived force field (QMDFF) for molecules and condensed phase simulations. J. Chem. Theory Comput. **10**(10), 4497–4514 (2014)
Gront, D., Kolinski, A.: Efficient scheme for optimization of parallel tempering Monte Carlo method. J. Phys. Condens. Matter **19**, 036225 (2007)
Gronwall, T.H., La Mer, V.K., Sandved, K.: Über den einfluss der sogenannten höheren glieder in der Debye-Hückelschen theorie der lösungen starker elektrolyte. Phys. Zeitschr. **29**, 358–393 (1928)
Grubmüller, H.: Predicting slow structural transitions in macromolecular systems: conformational flooding. Phys. Rev. E **52**, 2893 (1995)
Grycuk, T.: Deficiency of the Coulomb-field approximation in the generalised Born model: an improved formula for form radii evaluation. J. Chem. Phys. **119**, 4817–4826 (2003)
Hagler, A.T., Lifson, S.: Energy functions for peptides and proteins. II. Amide hydrogen bond and calculation of amide crystal properties. J. Am. Chem. Soc. **96**(17), 5327–5335 (1974)
Hagler, A.T., Huler, E., Lifson, S.: Energy functions for peptides and proteins. I. Derivation of a consistent force field including the hydrogen bond from amide crystals. J. Am. Chem. Soc. **96**(17), 5319–5327 (1974)
Hagler, A.T., Lifson, S., Dauber, P.: Consistent force field studies of intermolecular forces in hydrogen-bonded crystals. 1. Carboxylic acids, amides, and the C:O$\cdots$H-hydrogen bonds. J. Am. Chem. Soc. **101**(18), 5111–5121 (1979a)
Hagler, A.T., Lifson, S., Dauber, P.: Consistent force field studies of intermolecular forces in hydrogen-bonded crystals. 2. A benchmark for the objective comparison of alternative force fields. J. Am. Chem. Soc. **101**(18), 5122–5130 (1979b)
Hall, L.H., Kier, L.B.: Electrotopological state indices for atom types: a novel combination of electronic, topological, and valence state information. J. Chem. Inf. Comput. Sci. **35**, 1039–1045 (1995)
Halle, B., Andersson, T., Forsen, S., Lindman, B.: Protein hydration from water oxygen-17 magnetic relaxation. J. Am. Chem. Soc. **103**, 500–508 (1981)

Hänggi, P., Ingold, G.-L.: Fundamental aspects of quantum Brownian motion. Chaos **15**, 026105–1 (2005)
Hansen, J.P., McDonald, I.R.: Theory of Simple Liquids. Academic Press, London (1986)
Hansmann, U.H.E., Okamoto, Y.: Generalized-ensemble Monte Carlo method for systems with rough energy landscape. Phys. Rev. E **56**(2), 2228–2233 (1997)
Hansmann, U.H.E., Okamoto, Y.: Annual Reviews in Computational Physics VI. World Scientific, Singapore (1999)
Hansmann, U.H.E., Okamoto, Y., Eisenmenger, F.: Molecular dynamics, Langevin and hydrid Monte Carlo simulations in a multicanonical ensemble. Chem. Phys. Lett. **259**, 321–330 (1996)
Hardy, D.J., Stone, J.E., Vandivort, K.L., Gohara, D., Rodrigues, C., Schulten, K.: Fast molecular electrostatics algorithms on GPUs. In: Wen mei Hwu, W. (ed.) GPU Computing Gems, pp. 43–58. Morgan Kaufmann Publishers, San Francisco (2011)
Hastie, T., Tibshirani, R., Friedman, J.: The Elements of Statistical Learning, 2 edn. Springer, New York (2009)
Hawkins, G.D., Cramer, C.J., Truhlar, D.G.: Pairwise solute descreening of solute charges from a dielectric medium. Chem. Phys. Lett. **246**, 122–129 (1995)
Hawkins, G.D., Cramer, C.J., Truhlar, D.G.: Parametrized models of aqueous free energies of solvation based on pairwise descreening of solute atomic charges from a dielectric medium. J. Phys. Chem. **100**, 19824–19839 (1996)
Herr, J.E., Yao, K., McIntyre, R., Toth, D.W., Parkhill, J.: Metadynamics for training neural network model chemistries: a competitive assessment. J. Chem. Phys. **148**, 241710–9 (2018)
Hirata, F., Rossky, P.J.: An extended RISM equation for molecular polar fluids. Chem. Phys. Lett. **83**, 329–334 (1981)
Hockney, R., Eastwood, J.: Computer simulation using particles. McGraw-Hill, New York (1981)
Hockney, R.W., Eastwood, J.W.: Computer Simulation Using Particles. Adam Hilger, Bristol (1988)
Hoover, W.G.: Canonical dynamics: equilibrium phase-space distributions. Phys. Rev. A **31**, 1695–1697 (1985a)
Hoover, W.G.: Computational Statistical Mechanics, 1st edn. Elsevier Science, Burlington (1991)
Hoover, W.G., Ree, F.H.: Melting transition and communal entropy for hard spheres. J. Chem. Phys. **49**, 3609 (1968)
Hoover, W.G.: Canonical dynamics: equilibrium phase-space distributions. Phys. Rev. A **31**, 1695 (1985b)
Hu, M., Strachan, J.P., Li, Z., Grafals, E.M., Davila, N., Graves, C., Lam, S., Ge, N., Yang, J.J., Williams, R.S.: Dot-product engine for neuromorphic computing: programming 1t1m crossbar to accelerate matrix-vector multiplication. In: Proceedings of the 53rd annual design automation conference, p. 19 (2016)
Hukushima, K., Nemoto, K.: Exchange Monte Carlo method and application to spin glass simulations. J. Phys. Soc. Jpn. **65**, 1604–1608 (1996)
Humphrey, W., Dalke, A., Schulten, K.: VMD – Visual molecular dynamics. J. Mol. Graph. **14**, 33–38 (1996)
Hünenberger, P.H., van Gunsteren, W.F.: Computer Simulation of Biomolecular Systems, Theoretical and Experimental Applications. Kluwer, Dordrecht (1997)
Ichiye, T., Karplus, M.: Collective motions in proteins: a covariance analysis of atomic fluctuations in molecular-dynamics and normal mode simulations. Proteins **11**, 205–217 (1991)
Irbäck, A., Sjunnesson, F., Wallin, S.: Hydrogen bonds, hydrophobicity forces and the character of the collapse transition. Proc. Natl. Acad. Sci. U.S.A. **97**, 13614 (2000)
Islam, M.A.: Einstein - Smoluchowski diffusion equation: A discussion. Physica Scripta. **70**, 120 (2004)
Izaguirre, J.A., Catarello, D.P., Wozniak, J.M., Skeel, R.D.: Langevin stabilization of molecular dynamics. J. Chem. Phys. **114**(5), 2090–2098 (2001)
Jackson, J.D.: Classical Electrodynamics. Wiley, New York (1962)
Jain, M.K.: Introduction to Biological Membranes. Wiley-Interscience, New York (1988)

Janezic, D., Brooks, B.B.: Harmonic analysis of large systems: II. Comparison of different protein models. J. Comput. Chem. **16**, 1543–1553 (1995)

Jang, S., Voth, G.A.: Simple reversible molecular dynamics algorithms for Nosé-Hoover chain dynamics. J. Chem. Phys. **107**(22), 9514–9526 (1997)

Jang, S., Shin, S., Pak, Y.: Replica-exchange method using the generalized effective potential. Phys. Rev. Lett. **91**, 058305 (2003)

Jang, S., Kim, E., Pak, Y.: All-atom level direct folding simulation of $\beta\beta\alpha$ miniprotein. J. Chem. Phys. **128**, 105102 (2008)

Ji, J., Pettitt, B.M.: Phase transitions of water at constant excess chemical potential. An application of grand molecular dynamics. Mol. Phys. **82**, 67–83 (1994)

Ji, J., Cagin, T., Pettitt, B.M.: Dynamic simulations of water at constant chemical potential. J. Chem. Phys. **96**, 1333 (1992)

Joe, H.: Relative entropy measures of multivariate dependence. J. Am. Statist. Assoc. **84**, 157–164 (1989)

Jorgensen, W.L., Tirado-Rives, J.: The OPLS potential functions for proteins – energy minimizations for crystals of cyclic-peptides and crambin. J. Am. Chem. Soc. **110**, 1666–1671 (1988)

Kamath, A., Vargas-Hernández, R.A., Krems, R.V., Carrington, T. Jr., Manzhos, S.: Neural networks vs Gaussian process regression for representing potential energy surface: a comparative study of fit quality and vibrational spectrum accuracy. J. Chem. Phys. **148**, 241702–7 (2018)

Kamberaj, H.: A theoretical model for the collective motion of proteins by means of principal component analysis. Cent. Eur. J. Phys. **9**(1), 96–109 (2011)

Kamberaj, H.: Conformational sampling enhancement of replica exchange molecular dynamics simulations using swarm particle intelligence. J. Chem. Phys. **143**, 124105–8 (2015)

Kamberaj, H.: Sampling Convergence of collective motions in proteins. J. Appl. Phys. Sci. Int. **8**(3), 101–112 (2017)

Kamberaj, H.: Faster protein folding using enhanced conformational sampling of molecular dynamics simulation. J. Mol. Graph. Model. **81**, 32–49 (2018)

Kamberaj, H.: Advanced methods used in molecular dynamics simulation of macromolecules. In: Kale, S.A. (ed.) Mechanical Design, Materials and Manufacturing, pp. 57–134. Nova Science Publishers, Inc., New York (2019)

Kamberaj, H., van der Vaart, A.: Extracting the causality of correlated motions from molecular dynamics simulations. Biophys. J. **97**, 1747–1755 (2009a)

Kamberaj, H., van der Vaart, A.: Extracting the causality of correlated motions from molecular dynamic simulations. Biophys. J. **97**, 1747–1755 (2009b)

Kamberaj, H., van der Vaart, A.: Multiple scaling replica exchange for the conformational sampling of biomolecules in explicit water. J. Chem. Phys. **127**, 234102–234109 (2007)

Kamberaj, H., van der Vaart, A.: An optimised replica exchange method for molecular dynamics simulations. J. Chem. Phys. **130**, 074904 (2009)

Kamberaj, H., Low, R.J., Neal, M.P.: Time reversible and symplectic integrators for molecular dynamics simulations of rigid molecules. J. Chem. Phys. **122**(22), 224114 (2005)

Karhunen, K.: Über lineare Methoden in der Wahrscheinlichkeitsrechnung. Ann. Acad. Sci. Fenn. Ser. A1, **37**, 1–79 (1947)

Karolak, A., van der Vaart, A.: Importance of local interactions for the stability of inhibitory helix 1 of Ets-1 in the apo state. Biophys. Chem. **165–166**(3), 74–78 (2012)

Karplus, M., Jushick, J.N.: Method for estimating the configurational entropy of macromolecules. Macromolecules **14**, 325–332 (1981)

Karplus, M., Kuriyan, J.: Molecular dynamics and protein function. Proc. Natl. Acad. Sci. USA **102**(19), 6679–6685 (2005)

Karplus, M., McCammon, J.A.: Molecular dynamics simulations of biomolecules. Nat. Struct. Biol. **9**(9), 646–652 (2002); With corrigenda in Nat. Struct. Biol. **9**(10), 788 (2002)

Katchalski, A.: Polyelectrolytes. Pure Appl. Chem. **26**, 327–371 (1971)

Katzgraber, H.G., Trebst, S., Huse, D.A., Troyer, M.: Feedback-optimized parallel tempering Monte Carlo. J. Stat. Mech. **2006**(3), P03018 (2006)

Kennel, M.B., Brown, R., Abarbanel, H.D.I.: Determining embedding dimension for phase-space reconstruction using a geometrical construction. Phys. Rev. A **45**, 3403–3411 (1992)
Kim, J., Straub, J.E.: Optimal replica exchange method combined with Tsallis weight sampling. J. Chem. Phys. **130**, 144114–11 (2009)
Kirkwood, J.G.: On the theory of strong electrolyte solutions. J. Chem. Phys. **2**, 767–781 (1934a)
Kirkwood, J.G.: Theory of solutions of molecules containing widely separated charges with special applications to zwitterions. J. Chem. Phys. **7**, 351–361 (1934b)
Kitao, A., Hirata, F., Go, N.: The effect of solvent on the conformation and the collective motions of protein: normal mode analysis and molecular dynamics simulations of Melittin in water and in vacuum. Chem. Phys. **158**, 447–472 (1991)
Klimov, D.K., Thirumalai, D.: Viscosity dependence of the folding rates of proteins. Phys. Rev. Lett. **79**, 317–320 (1997)
Kollman, P.A.: Free energy calculations: applications to chemical and biochemical phenomena. Chem. Rev. **93**, 2395–2418 (1993)
Kollman, P.A., Massova, I., Reyes, C., Kuhn, B., Huo, S.H., Chong, L., Lee, M., Lee, T., Duan, Y., Wang, W., Donini, O., Cieplak, P., Srinivasan, J., Case, D.A., Cheatham, T.E.: Calculating structures and free energies of complex molecules: combining molecular mechanics and continuum models. Accounts Chem. Res. **33**(12), 889–897 (2000)
Kone, A., Kofke, D.A.: Selection of temperature intervals for parallel-tempering simulations. J. Chem. Phys. **122**, 206101 (2005)
Kraskov, A., Stögbauer, H., Grassberger, P.: Estimating mutual information. Phys. Rev. E **69**(6), 066138 (2004)
Kullback, S.: Information Theory and Statistics. Wiley, New York (1959)
Kullback, S.: The Kullback-Leibler distance. Am. Stat. **41**, 340–341 (1987)
Kullback, S., Leibler, R.A.: On information and sufficiency. Ann. Math. Stat. **22**, 79–86 (1951)
Lange, O.F., Grubmüller, H.: Can principal components yield a dimension reduced description of protein dynamics on long time scales? J. Phys. Chem. B **110**, 22842–22852 (2006)
Lange, O.F., Schäfer, L.V., Grubmüller, H.: Flooding in GROMACS: accelerated barrier crossing in molecular dynamics. J. Comput. Chem. **27**(14), 1693–1702 (2006)
Lazaridis, T.: Inhomogeneous fluid approach to solvation thermodynamics 2. Application to simple fluid. J. Phys. Chem. **102**, 3542–3550 (1998a)
Lazaridis, T.: Inhomogeneous fluid approach to solvation thermodynamics 1. Theory. J. Phys. Chem. **102**, 3531–3541 (1998b)
Lazaridis, T.: Solvent reorganisation energy and entropy in hydrophobic hydration. J. Phys. Chem. B **104**, 4964–4979 (2000)
Lazaridis, T.: Solvent size versus cohesive energy density as the origin of hydrophobicity. Acc. Chem. Res. **34**, 931–937 (2001)
Lazaridis, T., Karplus, M.: Effective energy function for proteins in solution. Proteins **35**, 133–152 (1999)
Lazaridis, T., Karplus, M.: Thermodynamics of protein folding: a microscopic view. Biophys. Chem. **100**, 367–395 (2003)
Lazaridis, T., Paulaitis, M.E.: Entropy of hydrophobic hydration: a new statistical mechanical formulation. J. Phys. Chem. **96**, 3847–3855 (1992)
Lazaridis, T., Paulaitis, M.E.: Simulation studies of the hydration entropy of simple hydrophobic solutes. J. Phys. Chem. **98**, 635–642 (1994)
Leach, A.R.: Molecular Modelling. Principles and Applications, 2nd edn. Pearson Education Limited, Edingburgh Gate/Prentice Hall (2001)
Lee, B., Richards, F.M.: The interpretation of protein structures: estimation of static accessibility. J. Mol. Biol. **55**, 379–400 (1971)
Lee, J., Bahri, Y., Novak, R., Schoenholz, S.S., Pennington, J., Sohl-Dickstein, J.: Deep neural networks as gaussian processes. In: Conference Proceedings in ICLR, pp. 1–8 (2018)
Lee, M.S., Salsbury, F.R. Jr., Brooks III, C.L.: Novel generalized Born methods. J. Chem. Phys. **116**, 10606–10614 (2002)

Lehrman, M., Rechester, A.B., White, R.B.: Symbolic analysis of chaotic signals and turbulent fluctuations. Phys. Rev. Lett **78**, 54–57 (1997)

Leimkuhler, B.J., Reich, S.: Geometric Numerical Methods for Hamiltonian Mechanics. Cambridge University Press, Cambridge (2004)

Levitt, M., Sander, C., Stern, P.S.: Protein normal-mode dynamics: trypsin inhibitor, cram bin, ribonuclease and lysozyme. J. Mol. Biol. **181**, 423–447 (1985)

Levy, R.M., Zhang, L.Y., Gallicchio, E., Felts, A.K.: On the nonpolar hydration free energy of proteins: surface area and continuum solvent models for the solute-solvent interaction energy. J. Am. Chem. Soc. **125**, 9523–9530 (2003)

Li, X., O'Brien, C.P., Collier, G., Vellore, N.A., Wang, F., Latour, R.A.: An improved replica-exchange sampling method: temperature intervals with global energy reassignment. J. Chem. Phys. **127**, 164116 (2007)

Liang, X.S.: The Liang-Kleeman information flow: theory and applications. Entropy **15**, 327–360 (2013)

Liang, X.S., Kleeman, R.: Information transfer between dynamical system components. Phys. Rev. Lett. **95**, 244101 (2005)

Lifson, S., Katchalski, A.: The electrostatic free energy of polyelectrolyte solutions. J. Polym. Sci. **13**, 43–55 (1954)

Lifson, S., Warshel, A.: Consistent force field for calculations of conformations, vibrational spectra, and enthalpies of cycloalkane and n-alkane molecules. J. Chem. Phys. **49**, 5116–5129 (1968)

Linderstrom-Lang, K.: On the ionisation of proteins. Compt. Rend. Trav. Lab. Carlsberg **15**, 1–29 (1924)

Liu, P., Kim, B., Friesner, R.A., Berne, B.J.: Replica exchange with solute tempering: a method for sampling biological systems in explicit water. Proc. Natl. Acad. Sci. USA **103**(39), 13749–13754 (2005)

Liu, Y., Tuckerman, M.E.: Generalized Gaussian moment thermostatting: a new continuous dynamical approach to the canonical ensemble. J. Chem. Phys. **112**(4), 1685 (2000)

Lo, C., Palmer, B.J.: Alternative Hamiltonian for molecular dynamics simulations in the grand canonical ensemble. J. Chem. Phys. **102**, 925 (1995)

Loncharich, R.J., Brooks, B.R., Pastor, R.W.: Langevin dynamics of peptides: the frictional dependence of isomerization rates of N-acetylalanyl-N′ – methylamide. Biopolymers **32**, 523–535 (1992)

London, F.: Zur Theorie und Systematik der Molekularkrafte. Zeitschrift für Physik **63**, 245–279 (1930)

Lubbers, N., Smith, J.S., Barros, K.: Hirarchical modeling of molecular energies using a deep neural network. J. Chem. Phys. **148**, 241715–8 (2018)

Luo, R., Moult, J., Gilson, M.K.: Dielectric screening treatment of electrostatic solvation. J. Phys. Chem. B **101**, 11216–11236 (1997)

Luty, B.A., David, M.E., Tironi, I.G., van Gunsteren., W.F.: A comparison of particle-particle, particle-mesh and Ewald methods for calculating electrostatic interactions in periodic molecular systems. Mol. Simul. **14**, 11–20 (1994)

Luty, B.A., Tironi, I.G., van Gunsteren, W.F.: Lattice-sum methods for calculating electrostatic interactions in molecular simulations. J. Chem. Phys. **103**, 3014–3021 (1995)

Lynch, G.C., Pettitt, B.M.: Grand canonical ensemble molecular dynamics simulations: reformulation of extended system dynamics approaches. J. Chem. Phys. **107**, 8594 (1997)

Ma, Q., Izaguirre, J.A.: Targeted mollified impulse: a multi scale stochastic integrator for long molecular dynamics simulations. Multiscale Model. Simul. **2**(1), 1–21 (2003)

Ma, Q., Izaguirre, J.A., Skeel, R.D.: Verlet-I/r-RESPA/Impulse is limited by nonlinear instabilities. SIAM J. Sci. Comput. **24**, 1951–1973 (2003)

Maciejczyk, M., Spasic, A., Liwo, A., Scheraga, H.A.: Coarse grained model of nucleic acid bases. J. Comput. Chem. **31**, 1644 (2010)

MacKerell, A., Banavali, N.: All-atom empirical force field for nucleic acids: II. Parameter optimization based on small molecule and condensed phase macromolecular target data. J. Comput. Chem. **21**, 105–120 (2000)

MacKerell, A.D., Bashford, D., Bellott, M., Dunbrack, R.L. Jr., Evanseck, J.D., Field, M.J., Fischer, S., Al Gao, J., Guo, H., Ha, S., McCarthy, D.J., Kuchnir, L., Kuczera, K., Lau, F.T.K., Mattos, C., Michnick, S., Ngo, T., Nguyen, D.T., Prodhom, B., Reiher, W.E., Toux, B., Schlenkrich, M., Smith, J.C., Stote, R., Straub, J., Watanabe, M., Kuczera, J.W., Yin, D., Karplus, M.: All atom empirical potential for molecular modelling and dynamics studies of protein. J. Phys. Chem. B **102**, 3586–3616 (1998)

MacKerell, A.D. Jr., Feig, M., Brooks III, C.L.: Extending the treatment of backbone energetics in protein force fields: limitations of gas-phase quantum mechanics in reproducing protein conformational distributions in molecular dynamics simulations. J. Comput. Chem. **25**, 1400–1415 (2004)

Majeux, N., Scarsi, M., Apostolakis, J., Ehrhardt, C., Caflisch, A.: Exhaustive docking of molecular fragments with electrostatic solvation. Proteins **37**, 88–105 (1999)

Mamuye, A.L., Rucco, M., Tesei, L., Merelli, E.: Persistent homology analysis of RNA. Mol. Based Math. Biol. **4**, 14–25 (2016)

Manning, G.S.: The molecular theory of polyelectrolyte solutions with applications to the electrostatic properties of polynucleotides. Q. Rev. Biophys. **11**, 179–246 (1978)

Marcus, R.A.: Calculation of thermodynamic properties of polyelectrolytes. J. Chem. Phys. **23**, 1057–1068 (1955)

Martyna, G.J., Tobias, D.J., Klein, M.L.: Constant pressure molecular dynamics algorithms. J. Chem. Phys. **101**(5), 4177–4189 (1994)

Martyna, G.J., Tuckerman, M., Tobias, D.J., Klein, M.L.: Explicit reversible integrators for extended systems dynamics. Mol. Phys. **87**(5), 1117–1157 (1996)

Massova, I., Kollman, P.A.: Combined molecular mechanical and continuum solvent approach (MM-PBSA/GBSA) to predict ligand binding. Persp. Drug Disc. Des. **18**, 113–135 (2000)

Matubayasi, N., Reed, L.H., Levy, R.M.: Thermodynamics of the hydration shell 1. Excess energy of a hydrophobic solute. J. Phys. Chem. **98**, 10640–10649 (1994)

McCammon, J.A., Northrup, S.H., Karplus, M., Levy, R.M.: Helixcoil transitions in a simple polypeptide model. Biopolymers **19**, 2033–2045 (1980)

McCulloch, W.S., Pitts, W.H.: A logical calculus of the ideas immanent in neural nets. Bull. Math. Biophys. **5**, 115–133 (1943)

McQuarrie, D.A.: Statistical Mechanics. Harper & Row, New York (1976)

McQuarrie, D.A.: Statistical Mechanics, 2nd edn. University Science Books, Sausalito (2000)

Mehler, E.L.: The Lorentz-Debye-Sack theory and dielectric screening of electrostatic effects in proteins and nucleic acids. In: Murray, J.S., Sen, K. (eds.) Molecular Electrostatic Potential: Concepts and Applications, vol. 3, pp. 371–405. Elsevier Science, Amsterdam (1996)

Mehler, E.L., Guarnieri, F.: A self-consistent, micro-environment modulated screened Coulomb potential approximation to calculate pH-dependent electrostatic effects in proteins. Biophys. J. **77**, 3–22 (1999)

Melchionna, S., Ciccotti, G., Holian, B.L.: Mol. Phys. **78**, 533 (1993)

Metz, M.P., Piszczatowski, K., Szalewicz, K.: Automatic generation of intermolecular potential energy surfaces. J. Chem. Theory Comput. **12**(12), 5895–5919 (2016)

Minary, P., Tuckerman, M.E., Martyna, G.J.: Long time molecular dynamics for enhanced conformational sampling in biomolecular systems. Phys. Rev. Lett. **93**, (150201-4) (2004)

Misquitta, A.J., Podeszwa, R., Jeziorski, B., Szalewicz, K.: Intermolecular potentials based on symmetry-adapted perturbation theory with dispersion energies from time-dependent density-functional calculations. J. Chem. Phys. **123**(21), 214103 (2005)

Misra, V.K., Sharp, K.A., Friedman, R.A., Honig, B.: Salt effects on ligand-DNA binding. Minor groove antibiotics. J. Mol. Biol. **238**, 245–263 (1994)

Mobley, D.L., Bannan, C.C., Rizzi, A., Bayly, C.I., Chodera, J.D., Lim, V.T., Lim, N.M., Beauchamp, K.A., Shirts, M.R., Gilson, M.K., Eastman, P.K.: Open force field consortium:

escaping atom types using direct chemical perception with SMIRNOFF. BioRxiv, Mar 21 (2018)
Mokshin, A.V., Yulmetyev, R.M., Hänggi, P.: Diffusion processes and memory effects. New J. Phys., **7**:9 (2005)
Moon, Y.I., Rajagopalam, B., Lall, U.: Estimation of mutual information using kernel density estimators. Phys. Rev. E **52**, 2318 (1995)
Murtola, T., Karttunen, M., Vattulainen, I.: Systematic coarse graining from structure using internal states: application to phospholipid/ cholesterol bilayer. J. Chem. Phys. **131**, 055101 (2009)
Nadler, W., Hansmann, U.H.E.: Generalized ensemble and tempering simulations: a unified view. Phys. Rev. E **75**, 026109 (2007)
Nakajima, N., Nakamura, H., Kidera, A.: Multicanonical ensemble generated by molecular dynamics simulation for enhanced conformational sampling of peptides. J. Phys. Chem. B **101**, 817–824 (1997)
Neal, R.M.: Sampling from multimodal distributions using tempered transitions. Stat. Comput. **6**(4), 353–366 (1996)
Nemethy, G., Scheraga, H.A.: The structure of water and hydrophobic bonding in proteins. III. The thermodynamic properties of hydrophobic bonds in proteins. J. Phys. Chem. **66**, 1773–1789 (1962)
Noakes, L.: The Takens embedding theorem. Int. J. Bifurcation Chaos Appl. Sci. Eng. **1**, 867–872 (1991)
Nosé, S.: A molecular dynamics method for simulation in the canonical ensemble. Mol. Phys. **52**, 255 (1984a)
Nosé, S.: A unified formulation of the constant temperature molecular dynamics methods. J. Chem. Phys. **81**, 511–519 (1984b)
Nosé, S.: A molecular dynamics method for simulation in the canonical ensemble. Mol. Phys. **52**, 255 (1984c)
Nosé, S.: Constant temperature molecular dynamics methods. Prog. Theor. Phys. Supp. **103**, 1–46 (1991)
Nosé, S., Klein, M.L.: Constant pressure molecular dynamics for molecular systems. Mol. Phys. **50**, 1055–1076 (1983)
Nozaki, Y., Tanford, C.: Examination of titration behaviour. Meth. Enzymol. **11**, 715–734 (1967)
Okamoto, Y., Hansmann, U.H.E.: Thermodynamics of helix-coil transitions studied by multicanonical algorithms. J. Phys. Chem. **99**, 11276–11287 (1995)
Okur, A., Wickstrom, L., Layten, M., Geney, R., Song, K., Hornak, V., Simmerling, C.J.: Improved efficiency of replica exchange simulations through use of hybrid explicit/implicit salvation model. J. Chem. Theory Comput. **2**, 420–433 (2006)
Oldziej, S., Liwo, A., Czaplewski, C., Pillardy, J., Scheraga, H.A.: Optimization of the UNRES force field by hierarchical design of the potential-energy landscape. 2. Off-lattice tests of the method with single proteins. J. Phys. Chem. B **108**, 16934–16949 (2004)
Onsager, L.: Theories of concentrated electrolytes. Chem. Rev. **13**, 73–89 (1933)
Ooi, T., Oobatake, M., Nemethy, G., Scheraga, H.A.: Accessible surface area as a measure of the thermodynamic parameters of hydration of peptides. Proc. Natl. Acad. Sci. USA **84**, 3086–3090 (1987)
Otting, G., Liepinsh, E., Wüthrich, K.: Protein hydration in aqueous solution. Science **254**, 974–980 (1991)
Paass, G.: Assessing and improving neural network predictions by the bootstrap algorithm. In: Hanson, S.J., Cowan, J.D., Giles, C.L. (eds.) Advances in Neural Information Processing Systems, vol. 5, pp. 196–203. Morgan-Kaufmann, San Francisco (1993)
Packard, N.H., Crutchfield, J.P., Farmer, J.D., Shaw, R.S.: Geometry from a time series. Phys. Rev. Lett. **45**(9), 712–716 (1980)
Pak, Y., Wang, S.: Folding of a 16-residue helical peptide using molecular dynamics simulation with Tsallis effective potential. J. Chem. Phys. **111**, 4359 (1999)

Pak, Y., Wang, S.: Application of a molecular dynamics simulation method with a generalized effective potential to the exible molecular docking problems. J. Phys. Chem. B **104**, 354–359 (2000)

Palmer, B.J., Lo, C.: Molecular dynamics implementation of the Gibbs-Ensemble calculation. J. Chem. Phys. **101**, 10899–10907 (1994)

Palmer, R.: Broken ergodicity. Adv. Phys. **32**, 669–735 (1982)

Park, S., Schulten, K.: Calculating potentials of mean force from steered molecular dynamics simulations. J. Chem. Phys. **120**(13), 5946 (2004)

Pastor, R.W.: Techniques and applications of Langevin dynamics simulations. In: The Molecular Dynamics of Liquid Crystals. Kluwer Academic, Dordrecht (1994)

Pastor, R.W., Brooks, B.R., Szabo, A.: An analysis of the accuracy of Langevin and molecular dynamics algorithms. Mol. Phys. **65**, 1409–1419 (1988)

Pearlman, D.A., Kollman, P.A.: A new method for carrying out free energy perturbation calculations: dynamically modified windows. J. Chem. Phys. **90**, 2460–2470 (1989)

Penna, T.J.P.: Traveling salesman problem and Tsallis statistics. Phys. Rev. E **51**, R1 (1995)

Perez, A., Morrone, J.A., Brini, E., MacCallum, J.L., Dill, K.A.: Blind protein structure prediction using accelerated free-energy simulations. Sci. Adv. **2**, e1601274 (2016)

Petitjean, M.: On the analysis calculation of van der Waals surfaces and volumes: some numerical aspects. J. Comput. Chem. **15**, 507–523 (1994)

Pettitt, B.M., Rossky, P.J.: Alkali halides in water: ion-solvent correlations and ion-ion potentials of mean force at infinity dilution. J. Chem. Phys. **84**, 5836–5844 (1986)

Phillips, J.C., Stone, J.E., Vandivort, K.L., Armstrong, T.G., Wozniak, J.M., Wilde, M., Schulten, K.: Petascale Tcl with NAMD, VMD, and Swift/T. In: SC'14 Workshop on High Performance Technical Computing in Dynamic Languages. IEEE Press (2014)

Piana, S., Klepeis, J.L., Shaw, D.E.: Assessing the accuracy of physical models used in protein-folding simulations: quantitative evidence from long molecular dynamics simulations. Curr. Opin. Struct. Biol. **24**, 98–105 (2014)

Pierotti, R.A.: A scaled particle theory of aqueous and nonaqueous solutions. Chem. Rev. **76**, 717–726 (1976)

Plastino, A.R., Anteneodo, C.: A dynamical thermostatting approach to nonextensive canonical ensembles. Ann. Phys. **255**, 250–269 (1997)

Poole, G.S.: Classical Mechanics, 3rd edn. Pearson Education Limited, Edinburgh (2001)

Posch, H.A., Hoover, W.G., Vesely, F.J.: Dynamics of the Nosé-Hoover oscillator: chaos, order, and stability. Phys. Rev. A **33**, 4253 (1986)

Potestio, R., Pontiggia, F., Micheletti, C.: Biophys. J. **96**, 4993 (2009)

Predescu, C., Predescu, M., Ciobanu, C.: The incomplete beta function law for parallel tempering sampling of classical canonical systems. J. Chem. Phys. **120**(9), 4119–4128 (2004)

Predescu, C., Predescu, M., Ciobanu, C.V.: On the efficiency of exchange in parallel tempering Monte Carlo simulations. J. Phys. Chem. B **109**, 4189–4196 (2005)

Prokopenko, M., Lizier, J.T.: Transfer entropy and transient limits of computation. Sci. Rep. **4**, 5394 (2014)

Prokopenko, M., Lizier, J.T., Price, D.C.: On thermodynamic interpretation of transfer entropy. Entropy **15**: 524–543 (2013)

Qian, N.: On the momentum term in gradient descent learning algorithms. Neural Netw. **12**, 145–151 (1999)

Raedt, H.D., Raedt, B.D.: Applications of the generalized Trotter formula. Phys. Rev. A **28**(6), 3575 (1983)

Rajagopal, G., Needs, R.: An optimized Ewald method for long-ranged potentials. J. Comput. Phys. **115**, 399–405 (1994)

Raschke, T.M., Tsai, J., Levitt, M.: Quantification of the hydrophobic interaction by simulations of the aggregation of small hydrophobic solutes in water. Proc. Natl. Acad. Sci. USA **98**, 5965–5969 (2001)

Rasmussen, C.E., Williams, C.K.: Gaussian Processes for Machine Learning, vol. 1. MIT Press, Cambridge (2006)

Rathore, N., Chopra, M., de Pablo, J.J.: Optimal allocation of replicas in parallel tempering simulations. J. Chem. Phys. **122**, 024111 (2005)

Rechester, A.B., White, R.B.: Symbolic kinetic equations for a chaotic attractor. Phys. Lett. A **156**, 419–424 (1991a)

Rechester, A.B., White, R.B.: Symbolic kinetic analysis of two-dimensional maps. Phys. Lett. A **158**, 51–56 (1991b)

Reiner, E.S., Radke, C.J.: Variational approach to the electrostatic free energy in charged colloidal suspensions: general theory for open systems. J. Chem. Soc. Faraday Trans. **86**, 3901–3912 (1990)

Reiss, H.: Scaled particle methods in the statistical thermodynamics of fluids. Adv. Chem. Phys. **9**, 1–84 (1965)

Richards, F.M.: Areas, volumes, packing, and protein structure. Annu. Rev. Biophys. Bioeng. **6**, 151–176 (1977)

Richards, F.M.: Optical matching of physical models and electron density maps: early developments. Methods Enzymol. **115**, 145–154 (1985)

Rogal, J., Bolhuis, P.G.: Multiple state transition path sampling. J. Chem. Phys. **129**, 224107 (2008)

Rogers, D., Hahn, M.: Extended-connectivity fingerprints. J. Chem. Inf. Model. **50**, 742–754 (2010)

Rotkiewicz, P., Skolnick, J.: Fast procedure for reconstruction of full-atom protein models from reduced representations. J. Comput. Chem. **29**, 1460–1465 (2008)

Roux, B., Simonson, T.: The implicit solvent models. Biophys. Chem. **78**, 1–20 (1999)

Rupp, M., Tkatchenko, A., Ḿuller, K.R., von Lilienfeld, O.A.: Fast and accurate modeling of molecular atomization energies with machine learning. Phys. Rev. Lett. **108**, 058301 (2012)

Ryckaert, J.P., Ciccotti, G., Berendsen, H.J.C.: Numerical integration of the Cartesian equations of motion of a system with constraints: molecular dynamics of n-alkanes. J. Comput. Phys. **23**, 327–341 (1977)

Sabo, D., Meuwly, M., Freeman, D.L., Doll, J.D.: A constant entropy increase model for the selection of parallel tempering ensembles. J. Chem. Phys. **128**, 174109 (2008)

Sanz-Serna, J.M., Calvo, M.P.: Numerical Hamiltonian Problems. Chapman and Hall, New York (1995)

Sauer, T., Yorke, J.A., Casdagli, M.: Embedology. J. Stat. Phys. **65**, 579–616 (1991)

Scarpazza, D.P., Ierardi, D.J., Lerer, A.K., Mackenzie, K.M., Pan, A.C., Bank, J.A., Chow, E., Dror, R.O., Grossman, J.P., Killebrew, D., Moraes, M.A., Predescu, C., Salmon, J.K., Shaw, D.E.: Extending the generality of molecular dynamics simulations on a special-purpose machine. In: Proceedings of the 27th IEEE International Parallel and Distributed processing Symposium, pp. 933–945. IEEE Computer Society, Boston (2013)

Schaefer, M., Froemmel, C.: A precise analytical method for calculating the electrostatic energy of macromolecules in aqueous solution. J. Mol. Biol. **216**, 1045–1066 (1990)

Schlick, T.: Mathematical Applications to Biomolecular Structure and Dynamics, IMA Volumes in Mathematics and its Applications, vol. 82. Springer, New York (1996)

Schlick, T.: Molecular Modeling and Simulation. An Interdisciplinary Guide, 2nd edn. Springer, New York (2010)

Schlick, T., Mandziuk, M., Skeel, R.D., Srinivas, K.: Nonlinear resonance artefacts in molecular dynamics simulations. J. Comput. Phys. **140**(1), 1–29 (1998)

Schneider, E., Dai, L., Topper, R.Q., Drechsel-Grau, C., Tuckerman, M.E.: Stochastic neural network approach for learning high-dimensional free energy surfaces. Phys. Rev. Lett. **119**, 150601 (2017)

Schreiber, T.: Measuring information transfer. Phys. Rev. Lett. **85**, 461–464 (2000)

Schutz, C.N., Warshel, A.: What are the dielectric "constants" of proteins and how to validate electrostatic models? Proteins Struct. Funct. Genet. **44**, 400–417 (2001)

Seyler, S.L., Beckstein, O.: Sampling large conformational transitions: adenylate kinase as a testing ground. Mol. Sim. **40**(10–11), 855–877 (2014)

Shannon, C.E., Weaver, W.: The Mathematical Theory of Information. University of Illinois Press, Urbana (1949)

Sharp, K.A., Honig, B.: Calculating total electrostatic energies with non-linear Poisson-Boltzmann equation. J. Phys. Chem. **94**, 7684–7692 (1990)
Sharp, K.A., Nicholis, A., Fine, R.F., Honig, B.: Reconciling the magnitude of the microscopic and macroscopic hydrophobic effects. Science **252**, 106–109 (1991)
Shi, Q., Izvekov, S., Voth, G.A.: Mixed atomistic and coarse-grained molecular dynamics: simulation of membrane-bound ion channel. J. Phys. Chem. B **110**, 15045–15048 (2006)
Shih, A.Y., Arkhipov, A., Freddolino, P.L., Schulten, K.: Coarse grained protein-lipid model with application to lipidprotein particles. J. Phys. Chem. B **110**, 3674–3684 (2006)
Shimada, J., Kaneko, H., Takada, T.: Efficient calculations of coulombic interactions in biomolecular simulations with periodic boundary conditions. J. Comput. Chem. **14**, 867–878 (1993)
Shimada, J., Kaneko, H., Takada, T.: Performance of fast multipole methods for calculating electrostatic interactions in biomacromolecular simulations. J. Comput. Chem. **15**, 28–43 (1994)
Simonson, T.: Accurate calculation of the dielectric constant of water from simulations of a microscopic droplet in vacuum. Chem. Phys. Lett. **250**, 450–454 (1996)
Simonson, T.: Dielectric constant of cytochrome c from simulations in water droplet including all electrostatic interactions. J. Am. Chem. Soc. **120**, 4875–4876 (1998)
Simonson, T.: Macromolecular electrostatics: continuum models and their growing pains. Curr. Opin. Struct. Biol. **11**, 243–252 (2001)
Simonson, T., Brunger, A.T.: Solvation free energies estimated from macroscopic continuum theory: an accuracy assessment. J. Phys. Chem. **98**, 4683–4694 (1994)
Sinitskiy, A.V., Saunders, M.G., Voth, G.A.: Optimal number of coarsegrained sites in different components of large biomolecular complexes. J. Phys. Chem. B **116**, 8363–8374 (2012)
Sitkoff, D., Sharp, K.A., Honig, B.: Accurate calculation of hydration free energies using macroscopic solvent models. J. Phys. Chem. **98**, 1978–1988 (1994)
Skeel, R.D., Zhang, G., Schlick, T.: A family of symplectic integrators: stability, accuracy, and molecular dynamics applications. SIAM J. Sci. Comput. **18**(1), 203–222 (1997)
Smith, A.V., Hall, C.K.: α-helix formation: discontinuous molecular dynamics on an intermediate-resolution protein model Proteins **44**, 344–360 (2001a)
Smith, A.V., Hall, C.K.: Assembly of a tetrameric α-helical bundle: computer simulations on an intermediate-resolution protein model. Proteins **44**, 376–391 (2001b)
Srinivasan, J., Trevathan, M.W., Beroza, P., Case, D.A.: Application of a pairwise generalized Born model to proteins and nucleic acids: inclusion of salt effects. Theor. Chem. Acc. **101**, 426–434 (1999)
Srivastava, N., Hinton, G.E., Krizhevsky, A., Sutskever, I., Salakhutdinov, R.: A simple way to prevent neural networks from overfitting. J. Mach. Learn. Res. **15**, 1929–1958 (2014)
Staniek, M., Lehnertz, K.: Symbolic transfer entropy. Phys. Rev. Lett. **100**, 158101 (2008)
Stepanova, M.: Dynamics of essential collective motions in proteins: theory. Phys. Rev. E **76**(5), 051918 (2007)
Still, W.C., Tempczyk, A., Hawley, R.C., Hendrickson, T.: Semi-analytical treatment of solvation for molecular mechanics and dynamics. J. Am. Chem. Soc. **112**, 6127–6129 (1990)
Stoffer, D.: Variable steps for reversible integration methods. Computing **55**, 1–22 (1995)
Stone, J.E., Hardy, D.J., Isralewitz, B., Schulten, K.: GPU algorithms for molecular modelling. In: Bader, D.A., Kurzak, J. (eds.) Scientific Computing with Multicore and Accelerators, pp. 351–371. Chapman & Hall, Boca Raton (2011)
Stone, J.E., Vandivort, K.L., Schulten, K.: Gpu-accelerated analysis and visualisation on petascale supercomputing platforms. In: Proceedings of the 8th International Workshop on Ultrascale Visualization, pp. 6:1–6:8, New York (2013) UltraVis'13
Strang, G.: On the construction and comparison of different schemes. SIAM J. Numer. Anal. **5**, 506–517 (1968)
Sturgeon, J.B., Laird, B.B.: Symplectic algorithm for constant-pressure molecular dynamics using a Nosé-Poincaré thermostat. J. Chem. Phys. **112**(8), 3474–3482 (2000)
Sugita, Y., Okamoto, Y.: Replica-exchange molecular dynamics method for protein folding. Chem. Phys. Lett. **314**, 141–151 (1999)

Sugita, Y., Okamoto, Y.: Replica-exchange multicanonical algorithm and multicanonical replica-exchange method for simulating systems with rough energy landscape. Chem. Phys. Lett. **329**, 261–270 (2000)

Suzuki, M.: General theory of fractal path integrals with applications to many-body theories and statistical physics. J. Math. Phys. **32**, 400 (1991)

Takahashi, M., Imada, M.: Monte Carlo calculation of quantum systems. II. Higher order correction. J. Phys. Soc. Jpn. **53**, 3765–3769 (1984)

Takens, F.: Detecting strange attractors in fluid turbulence, Dynamical Systems and Turbulence. Springer, Berlin (1981)

Tanford, C.: Interfacial free energy and the hydrophobic effect. Proc. Natl. Acad. Sci. USA **76**, 4175–4176 (1979)

Täuber, U.C.: Renormalization group: applications in statistical physics. Nucl. Phys. B Proc. Suppl. 00:1–28, (2011)

Thomas, M.C., Joy, A.T.: Elements of Information Theory. Wiley, Hoboken (2006)

Tjong, H., Zhou, H.X.: GBr^6: a parametrization-free, accurate, analytical generalised Born method. J. Phys. Chem. B **111**, 3055–3061 (2007a)

Tjong, H., Zhou, H.X.: GBr^6NL: a generalised Born method for accurately reproducing solvation energy of the nonlinear Poisson-Boltzmann equation. J. Chem. Phys. **126**, 195102 (2007b)

Tobias, D.J., Martyna, G.J., Klein, M.L.: An analysis of the accuracy of Langevin and molecular dynamics algorithms. J. Phys. Chem. **97**(49), 12959 (1993)

Toukmaji, A.Y., Board, J.A. Jr., Ewald summation techniques in perspective: a survey. Comput. Phys. Commun. **95**, 73–92 (1996)

Tozzini, V.: Coarse-grained models for proteins. Curr. Opin. Struct. Bio. **15**, 144–150 (2005)

Tozzini, V., McCammon, J.: A coarse grained model for the dynamics of the early stages of the binding mechanism of HIV-1 protease. Chem. Phys. Lett. **413**, 123–128 (2005)

Tozzini, V., Rocchia, W., McCammon, J.A.: Mapping all-atom models onto one-bead coarse-grained models: general properties and applications to a minimal polypeptide model. J. Chem. Theory Comput. **2**, 667–673 (2006)

Trebst, S., Huse, D.A., Troyer, M.: Optimizing the ensemble for equilibrium in broad-histogram Monte Carlo. Phys. Rev. E **70**, 046701 (2004)

Trebst, S., Troyer, M., Hansmann, U.H.E.: Optimized parallel tempering simulations of proteins. J. Chem. Phys. **124**, 174903 (2006)

Trotter, H.F.: On the product of semi-groups of operators. Proc. Am. Math. Soc. **10**, 545–551 (1959)

Tsallis, C.: Possible generalization of boltzmann-gibbs statistics. J. Stat. Phys. **52**, 479–487 (1988)

Tschöp, W., Kremer, K., Hahn, O., Batoulis, J., Bürger, T.: Simulation of polymer melts. II. From coarse-grained models back to atomistic description. Acta Polym. **49**, 75–79 (1998)

Tuckerman, M., Parrinello, M.: Integrating the Car-Parrinello equations. I. Basic integration techniques. J. Chem. Phys. **101**(2), 1302 (1994)

Tuckerman, M.E., Berne, B.J.: Molecular dynamics algorithm for multiple time scales: systems with long range forces. J. Chem. Phys. **94**, 6811 (1991a)

Tuckerman, M.E., Berne, B.J.: Molecular dynamics in systems with multiple time scales: Systems with stiff and soft degrees of freedom and with short and long range forces. J. Chem. Phys. **95**, 8362 (1991b)

Tuckerman, M.E., Martyna, G.J.: Understanding modern molecular dynamics: Techniques and Applications. J. Phys. Chem. B **104**, 159–178 (2000)

Tuckerman, M.E., Berne, B.J., Rossi, A.: Molecular dynamics algorithm for multiple time scales: Systems with disparate masses. J. Chem. Phys. **94**, 1465 (1991)

Tuckerman, M.E., Martyna, G.J., Berne, B.J.: Molecular dynamics algorithm for condensed systems with multiple time scales. J. Chem. Phys. **93**, 1287 (1990)

Tuckerman, M.E., Berne, B.J., Martyna, G.J.: Reversible multiple time step scale molecular dynamics. J. Chem. Phys. **97**(3), 1990–2001 (1992)

Tuckerman, M.E., Mundy, C.J., Martyna, G.J.: On the classical statistical mechanics of non-Hamiltonian systems. Europhys. Lett. **45**, 149–155 (1999)

Tuckerman, M.E., Liu, Y., Ciccotti, G., Martyna, G.J.: Non-Hamiltonian molecular dynamics: generalizing Hamiltonian phase space principles to non-Hamiltonian systems. J. Chem. Phys. **115**(4), 1678–1702 (2001)

Ueda, Y., Taketomi, H., Go, N.: Studies on protein folding, unfolding, and fluctuations by computer simulation. II. A. three-dimensional lattice model of lysozyme. Biopolymers **17**, 1531–1548 (1978)

Unke, O.T., Meuwly, M.: A reactive, scalable, and transferable model for molecular energies from a neural network approach based on local information. J. Chem. Phys. **148**, 241708–15 (2018)

Vandenbrande, S., Waroquier, M., Speybroeck, V.V., Verstraelen, T.: The monomer electron density force field (MEDFF): a physically inspired model for noncovalent interactions. J. Chem. Theory Comput. **13**(1), 161–179 (2017)

van Gunsteren, W.F., Bakowies, D., Baron, R., Chandrasekhar, I., Christen, M., Daura, X., Gee, P., Geerke, D.P., Glättli, A., Hünenberger, P.H., Kastenholz, M.A., Oostenbrink, C., Schenk, M., Trzesniak, D., van der Vegt, N.F.A., Yu, H.B.: Biomolecular modeling: goals, problems, perspectives. Angew. Chem. Int. Ed. **45**(25), 4064–4092 (2006)

Verwey, E.J.W., Overbeek, J.T.G.: Theory of the Stability of Lyophobic Colloids. Elsevier, Amsterdam (1948)

Van Vleet, M.J., Misquitta, A.J., Stone, A.J., Schmidt, J.: Beyond Borm-Mayer: improved models for short-range repulsion in ab initio force fields. J. Chem. Theory Comput. **12**, 3851–3870 (2016)

Vorobjev, Y.N., Hermans, J.: SIMS: computation of a smooth invariant molecular surface. Biophys. J. **73**, 722 (1997)

Voter, A.F.: Hyperdynamics: accelerated molecular dynamics of infrequent events. Phys. Rev. Lett. **78**, 3908–3911 (1997)

Voth, G.A. (ed.): Coarse-Graining of Condensed Phase and Biomolecular Systems. CRC Press, Boca Raton (2008)

Wade, R.C., Davis, M.E., Luty, B.A., Madura, J.D., McCammon, J.A.: Gating of the active site of triose phosphate isomerase: Brownian dynamics simulations of flexible peptide loops in the enzyme. Biophys. J. **64**, 9–15 (1993)

Wainwright, T., Alder, B.J., Gass, D.M.: Decay time correlations in two dimensions. Phys. Rev. A **4**, 233 (1971)

Wang, F., Landau, D.P.: Efficient, multiple-range random walk algorithm to calculate the density of states. Phys. Rev. Lett. **86**, 2050 (2001a)

Wang, F., Landau, D.P.: Efficient, multiple-range random walk algorithm to calculate the density of states. Phys. Rev. Lett. **86**(10), 2050–2053 (2001b)

Wang, F., Landau, D.P.: Determining the density of states for classical statistical models: a random walk algorithm to produce a flat histogram. Phys. Rev. E **64**, 056101–16 (2001c)

Wang, J.S., Swendsen, R.H.: Replica Monte Carlo simulation of spin glasses. Phys. Rev. Lett. **57**, 2607–2609 (1986)

Wang, W., Reyes, O.D.C., Kollman, P.A.: Biomolecular simulations: recent developments in force fields, simulations of enzyme catalysis, protein-ligand, protein-protein, and protein-nucleic acid noncovalent interactions. Annu. Rev. Bioph. Biom. **30**, 211–243 (2001)

Warshel, A., Lifson, S.: Consistent force field calculations. II. Crystal structures, sublimation energies, molecular and lattice vibrations, molecular conformations, and enthalpies of alkanes. J. Chem. Phys. **53**, 582 (1970)

Weerasinghe, S., Pettitt, B.M.: Ideal chemical potential contribution in molecular dynamics simulations of the grand canonical ensemble. Mol. Phys. **82**, 897 (1994)

Wehmeyer, C., Noé, F.: Time-lagged autoencoders: deep learning of slow collective variables for molecular kinetics. J. Chem. Phys. **148**, 241703–9 (2018)

Weininger, D.: SMILES, a chemical language and information system. 1. Introduction to methodology and encoding rules. J. Chem. Inf. Comput. Sci. **28**, 31–36 (1988)

Wetlaufer, D.B., Ristow, S.S.: Acquisition of three-dimensional structure of proteins. Annu. Rev. Biochem. **42**, 135–158 (1973)

Whitfield, T.W., Bu, L., Straub, J.E.: Generalized parallel sampling. Physica A: Statistical Mechanics and its Applications, **305**:157–171, (2002)

Widom, B.: Some topics in the theory of fluids. J. Chem. Phys. **39**, 2808–2812 (1963)

Wolfram, S.: A New Kind of Science. Wolfram Media, Inc., Champaign (2002)

Xia, K., Zhao, Z., Wei, G.W.: Multiresolution persistent homology for excessively large biomolecular datasets. J. Chem. Phys. **143**, 134103 (2015)

York, D.M., Wlodawer, A., Pedersen, L., Darden, T.A.: Atomic-level accuracy in simulations of large protein crystals. Proc. Natl. Acad. Sci. USA **91**, 8715–8718 (1994)

Yoshida, H.: Construction of higher order symplectic integrators. Phys. Lett. A. **150**, 262–268 (1990)

Yu, H.A., Karplus, M.: A thermodynamic analysis of solvation. J. Chem. Phys. **89**, 2366–2379 (1988)

Yulmetyev, R.M., Mokshin, A.V., Hänggi, P.: Diffusion time-scale invariance, randomization processes, and memory effects in lennard-jones liquids. Phys. Rev. E **68**, 051201 (2003)

Zacharias, M.: Protein-protein docking with a reduced protein model accounting for side-chain flexibility. Protein Sci. **12**, 1271 (2003)

Zare, K., Szebehely, V.: Time transformations in the extended phase-space. Celest. Mech. **11**, 469 (1975)

Zeni, C., Rossi, K., Glielmo, A., Fekete, Á., Gaston, N., Baletto, F., De Vita, A.: Building machine learning force fields for nanoclusters. J. Chem. Phys. **148**(24), 241739 (2018)

Zhang, G., Schlick, T.: LIN: a new algorithm combining implicit integration and normal mode techniques for molecular dynamics. J. Comput. Chem. **14**, 1212–1233 (1993)

Zhang, G., Schlick, T.: The Langevin/implicit-Euler/Normal-Mode scheme (LIN) for molecular dynamics at large time steps. J. Chem. Phys. **101**, 4995–5012 (1994)

Zhang, J., Muthukumar, M.: Simulations of nucleation and elongation of amyloid fibrils. J. Chem. Phys. **130**, 035102 (2009)

Zhou, H.X.: Macromolecular electrostatic energy within the nonlinear Poisson-Boltzmann equation. J. Chem. Phys. **100**, 3152–3162 (1994)

Zhou, R.: Free energy landscape of protein folding in water: explicit vs. implicit solvent. Proteins Struct. Funct. Bioinf. **53**(2), 148–161 (2003)

Zhou, R., Berne, B.J.: Can a continuum solvent model reproduce the free energy landscape of a β-hairpin folding in water? Proc. Natl. Acad. Sci. USA **99**, 12777–12782 (2002)

Zotin, A.I.: Thermodynamic Aspects of Developmental Biology. Monogr. Dev. Biol., **5**:1–59 (1972)

Zwanzig, R.W.: High-temperature equation of state by a perturbation method. 1. Nonpolar gases. J. Chem. Phys. **22**, 1420–1426 (1954)

Index

H. Kamberaj, *Molecular Dynamics Simulations in Statistical Physics: Theory and Applications*, Scientific Computation,
https://doi.org/10.1007/978-3-030-35702-3

MIX
Papier aus verantwortungsvollen Quellen
Paper from responsible sources
FSC® C105338
FSC
www.fsc.org

If you have any concerns about our products,
you can contact us on
ProductSafety@springernature.com

In case Publisher is established outside the EU,
the EU authorized representative is:
Springer Nature Customer Service Center GmbH
Europaplatz 3, 69115 Heidelberg, Germany

Printed by Libri Plureos GmbH
in Hamburg, Germany